Numerical Methods for Engineers and Scientists Using MATLAB®

Ramin S. Esfandiari

CRC Press is an imprint of the
Taylor & Francis Group, an **informa** business

CRC Press
Taylor & Francis Group
6000 Broken Sound Parkway NW, Suite 300
Boca Raton, FL 33487-2742

Printed on acid-free paper
Version Date: 20130426

International Standard Book Number-13: 978-1-4665-8569-0 (Paperback)

Visit the Taylor & Francis Web site at
http://www.taylorandfrancis.com

and the CRC Press Web site at
http://www.crcpress.com

To

My wife Haleh, my sisters Mandana and Roxana, and

my parents to whom I owe everything

Contents

Preface

The principal goal of this book is to provide the reader with a thorough knowledge of the fundamentals of numerical methods utilized in various disciplines in engineering and science. The very powerful software MATLAB® is introduced at the outset and is used throughout the book to perform symbolic, graphical, and numerical tasks. The textbook, written at the junior level, methodically covers a wide array of methods ranging from curve fitting a set of data to numerically solving ordinary and partial differential equations. Each method is accompanied by either a user-defined function or a MATLAB script file. MATLAB built-in functions are also presented for each main topic.

This book comprises 10 chapters. Chapter 1 presents the necessary background material and is divided into two parts. The first part covers differential equations, matrix analysis, and matrix eigenvalue problem. The second part introduces errors, approximations, iterative methods, and rates of convergence.

Chapter 2 gives a comprehensive introduction to the essentials of MATLAB as related to numerical methods. The chapter addresses fundamental features such as built-in functions and commands, formatting options, vector and matrix operations, program flow control, symbolic operations, and plotting capabilities. The reader also learns how to write a user-defined function or a MATLAB script file to perform specific tasks.

Chapters 3 and 4 introduce techniques for solving equations. Chapter 3 focuses on finding a single root or multiple roots of equations of a single variable. Chapter 4 covers methods for solving linear and nonlinear systems of equations.

Chapter 5 is entirely devoted to curve fitting and interpolation techniques, while Chapter 6 covers numerical differentiation and integration methods. Chapters 7 and 8 present numerical methods for solving initial-value problems and boundary-value problems, respectively.

Chapter 9 discusses the matrix eigenvalue problem, which entails numerical methods to approximate a few or all eigenvalues of a matrix.

Chapter 10 deals with the numerical solution of partial differential equations, specifically those that frequently arise in engineering and science.

Pedagogy of the Book

This book is written in a user-friendly manner that intends to make the material easy to understand by the reader. The materials are presented systematically using the following format:

- Each newly introduced method is accompanied by a user-defined function, or a script file, that utilizes the method to perform a desired task.
- This is followed by at least one fully worked-out example showing all details.
- The results are then confirmed through the execution of the user-defined function or the script file.
- When available, built-in functions are executed for reconfirmation.
- Plots are regularly generated to shed light on the soundness and significance of the numerical results.

Exercises

A large set of exercises, of different levels of difficulty, appears at the end of each chapter and can be worked out using either a

- ✍ Hand calculator, or
- MATLAB.

In many instances, the reader is asked to prepare a user-defined function, or a script file, that implements a specific technique. In many cases, these simply require revisions to those previously presented in the chapter.

Ancillary Material

The following will be provided to the instructors adopting this book:

- An instructor's solutions manual (in PDF format), featuring complete solution details of all exercises, prepared by the author, and
- A web download containing all user-defined functions and script files used throughout the book, available at www.crcpress.com/product/isbn/9781466585690.

An ample portion of the material in this book has been rigorously class tested over the past several years. In addition, the valuable remarks and suggestions made by students have greatly contributed to making this book as complete and user-friendly as possible.

MATLAB® and Simulink® are registered trademarks of The MathWorks, Inc. For product information, please contact:

The MathWorks, Inc.
3 Apple Hill Drive
Natick, MA 01760-2098 USA
Tel: 508 647 7000
Fax: 508-647-7001
E-mail: info@mathworks.com
Web: www.mathworks.com

Acknowledgments

The author expresses his deep gratitude to Jonathan Plant (senior editor, Mechanical, Aerospace, Nuclear and Energy Engineering) at Taylor & Francis/CRC Press for his assistance during various stages of the development of this project, as well as Dr. Nicholas R. Swart of the University of British Columbia for thoroughly reviewing the book and providing valuable input. The author also appreciates the feedback by his students that helped make this book as user-friendly as possible.

Author

Dr. Ramin S. Esfandiari is a professor of mechanical and aerospace engineering at California State University, Long Beach (CSULB), where he has served as a faculty member since 1989. He received his BS in mechanical engineering, as well as his MA and PhD in applied mathematics (optimal control), from the University of California, Santa Barbara.

He has authored several refereed research papers in high-quality engineering and scientific journals, including the *Journal of Optimization Theory and Applications*, the *Journal of Sound and Vibration*, *Optimal Control Applications and Methods*, and the *ASME Journal of Applied Mechanics*.

Dr. Esfandiari is the author of *Applied Mathematics for Engineers*, 4th edition (2007), *Matrix Analysis and Numerical Methods for Engineers* (2007), and *MATLAB Manual for Advanced Engineering Mathematics* (2007), all published by Atlantis Publishing Company, as well as *Modeling and Analysis of Dynamic Systems* (CRC Press, 2010) with Dr. Bei Lu. He is one of the select few contributing authors for the latest edition of the *Springer-Verlag Mechanical Engineering Handbook* (2009) and the coauthor (with Dr. H.V. Vu) of *Dynamic Systems: Modeling and Analysis* (McGraw-Hill, 1997).

Dr. Esfandiari is the recipient of several teaching and research awards, including two Meritorious Performance and Professional Promise Awards, the TRW Excellence in Teaching and Scholarship Award, and the Distinguished Faculty Teaching Award.

1

Background and Introduction

This chapter is essentially divided into two main parts. In the first part, we review some important mathematical concepts related to differential equations and matrix analysis. In the second part, fundamentals of numerical methods, such as computational errors and approximations, as well as iterations are introduced. The materials presented here will be fully integrated in the subsequent chapters.

1.1 Background

1.1.1 Differential Equations

Differential equations are divided into two categories: ordinary differential equations (ODEs) and partial differential equations (PDEs). An equation involving an unknown function and one or more of its derivatives is called a differential equation. When there is only one independent variable, the equation is called an ODE. If the unknown function is a function of several independent variables, the equation is a PDE. For example, $\dot{x} + 4x = \sin t$ is an ODE involving the unknown function $x(t)$, its first derivative $\dot{x} = dx/dt$, as well as a given function $\sin t$. Similarly, $t\ddot{x} - x\dot{x} = t + 1$ is an ODE relating $x(t)$ and its first and second derivatives with respect to t, as well as the function $t + 1$. The derivative of the highest order of the unknown function $x(t)$ with respect to t is the order of the ODE. For instance, $\dot{x} + 4x = \sin t$ is of order 1, while $t\ddot{x} - x\dot{x} = t + 1$ is of order 2.

Consider an nth-order ODE in the form

$$a_n x^{(n)} + a_{n-1} x^{(n-1)} + \cdots + a_1 \dot{x} + a_0 x = F(t) \tag{1.1}$$

where $x = x(t)$ and $x^{(n)} = d^n x/dt^n$. If all coefficients $a_0, a_1, \ldots, a_n$ are either constants or functions of the independent variable t, then the ODE is linear. Otherwise, the ODE is nonlinear. If $F(t) \equiv 0$, the ODE is homogeneous. Otherwise, it is nonhomogeneous. Therefore $\dot{x} + 4x = \sin t$ is linear, $t\ddot{x} - x\dot{x} = t + 1$ is nonlinear, and both are nonhomogeneous.

1.1.1.1 Linear First-Order ODEs

A linear, first-order ODE can be expressed as

$$a_1\dot{x} + a_0 x = F(t) \quad \overset{\text{Divide by } a_1}{\Rightarrow} \quad \dot{x} + g(t)x = f(t) \tag{1.2}$$

A general solution for Equation 1.2 is obtained as

$$x(t) = e^{-h}\left[\int e^{h} f(t)\mathrm{d}t + c\right], \quad h(t) = \int g(t)\mathrm{d}t, \quad c = \text{const} \tag{1.3}$$

To find a particular solution, an initial condition must be prescribed. Assuming t_0 is the initial time, a first-order initial-value problem (IVP) is described as

$$\dot{x} + g(t)x = f(t), \quad x(t_0) = x_0$$

Example 1.1: Linear First-Order IVP

Solve

$$2\dot{x} + x = e^{-2t}, \quad x(0) = \frac{2}{3}$$

SOLUTION

We first write the ODE in the standard form of Equation 1.2, as $\dot{x} + \frac{1}{2}x = \frac{1}{2}e^{-2t}$ so that $g(t) = \frac{1}{2}$, $f(t) = \frac{1}{2}e^{-2t}$. By Equation 1.3,

$$h = \int \frac{1}{2}\mathrm{d}t = \frac{1}{2}t, \quad x(t) = e^{-t/2}\left[\int e^{t/2}\frac{1}{2}e^{-2t}\mathrm{d}t + c\right] = -\frac{1}{3}e^{-2t} + ce^{-t/2}$$

Applying the initial condition, we find

$$x(0) = -\frac{1}{3} + c = \frac{2}{3} \quad \Rightarrow \quad c = 1$$

Therefore,

$$x(t) = -\frac{1}{3}e^{-2t} + e^{-t/2}$$

1.1.1.2 Second-Order ODEs with Constant Coefficients

Consider a second-order ODE in the form

$$\ddot{x} + a_1\dot{x} + a_0 x = f(t), \quad a_1, a_0 = \text{const} \tag{1.4}$$

The corresponding second-order IVP comprises this ODE accompanied by two initial conditions. A general solution of Equation 1.4 is a superposition of the homogeneous solution $x_h(t)$ and the particular solution $x_p(t)$.

1.1.1.2.1 *Homogeneous Solution*

The homogeneous solution is the solution of the homogeneous equation

$$\ddot{x} + a_1\dot{x} + a_0x = 0 \tag{1.5}$$

Assuming a solution in the form $x(t) = e^{\lambda t}$, substituting into Equation 1.5, and using the fact that $e^{\lambda t} \neq 0$, we find

$$\lambda^2 + a_1\lambda + a_0 = 0$$

known as the characteristic equation. The solution of Equation 1.5 is determined according to the nature of the two roots of the characteristic equation. These roots, labeled λ_1 and λ_2, are called the characteristic values.

1. When $\lambda_1 \neq \lambda_2$ (real), the homogeneous solution is

$$x_h(t) = c_1e^{\lambda_1 t} + c_2e^{\lambda_2 t}$$

2. When $\lambda_1 = \lambda_2$, we have

$$x_h(t) = c_1e^{\lambda_1 t} + c_2te^{\lambda_1 t}$$

3. When $\bar{\lambda}_1 = \lambda_2$ (complex conjugates), and $\lambda_1 = \sigma + i\omega$, we find

$$x_h(t) = e^{\sigma t}(c_1 \cos \omega t + c_2 \sin \omega t)$$

Example 1.2: Homogeneous, Second-Order ODE with Constant Coefficients

Consider the homogeneous equation

$$\ddot{x} + 4\dot{x} + 3x = 0$$

The characteristic equation is formed as $\lambda^2 + 4\lambda + 3 = 0$ so that the characteristic values are $\lambda_1 = -1$, $\lambda_2 = -3$, and

$$x_h(t) = c_1e^{-t} + c_2e^{-3t}$$

1.1.1.2.2 Particular Solution

The particular solution is determined by the function $f(t)$ in Equation 1.4 and how $f(t)$ is related to the independent functions that constitute the homogeneous solution. The particular solution is obtained by the method of undetermined coefficients. This method is limited in its applications only to cases where $f(t)$ is a polynomial, an exponential function, a sinusoidal function, or any of their combinations.

1.1.1.2.3 Method of Undetermined Coefficients

Table 1.1 considers different scenarios and the corresponding recommended $x_p(t)$. These recommended forms are subject to modification in some special cases as follows. If $x_p(t)$ contains a term that coincides with a solution of the homogeneous equation, and that the solution corresponds to a nonrepeated characteristic value, then the recommended $x_p(t)$ must be multiplied by t. If the characteristic value is repeated, then multiply $x_p(t)$ by t^2.

Example 1.3: Second-Order IVP

Solve the following second-order IVP:

$$\ddot{x} + 4\dot{x} + 3x = 2e^{-t}, \quad x(0) = 1, \quad \dot{x}(0) = -2$$

SOLUTION

The homogeneous solution was previously found in Example 1.2, as $x_h(t) = c_1 e^{-t} + c_2 e^{-3t}$. Since $f(t) = 2e^{-t}$, Table 1.1 recommends $x_p(t) = Ke^{-t}$. However, e^{-t} is one of the independent functions in the homogeneous solution, thus $x_p(t)$ must be modified. Since e^{-t} is associated with a nonrepeated characteristic value, the modification is $x_p(t) = Kte^{-t}$. Substitution into the ODE, and collecting like terms, yields

$$2Ke^{-t} = 2e^{-t} \Rightarrow k = 1$$

This implies $x_p(t) = te^{-t}$, and a general solution is found as $x(t) = c_1 e^{-t} + c_2 e^{-3t} + te^{-t}$. Applying the initial conditions,

TABLE 1.1

Method of Undetermined Coefficients

Term in $f(t)$	Recommended $x_p(t)$
$A_n t^n + A_{n-1} t^{n-1} + \cdots + A_1 t + A_0$	$K_n t^n + K_{n-1} t^{n-1} + \cdots + K_1 t + K_0$
Ae^{at}	Ke^{at}
$A \cos \alpha t$ or $A \sin \alpha t$	$K_1 \cos \alpha t + K_2 \sin \alpha t$
$Ae^{at} \cos \alpha t$ or $Ae^{at} \sin \alpha t$	$e^{at}(K_1 \cos \alpha t + K_2 \sin \alpha t)$

$$\begin{aligned} c_1 + c_2 &= 1 \\ -c_1 - 3c_2 + 1 &= -2 \end{aligned} \quad \overset{\text{Solve}}{\Rightarrow} \quad \begin{aligned} c_1 &= 0 \\ c_2 &= 1 \end{aligned}$$

Therefore, $x(t) = e^{-3t} + te^{-t}$.

1.1.2 Matrix Analysis

An n-dimensional vector $\mathbf{v}$ is an ordered set of n scalars, written as

$$\mathbf{v} = \begin{Bmatrix} v_1 \\ v_2 \\ \cdots \\ v_n \end{Bmatrix}$$

where each v_i $(i = 1, 2, \ldots, n)$ is a component of vector $\mathbf{v}$. A matrix is a collection of numbers (real or complex) or possibly functions, arranged in a rectangular array and enclosed by square brackets. Each of the elements in a matrix is an entry of the matrix. The horizontal and vertical lines are the rows and columns of the matrix, respectively. The number of rows and columns of a matrix determine its size. If a matrix $\mathbf{A}$ has m rows and n columns, then it is of size $m \times n$. A matrix is square if the number of its rows and columns are the same, and rectangular if different. Matrices are denoted by bold-faced capital letters, such as $\mathbf{A}$. The abbreviated form of an $m \times n$ matrix is

$$\mathbf{A} = [a_{ij}]_{m\times n}$$

where a_{ij} is known as the (i,j) entry of $\mathbf{A}$, located at the intersection of the ith row and the jth column of $\mathbf{A}$ so that a_{32}, for instance, is the entry at the intersection of the third row and the second column. In a square matrix $\mathbf{A}_{n\times n}$, the elements $a_{11}, a_{22}, \ldots, a_{nn}$ are the diagonal entries.

Two matrices $\mathbf{A} = [a_{ij}]$ and $\mathbf{B} = [b_{ij}]$ are equal if they have the same size and the same respective entries. A submatrix of $\mathbf{A}$ is generated by deleting some rows and/or columns of $\mathbf{A}$.

1.1.2.1 Matrix Operations

The sum of $\mathbf{A} = [a_{ij}]_{m\times n}$ and $\mathbf{B} = [b_{ij}]_{m\times n}$ is

$$\mathbf{C} = [c_{ij}]_{m\times n} = [a_{ij} + b_{ij}]_{m\times n}$$

The product of a scalar k and matrix $\mathbf{A} = [a_{ij}]_{m\times n}$ is

$$k\mathbf{A} = [ka_{ij}]_{m\times n}$$

Consider $\mathbf{A} = [a_{ij}]_{m\times n}$ and $\mathbf{B} = [b_{ij}]_{n\times p}$ so that the number of columns of **A** is equal to the number of rows of **B**. Then, their product $\mathbf{C} = \mathbf{AB}$ is $m \times p$ whose entries are obtained as

$$c_{ij} = \sum_{k=1}^{n} a_{ik} b_{kj}, \quad i = 1,2,\ldots,m, \quad j = 1,2,\ldots,p$$

1.1.2.1.1 Matrix Transpose

Given $\mathbf{A}_{m\times n}$, its transpose, denoted by $\mathbf{A}^T$, is an $n \times m$ matrix such that its first row is the first column of **A**, its second row is the second column of **A**, and so on. Provided all matrix operations are valid,

$$(\mathbf{A} + \mathbf{B})^T = \mathbf{A}^T + \mathbf{B}^T$$

$$(k\mathbf{A})^T = k\mathbf{A}^T \ (k = \text{scalar})$$

$$(\mathbf{AB})^T = \mathbf{B}^T\mathbf{A}^T$$

1.1.2.2 Special Matrices

A square matrix $\mathbf{A} = [a_{ij}]$ is symmetric if $\mathbf{A}^T = \mathbf{A}$, and skew-symmetric if $\mathbf{A}^T = -\mathbf{A}$. It is upper triangular if $a_{ij} = 0$ for all $i > j$, that is, all entries below the main diagonal are zeros. It is lower triangular if $a_{ij} = 0$ for all $i < j$, that is, all elements above the main diagonal are zeros. It is diagonal if $a_{ij} = 0$ for all $i \neq j$. In the upper and lower triangular matrices, the diagonal elements may be all zeros. However, in a diagonal matrix, at least one diagonal entry must be nonzero. The $n \times n$ identity matrix, denoted by **I**, is a diagonal matrix whose every diagonal entry is equal to 1.

1.1.2.3 Determinant of a Matrix

The determinant of a square matrix $\mathbf{A} = [a_{ij}]_{n\times n}$ is a real scalar denoted by $|\mathbf{A}|$ or det(**A**). For $n \geq 2$, the determinant may be calculated using any row or column—with preference given to the row or column with the most zeros. Using the ith row, the determinant is found as

$$|\mathbf{A}| = \sum_{k=1}^{n} a_{ik}(-1)^{i+k} M_{ik} \quad (i = 1,2,\ldots,n) \tag{1.6}$$

In Equation 1.6, M_{ik} is the minor of the entry a_{ik}, defined as the determinant of the $(n-1)\times(n-1)$ submatrix of **A** obtained by deleting the ith row and the kth column of **A**. The quantity $(-1)^{i+k}M_{ik}$ is the cofactor of a_{ik} and is denoted by C_{ik}. Also note that $(-1)^{i+k}$ is responsible for whether a term is multiplied by +1 or –1. A square matrix is nonsingular if its determinant is nonzero. Otherwise, it is called singular.

Example 1.4: Determinant

Calculate $|\mathbf{A}|$ if

$$\mathbf{A} = \begin{bmatrix} 1 & 2 & -1 & 3 \\ 2 & 0 & 1 & 4 \\ -1 & 1 & 5 & 2 \\ 3 & -4 & 2 & 3 \end{bmatrix}$$

SOLUTION

We will use the second row since it contains a zero entry.

$$|\mathbf{A}| = -2\begin{vmatrix} 2 & -1 & 3 \\ 1 & 5 & 2 \\ -4 & 2 & 3 \end{vmatrix} - \begin{vmatrix} 1 & 2 & 3 \\ -1 & 1 & 2 \\ 3 & -4 & 3 \end{vmatrix} + 4\begin{vmatrix} 1 & 2 & -1 \\ -1 & 1 & 5 \\ 3 & -4 & 2 \end{vmatrix}$$

$$= -2(99) - 32 + 4(55) = -10$$

Note that each of the 3×3 determinants is calculated via Equation 1.6.

1.1.2.3.1 *Properties of Determinant*

$$|\mathbf{AB}| = |\mathbf{A}||\mathbf{B}|$$

$$|\mathbf{A}^T| = |\mathbf{A}|$$

- The determinant of a lower triangular, upper triangular, or diagonal matrix is the product of the diagonal entries
- If any rows or columns of **A** are linearly dependent, then $|\mathbf{A}| = 0$

1.1.2.3.2 *Cramer's Rule*

Consider a linear system of n algebraic equations in n unknowns $x_1, x_2, \ldots, x_n$ in the form

$$\begin{cases} a_{11}x_1 + a_{12}x_2 + \cdots + a_{1n}x_n = b_1 \\ a_{21}x_1 + a_{22}x_2 + \cdots + a_{2n}x_n = b_2 \\ \quad\cdots \\ a_{n1}x_1 + a_{n2}x_2 + \cdots + a_{nn}x_n = b_n \end{cases} \quad (1.7)$$

where a_{ij} $(i, j = 1, 2, \ldots, n)$ and b_k $(k = 1, 2, \ldots, n)$ are known constants, and a_{ij}'s are the coefficients. Equation 1.7 can be expressed in matrix form, as

$$\mathbf{Ax} = \mathbf{b}$$

with

$$\mathbf{A} = \begin{bmatrix} a_{11} & a_{12} & \dots & a_{1n} \\ a_{21} & a_{22} & \dots & a_{2n} \\ \dots & \dots & \dots & \dots \\ a_{n1} & a_{n2} & \dots & a_{nn} \end{bmatrix}_{n\times n}, \quad \mathbf{x} = \begin{Bmatrix} x_1 \\ x_2 \\ \dots \\ x_n \end{Bmatrix}_{n\times 1}, \quad \mathbf{b} = \begin{Bmatrix} b_1 \\ b_2 \\ \dots \\ b_n \end{Bmatrix}_{n\times 1}$$

Assuming **A** is nonsingular, each unknown x_i ($i = 1, 2, \dots, n$) is uniquely determined via

$$x_i = \frac{\Delta_i}{\Delta}$$

where determinants Δ and Δ_i are described as

$$\Delta = \begin{vmatrix} a_{11} & a_{12} & \dots & a_{1n} \\ a_{21} & a_{22} & \dots & a_{2n} \\ \dots & \dots & \dots & \dots \\ a_{n1} & a_{n2} & \dots & a_{nn} \end{vmatrix}, \quad \Delta_i = \begin{vmatrix} a_{11} & \dots & \overset{i\text{th column of } \Delta}{b_1} & \dots & a_{1n} \\ a_{21} & \dots & b_2 & \dots & a_{2n} \\ \dots & \dots & \dots & \dots & \dots \\ \dots & \dots & \dots & \dots & \dots \\ a_{n1} & \dots & b_n & \dots & a_{nn} \end{vmatrix}$$

Example 1.5: Cramer's Rule

Solve the following system using Cramer's rule:

$$\begin{cases} 2x_1 + 3x_2 - x_3 = -3 \\ -x_1 + 2x_2 + x_3 = -6 \\ x_1 - 3x_2 - 2x_3 = 9 \end{cases}$$

SOLUTION

The determinants are calculated as

$$\Delta = \begin{vmatrix} 2 & 3 & -1 \\ -1 & 2 & 1 \\ 1 & -3 & -2 \end{vmatrix} = -6, \quad \Delta_1 = \begin{vmatrix} -3 & 3 & -1 \\ -6 & 2 & 1 \\ 9 & -3 & -2 \end{vmatrix} = -6,$$

$$\Delta_2 = \begin{vmatrix} 2 & -3 & -1 \\ -1 & -6 & 1 \\ 1 & 9 & -2 \end{vmatrix} = 12, \quad \Delta_3 = \begin{vmatrix} 2 & 3 & -3 \\ -1 & 2 & -6 \\ 1 & -3 & 9 \end{vmatrix} = 6$$

The unknowns are then found as

$$x_1 = \frac{\Delta_1}{\Delta} = 1, \quad x_2 = \frac{\Delta_2}{\Delta} = -2, \quad x_3 = \frac{\Delta_3}{\Delta} = -1$$

1.1.2.4 Inverse of a Matrix

The inverse of a square matrix $\mathbf{A}_{n\times n}$ is denoted by $\mathbf{A}^{-1}$ with the property $\mathbf{A}\mathbf{A}^{-1} = \mathbf{A}^{-1}\mathbf{A} = \mathbf{I}$ where $\mathbf{I}$ is the $n \times n$ identity matrix. The inverse of $\mathbf{A}$ exists only if $|\mathbf{A}| \neq 0$ and is obtained by using the adjoint matrix of $\mathbf{A}$, denoted by adj($\mathbf{A}$).

1.1.2.4.1 *Adjoint Matrix*

If $\mathbf{A} = [a_{ij}]_{n\times n}$, then the adjoint of $\mathbf{A}$ is defined as

$$\text{adj}(\mathbf{A}) = \begin{bmatrix} (-1)^{1+1}M_{11} & (-1)^{2+1}M_{21} & \cdots & (-1)^{n+1}M_{n1} \\ (-1)^{1+2}M_{12} & (-1)^{2+2}M_{22} & \cdots & (-1)^{n+2}M_{n2} \\ \cdots & \cdots & \cdots & \cdots \\ (-1)^{1+n}M_{1n} & (-1)^{2+n}M_{2n} & \cdots & (-1)^{n+n}M_{nn} \end{bmatrix} = \begin{bmatrix} C_{11} & C_{21} & \cdots & C_{n1} \\ C_{12} & C_{22} & \cdots & C_{n2} \\ \cdots & \cdots & \cdots & \cdots \\ C_{1n} & C_{2n} & \cdots & C_{nn} \end{bmatrix} \tag{1.8}$$

where M_{ij} is the minor of a_{ij} and $C_{ij} = (-1)^{i+j}M_{ij}$ is the cofactor of a_{ij}. Note that each minor M_{ij} (or cofactor C_{ij}) occupies the (j,i) position in the adjoint matrix. Then,

$$\mathbf{A}^{-1} = \frac{1}{|\mathbf{A}|}\text{adj}(\mathbf{A}) \tag{1.9}$$

Example 1.6: Inverse

Find the inverse of

$$\mathbf{A} = \begin{bmatrix} 3 & 1 & 1 \\ 1 & -1 & 1 \\ 0 & 2 & 1 \end{bmatrix}$$

SOLUTION

We first calculate $|\mathbf{A}| = -8$. Following the strategy outlined in Equation 1.8, the adjoint matrix of $\mathbf{A}$ is obtained as

$$\text{adj}(\mathbf{A}) = \begin{bmatrix} -3 & 1 & 2 \\ -1 & 3 & -2 \\ 2 & -6 & -4 \end{bmatrix}$$

Finally, by Equation 1.9, we have

$$\mathbf{A}^{-1} = \frac{1}{-8}\begin{bmatrix} -3 & 1 & 2 \\ -1 & 3 & -2 \\ 2 & -6 & -4 \end{bmatrix} = \begin{bmatrix} 0.3750 & -0.1250 & -0.2500 \\ 0.1250 & -0.3750 & 0.2500 \\ -0.2500 & 0.7500 & 0.5000 \end{bmatrix}$$

1.1.2.4.2 *Properties of Inverse*

$$(\mathbf{A}^{-1})^{-1} = \mathbf{A}$$

$$(\mathbf{AB})^{-1} = \mathbf{B}^{-1}\mathbf{A}^{-1}$$

$$(\mathbf{A}^{-1})^{p} = (\mathbf{A}^{p})^{-1}, \quad p = \text{integer} > 0$$

$$\left|\mathbf{A}^{-1}\right| = 1/\left|\mathbf{A}\right|$$

$$(\mathbf{A}^{-1})^{T} = (\mathbf{A}^{T})^{-1}$$

- Inverse of a symmetric matrix is symmetric
- Inverse of a diagonal matrix is diagonal whose entries are the reciprocals of the entries of the original matrix

1.1.2.4.3 *Solving a System of Equations*

A system of equations $\mathbf{Ax} = \mathbf{b}$ can be solved using $\mathbf{A}^{-1}$. Assuming $\mathbf{A}_{n\times n}$ is nonsingular, the solution vector $\mathbf{x}_{n\times 1}$ is found as follows:

$$\mathbf{Ax} = \mathbf{b} \quad \underset{\text{both sides by } \mathbf{A}^{-1}}{\overset{\text{Pre-multiply}}{\Longrightarrow}} \quad \mathbf{x} = \mathbf{A}^{-1}\mathbf{b}$$

1.1.3 Matrix Eigenvalue Problem

Consider an $n \times n$ matrix $\mathbf{A}$, a scalar λ (complex, in general), and a nonzero $n \times 1$ vector $\mathbf{v}$. The eigenvalue problem associated with matrix $\mathbf{A}$ is defined as

$$\mathbf{Av} = \lambda\mathbf{v}, \quad \mathbf{v} \neq \mathbf{0} \tag{1.10}$$

where λ is an eigenvalue of $\mathbf{A}$, and $\mathbf{v}$ is the eigenvector of $\mathbf{A}$ corresponding to λ. Note that an eigenvector cannot be the zero vector.

1.1.3.1 Solving the Eigenvalue Problem

Rewriting Equation 1.10, we have

$$\mathbf{Av} - \lambda\mathbf{v} = \mathbf{0} \quad \underset{\text{from the right side}}{\overset{\text{Factoring } \mathbf{v}}{\Rightarrow}} \quad [\mathbf{A} - \lambda\mathbf{I}]\mathbf{v} = \mathbf{0} \tag{1.11}$$

where the identity matrix **I** has been inserted to make the two terms in brackets size compatible. Recall that the solution vector **v** cannot be the zero vector, or the trivial solution. That said, Equation 1.11 has a nontrivial solution if and only if the coefficient matrix is singular, that is,

$$|\mathbf{A} - \lambda\mathbf{I}| = 0 \tag{1.12}$$

This is known as the characteristic equation for **A**. Since **A** is assumed $n \times n$, Equation 1.12 has n roots, which are the eigenvalues of **A**. The corresponding eigenvector for each λ is obtained by solving Equation 1.11. Since $\mathbf{A} - \lambda\mathbf{I}$ is singular, it has at least one row dependent on other rows [see Section 1.1.2.3.1]. Therefore, for each λ, Equation 1.11 has infinitely many solutions. A basis of solutions will then represent all eigenvalues associated with λ.

Example 1.7: Eigenvalue Problem

Find the eigenvalues and eigenvectors of

$$\mathbf{A} = \begin{bmatrix} 1 & 0 & 1 \\ 0 & 1 & 0 \\ 1 & 0 & 1 \end{bmatrix}$$

SOLUTION

The characteristic equation is formed as

$$|\mathbf{A} - \lambda\mathbf{I}| = \begin{vmatrix} 1-\lambda & 0 & 1 \\ 0 & 1-\lambda & 0 \\ 1 & 0 & 1-\lambda \end{vmatrix} = \lambda(\lambda - 1)(\lambda - 2) = 0 \Rightarrow \lambda = 0,1,2$$

Solving Equation 1.11 with $\lambda_1 = 0$, we have

$$[\mathbf{A} - \lambda_1\mathbf{I}]\mathbf{v}_1 = \mathbf{0} \quad \Rightarrow \quad \begin{bmatrix} 1 & 0 & 1 \\ 0 & 1 & 0 \\ 1 & 0 & 1 \end{bmatrix}\mathbf{v}_1 = \begin{Bmatrix} 0 \\ 0 \\ 0 \end{Bmatrix}$$

Let the three components of $\mathbf{v}_1$ be a, b, c. Then, the above system yields $b = 0$ and $a + c = 0$. Since the coefficient matrix is singular, by design, there is a free variable, which can be either a or c. Letting $a = 1$ leads to $c = -1$, and consequently the eigenvector associated with $\lambda_1 = 0$ is determined as

$$\mathbf{v}_1 = \begin{Bmatrix} 1 \\ 0 \\ -1 \end{Bmatrix}$$

Similarly, the eigenvectors associated with the other two eigenvalues ($\lambda_2 = 1$, $\lambda_3 = 2$) will be obtained as

$$\mathbf{v}_2 = \begin{Bmatrix} 0 \\ 1 \\ 0 \end{Bmatrix}, \quad \mathbf{v}_3 = \begin{Bmatrix} 1 \\ 0 \\ 1 \end{Bmatrix}$$

1.1.3.2 Similarity Transformation

Consider a matrix $\mathbf{A}_{n \times n}$ and a nonsingular matrix $\mathbf{S}_{n \times n}$ and suppose

$$\mathbf{S}^{-1}\mathbf{AS} = \mathbf{B} \tag{1.13}$$

We say $\mathbf{B}$ has been obtained through a similarity transformation of $\mathbf{A}$, and that matrices $\mathbf{A}$ and $\mathbf{B}$ are similar. The most important property of similar matrices is that they have the same set of eigenvalues.

1.1.3.2.1 Matrix Diagonalization

Suppose matrix $\mathbf{A}_{n \times n}$ has eigenvalues $\lambda_1, \lambda_2, \ldots, \lambda_n$ and linearly independent eigenvectors $\mathbf{v}_1, \mathbf{v}_2, \ldots, \mathbf{v}_n$. Then, the modal matrix $\mathbf{P} = [\mathbf{v}_1 \;\; \mathbf{v}_2 \;\; \ldots \;\; \mathbf{v}_n]_{n \times n}$ diagonalizes $\mathbf{A}$ by means of a similarity transformation:

$$\mathbf{P}^{-1}\mathbf{AP} = \mathbf{D} = \begin{bmatrix} \lambda_1 & & & \\ & \lambda_2 & & \\ & & \ldots & \\ & & & \lambda_n \end{bmatrix} \tag{1.14}$$

Example 1.8: Matrix Diagonalization

Consider the matrix in Example 1.7. The modal matrix is formed as

$$\mathbf{P} = \begin{bmatrix} \mathbf{v}_1 & \mathbf{v}_2 & \mathbf{v}_3 \end{bmatrix} = \begin{bmatrix} 1 & 0 & 1 \\ 0 & 1 & 0 \\ -1 & 0 & 1 \end{bmatrix}$$

Subsequently,

$$\mathbf{P}^{-1}\mathbf{AP} = \begin{bmatrix} 0 & 0 & 0 \\ 0 & 1 & 0 \\ 0 & 0 & 2 \end{bmatrix} = \begin{bmatrix} \lambda_1 & & \\ & \lambda_2 & \\ & & \lambda_3 \end{bmatrix}$$

1.1.3.2.2 Eigenvalue Properties of Matrices

- The determinant of a matrix is the product of the eigenvalues of that matrix.
- Eigenvalues of lower triangular, upper triangular, and diagonal matrices are the diagonal entries of that matrix.
- Similar matrices have the same set of eigenvalues.
- Eigenvalues of a symmetric matrix are all real.
- Every eigenvalue of an orthogonal matrix $[\mathbf{A}^{-1} = \mathbf{A}^T]$ has an absolute value of 1.

1.2 Introduction to Numerical Methods

1.2.1 Errors and Approximations

Numerical methods are procedures that allow for efficient solution of a mathematically formulated problem in a finite number of steps to within an arbitrary precision. Although scientific calculators can handle simple problems, computers are needed in most other cases. Numerical methods commonly consist of a set of guidelines to perform predetermined mathematical (algebraic and logical) operations leading to an approximate solution of a specific problem. Such set of guidelines is known as an algorithm. A very important issue here is the errors caused in computations.

1.2.1.1 Computational Errors

While investigating the accuracy of the result of a certain numerical method, two key questions arise: (1) what are the possible sources of error, and (2) to what degree do these errors affect the ultimate result? In numerical computations, there exist three possible sources of error:

1. Error in the initial model
2. Truncation error
3. Round-off error

The first source occurs in the initial model of the problem; for example, when simplifying assumptions are made in the derivation of the mathematical model of a physical system. Also, if mathematical numbers such as e or π are encountered, then we need to use their approximate values such as 2.7183 and 3.1416.

The second source is due to truncation errors, which occur whenever an expression is approximated by some type of a mathematical method. As an example, suppose we use the Maclaurin series representation of the sine function:

$$\sin\alpha = \sum_{n=odd}^{\infty} \frac{(-1)^{(n-1)/2}}{n!}\alpha^n = \alpha - \frac{1}{3!}\alpha^3 + \frac{1}{5!}\alpha^5 - \cdots + \frac{(-1)^{(m-1)/2}}{m!}\alpha^m + E_m$$

where E_m is the tail end of the expansion, neglected in the process, and known as the truncation error.

The third type of computational error is caused by the computer within the process of translating a decimal number to a binary number. This is because unlike humans who use the decimal number system (in base 10), computers mostly use the binary number system (in base 2, or perhaps 16). In this process, the number that is inputted to the computer is first converted to base 2, arithmetic is done in base 2, and finally the outcome is converted back to base 10.

1.2.1.2 Binary and Hexadecimal Numbers

For ordinary purposes, base 10 is used to represent numbers. For example, the number 147 is expressed as

$$147 = \left[1 \times 10^2 + 4 \times 10^1 + 7 \times 10^0\right]_{10}$$

where the subscript is usually omitted when the base is 10. This is known as decimal notation. The so-called normalized decimal form of a number is

$$\pm 0.d_1 d_2 \ldots d_m \times 10^p, \quad 1 \le d_1 \le 9, \quad 0 \le d_2, d_3, \ldots, d_m \le 9 \qquad (1.15)$$

Equation 1.15 is also known as the floating-point form, to be explained shortly. On the other hand, most computers use the binary system (in base 2). For instance, the number 147 is expressed in base 2 as follows. First, we readily verify that

$$147 = \left[1 \times 2^7 + 0 \times 2^6 + 0 \times 2^5 + 1 \times 2^4 + 0 \times 2^3 + 0 \times 2^2 + 1 \times 2^1 + 1 \times 2^0\right]_2$$

Then, in base 2, we have

$$147 = (10010011)_2$$

We refer to a binary digit as a bit. This last expression represents a binary number. Similarly, the same number can be expressed in base 16, as

$$147 = \left[9 \times 16^1 + 3 \times 16^0\right]_{16} \overset{\text{In base 16}}{\Rightarrow} 147 = (93)_{16}$$

This last expression represents a hexadecimal number. While the binary system consists of only two digits, 0 and 1, there are 16 digits in the hexadecimal system; 0, 1, 2, … , 9, A, B, … , F, where A–F represent 10–15. We then sense that the hexadecimal system is a natural extension of the binary system. Since $2^4 = 16$, for every group of four bits, there is one hexadecimal digit. Examples include $C = (1100)_2$, $3 = (0011)_2$, and so on.

1.2.1.3 Floating Point and Rounding Errors

Because only a limited number of digits can be stored in computer memory, a number must be represented in a manner that uses a somewhat fixed number of digits. Digital computers mostly represent a number in one of two ways: fixed point and floating point. In a fixed-point setting, a fixed number of decimal places are used for the representation of numbers. For instance, in a system using 4 decimal places, we encounter numbers like –2.0000, 131.0174, 0.1234. On the other hand, in a floating-point setting, a fixed number of significant digits* are used for representation. For instance, if four significant digits are used, then we will encounter numbers such as†

$$0.2501 \times 10^{-2}, \quad -0.7012 \times 10^5$$

Note that these two numbers fit the general form given by Equation 1.15. In the floating-point representation of a number, one position is used to identify its sign, a prescribed number of bits to represent its fractional part, known as the mantissa, and another prescribed number of bits for its exponential part, known as the characteristic. Computers that use 32 bits for single-precision representation of numbers, use 1 bit for the sign, 24 bits for the mantissa, and 8 bits for the exponent. Typical computers can handle wide

* Note that significant digits are concerned with the first nonzero digit and the ones to its right. For example, 4.0127 and 0.088659 both have five significant digits.

† Also expressed in scientific notation, as 0.2501E – 2 and –0.7012E + 5.

ranges of exponents. As one example, the IEEE* floating point standard range is between −38 and +38. Outside of this range, the result is an underflow if the number is smaller than the minimum and an overflow if the number is larger than the maximum.

1.2.1.3.1 *Round-Off: Chopping and Rounding*

Consider a positive real number N expressed as

$$N = 0.d_1d_2 \ldots d_m d_{m+1} \ldots \times 10^p$$

The floating-point form of N, denoted by FL(N), in the form of Equation 1.15, is obtained by terminating its fractional part at m decimal digits. There are two ways to do this. The first method is called chopping, and involves chopping of the digits to the right of d_m to get

$$\text{FL}(N) = 0.d_1d_2 \ldots d_m \times 10^p$$

The second method is known as rounding, and involves adding $5 \times 10^{p-(m+1)}$ to N and then chopping. In this process, if $d_{m+1} < 5$, then all that happens is that the first m digits are retained. This is known as rounding down. If $d_{m+1} \geq 5$, then FL(N) is obtained by adding one to d_m. This is called rounding up. It is clear that when a number is replaced with its floating-point form, whether through rounding down or up, an error results. This error is called round-off error.

Example 1.9: Chopping and Rounding

Consider $e = 2.71828182\ldots = 0.271828182\ldots \times 10^1$. If we use 5-digit chopping ($m = 5$), the floating-point form is $\text{FL}(e) = 0.27182 \times 10^1 = 2.7182$. We next use rounding. Since the digit immediately to the right of d_5 is $d_6 = 8 > 5$, we add 1 to d_5 to obtain

$$\text{FL}(e) = 0.27183 \times 10^1 = 2.7183$$

so that we have rounded up. The same result is obtained by following the strategy of adding $5 \times 10^{p-(m+1)}$ to e and chopping. Note that $p = 1$ and $m = 5$, so that $5 \times 10^{p-(m+1)} = 5 \times 10^{-5} = 0.00005$. Adding this to e, we have

$$e + 0.00005 = 2.71828182\ldots + 0.00005 = 2.71833\ldots = 0.271833\ldots \times 10^1$$

5-digit chopping yields $\text{FL}(e) = 0.27183 \times 10^1 = 2.7183$, which agrees with the result of rounding up.

* Institute of Electrical and Electronics Engineers.

1.2.1.4 Absolute and Relative Errors

In the beginning of this section, we discussed the three possible sources of error in computations. Regardless of what the source may be, computations generally yield approximations as their output. This output may be an approximation to a true solution of an equation, or an approximation of a true value of some quantity. Errors are commonly measured in one of two ways: absolute error and relative error. If $\tilde{x}$ is an approximation to a quantity whose true value is x, the absolute error is defined as

$$e_{\text{abs}} = x - \tilde{x} \tag{1.16}$$

On the other hand, the true relative error is given by

$$e_{\text{rel}} = \frac{\text{Absolute error}}{\text{True value}} = \frac{e_{\text{abs}}}{x} = \frac{x - \tilde{x}}{x}, \quad x \neq 0 \tag{1.17}$$

Note that if the true value happens to be zero, the relative error is regarded as undefined. The relative error is generally of more significance than the absolute error, as we will discuss in Example 1.10. And because of that, whenever possible, we will present bounds for the relative error in computations.

Example 1.10: Absolute and Relative Errors

Consider two different computations. In the first one, an estimate $\tilde{x}_1 = 0.003$ is obtained for the true value $x_1 = 0.004$. In the second one, $\tilde{x}_2 = 1238$ for $x_2 = 1258$. Therefore, the absolute errors are

$$(e_{\text{abs}})_1 = x_1 - \tilde{x}_1 = 0.001, \quad (e_{\text{abs}})_2 = x_2 - \tilde{x}_2 = 20$$

The corresponding relative errors are

$$(e_{\text{rel}})_1 = \frac{(e_{\text{abs}})_1}{x_1} = \frac{0.001}{0.004} = 0.25, \quad (e_{\text{rel}})_2 = \frac{(e_{\text{abs}})_2}{x_2} = \frac{20}{1258} = 0.0159$$

We notice that the absolute errors of 0.001 and 20 can be rather misleading, judging by their magnitudes. In other words, the fact that 0.001 is much smaller than 20 does not make the first error a smaller error relative to its corresponding computation. In fact, looking at the relative errors, we see that 0.001 is associated with a 25% error, while 20 corresponds to 1.59% error, much smaller than the first. Because they convey a more specific type of information, relative errors are considered more significant than absolute errors.

1.2.1.4.1 Error Bound

It is customary to use the absolute value of e_{abs} so that only the upper bound needs to be obtained, since the lower bound is clearly zero. We say that α is an upper bound for the absolute error if

$$|e_{\text{abs}}| = |x - \tilde{x}| \leq \alpha$$

Note that α does not provide an estimate for $|x - \tilde{x}|$, and is simply a bound. Similarly, we say that β is an upper bound for the relative error if

$$|e_{\text{rel}}| = \frac{|x - \tilde{x}|}{|x|} \leq \beta, \quad x \neq 0$$

Example 1.11: Error Bound

Find two upper bounds for the relative errors caused by the 5-digit chopping and rounding of e in Example 1.9.

SOLUTION

Using the results of Example 1.9, we have

$$|e_{\text{rel}}|_{\text{Chopping}} = \frac{|e - \text{FL}(e)|}{|e|} = \frac{0.000008182\ldots \times 10^1}{0.271828182\ldots \times 10^1} = \frac{0.8182\ldots}{0.271828182\ldots} \times 10^{-5} \leq 10^{-4}$$

Here, we have used the fact that the numerator is less than 1, while the denominator is greater than 0.1. It can be shown that in the general case, an m-digit chopping results in an upper bound relative error of 10^{1-m}. For the 5-digit rounding, we have

$$|e_{\text{rel}}|_{\text{Rounding}} = \frac{|e - \text{FL}(e)|}{|e|} = \frac{0.000001818\ldots \times 10^1}{0.271828182\ldots \times 10^1} = \frac{0.1818\ldots}{0.271828182\ldots} \times 10^{-5} \leq 0.5 \times 10^{-4}$$

where we used the fact that the numerator is less than 0.5 and the denominator is greater than 0.1. In general, an m-digit rounding corresponds to an upper-bound relative error of $0.5 \times 10^{1-m}$.

1.2.1.4.2 Transmission of Error from a Source to the Final Result

Now that we have learned about the sources of error, we need to find out about the degree to which these errors affect the outcome of a computation. Depending on whether addition (and/or subtraction) or multiplication (and/or division) is considered, definite conclusions may be drawn.

Theorem 1.1

Suppose in a certain computation the approximate values $\tilde{x}_1$ and $\tilde{x}_2$ have been generated for true values x_1 and x_2, respectively, with absolute and relative errors $(e_{\text{abs}})_i$ and $(e_{\text{rel}})_i$, $i = 1, 2$, and

$$\left|(e_{\text{abs}})_1\right| \le \alpha_1, \quad \left|(e_{\text{abs}})_2\right| \le \alpha_2, \quad \left|(e_{\text{rel}})_1\right| \le \beta_1, \quad \left|(e_{\text{rel}})_2\right| \le \beta_2$$

a. The upper bound for the absolute error e_{abs} in addition and subtraction is the sum of the upper bounds of the absolute errors associated with the quantities involved. That is,

$$\left|e_{\text{abs}}\right| = \left|(x_1 \pm x_2) - (\tilde{x}_1 \pm \tilde{x}_2)\right| \le \alpha_1 + \alpha_2$$

b. The upper bound for relative error e_{rel} in multiplication and division is approximately equal to the sum of the upper bounds of the relative errors associated with the quantities involved. That is,

Multiplication

$$\left|e_{\text{rel}}\right| = \left|\frac{x_1 x_2 - \tilde{x}_1 \tilde{x}_2}{x_1 x_2}\right| \le \beta_1 + \beta_2 \tag{1.18}$$

Division

$$\left|e_{\text{rel}}\right| = \left|\frac{x_1/x_2 - \tilde{x}_1/\tilde{x}_2}{x_1/x_2}\right| \le \beta_1 + \beta_2 \tag{1.19}$$

Proof

a. We have

$$\begin{aligned}\left|e_{\text{abs}}\right| &= \left|(x_1 \pm x_2) - (\tilde{x}_1 \pm \tilde{x}_2)\right| = \left|(x_1 - \tilde{x}_1) \pm (x_2 - \tilde{x}_2)\right| \\ &\le \left|x_1 - \tilde{x}_1\right| + \left|x_2 - \tilde{x}_2\right| \le \alpha_1 + \alpha_2\end{aligned}$$

b. We will prove Equation 1.18. Noting that $(e_{\text{abs}})_i = x_i - \tilde{x}_i$ for $i = 1, 2$, we have $\tilde{x}_i = x_i - (e_{\text{abs}})_i$. Insertion into the left side of Equation 1.18 yields

$$\begin{aligned}\left|e_{\text{rel}}\right| &= \left|\frac{x_1 x_2 - \tilde{x}_1 \tilde{x}_2}{x_1 x_2}\right| = \left|\frac{x_1 x_2 - \left[x_1 - (e_{\text{abs}})_1\right]\left[x_2 - (e_{\text{abs}})_2\right]}{x_1 x_2}\right| \\ &= \left|\frac{-(e_{\text{abs}})_1 (e_{\text{abs}})_2 + (e_{\text{abs}})_2 x_1 + (e_{\text{abs}})_1 x_2}{x_1 x_2}\right|\end{aligned}$$

But $(e_{abs})_1$ $(e_{abs})_2$ can be assumed negligible relative to the other two terms in the numerator. As a result,

$$|e_{rel}| \cong \left|\frac{(e_{abs})_2 x_1 + (e_{abs})_1 x_2}{x_1 x_2}\right| = \left|\frac{(e_{abs})_1}{x_1} + \frac{(e_{abs})_2}{x_2}\right|$$

$$\leq \left|\frac{(e_{abs})_1}{x_1}\right| + \left|\frac{(e_{abs})_2}{x_2}\right| \leq \beta_1 + \beta_2$$

as asserted.

1.2.1.4.3 Subtraction of Nearly Equal Numbers

There are two particular instances leading to unacceptable inaccuracies: division by a number that is very small in magnitude, and subtraction of nearly equal numbers. Naturally, if this type of subtraction takes place in the denominator of a fraction, the latter gives rise to the former. Consider two numbers N_1 and N_2 having the same first k decimal digits in their floating-point forms, that is,

$$\mathrm{FL}(N_1) = 0.d_1 d_2 \ldots d_k a_{k+1} \ldots a_m \times 10^p$$
$$\mathrm{FL}(N_2) = 0.d_1 d_2 \ldots d_k b_{k+1} \ldots b_m \times 10^p$$

The larger the value of k, the more "nearly equal" the two numbers are considered to be. Subtraction yields

$$\mathrm{FL}(\mathrm{FL}(N_1) - \mathrm{FL}(N_2)) = 0.c_{k+1} \ldots c_m \times 10^{p-k}$$

where $c_{k+1}, \ldots, c_m$ are constant digits. From this expression we see that there are only $m-k$ significant digits in the representation of the difference. In comparison with the m significant digits available in the original representations of the two numbers, some significant digits have been lost in the process. This is precisely what contributes to the round-off error, which will then be propagated throughout the subsequent computations. This can often be remedied by a simple reformulation of the problem, as illustrated in the following example.

Example 1.12: The Quadratic Formula When $b^2 >> 4ac$

Consider the quadratic equation $x^2 + 52x + 3 = 0$ with approximate roots $x_1 = -0.05775645785$, $x_2 = -51.94224354$. Recall that the quadratic formula generally provides the solution of $ax^2 + bx + c = 0$, as

$$x_1 = \frac{-b + \sqrt{b^2 - 4ac}}{2a}, \quad x_2 = \frac{-b - \sqrt{b^2 - 4ac}}{2a}$$

But in our example, $b^2 >> 4ac$ so that $\sqrt{b^2 - 4ac} \cong b$. This means that in the calculation of x_1 we are subtracting nearly equal numbers in the numerator. Now, let us use a 4-digit rounding for floating-point representation. Then,

$$\text{FL}(x_1) = \frac{-52.00 + \sqrt{(52.00)^2 - 4(1.000)(3.000)}}{2(1.000)} = \frac{-52.00 + 51.88}{2.000} = -0.0600$$

and

$$\text{FL}(x_2) = \frac{-52.00 - \sqrt{(52.00)^2 - 4(1.000)(3.000)}}{2(1.000)} = -51.94$$

The corresponding relative errors, in magnitude, are computed as

$$\left|e_{\text{rel}}\right|_{x_1} = \frac{\left|x_1 - \text{FL}(x_1)\right|}{\left|x_1\right|} \cong 0.0388 \quad \text{or} \quad 3.88\%$$

$$\left|e_{\text{rel}}\right|_{x_2} = \frac{\left|x_2 - \text{FL}(x_2)\right|}{\left|x_2\right|} \cong 0.0043 \quad \text{or} \quad 0.43\%$$

Thus, the error associated with x_1 is rather large compared to that for x_2. We anticipated this because in the calculation of x_2 nearly equal numbers are added, causing no concern. As mentioned above, reformulation of the problem often remedies the situation. Also note that the roots of $ax^2 + bx + c = 0$ satisfy $x_1x_2 = c/a$. We will retain the value of FL(x_2) and calculate FL(x_1) via

$$\text{FL}(x_1) = \frac{c}{a\text{FL}(x_2)} = \frac{3.000}{(1.000)(-51.94)} = -0.05775$$

The resulting relative error is

$$\left|e_{\text{rel}}\right|_{x_1} = \frac{\left|x_1 - \text{FL}(x_1)\right|}{\left|x_1\right|} \cong 0.00011 \quad \text{or} \quad 0.011\%$$

which shows a dramatic improvement compared to the result of the first trial.

1.2.2 Iterative Methods

Numerical methods generally consist of a set of directions to perform predetermined algebraic and logical mathematical operations leading to an approximate solution of a specific problem. These set of directions are known as algorithms. To effectively describe a certain algorithm, we will use a code. Based on the programming language or the software package used, a code can easily be modified to accomplish the task at hand. A code consists of a set of inputs, the required operations, and a list of outputs. It is standard practice to use two types of punctuation symbols in an algorithm: the period (.) proclaims that the current step is terminated, and the semicolon (;) indicates that the step is still in progress. An algorithm is stable if a small change in the initial data will correspond to a small change in the final result. Otherwise, it is unstable.

An iterative method is a process that starts with an initial guess and computes successive approximations of the solution of a problem until a reasonably accurate approximation is obtained. As we will demonstrate throughout the book, iterative methods are used to find roots of algebraic equations, solutions of systems of algebraic equations, solutions of differential equations, and much more. An important issue in an iterative scheme is the manner in which it is terminated. There are two ways to stop a procedure: (1) when a terminating condition is satisfied or (2) when the maximum number of iterations is exceeded. In principle, the terminating condition should check to see whether an approximation calculated in a step is within a prescribed tolerance of the true value. In practice, however, the true value is not available. As a result, one practical form of a terminating condition is whether the difference between two successively generated quantities by the iterative method is within a prescribed tolerance. The ideal scenario is when an algorithm meets the terminating condition, and at a reasonably fast rate. If it does not, then the total number of iterations performed should not exceed a prescribed maximum number of iterations.

Example 1.13: An Algorithm and Its Code

Approximate e^{-2} to seven significant digits with a tolerance of $\varepsilon = 10^{-6}$.

SOLUTION

Retaining the first $n+1$ terms in the Maclaurin series of $f(x) = e^x$ yields the nth-degree Taylor's polynomial

$$T_n(x) = \sum_{i=0}^{n} \frac{1}{i!} x^i \tag{1.20}$$

We want to evaluate e^{-2} by determining the least value of n in Equation 1.20 such that

$$\left| e^{-2} - T_n(-2) \right| < \varepsilon \tag{1.21}$$

TABLE 1.2

Algorithm in Example 1.13

Input:	$x = 2$, $\varepsilon = 10^{-6}$, $N = 20$	
Output:	An approximate value of e^{-2} accurate to within ε, or a message of "failure"	
Step 1:	Set $n = 0$	
	Tval = e^{-x}	
	Term = 1	True value
	Psum = 0	Initiate partial sum
	Sgn = 1	Initiate alternating signs
Step 2:	While $n \le N$, do Step 3–Step 5	
Step 3:	Psum = Psum + Sgn*Term/n!	
Step 4:	If |Psum − Tval| < ε, then Output(n)	Terminating condition
	Stop	
Step 5:	$n = n + 1$	Update n
	Sgn = −Sgn	Alternate sign
	Term = Term*x	Update Term
Step 6:	Output(failure)	
	Stop	
End		

Equation 1.21 is the terminating condition. To seven significant digits, the true value is $e^{-2} = 0.1353353$. Let us set the maximum number of iterations as $N = 20$, so that the program is likely to fail if the number of iterations exceeds 20 and the terminating condition is not met. As soon as an approximate value within the given tolerance is reached, the terminating condition is satisfied and the program is terminated. Then the outputs are n and the corresponding value for e^{-2}. We write the algorithm listed in Table 1.2. It turns out that 14 iterations are needed before the terminating condition is satisfied, that is, $n = 13$. The approximate value for e^{-2} is 0.1353351 with an absolute error of $0.2 \times 10^{-6} < \varepsilon$.

1.2.2.1 A Fundamental Iterative Method

A fundamental iterative method is the one that uses repeated substitutions. Suppose that a function $g(x)$ and a starting value x_0 are known. Let us generate a sequence of values $x_1, x_2, \ldots$ via an iteration defined by

$$x_{n+1} = g(x_n), \quad n = 0, 1, 2, \ldots, \quad x_0 \text{ is known} \tag{1.22}$$

There are a few possible scenarios. The iteration may exhibit convergence, either at a fast rate or a slow rate. It is also possible that the iteration does not converge at all. Again, its divergence may happen at a slow or a fast rate. All these depend on critical factors such as the nature of the function $g(x)$ and the starting value, x_0. We will analyze these in detail in Chapter 3.

Example 1.14: Iteration by Repeated Substitutions

Consider the sequence described by $x_n = \left(\frac{1}{2}\right)^n$, $n = 0, 1, 2, \ldots$. To generate the same sequence of elements using iteration by repeated substitutions, we need to reformulate it to agree with Equation 1.22. To that end, we propose

$$x_{n+1} = \left(\frac{1}{2}\right)x_n, \quad n = 0, 1, 2, \ldots, \quad x_0 = 1$$

This way, the sequence starts with $x_0 = 1$, which matches the first element of the original sequence. Next, we calculate

$$x_1 = \frac{1}{2}x_0 = \frac{1}{2}$$

which agrees with the respective element in the original sequence. All elements of the sequence generated in this manner agree with the original. Therefore,

$$x_{n+1} = g(x_n), \quad n = 0, 1, 2, \ldots, \quad x_0 = 1$$

where $g(x) = \frac{1}{2}x$.

1.2.2.2 Rate of Convergence of an Iterative Method

Consider a sequence $\{x_n\}$ that converges to x. The error at the nth iteration is then defined as

$$e_n = x - x_n, \quad n = 0, 1, 2, \ldots$$

If there exists a number R and a constant $K \neq 0$ such that

$$\lim_{n\to\infty} \frac{|e_{n+1}|}{|e_n|^R} = K \tag{1.23}$$

then we say that R is the rate of convergence of the sequence. There are two types of convergence that we often encounter: linear and quadratic. A convergence is linear if $R = 1$, that is,

$$\lim_{n\to\infty} \frac{|e_{n+1}|}{|e_n|} = K \neq 0 \tag{1.24}$$

A convergence is said to be quadratic if $R = 2$, that is,

$$\lim_{n\to\infty} \frac{|e_{n+1}|}{|e_n|^2} = K \neq 0 \tag{1.25}$$

Rate of convergence is not always an integer. We will see in Section 3.6, for instance, that the secant method has a rate of $\frac{1}{2}(1+\sqrt{5}) \cong 1.618$.

Example 1.15: Rate of Convergence

Determine the rate of convergence for the sequence in Example 1.14.

SOLUTION

Since $x_n = \left(\frac{1}{2}\right)^n \to 0$ as $n \to \infty$, the limit is $x = 0$. With that, the error at the nth iteration is

$$e_n = x - x_n = 0 - \left(\frac{1}{2}\right)^n = -\left(\frac{1}{2}\right)^n$$

We will first examine $R = 1$, that is, Equation 1.24:

$$\lim_{n\to\infty} \frac{|e_{n+1}|}{|e_n|} = \lim_{n\to\infty} \frac{\left|-\left(\frac{1}{2}\right)^{n+1}\right|}{\left|-\left(\frac{1}{2}\right)^n\right|} = \frac{1}{2} \neq 0$$

Therefore, $R = 1$ works and convergence is linear. Once a value of R satisfies the condition in Equation 1.23, no other values need be inspected.

PROBLEM SET

Differential Equations (Section 1.1.1)

✍ ***Hand Calculation***

In Problems 1 through 8, solve each initial-value problem.

1. $\dot{y} + y = t, \quad y(0) = 1$
2. $\dot{y} + ty = t, \quad y(0) = 0$
3. $\frac{1}{2}\dot{y} + y = 0, \quad y(0) = \frac{1}{2}$
4. $\dot{y} + y = e^{-2t}, \quad y(0) = 1$
5. $\ddot{y} + 2\dot{y} + 2y = 0, \quad y(0) = 0, \quad \dot{y}(0) = 1$
6. $\ddot{y} + 2\dot{y} + y = e^{-t}, \quad y(0) = 0, \quad \dot{y}(0) = 0$
7. $\ddot{y} + 2\dot{y} = \sin t, \quad y(0) = 0, \quad \dot{y}(0) = 0$
8. $\ddot{y} + 2\dot{y} = t, \quad y(0) = 1, \quad \dot{y}(0) = 1$

Matrix Analysis (Section 1.1.2)

✍ In Problems 9 through 12 calculate the determinant of the given matrix.

9. $\mathbf{A} = \begin{bmatrix} 3 & -4 & 1 \\ 1 & -1 & 2 \\ 0 & 2 & 6 \end{bmatrix}$

10. $\mathbf{A} = \begin{bmatrix} 8 & 2 & -1 \\ 1 & 0 & 4 \\ -3 & 4 & 5 \end{bmatrix}$

11. $\mathbf{A} = \begin{bmatrix} -1 & 3 & 1 & 2 \\ 1 & 0 & 4 & 3 \\ 1 & -1 & 1 & 0 \\ 2 & 3 & 4 & -5 \end{bmatrix}$

12. $\mathbf{A} = \begin{bmatrix} 0 & 2 & -3 & 0 \\ -1 & 2 & 4 & -3 \\ 2 & 0 & 1 & 1 \\ 4 & 5 & -3 & 1 \end{bmatrix}$

✍ In Problems 13 through 16 solve each system using Cramer's rule.

13. $\begin{cases} x_1 + x_2 - 4x_3 = -2 \\ -2x_1 + x_2 + 3x_3 = -1 \\ x_2 + 5x_3 = 5 \end{cases}$

14. $\begin{cases} 4x_1 + x_2 - 3x_3 = -13 \\ -x_1 + 2x_2 + 6x_3 = 13 \\ x_1 + 7x_3 = 4 \end{cases}$

15. $\mathbf{Ax} = \mathbf{b}, \quad \mathbf{A} = \begin{bmatrix} 2 & 1 & 0 & -1 \\ 1 & 3 & -1 & 3 \\ 0 & 1 & -3 & 2 \\ 2 & 0 & 1 & 4 \end{bmatrix}, \quad \mathbf{x} = \begin{Bmatrix} x_1 \\ x_2 \\ x_3 \\ x_4 \end{Bmatrix}, \quad \mathbf{b} = \begin{Bmatrix} -3 \\ 13 \\ 5 \\ 11 \end{Bmatrix}$

16. $\mathbf{Ax} = \mathbf{b}, \quad \mathbf{A} = \begin{bmatrix} -1 & 0 & 4 & 2 \\ 5 & 1 & 3 & -1 \\ 1 & 0 & 2 & 2 \\ -3 & 2 & 0 & -2 \end{bmatrix}, \quad \mathbf{x} = \begin{Bmatrix} x_1 \\ x_2 \\ x_3 \\ x_4 \end{Bmatrix}, \quad \mathbf{b} = \begin{Bmatrix} 3 \\ 22 \\ 5 \\ -7 \end{Bmatrix}$

✍ In Problems 17 through 20 find the inverse of each matrix.

17. $\mathbf{A} = \begin{bmatrix} 4 & 0 & 1 \\ 0 & 3 & -2 \\ -1 & -2 & -1 \end{bmatrix}$

18. $\mathbf{A} = \begin{bmatrix} 0 & 1 & 0 \\ 0 & 0 & 1 \\ 1 & 2 & 1 \end{bmatrix}$

19. $\mathbf{A} = \begin{bmatrix} 1 & 0 & 0 \\ 0 & 5 & 0 \\ 4 & 3 & -2 \end{bmatrix}$

20. $\mathbf{A} = \begin{bmatrix} \alpha & 0 & -1 \\ 0 & \alpha+1 & 2 \\ 1 & 0 & \alpha+2 \end{bmatrix}, \quad \alpha = \text{parameter}$

Matrix Eigenvalue Problem (Section 1.1.3)

✍ In Problems 21 through 24 find the eigenvalues and corresponding eigenvectors of each matrix.

21. $\mathbf{A} = \begin{bmatrix} -3 & 0 \\ -2 & 1 \end{bmatrix}$

22. $\mathbf{A} = \begin{bmatrix} 2 & 2 & 0 \\ 1 & 1 & 0 \\ 0 & 0 & 1 \end{bmatrix}$

23. $\mathbf{A} = \begin{bmatrix} 1 & 2 & 1 \\ 0 & 2 & -3 \\ 0 & 0 & -1 \end{bmatrix}$

24. $\mathbf{A} = \begin{bmatrix} 1 & 0 & 0 \\ 1 & 2 & 0 \\ 2 & 3 & 3 \end{bmatrix}$

25. ✍ Prove that a singular matrix has at least one zero eigenvalue.

✍ In Problems 26 through 28 diagonalize each matrix by using an appropriate modal matrix.

26. $\mathbf{A} = \begin{bmatrix} -2 & -1 & -1 \\ 3 & 2 & 1 \\ 1 & 1 & 0 \end{bmatrix}$

27. $\mathbf{A} = \begin{bmatrix} 3 & 2 & 1 \\ 0 & 2 & 0 \\ 0 & 0 & 2 \end{bmatrix}$

28. $\mathbf{A} = \begin{bmatrix} 1 & 2 & 1 \\ 0 & 2 & -3 \\ 0 & 0 & -1 \end{bmatrix}$

Errors and Approximations (Section 1.2.1)

✍ In Problems 29 through 32 convert each decimal number to a binary number.

29. 67
30. 234
31. 45.25
32. 1127

✍ In Problems 33 through 36 convert each decimal number to a hexadecimal number.

33. 596
34. 1327
35. 23.1875
36. 364.5

✍ In Problems 37 through 40 convert each hexadecimal number to a binary number.

37. $(2B5.4)_{16}$
38. $(143)_{16}$
39. $(3D.2)_{16}$
40. $(12F.11)_{16}$

✍ In Problems 41 through 45 write the floating-point form of each decimal number by m-digit rounding for the given m.

41. -0.00031676 ($m = 4$)
42. 11.893 ($m = 4$)
43. 200.346 ($m = 5$)
44. -1203.423 ($m = 6$)
45. 22318 ($m = 4$)

46. ✍ Suppose m-digit chopping is used to find the floating-point form of

$$N = 0.d_1 d_2 \ldots d_m d_{m+1} \ldots \times 10^p$$

Show that

$$\left|e_{\text{rel}}\right|_{\text{Chopping}} = \frac{|N - \text{FL}(N)|}{|N|} \le 10^{1-m}$$

47. ✍ Suppose in Problem 46, we use m-digit rounding. Show that

$$\left|e_{\text{rel}}\right|_{\text{Rounding}} = \frac{|N - \text{FL}(N)|}{|N|} \le 0.5 \times 10^{1-m}$$

Iterative Methods (Section 1.2.2)

48. ✍ Consider the sequence described by $x_n = \dfrac{1}{1+n}$, $n = 0, 1, 2, \ldots$.
    a. Find a suitable function $g(x)$ so that the sequence can be generated by means of repeated substitution in the form $x_{n+1} = g(x_n)$, $n = 0, 1, 2, \ldots$
    b. Determine the rate of convergence of the sequence to its limit.

2

Introduction to MATLAB®

This chapter introduces fundamental features and capabilities of MATLAB® as related to numerical methods. These range from vector and matrix operations to plotting functions and sets of data. We will discuss several MATLAB built-in functions (commands) and their applications, as well as preparing user-defined functions to perform specific tasks.

2.1 MATLAB® Built-In Functions

MATLAB has a large number of built-in elementary functions, each accompanied by a brief but sufficient description through the help command. For example,

```
>> help sqrt
 sqrt   Square root.
    sqrt(X) is the square root of the elements of X. Complex
    results are produced if X is not positive.

    See also sqrtm, realsqrt, hypot.

    Reference page in Help browser
       doc sqrt

>> (1+sqrt(5))/2   % Calculate the golden ratio

ans =

    1.6180
```

The outcome of a calculation can be stored under a variable name, and suppressed by using a semicolon at the end of the statement:

```
>> g_ratio= (1+sqrt(5))/2;
```

Other elementary functions—assuming the variable name is a—include `abs(a)` for $|a|$, `sin(a)` for $\sin(a)$, `log(a)` for $\ln a$, `log10(a)` for $\log_{10}(a)$, `exp(a)` for e^a, and many more. Descriptions of these functions are available through the help command.

TABLE 2.1

MATLAB Rounding Functions

MATLAB Function	Example
Round(a) Round to the nearest integer	round(1.65) = 2, round(–4.7) = –5
Fix(a) Round toward zero	fix(1.65) = 1, fix(–4.7) = –4
Ceil(a) Round up toward infinity	ceil(1.65) = 2, ceil(–4.7) = –4
Floor(a) Round down toward minus infinity	floor(1.65) = 1, floor(–4.7) = –5

2.1.1 Rounding Commands

MATLAB has four built-in functions that round decimal numbers to the nearest integer via different rounding techniques. These are listed in Table 2.1.

2.1.2 Relational Operators

Table 2.2 gives a list of the relational and logical operators used in MATLAB.

2.1.3 Format Options

The `format` built-in function offers several options for displaying output. The preferred option can be chosen by selecting the following in the pull-down menu: File → Preferences → Command Window. A few of the format options are listed in Table 2.3.

TABLE 2.2

MATLAB Relational Operators

Mathematical Symbol	MATLAB Symbol
$=$	==
$\neq$	~=
$<$	<
$>$	>
$\leq$	<=
$\geq$	>=
AND	& or &&
OR	| or ||
NOT	~

TABLE 2.3
MATLAB Format Options

Format Option	Description	Example: 73/7
format short (default)	Fixed point with 4 decimal digits	10.4286
format long	Fixed point with 14 decimal digits	10.428571428571429
format short e	Scientific notation with 4 decimal digits	1.0429e+001
format long e	Scientific notation with 14 decimal digits	1.042857142857143e+001
format bank	Fixed point with 2 decimal digits	10.43

2.2 Vectors and Matrices

Vectors can be created in several ways in MATLAB. The row vector **v** = [1 4 6 7 10] is created as

```
>> v = [1 4 6 7 10]

v =
     1     4     6     7    10
```

Commas may be used instead of spaces between elements. For column vectors, the elements must be separated by semicolons.

```
>> v = [1;4;6;7;10]

v =
     1
     4
     6
     7
    10
```

The length of a vector is determined by using the `length` command:

```
>> length(v)

ans =

     5
```

The size of a vector is determined by the `size` command. For the last (column) vector defined above, we find

```
>> size(v)

ans =

     5     1
```

Arrays of numbers with equal spacing can be created more effectively. For example, a row vector whose first element is 2, its last element is 17, with a spacing of 3 is created as

```
>> v = [2:3:17]    or    >> v = 2:3:17

v =
      2    5    8    11    14    17
```

To create a column vector with the same properties:

```
>> v = [2:3:17]'

v =
      2
      5
      8
     11
     14
     17
```

Any component of a vector can be easily retrieved. For example, the third component of the above vector is retrieved by typing

```
>> v(3)

ans =

     8
```

A group of components may be retrieved as well. For example, the last three components of the row vector defined earlier are recovered as

```
>> v = 2:3:17;
>> v(end-2:end)

ans =

    11    14    17
```

2.2.1 Linspace

Another way to create vectors with equally spaced elements is by using the `linspace` command.

```
>> x = linspace(1,5,6) % 6 equally-spaced points between 1 and 5

x =
     1.0000    1.8000    2.6000    3.4000    4.2000    5.0000
```

The default value for the number of points is 100. Therefore, if we use `x=linspace(1,5)`, then 100 equally spaced points will be generated between 1 and 5.

2.2.2 Matrices

A matrix can be created by using brackets enclosing all of its elements, rows separated by a semicolon.

```
>> A=[1 -2 3;-3 0 1;5 1 4]

A =
     1    -2     3
    -3     0     1
     5     1     4
```

An entry can be accessed by using the row and column number of the location of that entry.

```
>> A(3,2)
% Entry at the intersection of the 3rd row and 2nd column

ans =

     1
```

An entire row or column of a matrix is accessed by using a colon.

```
>> Row_2=A(2,:)      % 2nd row of A

Row_2 =

    -3     0     1

>> Col_3=A(:,3)      % 3rd column of A

Col_3 =

     3
     1
     4
```

To replace an entire column of matrix `A` by a given vector v, we proceed as follows.

```
>> v=[1;0;1];
>> A_new=A;          % Pre-allocate the new matrix
>> A_new(:,2)=v      % Replace the 2nd column with v

A_new =

     1     1     3
    -3     0     1
     5     1     4
```

The $m \times n$ zero matrix is created by using `zeros(m,n)`; for instance, the 3×2 zero matrix:

```
>> A = zeros(3,2)

A =
     0     0
     0     0
     0     0
```

The $m \times n$ zero matrix is commonly used for pre-allocation of matrices to save memory space. In the matrix `A` defined above, any entry can be altered while others remain unchanged.

```
>> A(2,1) =-3; A(3,2) =-1

A =
     0     0
    -3     0
     0    -1
```

Size of a matrix is determined by using the `size` command:

```
>> size(A)

ans =

     3     2
```

The $n \times n$ identity matrix is created by `eye(n)`:

```
>> I = eye(3)

I =

     1     0     0
     0     1     0
     0     0     1
```

Matrix operations (Section 1.1.2) can be easily performed in MATLAB. If the sizes are not compatible, or the operations are not defined, MATLAB returns an error message to that effect.

```
>> A = [1 2;2 -2;4 0]; B = [-1 3;2 1];
% A is 3-by-2, B is 2-by-2
>> C = A*B    % Operation is valid

C =
     3     5
    -6     4
    -4    12
```

2.2.3 Determinant, Transpose, and Inverse

The determinant of an $n \times n$ matrix is calculated by the det command.

```
>> A = [1 2 -3;0 2 1;1 2 5]; det(A)

ans =

    16
```

The transpose of a matrix is found as

```
>> At = A.'

At =

     1     0     1
     2     2     2
    -3     1     5
```

The inverse of a (nonsingular) matrix is calculated by the inv command:

```
>> Ai = inv(A)

Ai =

    0.5000   -1.0000    0.5000
    0.0625    0.5000   -0.0625
   -0.1250         0    0.1250
```

2.2.4 Slash Operators

There are two slash operators in MATLAB: backslash (\) and slash (/).

```
>> help \

 \ Backslash or left matrix divide.

   A\B is the matrix division of A into B, which is roughly the
   same as INV(A)*B, except it is computed in a different way.
   If A is an N-by-N matrix and B is a column vector with N
   components, or a matrix with several such columns, then
   X = A\B is the solution to the equation A*X = B. A warning
   message is printed if A is badly scaled or nearly singular.
   A\EYE(SIZE(A)) produces the inverse of A.
```

The backslash (\) operator is employed for solving a linear system of algebraic equations $\mathbf{Ax} = \mathbf{b}$, whose solution vector is obtained as $\mathbf{x} = \mathbf{A}^{-1}\,\mathbf{b}$. However, instead of performing x = inv(A)*b, it is most efficient to find it as follows:

```
>> A = [1 -1 2;2 0 3;1 -2 1]; b = [2;8;-3];
>> x = A\b
```

```
x =
     1
     3
     2
```

The description of the slash (/) operator is given below.

```
>> help /

 /  Slash or right matrix divide.

    A/B is the matrix division of B into A, which is roughly the
    same as A*INV(B), except it is computed in a different way.
    More precisely, A/B = (B'\A')'. See MLDIVIDE for details.
```

2.2.5 Element-by-Element Operations

Element-by-element operations are summarized in Table 2.4. These are used when operations are performed on each element of a vector or matrix.

For example, suppose we want to raise each element of a vector to power of 2.

```
>> x = 0:2:10

x =
     0     2     4     6     8    10

>> x.^2

% If we use x^2 instead, an error is returned by MATLAB

ans =

     0     4    16    36    64   100
```

Now consider $(1 + x)/(2 + x)$ where x is the vector defined above. This fraction is to be evaluated for each value of x.

```
>> (1.+x)./(2.+x)

ans =

    0.5000    0.7500    0.8333    0.8750    0.9000    0.9167
```

TABLE 2.4

Element-by-Element Operations

MATLAB Symbol	Description
.*	Multiplication
./	(right) Division
.^	Exponentiation

If two arrays are involved in the element-by-element operation, they must be of the same size.

```
>> v = [1;2;3];
>> w = [2;3;4];
>> v.*w

ans =

     2
     6
    12
```

2.3 User-Defined Functions and Script Files

User-defined M file functions and scripts may be created, saved, and edited in MATLAB using the `edit` command. For example, suppose we want to create a function (say, `Circ_area`) that returns the area of a circle with a given radius. The function can be saved in a folder on the MATLAB path or in the current directory. The current directory can be viewed and/or changed using the drop-down menu at the top of the MATLAB command window. Once the current directory has been properly selected, type

```
>> edit Circ_area
```

A new window (Editor Window) will be opened where the function can be created. The generic structure of a function is

```
function [output variables] = FunctionName(input variables)
% Comments
Expressions/statements
Calculation of all output variables
```

In the case of our function, `Circ_area`, this translates to:

```
function A = Circ_area(r)
%
% Circ_area calculates the area of a circle of a given radius.
%
%   A = Circ_area(r) where
%
%     r is the radius of the circle,
%     A is the area.

A = pi*r^2;
```

To compute the area of a circle with radius 1.3, we simply type

```
>> A=Circ_area(1.3)

A =
    5.3093
```

Often, functions with multiple outputs are desired. For example, suppose our `Circ_area` function is to return two outputs: area of the circle, and the perimeter of the circle. We create this as follows:

```
function [A P] = Circ_area(r)
%
%  Circ_area calculates the area and the perimeter of a
%  circle of a given radius.
%
%      [A P] = Circ_area(r) where
%
%        r is the radius of the circle,
%        A is the area,
%        P is the perimeter.

A=pi*r^2;
P=2*pi*r;
```

Executing this function for the case of radius 1.3, we find

```
>> [A P] = Circ_area(1.3)

A =
    5.3093

P =
    8.1681
```

2.3.1 Setting Default Values for Input Variables

Sometimes, default values are declared for one or more of the input variables of a function. As an example, consider a function `y_int` that returns the y-intercept of a straight line that passes through a specified point with a given slope. Suppose the slope is 1 unless specified otherwise. If the specified point has coordinates (x_0, y_0) and the slope is m, then the y-intercept is found as $y = y_0 - mx_0$. Based on this we write the function as follows:

```
function y=y_int(x0,y0,m)
%
%   y_int finds the y-intercept of a line passing through
%   a point (x0,y0) with slope m.
%
%     y=y_int(x0,y0,m) where
%
%       x0, y0 are the coordinates of the given point,
%       m is the slope of the line (default=1),
%       y is the y-intercept of the line.

if nargin<3 || isempty(m)
    m=1;
end
y=y0 - m*x0;
```

To find the y-intercept of the line going through (–1, 2) with slope 1, we proceed as follows:

```
>> y=y_int(-1,2)

% Omitting the 3rd input variable, thus using default value

y =
    3
```

2.3.2 Creating Script Files

A script file comprises a list of commands as if they were typed at the command line. Script files can be created in the MATLAB Editor, and saved as an M file. For example, typing

```
>> edit My_script
```

opens the Editor Window, where the script can be created and saved under the name `My_script`. It is recommended that a script start with the functions `clear` and `clc`. The first one clears all the previously generated variables, and the second one clears the Command Window. Suppose we type the following lines in the Editor Window:

```
clear
clc
x=2; N=10;
a=cos(x)*N^2;
```

While in the Editor Window, select "Run My_script.m" under the Debug pull-down menu. This will execute the lines in the script file and return the Command Prompt. Simply type `a` at the prompt to see the result.

```
>> My_script
>> a

a =

  -41.6147
```

This can also be done by highlighting the contents and selecting "Evaluate Selection." An obvious advantage of creating a script file is that it allows us to simply make changes to the contents without having to retype all the commands.

2.3.3 Anonymous Functions

Another, more simple and more limited, way to create a function is to create an anonymous function. An anonymous function offers a way to create a function for simple expressions without creating an M file. Anonymous functions can only contain one expression and cannot return more than one output variable. They can either be created in the Command Window or as a script. The generic form is

```
My_function=@(arguments)(expression)
```

The function handle `My_function` is used the same way as one created as a user-defined function, described earlier. As an example, we will create an anonymous function (in the Command Window) to evaluate $\alpha = \sqrt{1+e^{-bx/2}}$ when $b = 1$ and $x = 2$:

```
>> alfa=@(b,x)(sqrt(1+exp(-b*x/2)));
% This creates alfa(b,x), which is then evaluated for specific
% values of b and x
>> alfa(1,2)

ans =

     1.1696
```

An anonymous function can be used in another anonymous function. For example, to evaluate $\beta = \ln\sqrt{1+e^{-bx/2}}$:

```
>> alfa=@(b,x)(sqrt(1+exp(-b*x/2)));
>> betta=@(b,x)(log(alfa(b,x)));
>> betta(1,2)

ans =

   0.1566
```

2.3.4 Inline

The built-in function `inline` is ideal for defining and evaluating functions of one or more variables. As with the anonymous function, `inline` function does not have to be saved in a separate file, and cannot return more than one output variable. Furthermore, it cannot call another `inline` but can use a user-defined function that has already been created. The generic form is

```
My_function=inline('expression','var1','var2', ...)
```

Let us consider the example involving evaluation of $\alpha = \sqrt{1+e^{-bx/2}}$ when $b=1$ and $x=2$.

```
>> alfa=inline('sqrt(1+exp(-b*x/2))','b','x');
>> alfa(1,2)

ans =

    1.1696
```

This agrees with the earlier result using the anonymous function. Note that if the desired order of variables is not specified by the user, MATLAB will list them in alphabetical order. In the above example, omitting the list of variables would still result in `alfa(b,x)`.

In this last example, suppose `b` is a scalar and `x` is a vector. Then, the function must be defined to include the "dot" to accommodate the element-by-element operation. Let $b=1$ and $\mathbf{x} = [1\ 2\ 3]$. Then,

```
>> alfa=inline('sqrt(1+exp(-b*x./2))','b','x');
>> x=[1 2 3];
>> alfa(1,x)

ans =

    1.2675    1.1696    1.1060
```

Note that, as expected, the second returned output matches what we got earlier for the case of $b=1$ and $x=2$.

2.4 Program Flow Control

Program flow can be controlled with the following three commands: `for`, `if`, and `while`.

2.4.1 for Loop

A for/end loop repeats a statement, or a group of statements, a specific number of times. Its generic form is

```
for i = first:increment:last,
  statements
end
```

The loop is executed as follows. The index i assumes its first value, all statements in the subsequent lines are executed with i = first, then the program goes back to the for command and i assumes the value i = first + increment and the process continues until the very last run corresponding to i = last.

As an example, suppose we want to generate a 5×5 matrix **A** with diagonal entries all equal to 1, and super diagonal entries all equal to 2, while all other entries are zero.

```
A = zeros(5,5);      % Pre-allocate
for i = 1:5,
  A(i,i) = 1;     % Diagonal entries
end
for i = 1:4,
  A(i,i+1) = 2;    % Super diagonal entries
end
```

Execution of this script returns

```
>> A

A =

     1     2     0     0     0
     0     1     2     0     0
     0     0     1     2     0
     0     0     0     1     2
     0     0     0     0     1
```

2.4.2 if Command

The most general form of the if command is

```
if condition 1
  set of expressions 1
else if condition 2
  set of expressions 2
else
  set of expressions 3
end
```

The simplest form of a conditional statement is the if/end structure. For example,

```
f = inline('log(x/3)'); x = 1;
if f(x) ~= 0,
error('x is not a root')
end
```

Execution of this script returns

```
x is not a root
```

The if/else/end structure allows for choosing one group of expressions from two groups. The most complete form of the conditional statement is the if/elseif/else/end structure. Let us create the same 5 × 5 matrix **A** as above, this time employing the if/elseif/else/end structure.

```
A = zeros(5,5);
for i = 1:5,
 for j = 1:5,
   if j == i+1,
     A(i,j) = 2;
elseif j == i,
   A(i,j) = 1;
end
end
end
```

Note that each for statement is accompanied by an end statement. Execution of this script returns

```
>> A

A =

     1     2     0     0     0
     0     1     2     0     0
     0     0     1     2     0
     0     0     0     1     2
     0     0     0     0     1
```

2.4.3 while Loop

A while/end loop repeats a statement, or a group of statements, until a specific condition is met. Its generic form is

```
while condition
  statements
end
```

We will generate the same 5×5 matrix **A** as before, this time with the aid of the `while` loop.

```
A = eye(5); i = 1;
while i < 5,
    A(i,i+1) = 2;
    i = i+1;
end
```

Executing this script returns the same matrix as above.

```
>> A

A =

     1     2     0     0     0
     0     1     2     0     0
     0     0     1     2     0
     0     0     0     1     2
     0     0     0     0     1
```

2.5 Displaying Formatted Data

Formatted data can be displayed by using either `disp` or `fprintf`. An example of how the `disp` command is used is

```
>> v = [1.2 -9.7 2.8];
>> disp(v)

    1.2000   -9.7000    2.8000
```

For better control of the formatted data, `fprintf` is used. Consider the script below. A function $f(x) = x\cos x + 1$ is defined. For $k = 1,2,3,4,5$ we want to calculate each $c = (1/2)^k$ as well as the corresponding function value $f(c)$. The output is to be displayed in tabulated form containing the values of k, c, and $f(c)$ for each $k = 1,2,3,4,5$.

```
f = inline('x*cos(x)+1');
disp('  k    c     f(c)')
for k = 1:5,
    c = (1/2)^k;
    fc = f(c);
    fprintf('%2i    %6.4f    %6.4f\n',k,c,fc)
end
```

Execution of this script returns

```
k      c        f(c)
1   0.5000    1.4388
2   0.2500    1.2422
3   0.1250    1.1240
4   0.0625    1.0624
5   0.0313    1.0312
```

The disp command simply displays all contents inside the single quotes. The fprintf command is used inside for loop. For each k in the loop, fprintf writes the value of k, the calculated value of c, as well as f(c). The format %2i means integer of length 2, which is being used for displaying the value of k. In %6.4f, the letter f represents the fixed-point format, 6 is the length, and the number 4 is the number of digits to the right of the decimal. Finally, \n means new line. A more detailed description is available through the help command.

2.6 Symbolic Toolbox

The Symbolic Math Toolbox allows for manipulation of symbols to perform operations symbolically. Symbolic variables are created by using the syms command:

```
>> syms a b
```

Consider a function $g = 4.81 \sin(a/3) + 3e^{-b/c}$ where $c = 2.1$. This function can be defined symbolically as follows:

```
>> syms a b
>> c = 2.1;
>> g = 4.81*sin(a/3) + 3*exp(-b/c)

g =

3/exp((10*b)/21) + (481*sin(a/3))/100
```

In symbolic expressions, numbers are always converted to the ratio of two integers, as it is observed here as well. For decimal representation of numbers, we use the vpa (variable precision arithmetic) command. The syntax is

```
>> vpa(g,n)
```

where n is the number of desired digits. For example, if four digits are desired in our current example, then

```
>> g = vpa(4.81*sin(a/3) + 3*exp(-b/c),4)

g =

4.81*sin(0.3333*a) + 3.0/exp(0.4762*b)
```

To evaluate this symbolic function for specified values of a and b, we use the subs command. For instance, to evaluate g when a = 1 and b = 2,

```
>> a = 1;b = 2;subs(g)

ans =

    2.7311
```

Symbolic functions can be converted to inline functions. In our current example,

```
>> gg = inline(char(g))

gg =

    Inline function:
    gg(a,b) = 4.81*sin(0.3333*a) + 3.0/exp(0.4762*b)
```

This function can easily be evaluated for the specified values of its variables:

```
>> gg(1,2)

ans =

   2.7311   % Same as before
```

2.6.1 Differentiation

Consider $f = t^3 - \sin t$, which may be defined symbolically as

```
>> f = sym('t^3-sin(t)');
```

To find df/dt we use the diff command:

```
>> dfdt = diff(f)

dfdt =

3*t^2 - cos(t)
```

The second derivative d^2f/dt^2 is found as

```
>> dfdt2 = diff(f,2)

dfdt2 =

6*t + sin(t)
```

The symbolic derivatives can always be converted to inline functions for convenient evaluation.

2.6.2 Integration

Indefinite and definite integrals are calculated symbolically via the `int` command. For example, the indefinite integral $\int (t + \frac{1}{2}\sin 2t)\,dt$ is calculated as

```
>> syms t
>> int(t+(1/2)*sin(2*t))

ans =

t^2/2 - cos(t)^2/2
```

The definite integral $\int_1^3 (at - e^{t/2})dt$, where $a = const$, is calculated as follows:

```
>>  syms a t
>>  int(a*t-exp(t/2),t,1,3)

ans =

4*a - 2*exp(1/2)*(exp(1) - 1)
```

To convert all constants to decimals with two digits, we use `vpa`:

```
>> vpa(int(a*t-exp(t/2),t,1,3),2)

ans =

4.0*a - 5.7
```

2.6.3 Differential Equations

Differential equations and initial-value problems can be solved by the `dsolve` function. For example, the solution of the differential equation $y' + (x+1)y = 0$ is obtained as

```
>> y = dsolve('Dy+(x+1)*y=0','x')

y =

C4/exp((x+1)^2/2)
```

Note that the default independent variable in `dsolve` is `t`. Since in our example, the independent variable is `x`, we needed to specify that in single quotes. The initial-value problem $\ddot{x} + 2\dot{x} + 2x = e^{-t}$, $x(0) = 0$, $\dot{x}(0) = 1$ is solved as

```
>> x = dsolve('D2x+2*Dx+2*x=exp(-t)','x(0)=0,Dx(0)=1')

x =

1/exp(t) - cos(t)/exp(t) + sin(t)/exp(t)
```

2.7 Plotting

Plotting a vector of values versus another vector of values is done by using the `plot` command. For example, to plot the function $x(t) = e^{-t}(\cos t + \sin t)$ over the interval [0,5] using 100 points:

```
>> t = linspace(0,5);
>> x = exp(-t).*(cos(t)+sin(t));
>> plot(t,x)   % Figure 2.1
```

The Figure Window can be used to edit the figure. These include adding grid, adjusting thickness of lines and curves, adding text and legend, axes titles, and much more.

2.7.1 Subplot

The built-in function `subplot` is designed to create multiple figures in tiled positions. Suppose we want to plot the function $z(x,t) = e^{-x}\sin(t + 2x)$ versus $0 \le x \le 5$ for four values of $t = 0,1,2,3$. Let us generate the four plots and arrange them in a 2×2 formation.

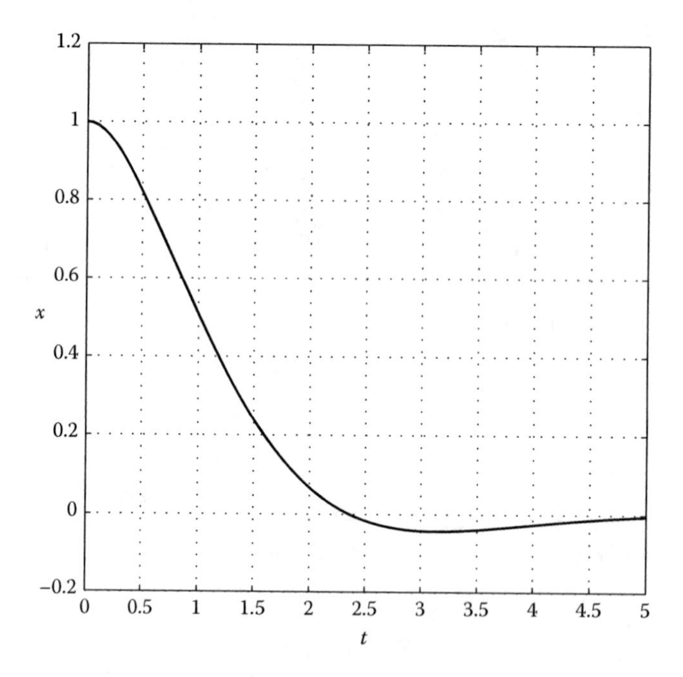

FIGURE 2.1
Plot of a function versus its variable.

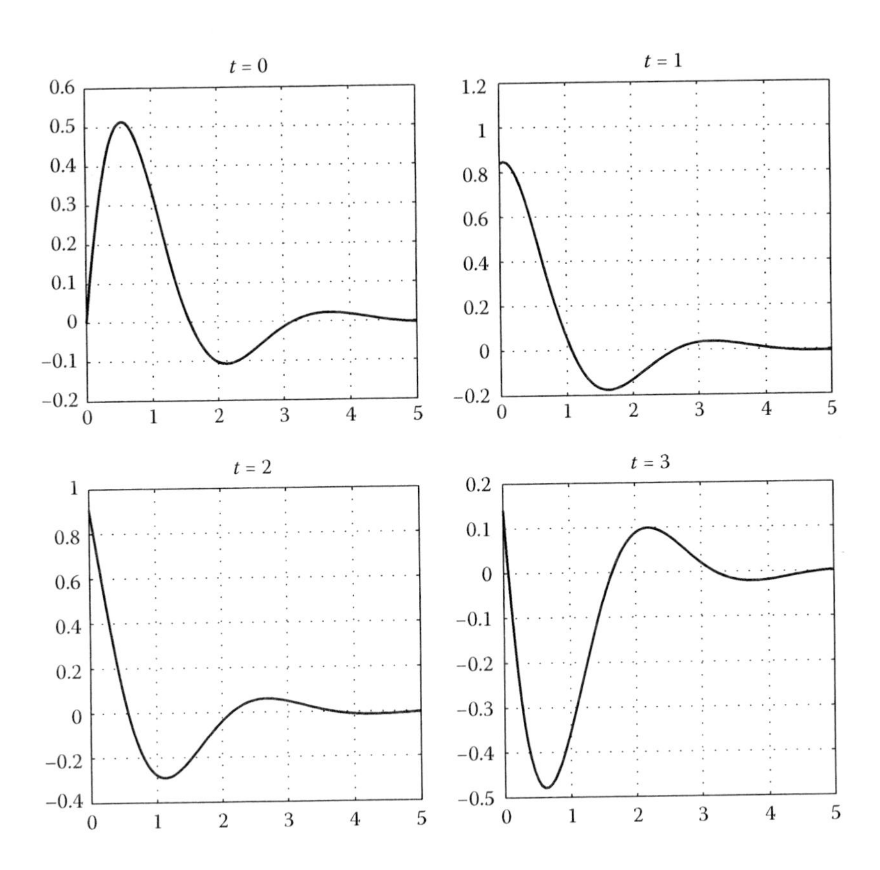

FIGURE 2.2
Subplot in a 2 × 2 formation.

```
x=linspace(0,5); t=0:1:3;
for i=1:4,
    for j=1:100,
    z(j,i)=exp(-x(j))*sin(t(i)+2*x(j));
    % Generate 100 values of z for each t
    end
end

% Initiate Figure 2.2
subplot(2,2,1), plot(x,z(:,1)), title('t=0')
subplot(2,2,2), plot(x,z(:,2)), title('t=1')
subplot(2,2,3), plot(x,z(:,3)), title('t=2')
subplot(2,2,4), plot(x,z(:,4)), title('t=3')
```

2.7.2 Plotting Analytical Expressions

An alternative way to plot a function is to use the `ezplot` command, which plots the function without requiring data generation. As an example, consider

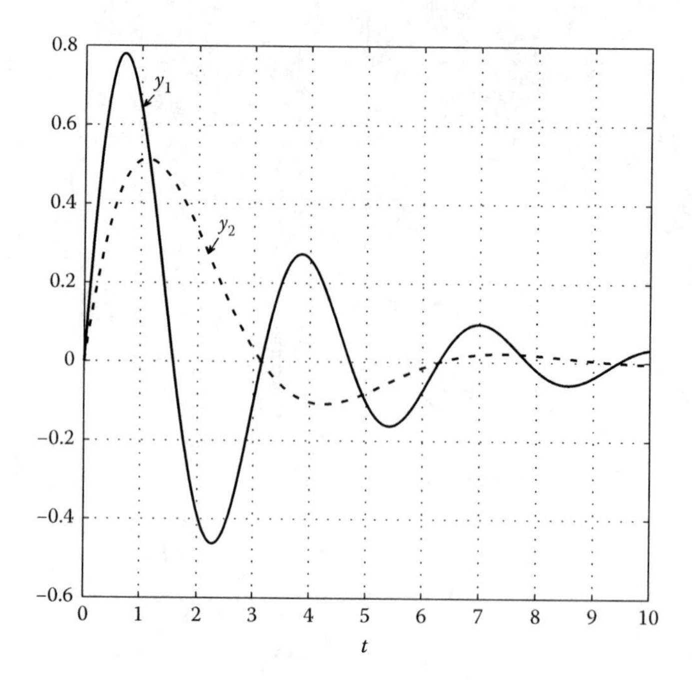

FIGURE 2.3
Multiple plots.

the function $x(t) = e^{-t}(\cos t + \sin t)$ that we previously plotted over the interval [0,5]. The plot in Figure 2.1 can be regenerated using `ezplot` as follows:

```
>> x = inline('exp(-t)*(cos(t)+sin(t))');
>> ezplot(x, [0,5])
```

2.7.3 Multiple Plots

Multiple plots can also be created using `ezplot`. Suppose the two functions $y_1 = e^{-t/3} \sin 2t$ and $y_2 = e^{-t/2} \sin t$ are to be plotted versus $0 \le t \le 10$ in the same graph.

```
syms t
y1 = exp(-t/3)*sin(2*t);
y2 = exp(-t/2)*sin(t);
ezplot(y1, [0,10])    % Initiate Figure 2.3
hold on
ezplot(y2, [0,10])    % Complete plot
```

PROBLEM SET

1. Write a user-defined function with function call `val = function_eval(f,a,b)` where f is an inline function, and a and b are constants such that $a < b$. The function calculates the midpoint m of the interval

[a,b] and returns the value of $f(a) + (1/2)\, f(m) + f(b)$. Execute the function for $f = e^{x/2}$, $a = -2$, $b = 4$.

2. Write a user-defined function with function call `m = midpoint_seq(a,b,tol)` where a and b are constants such that $a < b$, and `tol` is a specified tolerance. The function first calculates the midpoint m_1 of the interval $[a, b]$, then the midpoint m_2 of $[a, m_1]$, then the midpoint m_3 of $[a, m_2]$, and so on. The process terminates when two successive midpoints are within `tol` of each other. Allow a maximum of 20 iterations. The output of the function is the sequence $m_1, m_2, m_3, \ldots$. Execute the function for $a = -4$, $b = 10$, tol $= 10^{-3}$.
3. Write a user-defined function with function call `C = temp_conv(F)` where F is temperature in Fahrenheit, and C is the corresponding temperature in Celsius. Execute the function for $F = 87$.
4. Write a user-defined function with function call `P = partial_eval(f,a)` where f is a function defined symbolically, and a is a constant. The function returns the value of $f' + f''$ at $x = a$. Execute the function for $f = 3x^2 - e^x/3$, and $a = 1$.
5. Write a user-defined function with function call `P = partial_eval(f,g,a)` where f and g are functions defined symbolically, and a is a constant. The function returns the value of $f' + g'$ at $x = a$. Execute the function for $f = x^2 + e^{-x}$, $g = \sin(0.3x)$, and $a = 0.8$.
6. Write a script file that employs any combination of the flow control commands to generate

$$\mathbf{A} = \begin{bmatrix} 1 & 0 & -1 & 0 & 0 & 0 \\ 0 & 2 & 0 & -1 & 0 & 0 \\ 2 & 0 & 3 & 0 & -1 & 0 \\ 0 & 2 & 0 & 4 & 0 & -1 \\ 0 & 0 & 2 & 0 & 5 & 0 \\ 0 & 0 & 0 & 2 & 0 & 6 \end{bmatrix}$$

7. Write a script file that employs any combination of the flow control commands to generate

$$\mathbf{A} = \begin{bmatrix} 4 & 1 & -2 & 3 & 0 & 0 \\ 0 & 4 & -1 & 2 & 3 & 0 \\ 0 & 0 & 4 & 1 & -2 & 3 \\ 0 & 0 & 0 & 4 & -1 & 2 \\ 0 & 0 & 0 & 0 & 4 & 1 \\ 0 & 0 & 0 & 0 & 0 & 4 \end{bmatrix}$$

8. Plot $\int_1^t e^{t-2x}\sin x\,dx$ versus $-1 \le t \le 1$, add grid and label.
9. Plot $\int_0^t (x+t)^2 e^{-(t-x)}dx$ versus $-2 \le t \le 1$, add grid and label.
10. Evaluate $\int\limits_0^{\infty} \frac{\sin\omega}{\omega}\,d\omega$.
11. Differentiate $f(t) = (t-1)\ln(t+1) - t\cos(t/2)$ with respect to t, make the outcome an inline function, and evaluate at $t = 1.3$.
12. Differentiate $g(x) = 2^{x-2}\sin x - e^{3-2x}$ with respect to x, make the outcome an inline function, and evaluate at $x = 0.9$.
13. Plot $y_1 = \frac{1}{3}e^{-t}\sin(t\sqrt{2})$ and $y_2 = e^{-t/2}$ versus $0 \le t \le 5$ in the same graph. Add grid, and label.
14. Generate 100 points for each of the two functions in Problem 13 and plot versus $0 \le t \le 5$ in the same graph. Add grid, and label.
15. Plot $u(x, t) = \cos(1.7x)\sin(3.2t)$ versus $0 \le x \le 5$ for four values of $t = 1$, 1.5, 2, 2.5 in a 2 × 2 tile. Add grid and title.
16. Plot $u(x, t) = (1 - \sin x)e^{-(t+1)}$ versus $0 \le x \le 5$ for values of $t = 1,3$ in a 1 × 2 tile. Add grid and title.
17. Write a user-defined function with function call `[r k] = root_finder(f,x0,kmax,tol)` where f is an inline function, x_0 is a specified value, kmax is the maximum number of iterations, and tol is a specified tolerance. The function sets $x_1 = x_0$, calculates $|f(x_1)|$, and if it is less than the tolerance, then x_1 approximates the root r. If not, it will increment x_1 by 0.01 to obtain x_2, repeat the procedure, and so on. The process terminates as soon as $|f(x_k)| <$ tol for some k. The outputs of the function are the approximate root and the number of iterations it took to find it. Execute the function for $f(x) = x^2 - 3.3x + 2.1$, $x_0 = 0.5$, kmax = 50, tol = 10^{-2}.
18. Write a user-defined function with function call `[opt k] = opt_finder(fp,x0,kmax,tol)` where fp is the derivative (as an inline function) of a given function f, x_0 is a specified value, kmax is the maximum number of iterations, and tol is a specified tolerance. The function sets $x_1 = x_0$, calculates $|\text{fp}(x_1)|$, and if it is less than the tolerance, then x_1 approximates the critical point at which the derivative is near zero. If not, it will increment x_1 by 0.1 to obtain x_2, repeat the procedure, and so on. The process terminates as soon as $|\text{fp}(x_k)| <$ tol for some k. The outputs are the approximate critical point and the number of iterations it took to find it. Execute the function for $f(x) = x + (x-2)^2$, $x_0 = 1$, kmax = 50, tol = 10^{-2}.
19. Evaluate $[x^2 + e^{-a(x+1)}]^{1/3}$ when $a = -1$, $x = 3$ by using an anonymous function in another anonymous function.
20. Evaluate $\sqrt{\left|x + \ln\left|1 - e^{(a+2)x/3}\right|\right|}$ when $a = -3$, $x = 1$ by using an anonymous function in another anonymous function.

3

Solution of Equations of a Single Variable

The focus of this chapter is on numerical solution of equations in the general form

$$f(x) = 0 \tag{3.1}$$

Graphically, a solution, or a root, of Equation 3.1 refers to the point of intersection of $f(x)$ and the x-axis. Therefore, depending on the nature of the curve of $f(x)$ in relation to the x-axis, Equation 3.1 may have a unique solution, multiple solutions, or no solution. A root of an equation can sometimes be determined analytically resulting in an exact solution. For instance, the equation $e^{2x} - 3 = 0$ can be solved analytically to obtain a unique solution $x = \frac{1}{2}\ln 3$. In most situations, however, this is not possible and the root(s) must be found using a numerical procedure. An example would be an equation in the form $1 - x + \sin x = 0$. Figure 3.1 shows that this equation has one solution only, slightly smaller than 2, which may be approximated to within a desired accuracy with the aid of a numerical method.

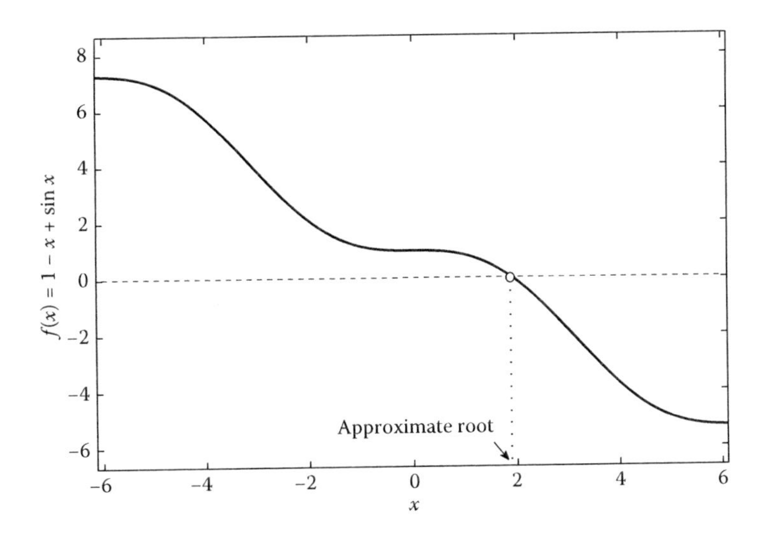

FIGURE 3.1
Approximate solution of $1 - x + \sin x = 0$.

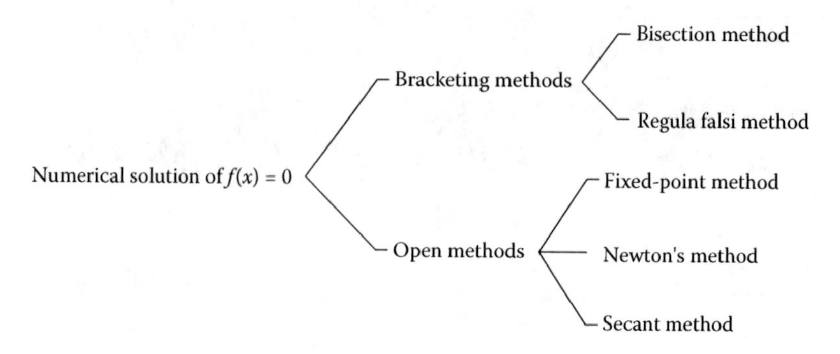

FIGURE 3.2
Classification of methods to solve an equation of one variable.

3.1 Numerical Solution of Equations

As described in Figure 3.2, numerical methods for solving an equation are divided into two main categories: bracketing methods and open methods.

Bracketing methods require that an initial interval containing the root be identified. Referring to Figure 3.3, this means an interval $[a,b]$ such that $f(a)f(b) < 0$. The length of the interval is then reduced in succession until it satisfies the desired accuracy. Exactly how this interval gets narrowed in each step depends on the specific method used. It is readily seen that bracketing methods always converge to the root. Open methods require an initial estimate of the solution, close to the actual root. Then, more accurate estimates are successively generated by a specific method (Figure 3.4). Open methods are more efficient than bracketing methods, but do not always generate a sequence that converges to the root.

3.2 Bisection Method

Bisection method is the simplest bracketing method to find a root of $f(x) = 0$. It is assumed that $f(x)$ is continuous on an interval $[a,b]$ and has a root there so that $f(a)$ and $f(b)$ have opposite signs, hence $f(a)f(b) < 0$. The procedure goes as follows: Locate the midpoint of $[a,b]$, that is, $c_1 = \frac{1}{2}(a+b)$ (Figure 3.5). If $f(a)$ and $f(c_1)$ have opposite signs, the interval $[a,c_1]$ contains the root and will be retained for further analysis. If $f(b)$ and $f(c_1)$ have opposite signs, we continue with $[c_1,b]$. In Figure 3.5, it so happens that the interval $[c_1,b]$ brackets the root and is retained. Since the right endpoint is unchanged, we update the interval $[a,b]$ by resetting the left endpoint $a = c_1$. The process is repeated until the length of the most recent interval $[a,b]$ satisfies the desired accuracy.

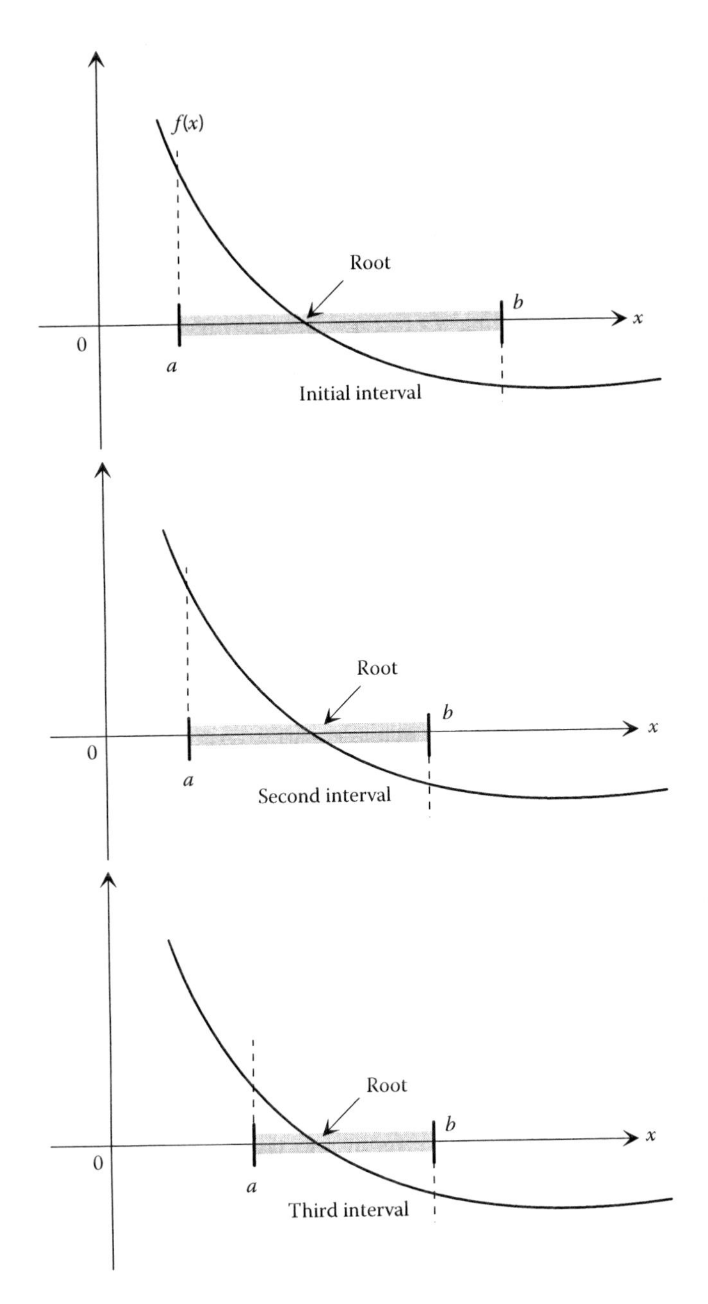

FIGURE 3.3
The philosophy of bracketing methods.

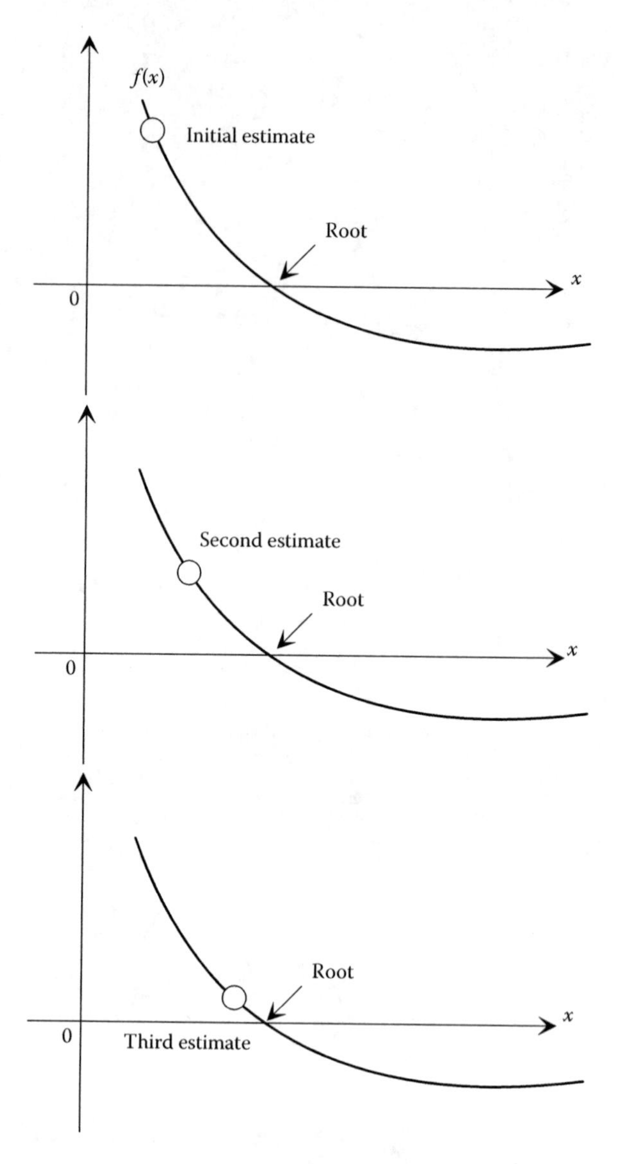

FIGURE 3.4
The philosophy of open methods.

The initial interval $[a,b]$ has length $b - a$. Beyond that, the first generated interval has length $\frac{1}{2}(b - a)$, the next interval $\frac{1}{4}(b - a)$, and so on. Thus, the nth interval constructed in this manner has length $(b - a)/2^{n-1}$, and because it brackets the root, the absolute error associated with the nth iteration satisfies

$$|e_n| \le \frac{b - a}{2^{n-1}} \quad (b > a)$$

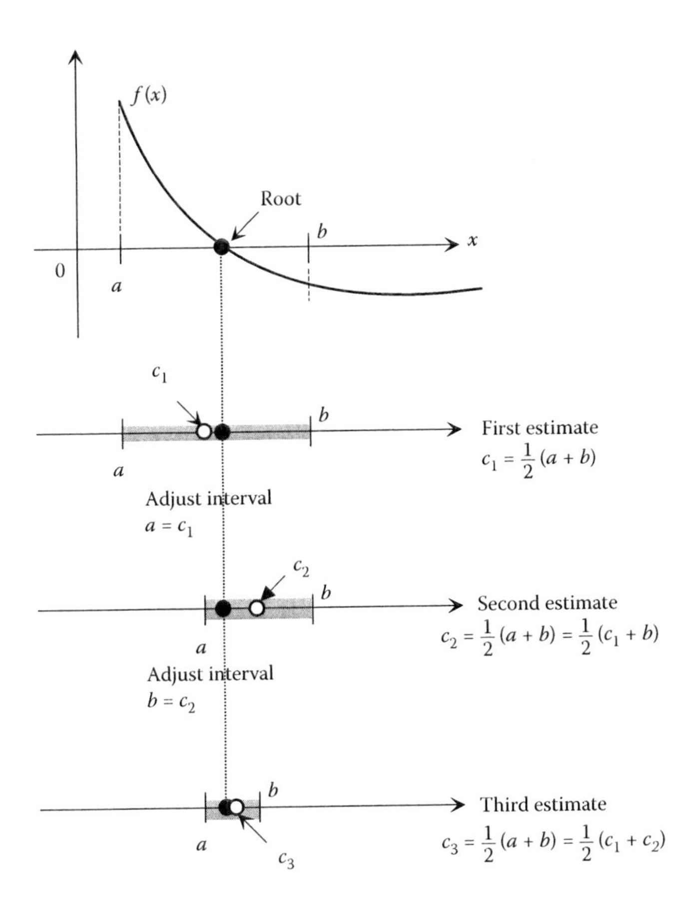

FIGURE 3.5
Bisection method: three iterations shown.

This upper bound is usually larger than the actual error at the nth step. If the bisection method is used to approximate the root of $f(x) = 0$ within a prescribed tolerance $\varepsilon > 0$, then it can be shown that the number N of iterations needed to meet the tolerance condition satisfies

$$N > \frac{\ln(b - a) - \ln \varepsilon}{\ln 2} \tag{3.2}$$

The user-defined function `Bisection` shown below generates a sequence of values (midpoints) that ultimately converges to the true solution. The iterations terminate when $\frac{1}{2}(b - a) < \varepsilon$, where ε is a prescribed tolerance. The output of the function is the last generated estimate of the root at the time the tolerance was met. It also returns a table that comprises iteration counter, interval endpoints and interval midpoint per iteration, as well as the value of $\frac{1}{2}(b - a)$ to see when the terminating condition is satisfied.

```
function c = Bisection(f, a, b, kmax, tol)
%
% Bisection uses the bisection method to approximate a
% root of f(x) = 0 in the interval [a,b].
%
%    c = Bisection(f, a, b, kmax, tol) where
%
%      f is an inline function representing f(x),
%      a and b are the limits of the interval [a,b],
%      kmax is the maximum number of iterations
%      (default 20),
%      tol is the scalar tolerance for convergence
%      (default 1e-4),
%
%      c is the approximate root of f(x) = 0.
%
if nargin < 5, tol = 1e-4; end
if nargin < 4, kmax = 20; end
if f(a)*f(b) > 0
   c ='failure';
   return
end
disp(' k        a          b          c          (b-a)/2')
for k = 1:kmax,
    c = (a+b)/2;     % Find the first midpoint
    if f(c) == 0       % Stop if a root has been found
       return
    end
fprintf('%3i %11.6f%11.6f%11.6f%11.6f\n',k,a,b,c,(b-a)/2)
   if (b-a)/2 < tol,      % Stop if tolerance is met
       return
   end
   if f(b)*f(c) > 0        % Check sign changes
       b = c;            % Adjust the endpoint of interval
   else a = c;
   end
end
c = 'failure';
```

Example 3.1: Bisection Method

The equation $x \cos x = -1$ has a root in the interval [–2,4], as shown in Figure 3.6. The figure is produced by

```
>> syms x
>> ezplot(x*cos(x)+1,[-2 4])
```

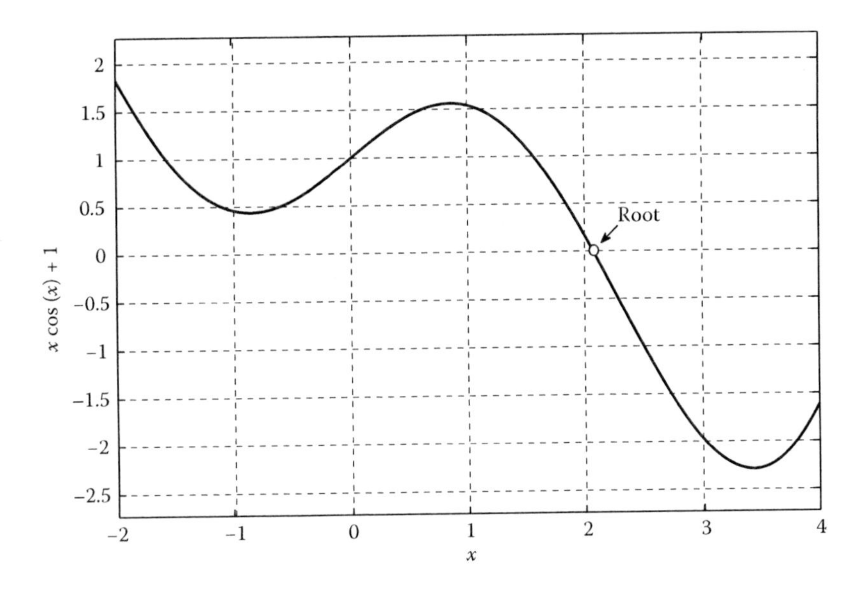

FIGURE 3.6
Location of the root of $x \cos x = -1$ in [–2,4].

Execute the user-defined function `Bisection` with $\varepsilon = 10^{-2}$ and no more than 20 iterations.

```
>> f = inline('x*cos(x)+1');
>> c = Bisection(f, -2, 4, 20, 1e-2)

k        a          b          c          (b-a)/2

1   -2.000000   4.000000   1.000000   3.000000
2    1.000000   4.000000   2.500000   1.500000
3    1.000000   2.500000   1.750000   0.750000
4    1.750000   2.500000   2.125000   0.375000
5    1.750000   2.125000   1.937500   0.187500
6    1.937500   2.125000   2.031250   0.093750
7    2.031250   2.125000   2.078125   0.046875
8    2.031250   2.078125   2.054688   0.023438
9    2.054688   2.078125   2.066406   0.011719
10   2.066406   2.078125   2.072266   0.005859

c =

    2.0723
```

Iterations stopped when $\frac{1}{2}(b-a) = 0.005859 < \varepsilon = 0.01$. Note that by Equation 3.2,

$$N > \frac{\ln(b-a) - \ln\varepsilon}{\ln 2} \overset{a=-2,\, b=4}{\underset{\varepsilon=0.01}{=}} \frac{\ln 6 - \ln 0.01}{\ln 2} = 9.23$$

which means at least 10 iterations are required for convergence. This is in agreement with the findings here, as we saw that tolerance was met after 10 iterations. The accuracy of the solution estimate will improve if a smaller tolerance is imposed.

3.2.1 MATLAB® Built-In Function fzero

The fzero function in MATLAB® finds the roots of $f(x) = 0$ for a real function $f(x)$.

```
FZERO Scalar nonlinear zero finding.

   X = FZERO(FUN,X0) tries to find a zero of the function FUN
   near X0,
   if X0 is a scalar.
```

The fzero function uses a combination of the bisection, secant, and inverse quadratic interpolation methods. If we know two points where the function value differs in sign, we can specify this starting interval using a two-element vector for x0. This algorithm is guaranteed to return a solution. If we specify a scalar starting point x0, then fzero initially searches for an interval around this point where the function changes sign. If an interval is found, then fzero returns a value near where the function changes sign. If no interval is found, fzero returns a NaN value.

The MATLAB built-in function fzero can be used to confirm the approximate root in Example 3.1:

```
>> fzero(f,1)

ans =

   2.0739
```

3.3 Regula Falsi Method (Method of False Position)

The regula falsi method is another bracketing method to find a root of $f(x) = 0$. Once again, it is assumed that $f(x)$ is continuous on an interval $[a,b]$ and has a root there so that $f(a)$ and $f(b)$ have opposite signs, $f(a)f(b) < 0$. The technique is geometrical in nature and described as follows. Let $[a_1,b_1] = [a,b]$. Connect points $A{:}(a_1, f(a_1))$ and $B{:}(b_1, f(b_1))$ by a straight line as in Figure 3.7 and let c_1 be its x-intercept. If $f(a_1)f(c_1) < 0$, then $[a_1,c_1]$ brackets the root. Otherwise, the root is in $[c_1,b_1]$. In Figure 3.7, it just so happens that $[a_1,c_1]$ brackets the root. Continuing this process generates a sequence c_2, c_3, … that eventually converges to the root. In the case shown in Figure 3.7, the curve of $f(x)$ is concave up and the left end of the interval remains fixed throughout the process. This issue will be addressed shortly.

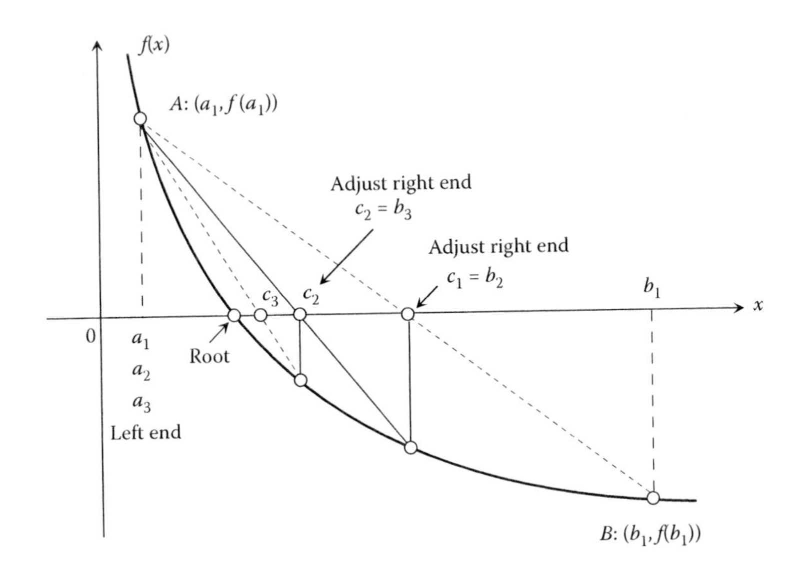

FIGURE 3.7
Method of false position.

Analytically, the procedure is illustrated as follows. The equation of the line connecting points A and B is

$$y - f(b_1) = \frac{f(b_1) - f(a_1)}{b_1 - a_1}(x - b_1)$$

To find the x-intercept, set $y = 0$ and solve for $x = c_1$:

$$c_1 = b_1 - \frac{b_1 - a_1}{f(b_1) - f(a_1)} f(b_1) \overset{\text{Simplify}}{=} \frac{a_1 f(b_1) - b_1 f(a_1)}{f(b_1) - f(a_1)}$$

Generalizing this result, the sequence of points that converges to the root is generated via

$$c_n = \frac{a_n f(b_n) - b_n f(a_n)}{f(b_n) - f(a_n)}, \quad n = 1, 2, 3, \ldots \tag{3.3}$$

The user-defined function `RegulaFalsi` generates a sequence of elements that eventually converges to the root of $f(x) = 0$. The iterations stop when two consecutive x-intercepts are close to one another. That is, the terminating condition is $|c_{n+1} - c_n| < \varepsilon$, where ε is the imposed tolerance. The outputs are the approximate root and the number of iterations required to meet the tolerance. The function also returns a table comprised of the intervals containing the root in all iterations performed.

```
function [r k] = RegulaFalsi(f, a, b, kmax, tol)
%
% RegulaFalsi uses the regula falsi method to approximate
% a root of f(x) = 0 in the interval [a,b].
%
%   [r k] = RegulaFalsi(f, a, b, kmax, tol), where
%
%     f is an inline function representing f(x),
%     a and b are the limits of interval [a,b],
%     kmax is the maximum number of iterations (default 20),
%     tol is the scalar tolerance for convergence
%     (default 1e-4),
%
%     r is the approximate root of f(x) = 0,
%     k is the number of iterations needed for convergence.
%
if nargin < 5, tol = 1e-4; end
if nargin < 4, kmax = 20; end

c = zeros(1, kmax); % Pre-allocate

if f(a)*f(b) > 0
   r = 'failure';
   return
end

disp(' k      a       b')

for k = 1:kmax,
   c(k) = (a*f(b)-b*f(a))/(f(b)-f(a)); % Find the x-intercept
   if f(c(k)) == 0  % Stop if a root has been found
      return
   end

  fprintf('%2i  %11.6f%11.6f\n',k,a,b)

   if f(b)*f(c(k)) > 0   % Check sign changes
      b = c(k);          % Adjust the endpoint of interval
   else a = c(k);
   end

   c(k+1) = (a*f(b)-b*f(a))/(f(b)-f(a));
   % Find the next x-intercept

   if abs(c(k+1)-c(k)) < tol,  % Stop if tolerance is met
      r = c(k+1);
      return
   end
end
r = 'failure';
```

Example 3.2: Regula Falsi Method

We will find an approximate root of $x \cos x = -1$ in the interval $[-2,4]$ by executing the `RegulaFalsi` function with tolerance $\varepsilon = 10^{-4}$. Since this happens to be the default value for tolerance in `RegulaFalsi`, we simply execute

```
>> [r k] =RegulaFalsi(f, -2, 4)

  k        a            b
  1    -2.000000    4.000000
  2     1.189493    4.000000
  3     1.189493    2.515720
  4     1.960504    2.515720
  5     2.069986    2.515720

r =
    2.0739

k =
     5
```

Therefore, the approximate root with the desired accuracy is obtained after five iterations.

3.3.1 Modified Regula Falsi Method

In many cases, the graph of the function $f(x)$ happens to be concave up or concave down. In these situations, when regula falsi is employed, one of the endpoints of the interval remains the same through all iterations, while the other endpoint advances in the direction of the root. For instance, in Figure 3.7,

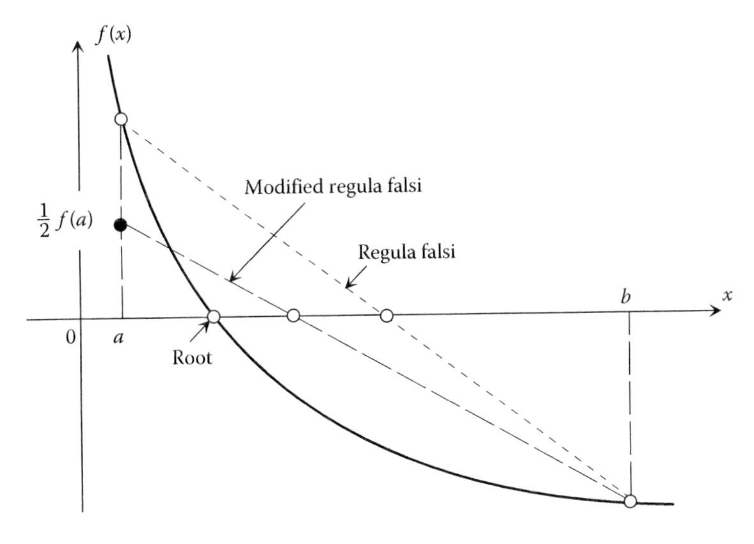

FIGURE 3.8
Modified regula falsi method.

the function is concave up, the left endpoint remains unchanged, and the right endpoint moves toward the root. The regula falsi method can be modified such that both ends of the interval move toward the root, thus improving the rate of convergence. Among many proposed modifications, there is one that is presented here. Reconsider the scenario portrayed in Figure 3.7 now shown in Figure 3.8. If endpoint a stays the same after, say, three consecutive iterations, the usual straight line is replaced with one that is less steep, going through the point at $\frac{1}{2}f(a)$ instead of $f(a)$. This causes the x-intercept to be closer to the actual root. It is possible that this still does not force the endpoint a to move toward the root. In that event, if endpoint a remains the same after three more iterations, the modified line will be replaced with yet a less steep line going through $\frac{1}{4}f(a)$, and so on; see the Problem Set for Section 3.3 at the end of this chapter.

3.4 Fixed-Point Method

The fixed-point method is an open method to find a root of $f(x) = 0$. The idea is to rewrite $f(x) = 0$ as $x = g(x)$, where $g(x)$ is called the iteration function.

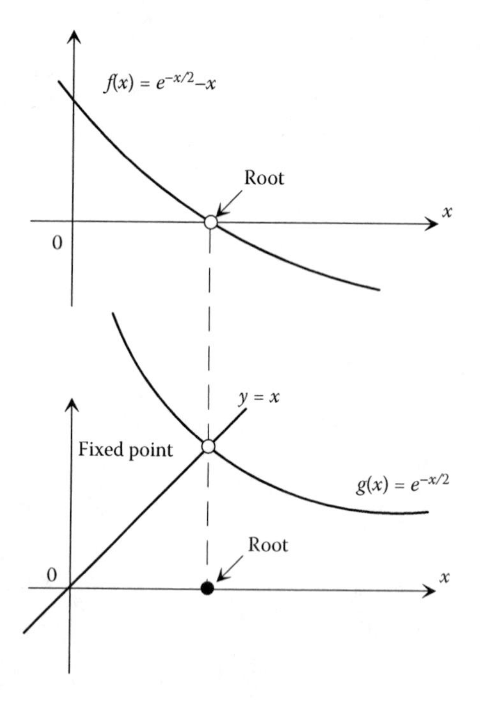

FIGURE 3.9
Root of an equation interpreted as a fixed point of an iteration function.

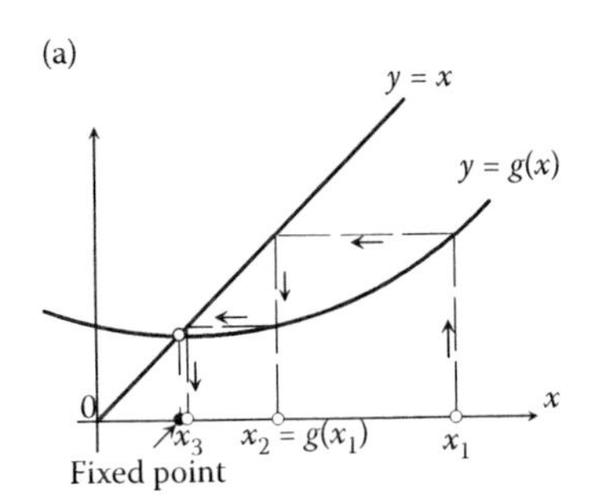

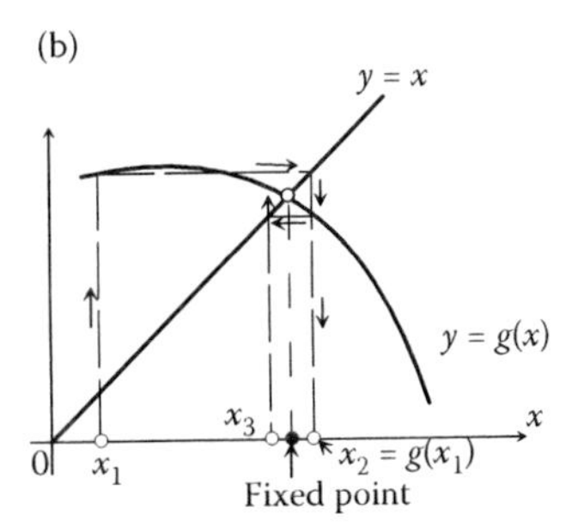

FIGURE 3.10
Fixed-point iteration: (a) monotone convergence and (b) oscillatory convergence.

Consequently, a point of intersection of $y = g(x)$ and $y = x$, known as a fixed point of $g(x)$, is also a root of $f(x) = 0$. As an example, consider $e^{-x/2} - x = 0$ and its root as shown in Figure 3.9. The equation is rewritten as $x = e^{-x/2}$ so that $g(x) = e^{-x/2}$ is the iteration function. It is observed that $g(x)$ has only one fixed point, which is the only root of the original equation. It should be noted that for a given equation $f(x) = 0$ there usually exist more than one iteration function. For instance, $e^{-x/2} - x = 0$ can also be rewritten as $x = -2 \ln x$ so that $g(x) = -2 \ln x$.

The fixed point of $g(x)$ is found numerically via the fixed-point iteration:

$$x_{n+1} = g(x_n), \quad n = 1, 2, 3, \ldots, \quad x_1 = \text{initial guess} \tag{3.4}$$

The procedure begins with an initial guess x_1 near the fixed point. The next point x_2 is found by evaluating $g(x_1)$, and so on. This continues until convergence is observed, that is, until two successive points are within a prescribed distance of each other, or

$$|x_{n+1} - x_n| < \varepsilon$$

Two types of convergence can be exhibited by the fixed-point iteration: monotone and oscillatory, as illustrated in Figure 3.10. In a monotone convergence, the elements of the generated sequence converge to the fixed point from one side, while in an oscillatory convergence, the elements bounce from one side of the fixed point to the other as they approach it.

3.4.1 Selection of a Suitable Iteration Function

As mentioned above, there is usually more than one way to rewrite a given equation $f(x) = 0$ as $x = g(x)$. The iteration function $g(x)$ must be suitably selected so that when used in Equation 3.4, the iterations converge to the fixed point. In some cases, more than one of the possible forms can be successfully used. Sometimes, none of the forms is suitable, which means that the root cannot be found by the fixed-point method. When there are multiple roots, one possible form may be used to find one root, while another form leads to another root. As demonstrated in Theorem 3.1, there is a way to decide whether a fixed-point iteration converges or not for a specific choice of iteration function.

Theorem 3.1

Suppose $r \in I$ is a fixed point of $g(x)$. Assume that $g(x)$ has a continuous derivative in interval I, and $|g'(x)| \le K < 1$ for all $x \in I$. Then, for any initial point $x_1 \in I$, the fixed-point iteration in Equation 3.4 generates a sequence $\{x_n\}$ that converges to r. Furthermore, if $e_1 = x_1 - r$ and $e_n = x_n - r$ denote the initial error and the error at the nth iteration, we have

$$|e_n| \le K^n \, |e_1| \tag{3.5}$$

Proof

Suppose $x \in I$. Then, by the mean value theorem for derivatives (MVT), there exists a point $\xi \in (x,r)$ such that

$$g(x) - g(r) = g'(\xi)(x - r)$$

Next, let us consider the left side of Equation 3.5. Noting that $r = g(r)$ and $x_n = g(x_{n-1})$, we have

$$\begin{aligned}
|e_n| = |x_n - r| = |g(x_{n-1}) - g(r)| &\overset{\text{MVT}}{=} |g'(\xi)||x_{n-1} - r| \le K|x_{n-1} - r| \\
&= K|g(x_{n-2}) - g(r)| \\
&\overset{\text{MVT}}{=} K|g'(\eta)||x_{n-2} - r| \le K^2|x_{n-2} - r| \le \cdots \le K^n|x_1 - r| \\
&= K^n|e_1|
\end{aligned}$$

Since $K < 1$ by assumption, $|e_n| = |x_n - r| \to 0$ as $n \to \infty$. That completes the proof.

3.4.2 A Note on Convergence

Following Theorem 3.1, if $|g'(x)| < 1$ near a fixed point of $g(x)$, convergence is guaranteed. In other words, if in a neighborhood of a root, the curve of $g(x)$ is less steep than the line $y = x$, the fixed-point iteration converges. Note that this is a sufficient, and not necessary, condition for convergence.

The user-defined function FixedPoint uses an initial x_1 and generates a sequence of elements $\{x_n\}$ that eventually converges to the fixed point of $g(x)$. The iterations stop when two consecutive elements are sufficiently close to one another, that is, $|x_{n+1} - x_n| < \varepsilon$, where ε is the imposed tolerance. The outputs are the approximate value of the fixed point and the number of iterations needed to meet the tolerance.

```
function [r n] = FixedPoint(g, x1, kmax, tol)
%
% FixedPoint uses the fixed-point method to approximate a
% fixed point of g(x), that is, x=g(x).
%
%  [r n] = FixedPoint(g, x1, kmax, tol), where
%
%    g is an inline function representing g(x),
%    x1 is the initial point,
%    kmax is the maximum number of iterations (default 20),
%    tol is the scalar tolerance for convergence
%    (default 1e-4),
%
%    r is the approximate fixed point of g(x),
%    n is the number of iterations needed for convergence.

if nargin<4, tol=1e-4;end
if nargin<3, kmax=20;end
x(1)=x1;
for n=1:kmax,
    x(n+1)=g(x(n));        % Generate the next point
    if abs(x(n+1)-x(n))<tol,    % Stop if tolerance is met
        r=x(n+1);
        return
    end
end
r='failure';
```

Example 3.3: Fixed-Point Method

The objective is to find the root of $x - 2^{-x} = 0$ using the fixed-point method. Rewrite the equation as $x = 2^{-x}$ so that $g(x) = 2^{-x}$. The fixed point can be roughly located as in Figure 3.11.

```
>> g=inline('2^(-x)');
>> syms x
>> ezplot(g, [0,2])
>> hold on
>> ezplot(x, [0,2])        % Figure 3.11
```

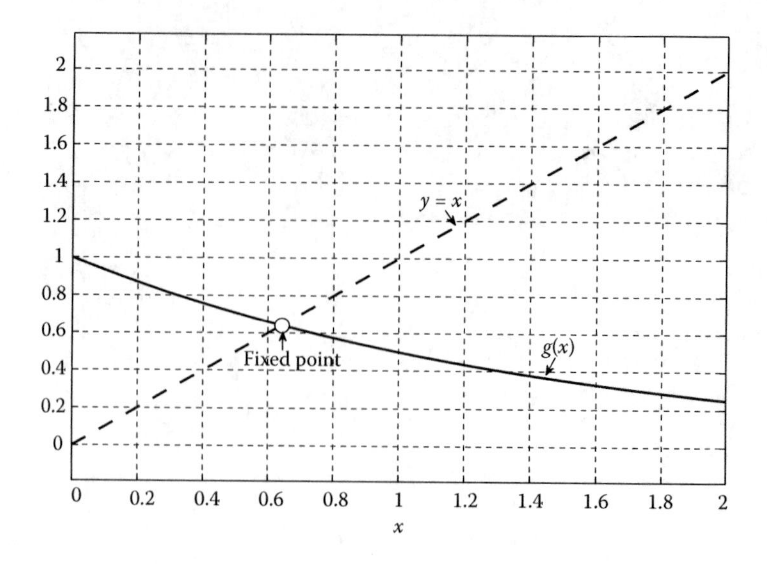

FIGURE 3.11
Location of the fixed point of $g(x) = 2^{-x}$.

Before applying the fixed-point iteration, we need to check the condition of convergence, Theorem 3.1, as follows:

$$|g'(x)| = |-2^{-x}\ln 2| = 2^{-x}\ln 2 < 1 \quad \Rightarrow \quad x > 0.5288$$

This means if the fixed point is in an interval comprised of values of x larger than 0.5288, the fixed-point iteration is guaranteed to converge. Figure 3.11 confirms that this condition is satisfied.

We will execute the user-defined function `FixedPoint` with $x_1 = 0$ and default values for `kmax` and `tol`.

```
>> [r n] = FixedPoint(g, 0)

r =
    0.6412
n =
    13
```

Therefore, the fixed point of $g(x) = 2^{-x}$, which is the root of $x - 2^{-x} = 0$, is found after 13 iterations. The reader may verify that the convergence is oscillatory. In fact, the sequence of elements generated by the iteration is

```
0.0000  1.0000  0.5000  0.7071  0.6125  0.6540  0.6355  0.6437
0.6401  0.6417  0.6410  0.6413  0.6411  0.6412
```

Example 3.4: Selection of a Suitable Iteration Function

Consider the quadratic equation $x^2 - 4x + 1 = 0$. As stated earlier, there is more than one way to construct an iteration function $g(x)$. For instance, two such forms are

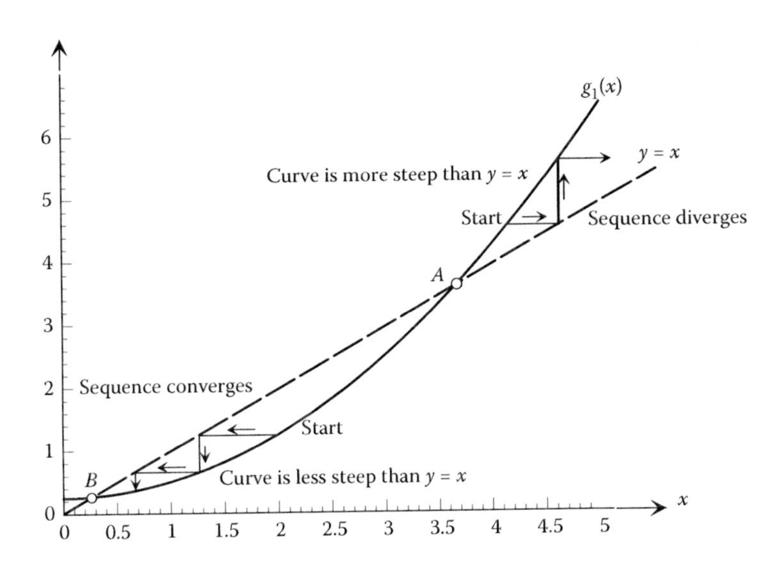

FIGURE 3.12
Fixed points of $g_1(x) = \frac{1}{4}(x^2 + 1)$.

$$g_1(x) = \frac{1}{4}(x^2 + 1), \quad g_2(x) = 4 - \frac{1}{x}$$

Let us first consider $g_1(x)$, Figure 3.12. The portion of the curve of $g_1(x)$ below point A is less steep than $y = x$. Starting at any arbitrary point in that region, we see that the iteration always converges to the smaller root B. On the other hand, above point A, the curve is steeper than the line $y = x$. Starting at any point there, the iteration will diverge. Thus, only the smaller of the two roots can be approximated if $g_1(x)$ is used. Now, referring to Figure 3.13, the curve of $g_2(x)$ is much less steep near A than it is near B. Also it appears that starting at any point above or below A (also above B), the iteration converges to the larger root.

Let us inspect the condition of convergence as stated in Theorem 3.1. In relation to $g_1(x)$, we have

$$|g_1'(x)| = \left|\frac{1}{2}x\right| < 1 \;\Rightarrow\; |x| < 2 \;\Rightarrow\; -2 < x < 2$$

The fixed point B falls inside this interval, and starting at around $x_1 = 2$ (Figure 3.12), the sequence did converge toward B. When we started at around $x_1 = 4$, however, the sequence showed divergence. In relation to $g_2(x)$,

$$|g_2'(x)| = \left|\frac{1}{x^2}\right| < 1 \;\Rightarrow\; |x^2| > 1 \;\Rightarrow\; x < -1 \quad \text{or} \quad x > 1$$

Figure 3.13 shows that the fixed point A certainly falls in the interval $x > 1$. However, we see that our starting choice of around $x_1 = 0.3$ led to

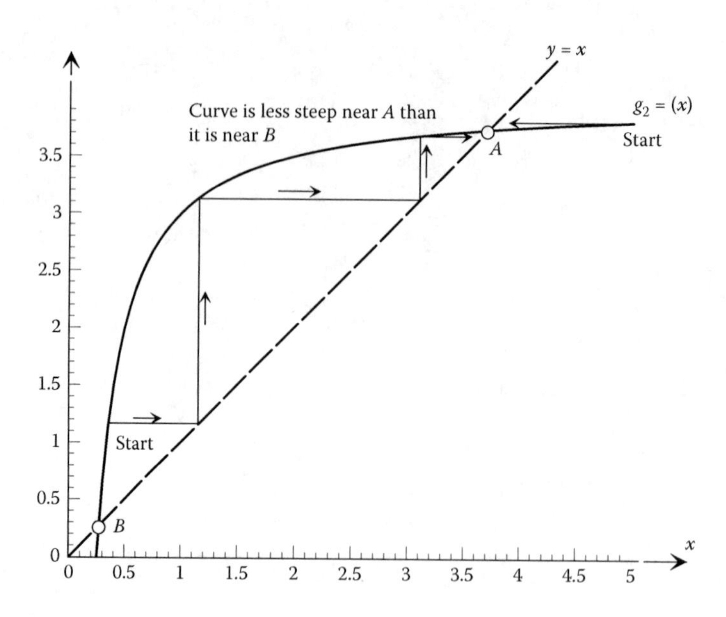

FIGURE 3.13
Fixed points of $g_2(x) = 4 - \dfrac{1}{x}$.

convergence, even though it was not inside the required interval. This of course is due to the fact that the condition of convergence is only sufficient and not necessary.

```
>> g1 = inline('(x^2+1)/4');
>> g2 = inline('4-1/x');
>> [r1 n] = FixedPoint(g1, 2)

r1 =
    0.2680  % First root found by using iteration function g1

n =
    8

>> [r2 n] = FixedPoint(g2, 0.3)

r2 =
    3.7320  % Second root found by using iteration function g2

n =
    7
```

The original equation is $x^2 - 4x + 1 = 0$ so that we are looking for the roots of a polynomial. MATLAB has a built-in function `roots`, which performs this task:

```
>> roots([1 -4 1])

ans =

    3.7321
    0.2679
```

3.4.3 Rate of Convergence of the Fixed-Point Iteration

Suppose r is a fixed point of $g(x)$, and that $g(x)$ satisfies the hypotheses of Theorem 3.1 in some interval I. Also assume the $(k + 1)$th derivative of $g(x)$ is continuous in I. Expanding $g(x)$ in a Taylor's series about $x = r$, and noting that $r = g(r)$, $x_{n+1} = g(x_n)$, and $e_n = x_n - r$, the error at the $(n + 1)$th iteration is obtained as

$$\begin{aligned} e_{n+1} &= x_{n+1} - r = g(x_n) - g(r) \\ &= g'(r)e_n + \frac{g''(r)}{2!}e_n^2 + \cdots + \frac{g^{(k)}(r)}{k!}e_n^k + E_{k,n} \end{aligned} \tag{3.6}$$

where $E_{k,n}$, the error due to truncation, is given by

$$E_{k,n} = \frac{g^{(k+1)}(\xi_n)}{(k+1)!}e_n^{k+1} \quad \text{for some} \quad \xi_n \in (r, x_n)$$

Assume $g'(x) \neq 0 \ \forall x \in I$. Then, for $k = 0$, Equation 3.6 yields $e_{n+1} = g'(\xi_n)e_n$. But since $x_n \to r$ as $n \to \infty$ (by Theorem 3.1), we have $\xi_n \to r$ as well. Consequently,

$$\lim_{n\to\infty} \frac{e_{n+1}}{e_n} = \lim_{n\to\infty} g'(\xi_n) = g'(r) \neq 0$$

Therefore, convergence is linear. The rate of convergence will be improved if $g'(r) = 0$ and $g''(x) \neq 0 \ \forall x \in I$. In that case, it can be shown that

$$\lim_{n\to\infty} \frac{e_{n+1}}{e_n^2} = \frac{g''(r)}{2!} \neq 0$$

so that convergence is quadratic. We will see shortly that Newton's method falls in this category. From the foregoing analysis it is evident that the more derivatives of $g(x)$ vanish at the root, the faster the rate of the fixed-point iteration.

3.5 Newton's Method (Newton–Raphson Method)

The most commonly used open method to solve $f(x) = 0$, where $f'(x)$ is continuous, is Newton's method. Consider the graph of $f(x)$ in Figure 3.14. Start with an initial point x_1 and locate the point $(x_1, f(x_1))$ on the curve. Draw the tangent line to the curve at that point, and let its x-intercept be x_2. Locate

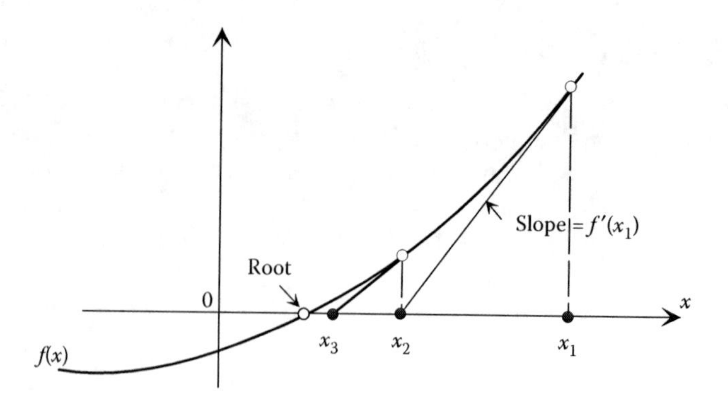

FIGURE 3.14
Geometry of Newton's method.

$(x_2, f(x_2))$, draw the tangent line to the curve there, and let x_3 be its x-intercept. Repeat this until convergence $x_n \to r$ is observed. In general, two consecutive elements x_n and x_{n+1} are related via

$$x_{n+1} = x_n - \frac{f(x_n)}{f'(x_n)}, \quad n = 1, 2, 3, \ldots, \quad x_1 = \text{initial point} \tag{3.7}$$

The user-defined function `Newton` uses an initial x_1 and generates a sequence of elements $\{x_n\}$ via Equation 3.7 that eventually converges to the root of $f(x) = 0$. The iterations stop when two consecutive elements are sufficiently close to one another, that is, $|x_{n+1} - x_n| < \varepsilon$, where ε is the prescribed tolerance. The outputs are the approximate value of the root and the number of iterations needed to meet the tolerance.

```
function [r n] = Newton(f, fp, x1, tol, N)
%
% Newton uses Newton's method to approximate a root of
% f(x) = 0.
%
%   [r n] = Newton(f, fp, x1, tol, N), where
%
%     f is an inline function representing f(x),
%     fp is an inline function representing f'(x),
%     x1 is the initial point,
%     tol is the scalar tolerance for convergence
%     (default 1e-4),
%     N is the maximum number of iterations (default 20),
%
%     r is the approximate root of f(x) = 0,
```

```
%   n is the number of iterations required for convergence.
if nargin<5, N=20;end
if nargin<4,tol=1e-4;end
x=zeros(1, N+1);  % Pre-allocate
x(1)=x1;
for n=1:N,
    if fp(x(n))==0
       r='failure';
       return
    end
    x(n+1)=x(n)-f(x(n))/fp(x(n));
    if abs(x(n+1)-x(n))<tol,
       r=x(n+1);
       return
    end
end
r='failure';
```

Example 3.5: Newton's Method

Find the first positive root of $x\cos x = -1$ using Newton's method with $\varepsilon = 10^{-4}$ and maximum 20 iterations.

SOLUTION

This equation was previously tackled in Examples 3.1 and 3.2. According to Figure 3.6, the first positive root is located around $x = 2$, thus we will execute the user-defined function `Newton` with initial point $x_1 = 1$. In the meantime, we note that $f(x) = x\cos x + 1$ and $f'(x) = \cos x - x\sin x$.

```
>> f=inline('x*cos(x)+1');
>> fp=inline('cos(x)-x*sin(x)');
>> [r n]=Newton(f, fp, 1)

r =
    2.0739

n =
    6
```

The result fully agrees with those of Examples 3.1 and 3.2.

Example 3.6: Newton's Method

Find the roots of $x^2 - 3x - 7 = 0$ using Newton's method with $\varepsilon = 10^{-4}$ and maximum 20 iterations.

SOLUTION

We first plot $f(x) = x^2 - 3x - 7$ to find approximate locations of its roots.

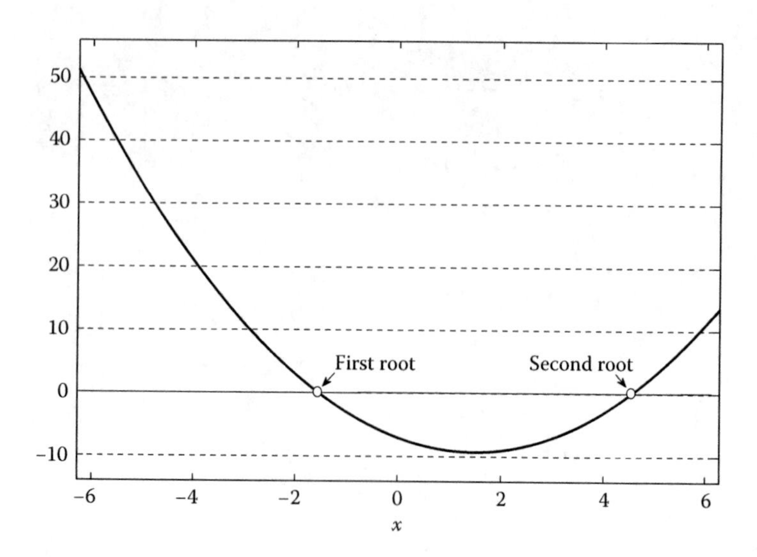

FIGURE 3.15
Location of the two roots of $x^2 - 3x - 7 = 0$.

```
>> f = inline('x^2-3*x-7');
>> ezplot(f)  % Figure 3.15
```

Inspired by Figure 3.15, we will execute the user-defined function `Newton` two separate times, once with initial point $x_1 = -2$ and a second time with $x_1 = 4$.

```
>> fp = inline('2*x-3');
>> [r1 n1] = Newton(f, fp, -2)

r1 =
   -1.5414    % First root

n1 =
   4

>> [r2 n2] = Newton(f, fp, 4)

r2 =
   4.5414     % Second root

n2 =
   4
```

In both applications, convergence was achieved after four iterations. Since $f(x)$ is a polynomial, the built-in MATLAB function `roots` can be used to find its roots.

```
>> roots([1 -3 -7])

ans =

    4.5414
   -1.5414
```

3.5.1 Rate of Convergence of Newton's Method

It turns out that the speed of convergence of Newton's method depends on the multiplicity of a certain root of $f(x) = 0$. We say that a root r of $f(x) = 0$ is of multiplicity (or order) m if and only if

$$f(r) = 0, \quad f'(r) = 0, \quad f''(r) = 0, \ldots, \ f^{(m-1)}(r) = 0, \quad f^{(m)}(r) \neq 0$$

A root of order 1 is commonly known as a simple root.

Theorem 3.2

Let r be a root of $f(x) = 0$, and in Newton's iteration, Equation 3.7, let x_1 be sufficiently close to r.

a. If $f''(x)$ is continuous and r is a simple root, then

$$|e_{n+1}| \cong \frac{1}{2}\frac{|f''(r)|}{|f'(r)|}|e_n|^2 \Rightarrow \lim_{n\to\infty}\frac{|e_{n+1}|}{|e_n|^2} = \frac{1}{2}\frac{|f''(r)|}{|f'(r)|} \neq 0 \tag{3.8a}$$

and convergence $\{x_n\} \to r$ is quadratic.

b. If $\{x_n\} \to r$, where r is root of order $m > 1$, then

$$|e_{n+1}| \cong \frac{m-1}{m}|e_n| \Rightarrow \lim_{n\to\infty}\frac{|e_{n+1}|}{|e_n|} = \frac{m-1}{m} \neq 0 \tag{3.8b}$$

and convergence is linear.

Example 3.7: Quadratic Convergence; Newton's Method

Suppose Newton's method is to be used to find the two roots of $x^2 - 3x - 4 = 0$, which are $r = -1, 4$. Let us focus on the task of finding the larger root, $r = 4$. Since the root is simple, by Equation 3.8a we have

$$\lim_{n\to\infty}\frac{|e_{n+1}|}{|e_n|^2} \cong \frac{1}{2}\frac{|f''(4)|}{|f'(4)|} = \frac{1}{2}\left(\frac{2}{5}\right) = 0.2$$

This indicates that convergence is quadratic, as stated in Theorem 3.2. While finding the smaller root $r = -1$, this limit is once again 0.2, thus confirming that convergence is quadratic. The reader can readily verify this by tracking the ratio $|e_{n+1}| / |e_n|^2$ while running Newton's method.

3.5.2 A Few Notes on Newton's Method

- When Newton's method works, it generates a sequence that converges rapidly to the intended root.
- Several factors may cause Newton's method to fail. Usually, it is because the initial point x_1 is not sufficiently close to the intended root. Another one is that at some point in the iterations, $f'(x_n)$ may be close to or equal to zero. Other scenarios, where the iteration simply halts or the sequence diverges, are shown in Figure 3.16 and explained in Example 3.8.
- If $f(x)$, $f'(x)$, and $f''(x)$ are continuous, $f'(\text{root}) \neq 0$, and the initial point x_1 is close to the root, then the sequence generated by Newton's method converges to the root.
- A downside of Newton's method is that it requires the expression for $f'(x)$. However, finding $f'(x)$ can at times be difficult. In these cases, the secant method (described below in Section 3.6) can be used instead.

Example 3.8: Newton's Method

Apply Newton's method to find the root of $2/(x+1) = 1$. For the initial point use (a) $x_1 = 3$, and (b) $x_1 = 4$.

SOLUTION

$$f(x) = \frac{2}{x+1} - 1 \quad \text{so that } f'(x) = -\frac{2}{(x+1)^2}$$

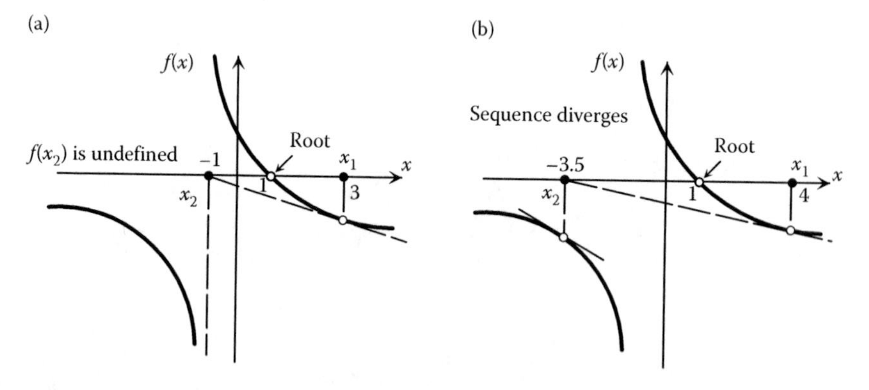

FIGURE 3.16
Newton's method (a) halts, (b) diverges.

a. Starting with $x_1 = 3$, we find

$$x_2 = x_1 - \frac{f(x_1)}{f'(x_1)} = 3 - \frac{-\frac{1}{2}}{-\frac{1}{8}} = -1$$

The iterations halt at this point because $f(-1)$ is undefined. This is illustrated in Figure 3.16a.

b. Starting with $x_1 = 4$, we find

$$x_2 = x_1 - \frac{f(x_1)}{f'(x_1)} = 4 - \frac{-\frac{3}{5}}{-\frac{2}{25}} = -3.5, \quad x_3 = -9.1250, \quad x_4 = -50.2578, \ldots$$

The sequence clearly diverges. This is illustrated in Figure 3.16b.

3.5.3 Modified Newton's Method for Roots with Multiplicity 2 or Higher

If r is a root of $f(x)$ and r has a multiplicity 2 or higher, then convergence of the sequence generated by Newton's method is linear; see Theorem 3.2. In these situations, Newton's method may be modified to improve the efficiency. The modified Newton's method designed for roots of multiplicity 2 or higher is described as

$$x_{n+1} = x_n - \frac{f(x_n)f'(x_n)}{\left[f'(x_n)\right]^2 - f(x_n)f''(x_n)}, \quad n = 1, 2, 3, \ldots, \quad x_1 = \text{initial point} \tag{3.9}$$

The user-defined function `NewtonMod` uses an initial x_1 and generates a sequence of elements $\{x_n\}$ via Equation 3.9 that eventually converges to the root of $f(x) = 0$, where the root has multiplicity 2 or higher. The iterations stop when two consecutive elements are sufficiently close to one another, that is, $|x_{n+1} - x_n| < \varepsilon$, where ε is the prescribed tolerance. The outputs are the approximate value of the root and the number of iterations needed to meet the tolerance.

```
function [r n] = NewtonMod(f, fp, f2p, x1, tol, N)
%
% NewtonMod uses modified Newton's method to approximate
% a root with multiplicity 2 or higher of f(x) = 0.
%
%   [r n] = NewtonMod(f, fp, f2p, x1, tol, N), where
%
%     f is an inline function representing f(x),
%     fp is an inline function representing f'(x),
%     f2p is an inline function representing f"(x),
```

```
%       x1 is the initial point,
%       tol is the scalar tolerance for convergence
%       (default 1e-4),
%       N is the maximum number of iterations (default 20),
%
%       r is the approximate root of f(x) = 0,
%       n is the number of iterations required for
%       convergence.
if nargin < 6, N = 20;end
if nargin < 5,tol = 1e-4;end
x = zeros(1,N+1);  % Pre-allocate
x(1) = x1;
for n = 1:N,
    x(n+1) = x(n)-(f(x(n))*fp(x(n)))/(fp(x(n))^2-f(x(n))
    *f2p (x(n)));
    if abs(x(n+1)-x(n)) < tol,
        r = x(n+1);
        return
    end
end
r = 'failure';
```

Example 3.9: Modified Newton's Method

Find the roots of $4x^3 + 4x^2 - 7x + 2 = 0$ using Newton's method and its modified version. Discuss the results.

SOLUTION

Figure 3.17 reveals that $f(x) = 4x^3 + 4x^2 - 7x + 2$ has a simple root at −2 and a double root (multiplicity 2) at 0.5 since it is tangent to the x-axis at that point. We will execute `NewtonMod` with default parameter values and $x_1 = 0$.

```
>> f = inline('4*x^3+4*x^2-7*x+2');
>> fp = inline('12*x^2+8*x-7');
>> f2p = inline('24*x+8');
>> [r n] = NewtonMod(f, fp, f2p, 0)

r =
    0.5000

n =
    4
```

Executing `Newton` with default parameters and $x_1 = 0$ yields

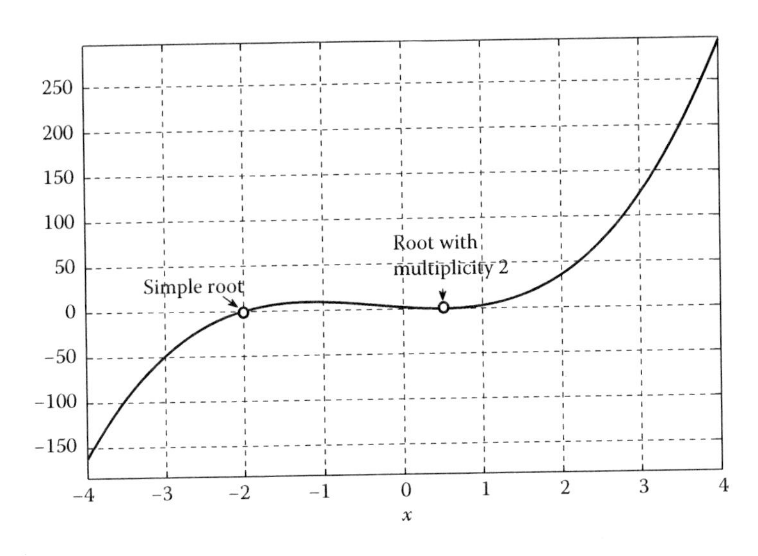

FIGURE 3.17
A simple and a multiple root of $4x^3 + 4x^2 - 7x + 2 = 0$.

```
>> [r n]=Newton(f, fp, 0)

r =
    0.4999

n =
    12
```

The modified Newton's method is clearly superior to the standard Newton's method when approximating a multiple root. The same, however, is not true for simple roots. Applying both methods with $x_1 = -3$ and default parameters yields

```
>> [r n]=NewtonMod(f, fp, f2p, -3)

r =
    -2.0000

n =
     6

>> [r n]=Newton(f, fp, -3)

r =
    -2.0000

n =
     5
```

The standard Newton's method generally exhibits a faster convergence (quadratic, by Theorem 3.2) to the simple root. The modified Newton's method outperforms the standard one when finding multiple roots. But for simple roots, it is not as efficient as Newton's method and requires more computations.

3.6 Secant Method

The secant method is another open method to solve $f(x) = 0$. Consider the graph of $f(x)$ in Figure 3.18. Start with two initial points x_1 and x_2, locate the points $(x_1, f(x_1))$ and $(x_2, f(x_2))$ on the curve, and draw the secant line connecting them. The x-intercept of this secant line is x_3. Next, use x_2 and x_3 to define a secant line and let the x-intercept of this line be x_4. Continue the process until the sequence converges to the root. In general, two consecutive elements x_n and x_{n+1} generated by the secant method are related via

$$x_{n+1} = x_n - \frac{x_n - x_{n-1}}{f(x_n) - f(x_{n-1})} f(x_n), \quad n = 2, 3, 4, \ldots, \quad x_1, x_2 = \text{initial points} \tag{3.10}$$

Comparing with Newton's method, we see that $f'(x_n)$ in Equation 3.7 is essentially approximated by, and replaced with, the difference quotient

$$\frac{f(x_n) - f(x_{n-1})}{x_n - x_{n-1}}$$

The user-defined function `Secant` uses initial points x_1 and x_2 and generates a sequence of elements $\{x_n\}$ that eventually converges to the root of $f(x) = 0$. The iterations stop when two consecutive elements are sufficiently close to one another, that is, $|x_{n+1} - x_n| < \varepsilon$, where ε is the prescribed tolerance. The outputs are the approximate value of the root and the number of iterations needed to meet the tolerance.

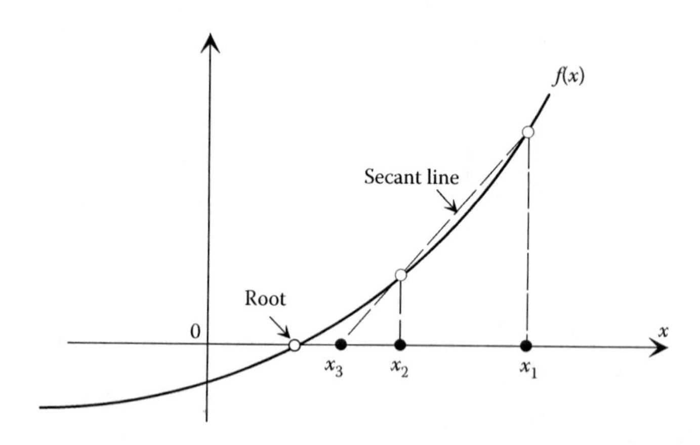

FIGURE 3.18
Geometry of the secant method.

```
function [r n] = Secant(f, x1, x2, tol, N)
%
% Secant uses secant method to approximate roots of
% f(x) = 0.
%
%   [r n] = Secant(f, x1, x2, tol, N), where
%
%     f is an inline function which represents f(x),
%     x1 and x2 are the initial values of x,
%     tol is the scalar tolerance of convergence (default
%     is 1e-4),
%     N is the maximum number of iterations (default is 20),
%
%     r is the approximate root of f(x) = 0,
%     n is the number of iterations required for
%     convergence.

if nargin<5, N=20;end
if nargin<4,tol=1e-4;end

x=zeros(1, N+1); % Pre-allocate

for n=2:N,
    if x1==x2
       r='failure';
       return
    end

     x(1) = x1; x(2) = x2;
     x(n+1) = x(n)-((x(n)-x(n-1))/(f(x(n))-f(x(n-1))))
     * f(x(n));
     if abs(x(n+1)-x(n))<tol,
         r=x(n+1);
          return
     end
end
r='failure';
```

Example 3.10: Secant Method

Find the first positive root of $x \cos x = -1$ using the secant method with $\varepsilon = 10^{-4}$ and maximum 20 iterations.

SOLUTION

We have worked with this equation on a few occasions so far, the last time in Example 3.5 while using Newton's method. According to Figure 3.6, the first positive root is around $x = 2$, so we will execute the user-defined function `Secant` with initial points $x_1 = 1$ and $x_2 = 1.5$.

```
>> f = inline('x*cos(x)+1');
>> [r n] = Secant(f, 1, 1.5)

r =
   2.0739

n =
   6
```

Recalling that Newton's method starting with $x_1 = 1$ also required six iterations, we see that the secant method in this case has a similar rate of convergence.

3.6.1 Rate of Convergence of Secant Method

Assuming a simple root r, the rate of convergence of the secant method is $\frac{1}{2}\left(1+\sqrt{5}\right) \cong 1.618$. More exactly,

$$\lim_{n\to\infty} \frac{|e_{n+1}|}{|e_n|^{1.618}} = \left|\frac{f''(r)}{2f'(r)}\right|^{0.618} \neq 0 \tag{3.11}$$

3.6.2 A Few Notes on Secant Method

- As with Newton's method, because the secant method does not bracket the root in each iteration, the sequence that it generates is not guaranteed to converge to the root.
- For the case of a simple root, the rate of convergence for the secant method is 1.618, thus the generated sequence converges faster than linear, but slower than quadratic. Therefore, it is slower than Newton's method—which has quadratic convergence for simple root—but the expression of $f'(x)$ does not need to be obtained.
- If $f(x)$, $f'(x)$, and $f''(x)$ are continuous on an interval I, which contains the root, $f'(\text{root}) \neq 0$, and the initial points x_1 and x_2 are close to the root, then the secant method converges to the root.

3.7 Equations with Several Roots

All the bracketing and open methods presented in this chapter are capable of finding one root of $f(x) = 0$ at a time. The built-in MATLAB function `fzero` also finds a root nearest a specified point, with syntax `fzero(f,x0)`.

Sometimes several roots of an equation are desired. To find zeros of a function, we must start with an initial point, increment the variable in either one or both directions, look for sign changes in the function value, and finally zoom in onto a root with a desired accuracy.

3.7.1 Finding Zeros to the Right of a Specified Point

The user-defined function `NZeros` finds n roots of $f(x) = 0$ to the right of the specified initial point x_0 by incrementing x and inspecting the sign of the corresponding $f(x)$. A root is detected when $|\Delta x / x| < \varepsilon$, where ε is the prescribed tolerance. The output is the list of the desired number of approximate roots.

```
function NRoots = NZeros(f, n, x0, tol, delx)
%
%  NZeros approximates a desired number of roots of f(x)
%  on the right side of a specified point.
%
%    NRoots = NZeros(f, n, x0, tol, delx) where
%
%       f is an inline function representing f(x),
%       n is the number of desired roots,
%       x0 is the starting value,
%       tol is the scalar tolerance (default is 1e-6),
%       delx is the increment in x (default is 0.1),
%
%       NRoots is the list of n roots of f(x) to the
%       right of x0.

if nargin < 5, delx = 0.1;end
if nargin < 4, tol = 1e-6;end

x = x0;
dx = delx;
NRoots = zeros(n,1);              % pre-allocate
for m = 1:n,
    sgn1 = sign((f(x)));
    while abs(dx/x) > tol,
       if sgn1~= sign((f(x+dx))),
         dx = dx/2;
       else
         x = x+dx;
       end
    end
 NRoots(m) = x;
```

```
dx = delx;
x = x + abs(0.05*x);
end
```

Example 3.11: Several Roots

Find the roots of $x \cos x + 1 = 0$ in [−6,4].

SOLUTION

We first plot the function to see how many roots are there in the given interval.

```
>> f = inline('x*cos(x)+1');
>> ezplot(f)     % Figure 3.19
```

Figure 3.19 shows that there are two roots in [−6,4]. Since both roots (n = 2) are located to the right of −6, we set `x0 = -6` and execute `NZeros` with default values for the remaining input parameters.

```
>> NRoots = NZeros(f, 2, -6)

NRoots =

  -4.9172
   2.0739
```

The second root is what we have repeatedly found using various techniques in this chapter. The first root completely agrees with Figure 3.19.

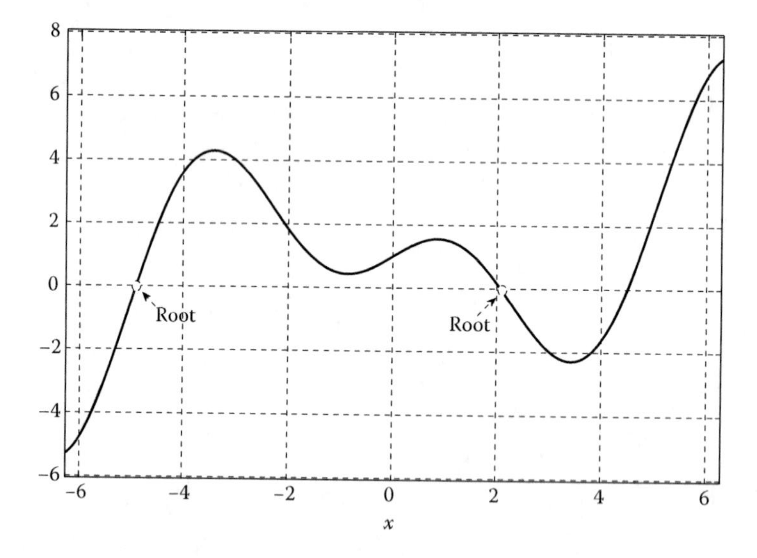

FIGURE 3.19
The two roots of $x \cos x = -1$ in [−6,4].

3.7.2 Finding Zeros on Two Sides of a Specified Point

The user-defined function NZerosMod finds n roots of $f(x) = 0$ on both sides of the specified initial point x_0 by incrementing x in both directions and inspecting the sign of the corresponding $f(x)$. A root is detected when $|\Delta x / x| < \varepsilon$, where ε is the prescribed tolerance. The output is the list of the desired number of approximate roots.

```
function NRoots = NZerosMod(f, n, x0, tol, delx)
%
%  NZerosMod approximates a desired number of roots of
%  f(x) on both sides of a specified point.
%
%    NRoots = NZerosMod(f, n, x0, tol, delx) where
%
%       f is an inline function representing f(x),
%       n is the number of desired roots,
%       x0 is the starting value,
%       tol is the scalar tolerance (default is 1e-6),
%       delx is the increment in x (default is 0.1),
%
%       NRoots is the list of n roots of f(x) on two
%       sides of x0.
if nargin < 5, delx = 0.1;end
if nargin < 4,tol = 1e-6;end

x = x0;
xtrack = x0;
% Vector that tracks which intervals have been searched
dx = delx;
NRoots = zeros(n,1);     % pre-allocate
for m = 1:n
    sgn1 = sign((f(x)));
    while abs(dx/x) > tol
       if sgn1~= sign((f(x+ (-1)^(m+1)*dx)))
          dx = dx/2;
       else
         x = x+ (-1)^(m+1)*dx; % Changes direction of search
       end
    end
NRoots(m) = x;
xtrack(m+1) = x;
xtrack = sort(xtrack);      % Sorting searched intervals
dx = delx;
```

```
% Setting x to the appropriate value so intervals are not
% searched twice
if mod(m,2) == 1
      x=xtrack(1);
else
      x=xtrack(end);
end
x=x+(-1)^m*abs(0.05*x);
end
NRoots=sort(NRoots);
```

Example 3.12: Several Roots

Find the roots of $\cos(\pi x/3) = \frac{1}{2}$ in [−6,4].

SOLUTION

Figure 3.20 shows there are three roots in [−6,4]. We start with `x0 = -2` and execute `NzerosMod` with default values for the remaining input parameters.

```
>> f=inline('cos(pi*x/3)-0.5');
>> NRoots=NzerosMod(f, 3, -2)
```

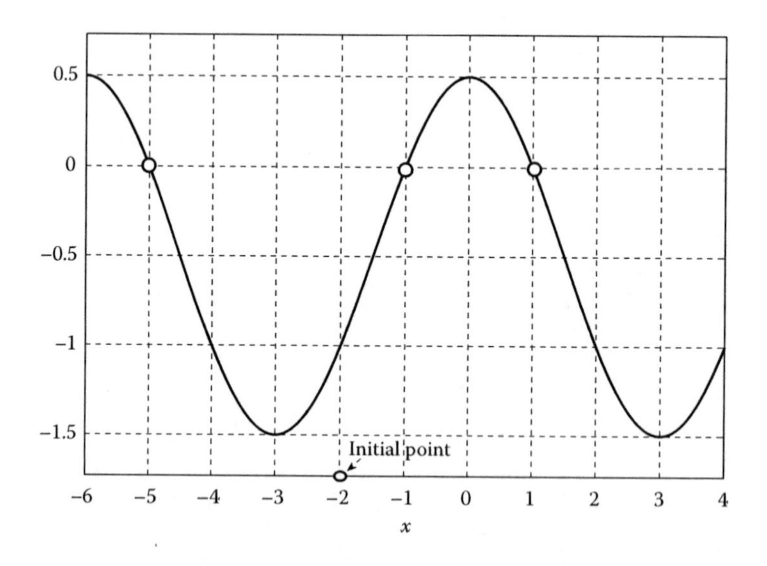

FIGURE 3.20
The three roots of $\cos(\pi x/3) - 1/2 = 0$ in [−6,4].

```
NRoots =

   -5.0000
   -1.0000
    1.0000
```

We see that these are the three roots closest to the starting point.

3.7.3 Using `fzero` to Find Several Roots

The following example demonstrates how the built-in function `fzero` can be utilized to find several roots of an equation in a given interval.

Example 3.13: Several Roots Using `fzero`

Find all roots of $x \sin x = 0$ in $[-10,10]$.

SOLUTION

There are several roots in the given range. We must solve for each root individually using an appropriate initial guess. These initial guesses can be generated by evaluating the function at a few points in the given range, and identifying any sign changes.

```
fun=inline('x.*sin(x)','x');
x=linspace(-10,10,20);
% Generate 20 points in the given range
f=fun(x);      % Evaluate function
plot(x, f)     % Figure 3.21
```

Any sign changes in `f` can be identified by

```
I=find(sign(f(2:end))~=sign(f(1:end-1)));
```

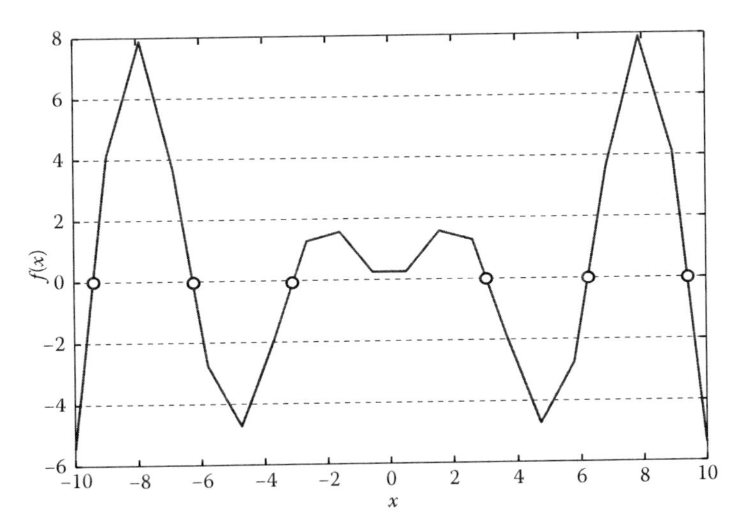

FIGURE 3.21
Several roots of $x \sin x = 0$ in $[-10,10]$.

Now, we can iterate through each sign change, and use fzero with a two-element initial guess.

```
for n=1:length(I)
    r(n)=fzero(fun, x([I(n) I(n)+1]));   % Approximate roots
end
```

```
>> disp(r)    % Display roots
r =

   -9.4248   -6.2832   -3.1416    3.1416    6.2832    9.4248
```

We note that this method missed the obvious root at $x = 0$. This is because the function never changes sign around this root; see Figure 3.22, which is the plot of $x \sin x$ using 101 points for x. The fzero function only finds roots where the function changes sign.

The user-defined function NZeros also skips the root at $x = 0$, again because the function does not exhibit a sign change at that point. Based on Figure 3.22, we will try to find seven roots of the function to the right of –10.

```
>> NRoots=NZeros(fun, 7, -10)

NRoots =

   -9.4248
   -6.2832
   -3.1416
    3.1416
    6.2832
    9.4248
   12.5664    % Root outside range, not accepted
```

The root at 0 is missed by this function as well. Instead, the seventh root is a root outside of the given range.

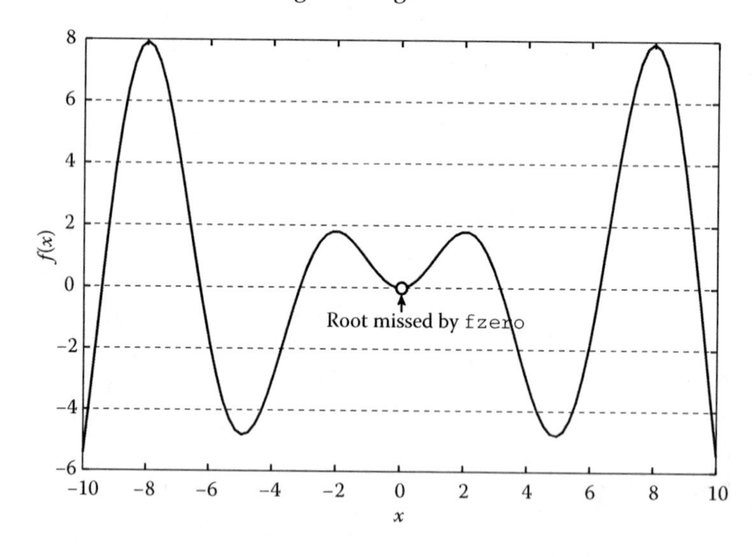

FIGURE 3.22
Root of $x \sin x = 0$ at $x = 0$ missed by fzero.

3.7.4 Points of Discontinuity Mistaken for Roots

If a function is not continuous, fzero may return values that are points of discontinuity rather than roots. In general, it can be difficult to find all roots of any arbitrary nonlinear function unless we have knowledge of the behavior of the function, and an approximate location of all desired roots.

Example 3.14: Points of Discontinuity

Find the roots of tan x = tanh x in [–2,2].

SOLUTION

```
>> fun = inline('tan(x)-tanh(x)','x');
>> x = linspace(-2,2,100);
>> f = fun(x);
>> I = find(sign(f(2:end)) ~= sign(f(1:end-1)));
>> for n = 1:length(I)
     r(n) = fzero(fun, x([I(n) I(n)+1]));
end
>> r

r =

   -1.5708    -0.0000    1.5708
```

We realize that the only legitimate root is the one at 0, while the other two are merely points of discontinuity; Figure 3.23. The user-defined function NZeros also mistakes such points for roots.

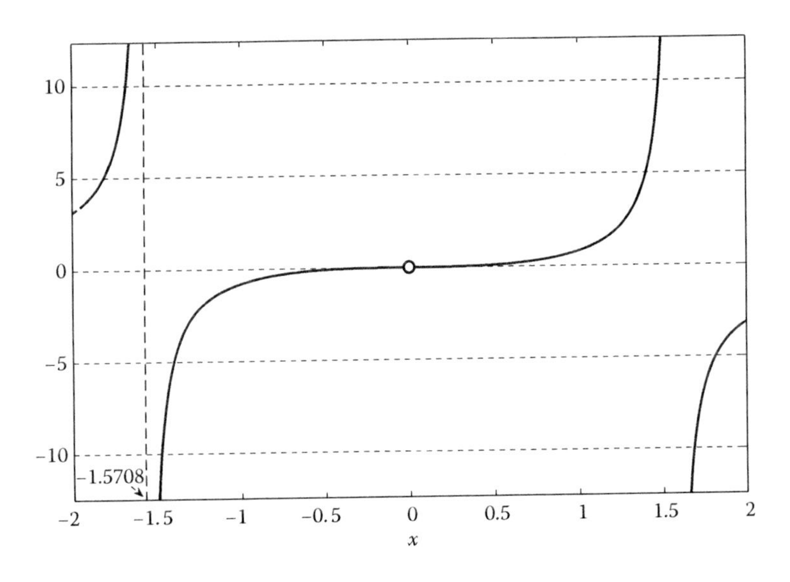

FIGURE 3.23
Discontinuity points mistaken for roots.

PROBLEM SET

Bisection Method (Section 3.2)

✍ In Problems 1 through 6, the given equation has a root in the indicated interval. Using bisection method, generate the first four midpoints and intervals (besides the original interval given) containing the root.

1. $\cos x + \sin x = 0$, [2,4]
2. $e^x - 2x = 2$, [0,2]
3. $x^2 - 4x + 2 = 0$, [3,4]
4. $\cos x \cosh x = -1$, $[-5, -4]$
5. $\tan(x/2) + x + 1 = 0$, $[-1,0]$
6. $\frac{1}{3}x + \ln x = 1, [1,2]$
7. Modify the user-defined function `Bisection` (Section 3.2) so that the table is not generated and the outputs are the approximate root and the number of iterations needed for the tolerance to be met. All other parameters, including default values, are to remain unchanged. Save this function as `Bisection_New`. Apply `Bisection_New` to the equation in Problem 1, using the given tolerance:

$$\cos x + \sin x = 0, [2,4], \quad \varepsilon = 10^{-4}$$

8. Apply the user-defined function `Bisection_New` (Problem 7) to

$$2^{-x} = x, [0,2], \quad \varepsilon = 10^{-4}$$

9. The objective is to find all roots of $e^{-x/2} + \frac{1}{2}x = 3$, one at a time, using the bisection method, as follows: first, locate the roots graphically and identify the intervals containing the roots. The endpoints of each interval must be chosen as the integers closest to a root on each side of that root. Then apply the user-defined function `Bisection_New` (Problem 7) with default values for tolerance and maximum number of iterations allowed to find one root at a time.

10. The goal is to find an approximate value for $\sqrt{7}$ using the bisection method. Define a suitable function $f(x)$, plot to locate the root of $f(x) = 0$ and identify the interval containing the root. The endpoints of the interval must be the integers closest to the intended root on each side of the root. Then apply the user-defined function `Bisection_New` (Problem 7) with default values for tolerance and maximum number of iterations allowed.

Regula Falsi Method (Section 3.3)

✍ In Problems 11 through 16, the given equation has a root in the indicated interval. Using regula falsi method, generate the first four elements in the sequence that eventually converges to the root.

11. $2e^{-x/2} + \ln x = 1,\ [0.1,1]$
12. $\sin x + \sin 2x = 1$, [1,2]
13. $\cos x \cosh x = 1$, [4,5]
14. $x^3 - 5x + 3 = 0$, [1,2]
15. $2x^2 - 3 = 0$, [−2, − 1]
16. $e^{-x} = x^2$, [0,2]
17. Modify the user-defined function `RegulaFalsi` (Section 3.3) so that the table is not generated and the terminating condition is $|f(c_n)| < \varepsilon$, where ε is an imposed tolerance. The outputs are the approximate root and the number of iterations needed for the tolerance to be met. All other parameters, including default values, are to remain unchanged. Save this function as `RegulaFalsi_New`. Apply `RegulaFalsi_New` to the equation in Problem 11, using the given tolerance:

$$2e^{-x/2} + \ln x = 1, [0.1,1], \quad \varepsilon = 10^{-4}$$

18. Apply the user-defined function `RegulaFalsi_New` (Problem 17) to

$$3^{-x} = x^3, [0,1], \quad \varepsilon = 10^{-4}$$

19. The goal is to find all three roots of $2x^3 - 2.5x^2 - 8.01x + 2.5740 = 0$, one at a time, using regula falsi method, as follows: first, locate the roots graphically and identify the intervals containing the roots. The endpoints of each interval must be chosen as the integers closest to a root on each side of that root. Then apply the user-defined function `RegulaFalsi_New` (Problem 17) with default values for tolerance and maximum number of iterations allowed to find one root at a time.
20. The objective is to find the two roots of $e^{-x^2} + 2\sin x = 0.4$ in [−2,4], one at a time, using regula falsi method, as follows: first, locate the roots graphically and identify the intervals containing the roots. The endpoints of each interval must be selected as the integers closest to a root on each side of that root. Apply the user-defined function `RegulaFalsi_New` (Problem 17) with default values for tolerance and maximum number of iterations allowed to find one root at a time.

21. The goal is to find an approximate value for $\sqrt[3]{15}$ using regula falsi method. Define a suitable function $f(x)$, plot to locate the root of $f(x) = 0$, and identify the interval containing the root. The endpoints of the interval must be the integers closest to the intended root on each side of the root. Then apply the user-defined function `RegulaFalsi_New` (Problem 17) with default values for tolerance and maximum number of iterations allowed.

Modified Regula Falsi

22. Modify the user-defined function `RegulaFalsi` (Section 3.3) so that if an endpoint remains stationary for three consecutive iterations, $\frac{1}{2}f(\text{endpoint})$ is used in the calculation of the next x-intercept, and if the endpoint still does not move, $\frac{1}{4}f(\text{endpoint})$ is used, and so on. All other parameters, including the default values are to remain the same as before. Save this function as `RegulaFalsi_Mod`.

    Apply `RegulaFalsi` to find a root of $\frac{1}{3}(x-2)^2 - 3 = 0$ in [−6,2]. Next, apply `RegulaFalsi_Mod` and compare the results.

Fixed-Point Method (Section 3.4)

23. The two roots of $3^{-x} = 4 - x$ are to be found by the fixed-point method as follows: Define two iteration functions $g_1(x) = 4 - 3^{-x}$ and $g_2(x) = -\log_3(4 - x)$.
    a. Locate the fixed points of $g_1(x)$ and $g_2(x)$ graphically.
    b. Referring to the figure showing the fixed points of g_1, set x_1 to be the nearest integer to the left of the smaller fixed point and perform four iterations using the fixed-point method. Next, set x_1 to be the nearest integer to the right of the same fixed point and perform four iterations. If both fixed points were not found, repeat the process applied to g_2. Discuss any convergence issues as related to Theorem 3.1.

24. The two roots of $3x^2 + 2.72x - 1.24 = 0$ are to be found using the fixed-point method as follows: Define iteration functions

$$g_1(x) = \frac{-3x^2 + 1.24}{2.72}, \quad g_2(x) = \frac{-2.72x + 1.24}{3x}$$

    a. Locate the fixed points of $g_1(x)$ and $g_2(x)$ graphically.
    b. Focus on g_1 first. Execute the user-defined function `FixedPoint` (Section 3.4) with initial point x_1 chosen as the nearest integer to the left of the smaller fixed point. Execute a second time with x_1

an integer between the two fixed points. Finally, with x_1 to the right of the larger fixed point. In all cases, use the default tolerance, but increase kmax if necessary. Discuss all convergence issues as related to Theorem 3.1.

c. Repeat Part (b), this time focusing on g_2.

25. Consider the fixed-point iteration described by

$$x_{n+1} = g(x_n) = \frac{1}{2}\left(x_n + \frac{a}{x_n}\right), \quad n = 1, 2, 3, \ldots, \quad a > 0$$

a. ✍ Show that the iteration converges to $\sqrt{a}$ for any initial point $x_1 > 0$, and that the convergence is quadratic.

b. Apply this iteration function $g(x)$ to approximate $\sqrt{5}$. Execute the user-defined function FixedPoint (Section 3.4) using default values for kmax and tol, and x_1 chosen as the nearest integer on the left of the fixed point.

26. The goal is to find the root of $0.3x^2 - x^{1/3} = 1.4$ using the fixed-point method.

a. As a potential iteration function, select $g_1(x) = (x^{1/3} + 1.4)/0.3x$. Graphically locate the fixed point of $g_1(x)$. Execute the user-defined function FixedPoint (Section 3.4) twice, once with initial point chosen as the integer nearest the fixed point on its left, and a second time with the nearest integer on its right. Use default values for kmax and tol, but increase kmax if necessary. Fully discuss convergence issues as related to Theorem 3.1.

b. Next, as the iteration function select $g_2(x) = (0.3x^2 - 1.4)^3$ and repeat all steps in Part (a).

27. The two roots of $x^2 - 3.13x + 2.0332 = 0$ are to be found using the fixed-point method as follows: Define iteration functions

$$g_1(x) = \frac{3.13x - 2.0332}{x}, \quad g_2(x) = \frac{x^2 + 2.0332}{3.13}$$

a. Locate the fixed points of $g_1(x)$ and $g_2(x)$ graphically.

b. Focus on g_1 first. Execute the user-defined function FixedPoint (Section 3.4) with initial point x_1 chosen as the nearest integer to the left of the smaller fixed point. Execute a second time with x_1 an integer between the two fixed points. Finally, with x_1 to the right of the larger fixed point. In all cases, use the default tolerance, but increase kmax if necessary. Discuss all convergence issues as related to Theorem 3.1.

c. Repeat Part (b), this time focusing on g_2.

28. The two roots of $2^{-x/3} + e^x = 2.2$ are to be found by the fixed-point method as follows: Define two iteration functions $g_1(x) = -3\log_2(2.2 - e^x)$ and $g_2(x) = \ln(2.2 - 2^{-x/3})$.
    a. Locate the fixed points of $g_1(x)$ and $g_2(x)$ graphically.
    b. Referring to the figure showing the fixed points of g_1, set x_1 to the nearest integer to the left of the smaller fixed point and perform four iterations using the fixed-point method. Next, set x_1 to the nearest integer to the right of the same fixed point and perform four iterations. If both fixed points were not found, repeat the process applied to g_2. Discuss any convergence issues as related to Theorem 3.1.

Newton's Method (Section 3.5)

In Problems 29 through 34, the given equation has a root in the indicated interval. Using Newton's method, with the initial point set to be the left end of the interval, generate the first four elements in the sequence that eventually converges to the root.

29. $3x^2 - x - 4 = 0$, $[-2,0]$
30. $x^3 + 2x^2 + x + 2 = 0$, $[-3, -1]$
31. $\sin x = x - 1$, $[1,3]$
32. $\ln(x+1) = \frac{1}{2}x$, $[2,4]$
33. $e^{-(x-1)} - \cos(x+1) = 2.3$, $[-1,1]$
34. $\sin x \sinh x + 1 = 0$, $[3,4]$
35. Determine graphically how many roots the equation $0.2x^3 - x^{1/2} = 0.9$ has. Then find each root by executing the user-defined function `Newton` (Section 3.5) with default parameter values and x_1 chosen as the closest integer on the left of the root.
36. The goal is to find an approximate value for $\sqrt{6}$ using Newton's method. Define a suitable function $f(x)$, plot to locate the root of $f(x) = 0$ and identify the interval containing the root. Then apply the user-defined function `Newton` (Section 3.5) with default parameter values and x_1 chosen as the closest integer on the left of the root.
37. The goal is to find two roots of $\cos x \cosh x = -1.3$ in $[-4,4]$.
    a. Graphically locate the roots of interest,
    b. To approximate each root, execute the user-defined function `Newton` with default parameter values and x_1 chosen as the closest integer on the left of the root. If the intended root is not found this way, set x_1 to be the integer closest to the root on its right and reexecute `Newton`. Discuss the results.

38. All three roots of the equation $x^3 - 0.8x^2 - 1.12x - 0.2560 = 0$ are inside [–2,2].
    a. Graphically locate the roots, and decide whether each root is simple or of higher multiplicity,
    b. Approximate the root with higher multiplicity by executing the user-defined function `NewtonMod` (Section 3.5), and the simple root by executing `Newton`. In both cases use default parameter values, and x_1 chosen as the closest integer on the left of the root.
39. Roots of $x^3 - 0.9x^2 + 0.27x - 0.027 = 0$ are inside [–1,1].
    a. Graphically locate the roots, and determine if a root is simple or of higher multiplicity,
    b. Approximate the root with higher multiplicity by executing the user-defined function `NewtonMod` (Section 3.5), and the simple root by executing `Newton`. Use default parameter values, and let x_1 be the closest integer on the left of the root.
40. Locate the root(s) of $0.6(x - 2\sin x) = 3$ graphically, and depending on multiplicity, use Newton's method or the modified Newton's method to find the root(s). Use default parameter values, and let x_1 be the closest integer on the left of the root. Verify the result by using the built-in `fzero` function.

Secant Method (Section 3.6)

In Problems 41 through 46, apply the secant method, with the given initial points x_1 and x_2, to generate the next four elements in the sequence that eventually converges to the root.

41. $x^3 - x^{1/4} = 2.45,\ x_1 = 5,\ x_2 = 4.5$
42. $x^2 + 1.3x - 1.4 = 0,\ x_1 = -5,\ x_2 = -4$
43. $\sin 3x = x - 2,\ x_1 = 0,\ x_2 = 1$
44. $e^{-x} + \sqrt{2}x^3 = 1,\ x_1 = -2,\ x_2 = -1.8$
45. $\ln(0.2x - 1) + 0.3x + 1 = 0,\ x_1 = 6,\ x_2 = 5.8$
46. $\cosh\left(\frac{1}{2}x\right) = x,\ x_1 = -4,\ x_2 = -3$
47. An approximate value for $\sqrt{7}$ using the secant method is desired. Define a suitable function $f(x)$, plot to locate the root of $f(x) = 0$ and identify the interval containing the root. Then apply the user-defined function `Secant` (Section 3.6) with default parameter values and x_1 and x_2 chosen as the two integers closest to the root on the right side of the root.
48. Graphically locate the root of $x^3 + 0.7x^2 + 2x + 1.4 = 0$. Find the root numerically by applying the user-defined function `Secant`

[with $x_1 = 4$, $x_2 = 3.9$]. Next, apply the function `Newton` (Section 3.5) with $x_1 = 4$. In both cases, use default parameter values. Compare the results.

49. Graphically locate the root of $x^2 + x + 0.1x^{-1/3} = 1$. Find the root numerically by applying the user-defined function `Secant` [with $x_1 = 3$, $x_2 = 2.9$]. Next, apply the function `Newton` (Section 3.5) with $x_1 = 3$. In both cases, use default parameter values. Compare the results.

50. Locate the root of $x\sqrt{x^2 + 250} = 420$ graphically. Find the approximate value of the root by applying the user-defined function `Secant` [with $x_1 = -5$, $x_2 = 0$]. Next, apply the function `Newton` (Section 3.5) with $x_1 = -5$. In both cases, use default parameter values. Compare the results.

Equations with Several Roots (Section 3.7)

In each problem, the user-defined functions `NZeros` or `NzerosMod` must be executed. Use the default parameter values unless otherwise specified.

51. Find the first four positive roots of $\sin x + \cos 2x = 0.2$.
52. Find all roots of $\sin(\pi x/2) = \frac{1}{3}$ in $[-4,4]$.
53. A very important function in engineering applications is the Bessel function of the first kind. One specific such function is of order 0, denoted by $J_0(x)$, represented in MATLAB by `besselj(0,x)`. The zeros of Bessel functions arise in applications such as vibration analysis of circular membranes. Find the first four positive zeros of $J_0(x)$, and verify them graphically.
54. The natural frequencies of a beam are the roots of the so-called frequency equation. For a beam fixed at both of its ends, the frequency equation is derived as $\cos x \cosh x = 1$. Find the first five (positive) natural frequencies (rad/s) of this particular beam.
55. Find all roots of $0.2x^4 + 0.58x^3 - 12.1040x^2 + 20.3360x - 6.24 = 0$. Verify the findings by using the MATLAB built-in function `roots`.

4

Solution of Systems of Equations

This chapter is concerned with the numerical solution of systems of equations, and is composed of two main parts: linear systems and nonlinear systems.

4.1 Linear Systems of Equations

A linear system of n algebraic equations in n unknowns $x_1, x_2, \ldots, x_n$ is in the form

$$\begin{cases} a_{11}x_1 + a_{12}x_2 + \cdots + a_{1n}x_n = b_1 \\ a_{21}x_1 + a_{22}x_2 + \cdots + a_{2n}x_n = b_2 \\ \quad \ldots \\ a_{n1}x_1 + a_{n2}x_2 + \cdots + a_{nn}x_n = b_n \end{cases} \tag{4.1}$$

where a_{ij} $(i, j = 1, 2, \ldots, n)$ and b_k $(k = 1, 2, \ldots, n)$ are known constants and a_{ij}'s are the coefficients. If every b_k is zero, the system is homogeneous; otherwise it is nonhomogeneous. Equation 4.1 can be expressed in matrix form as

$$\mathbf{Ax} = \mathbf{b} \tag{4.2}$$

with

$$\mathbf{A} = \begin{bmatrix} a_{11} & a_{12} & \ldots & a_{1n} \\ a_{21} & a_{22} & \ldots & a_{2n} \\ \ldots & \ldots & \ldots & \ldots \\ a_{n1} & a_{n2} & \ldots & a_{nn} \end{bmatrix}_{n\times n}, \quad \mathbf{x} = \begin{Bmatrix} x_1 \\ x_2 \\ \ldots \\ x_n \end{Bmatrix}_{n\times 1}, \quad \mathbf{b} = \begin{Bmatrix} b_1 \\ b_2 \\ \ldots \\ b_n \end{Bmatrix}_{n\times 1}$$

where $\mathbf{A}$ is the coefficient matrix. A set of values for $x_1, x_2, \ldots, x_n$ satisfying Equation 4.1 forms a solution of the system. The vector $\mathbf{x}$ with components $x_1, x_2, \ldots, x_n$ is the solution vector for Equation 4.2. If $x_1 = 0 = x_2 = \cdots = x_n$, the solution $\mathbf{x} = \mathbf{0}_{n\times 1}$ is called the trivial solution. The augmented matrix for Equation 4.2 is defined as

$$\left[\mathbf{A} \middle| \mathbf{b}\right] = \begin{bmatrix} a_{11} & a_{12} & \dots & a_{1n} & b_1 \\ a_{21} & a_{22} & \dots & a_{2n} & b_2 \\ \dots & \dots & \dots & \dots & \dots \\ a_{n1} & a_{n2} & \dots & a_{nn} & b_n \end{bmatrix}_{n\times(n+1)} \tag{4.3}$$

4.2 Numerical Solution of Linear Systems

As described in Figure 4.1, numerical methods for solving linear systems of equations are divided into two categories: direct methods and indirect methods.

A direct method computes the solution of Equation 4.2 by performing a predetermined number of operations. These methods transform the original system into an equivalent system in which the coefficient matrix is upper-triangular, lower-triangular, or diagonal, making the system much easier to solve. Indirect methods use iterations to approximate the solution. The iteration process begins with an initial vector and generates successive approximations that eventually converge to the actual solution. Unlike direct methods, the number of operations required by iterative methods is not known in advance.

4.3 Gauss Elimination Method

Gauss elimination is a procedure that transforms a given linear system of equations into upper-triangular form, the solution of which is found by back substitution. It is important to note that the augmented matrix $[\mathbf{A}|\mathbf{b}]$ completely

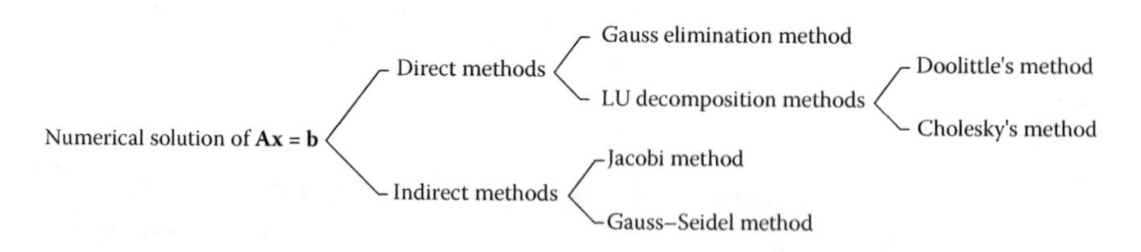

FIGURE 4.1
Classification of methods to solve a linear system of equations.

represents the linear system $\mathbf{Ax} = \mathbf{b}$; therefore, all modifications must be applied to the augmented matrix and not just matrix $\mathbf{A}$. The transformation into upper-triangular form is achieved by using elementary row operations (EROs):

- ERO_1 Multiply a row by a nonzero constant.
- ERO_2 Interchange any two rows.
- ERO_3 Multiply the ith row by a constant $\alpha \neq 0$ and add the result to the kth row, then replace the kth row with the outcome. The ith row is called the pivot row.

The nature of a linear system is preserved under EROs. If a linear system undergoes a finite number of EROs, then the new system and the original one are called row-equivalent.

Consider the system given in Equation 4.1. The first objective is to eliminate x_1 in all equations below the first; thus, the first row is the pivot row. The entry that plays the most important role here is a_{11}, the coefficient of x_1 in the first row, known as the pivot. If $a_{11} = 0$, the first row must be interchanged with another row (ERO_2) to ensure that x_1 has a nonzero coefficient. This is called partial pivoting. Another situation that may lead to partial pivoting is when a pivot is very small in magnitude, with a potential to cause round-off errors. Suppose x_1 has been eliminated via EROs, so that we now have a new system in which the first equation is as in the original, while the rest are generally changed, and are free of x_1. The next step is to focus on the coefficient of x_2 in the second row of this new system. If it is nonzero, and not very small, we use it as the pivot and eliminate x_2 in all the lower-level equations. Here, the second row is the pivot row and remains unchanged. This continues until an upper-triangular system is formed. Finally, back substitution is used to find the solution.

Example 4.1: Gauss Elimination with Partial Pivoting

Using Gauss elimination, find the solution x_1, x_2, x_3, x_4 of the system whose augmented matrix is

$$\left[\begin{array}{cccc|c} -1 & 2 & 3 & 1 & 3 \\ 2 & -4 & 1 & 2 & -1 \\ -3 & 8 & 4 & -1 & 6 \\ 1 & 4 & 7 & -2 & -4 \end{array}\right]$$

SOLUTION

Because the (1,1) entry is nonzero, we use it as the pivot to eliminate the entries directly below it. For instance, multiply the first row (pivot row) by 2 and add the result to the second row, then replace the second row by the outcome; ERO_3. All details are shown in Figure 4.2. Next, we focus on the (2,2) entry in the second row of the new system. Because it is zero, the second row must be switched with any other row below it, say, the third

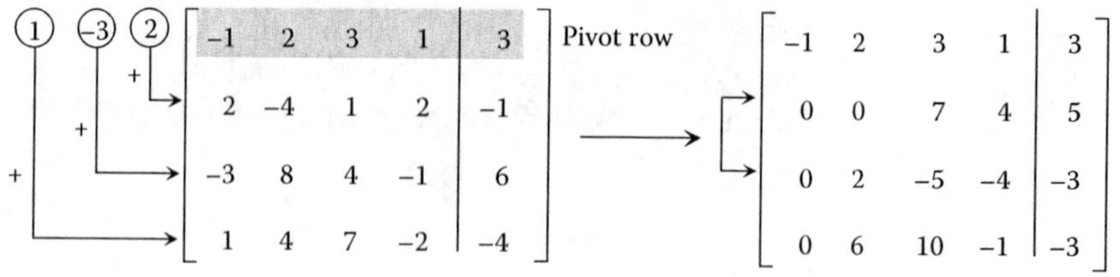

$$\begin{array}{c} \textcircled{1}\ \textcircled{-3}\ \textcircled{2} \end{array}\left[\begin{array}{cccc|c} -1 & 2 & 3 & 1 & 3 \\ 2 & -4 & 1 & 2 & -1 \\ -3 & 8 & 4 & -1 & 6 \\ 1 & 4 & 7 & -2 & -4 \end{array}\right] \text{Pivot row} \longrightarrow \left[\begin{array}{cccc|c} -1 & 2 & 3 & 1 & 3 \\ 0 & 0 & 7 & 4 & 5 \\ 0 & 2 & -5 & -4 & -3 \\ 0 & 6 & 10 & -1 & -3 \end{array}\right]$$

FIGURE 4.2
The first three operations in Example 4.1.

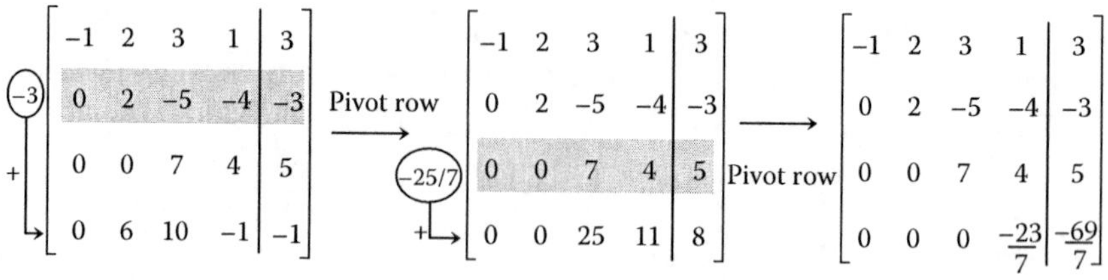

$$\left[\begin{array}{cccc|c} -1 & 2 & 3 & 1 & 3 \\ 0 & 2 & -5 & -4 & -3 \\ 0 & 0 & 7 & 4 & 5 \\ 0 & 6 & 10 & -1 & -1 \end{array}\right] \xrightarrow{\text{Pivot row}} \left[\begin{array}{cccc|c} -1 & 2 & 3 & 1 & 3 \\ 0 & 2 & -5 & -4 & -3 \\ 0 & 0 & 7 & 4 & 5 \\ 0 & 0 & 25 & 11 & 8 \end{array}\right] \text{Pivot row} \longrightarrow \left[\begin{array}{cccc|c} -1 & 2 & 3 & 1 & 3 \\ 0 & 2 & -5 & -4 & -3 \\ 0 & 0 & 7 & 4 & 5 \\ 0 & 0 & 0 & -\frac{23}{7} & -\frac{69}{7} \end{array}\right]$$

FIGURE 4.3
Transformation into upper-triangular form in Example 4.1.

row. As a result, the (2,2) element is now 2, and is used as the pivot to zero out the entries below it. Since the one directly beneath it is already zero, by design, only one ERO_3 is needed. Finally, the (3,3) entry in the latest system is 7, and applying one last ERO_3 yields an upper-triangular system as shown in Figure 4.3.

The solution is then found by back substitution as follows. The last row gives

$$-\left(\frac{23}{7}\right)x_4 = -\left(\frac{69}{7}\right) \quad \Rightarrow \quad x_4 = 3$$

Moving up one row at a time, each time using the latest information on the unknowns, we find

$$\begin{aligned} x_3 &= (5 - 4x_4)/7 = -1 \\ x_2 &= (5x_3 + 4x_4 - 3)/2 = 2 \\ x_1 &= 2x_2 + 3x_3 + x_4 - 3 = 1 \qquad x_3 = -1 \end{aligned}$$

Therefore, the solution is $x_1 = 1$, $x_2 = 2$, $x_3 = -1$, $x_4 = 3$.

4.3.1 Choosing the Pivot Row: Partial Pivoting with Row Scaling

When using partial pivoting, in the first step of the elimination process, it is common to choose as the pivot row the row in which x_1 has the largest (in absolute value) coefficient. The subsequent steps are treated in a similar manner. This is mainly to deal with the problems that round-off can cause while dealing with large matrices. There is also total pivoting where the idea is to locate the entry of the coefficient matrix **A** that is the largest in absolute value among all entries. This entry corresponds to one of the unknowns, say, x_m. Then, the first variable to be eliminated is x_m. A similar logic applies to the new system to decide which variable has to be eliminated next. Because total pivoting requires much more computational effort than partial pivoting, it is not implemented in practice. Instead, partial pivoting with scaling is used where we choose the pivot row as the row in which x_1 has the largest (in absolute value) coefficient relative to the other entries in that row. More specifically, consider the first step, where x_1 is to be eliminated. We will choose the pivot row as follows. Assume **A** is $n \times n$.

1. In each row i of **A**, find the entry with the largest absolute value. Call it M_i.
2. In each row i, find the ratio of the absolute value of the coefficient of x_1 to the absolute value of M_i, that is

$$r_i = \frac{|a_{i1}|}{|M_i|}$$

3. Among r_i ($i = 1, 2, \ldots, n$), pick the largest. Whichever row is responsible for this maximum value is picked as the pivot row. Eliminate x_1 to obtain a new system.
4. In the new system, consider the $(n-1) \times (n-1)$ submatrix of the coefficient matrix occupying the lower right corner. In this matrix, use the same logic as above to choose the pivot row to eliminate x_2, and so on.

Example 4.2: Partial Pivoting with Scaling

Use partial pivoting with scaling to solve the 3×3 system with the augmented matrix

$$\left[\mathbf{A} | \mathbf{b}\right] = \left[\begin{array}{ccc|c} -4 & -3 & 5 & 0 \\ 6 & 7 & -3 & 2 \\ 2 & -1 & 1 & 6 \end{array}\right]$$

SOLUTION

The three values of r_i are found as

$$r_1 = \frac{|-4|}{|5|} = \frac{4}{5}, \quad r_2 = \frac{|6|}{|7|} = \frac{6}{7}, \quad r_3 = \frac{|2|}{|2|} = 1$$

Since r_3 is the largest, it is the third row that produces the maximum value; hence, it is chosen as the pivot row. Switch the first and the third row in the original system and eliminate x_1 using EROs to obtain Figure 4.4.

$$\left[\begin{array}{ccc|c} 2 & -1 & 1 & 6 \\ 6 & 7 & -3 & 2 \\ -4 & -3 & 5 & 0 \end{array}\right] \xrightarrow{\text{Eliminate } x_1} \left[\begin{array}{ccc|c} 2 & -1 & 1 & 6 \\ 0 & 10 & -6 & -16 \\ 0 & -5 & 7 & 12 \end{array}\right]$$

B

FIGURE 4.4
Partial pivoting with scaling.

To eliminate x_2, consider the 2×2 submatrix **B** and compute the corresponding ratios

$$\frac{|10|}{|10|} = 1, \quad \frac{|-5|}{|7|} = \frac{5}{7}$$

so that the 1st row is picked as the pivot row. Row operations yield

$$\left[\begin{array}{ccc|c} 2 & -1 & 1 & 6 \\ 0 & 10 & -6 & -16 \\ 0 & 0 & 4 & 4 \end{array}\right]$$

and back substitution gives the solution; $x_3 = 1, x_2 = -1, x_1 = 2$.

4.3.2 Permutation Matrices

In the foregoing analysis, a linear $n \times n$ system was solved by Gauss elimination via EROs. In the process, the original system $\mathbf{Ax} = \mathbf{b}$ was transformed into $\mathbf{Ux} = \tilde{\mathbf{b}}$, where $\mathbf{U}$ is an upper-triangular matrix with nonzero diagonal entries. So, there must exist an $n \times n$ matrix $\mathbf{P}$ so that premultiplication of the original system by $\mathbf{P}$ yields

$$\mathbf{P}[\mathbf{Ax}] = \mathbf{Pb} \quad \Rightarrow \quad [\mathbf{PA}]\mathbf{x} = \mathbf{Pb} \quad \Rightarrow \quad \mathbf{Ux} = \tilde{\mathbf{b}} \tag{4.4}$$

To identify this matrix **P**, we need what are known as permutation matrices. The simplest way to describe these matrices is to go through an example. Let us refer to the 4×4 system in Example 4.1. Because the size is 4, we start with the 4×4 identity matrix

$$\mathbf{I} = \begin{bmatrix} 1 & 0 & 0 & 0 \\ 0 & 1 & 0 & 0 \\ 0 & 0 & 1 & 0 \\ 0 & 0 & 0 & 1 \end{bmatrix}$$

Consider the three EROs in the first step of Example 4.1; see Figure 4.2. Apply them to **I** to get the matrix $\mathbf{P}_1$ (shown below). Next, focus on the second step, where there was only one ERO; the second and third rows were switched. Apply that to **I** to obtain $\mathbf{P}_2$. The third step also involved one ERO only. Apply to **I** to get $\mathbf{P}_3$. Finally, application of the operation in the last step to **I** gives $\mathbf{P}_4$.

$$\mathbf{P}_1 = \begin{bmatrix} 1 & 0 & 0 & 0 \\ 2 & 1 & 0 & 0 \\ -3 & 0 & 1 & 0 \\ 1 & 0 & 0 & 1 \end{bmatrix}, \quad \mathbf{P}_2 = \begin{bmatrix} 1 & 0 & 0 & 0 \\ 0 & 0 & 1 & 0 \\ 0 & 1 & 0 & 0 \\ 0 & 0 & 0 & 1 \end{bmatrix},$$

$$\mathbf{P}_3 = \begin{bmatrix} 1 & 0 & 0 & 0 \\ 0 & 1 & 0 & 0 \\ 0 & 0 & 1 & 0 \\ 0 & -3 & 0 & 1 \end{bmatrix}, \quad \mathbf{P}_4 = \begin{bmatrix} 1 & 0 & 0 & 0 \\ 0 & 1 & 0 & 0 \\ 0 & 0 & 1 & 0 \\ 0 & 0 & -\frac{25}{7} & 1 \end{bmatrix}$$

Each $\mathbf{P}_i$ is called a permutation matrix, reflecting the operations in each step of Gauss elimination.

Then

$\mathbf{P}_1\,\mathbf{A}$ yields exactly the coefficient matrix at the conclusion of the first step in Example 4.1.

$\mathbf{P}_2\,(\mathbf{P}_1\,\mathbf{A})$ gives the coefficient matrix at the end of the second step.

$\mathbf{P}_3\,(\mathbf{P}_2\,\mathbf{P}_1\,\mathbf{A})$ produces the coefficient matrix at the end of the third step.

$\mathbf{P}_4\,(\mathbf{P}_3\,\mathbf{P}_2\,\mathbf{P}_1\,\mathbf{A})$ gives the upper-triangular coefficient matrix **U** at the end of the fourth step.

Letting $\mathbf{P} = \mathbf{P}_4\,\mathbf{P}_3\,\mathbf{P}_2\,\mathbf{P}_1$

$$\mathbf{PA} = \begin{bmatrix} -1 & 2 & 3 & 1 \\ 0 & 2 & -5 & -4 \\ 0 & 0 & 7 & 4 \\ 0 & 0 & 0 & -\frac{23}{7} \end{bmatrix} \quad \text{and} \quad \mathbf{Pb} = \tilde{\mathbf{b}} = \begin{Bmatrix} 3 \\ -3 \\ 5 \\ -\frac{69}{7} \end{Bmatrix}$$

Subsequently, the final triangular system has the augmented matrix $[\mathbf{U}|\tilde{\mathbf{b}}]$ as suggested by Equation 4.4.

The user-defined function `GaussPivotScale` uses Gauss elimination with partial pivoting and row scaling to transform a linear system $\mathbf{Ax} = \mathbf{b}$ into an upper-triangular system, and subsequently finds the solution vector by back substitution. The user-defined function `BackSub` performs the back substitution portion and is given below.

```
function x = BackSub(Ab)
%
% BackSub returns the solution vector of the upper
% triangular augmented matrix Ab using back substitution.
%
%   x = BackSub(Ab) where
%
%     Ab is the n-by-(n+1) augmented matrix,
%
%     x is the n-by-1 solution vector.
n = size(Ab, 1);
for k = n:-1:1,
    Ab(k,:) = Ab(k,:)./Ab(k, k); % Construct multipliers
    Ab(1:k-1, n+1) = Ab(1:k-1, n+1)-Ab(1:k-1, k)*Ab(k, n+1);
    % Adjust rows
end
x = Ab(:, end);
```

```
function x = GaussPivotScale(A,b)
%
% GaussPivotScale uses Gauss elimination with partial
% pivoting and row scaling to solve the linear system
% Ax = b.
%
%   x = GaussPivotScale(A,b) where
%
%     A is the n-by-n coefficient matrix,
%     b is the n-by-1 result vector,
%
%     x is the n-by-1 solution vector.
n = length(b);
A = [A b];      % Augmented matrix
for k = 1:n-1,
    M = max(abs(A(k:end, k:end-1)), [], 2);
    % Find maximum for each row
    a = abs(A(k:end, k)); % Find maximum for kth column
    [ignore I] = max(a./M);
    % Find relative row with maximum ratio
    I = I + k-1;      % Adjust relative row to actual row
    if I > k
```

```
        % Pivot rows
        A([k I],:)=A([I k],:);
    end
    m=A(k+1:n, k)/A(k, k);         % Construct multipliers
    [Ak M]=meshgrid(A(k,:), m);        % Create mesh
    A(k+1:n,:)=A(k+1:n,:) - Ak.*M;
end
Ab=A;
% Find the solution vector using back substitution
x=BackSub(Ab);
```

Example 4.3: Partial Pivoting with Scaling

The linear system in Example 4.2 can be solved using the under-defined function `GaussPivotScale` as follows:

```
>> A=[-4 -3 5;6 7 -3;2 -1 1];
>> b=[0;2;6];
>> x=GaussPivotScale(A,b)

x=
     2
    -1
     1
```

4.3.3 Counting the Number of Operations

The objective is to determine approximately the total number of operations required by Gauss elimination for solving an $n \times n$ system. We note that the entire process consists of two parts: (1) elimination and (2) back substitution.

4.3.3.1 Elimination

Suppose the first $k-1$ steps of elimination have been performed, and we are in the kth step. This means that the coefficients of x_k must be made into zeros in the remaining $n-k$ rows of the augmented matrix. There

$n-k$ divisions are needed to figure out the multipliers

$(n-k)(n-k+1)$ multiplications

$(n-k)(n-k+1)$ additions

Noting that the elimination process consists of $n-1$ steps, the total number of operations N_e is

$$N_e = \underbrace{\sum_{k=1}^{n-1}(n-k)}_{\text{Divisions}} + \underbrace{\sum_{k=1}^{n-1}(n-k)(n-k+1)}_{\text{Multiplications}} + \underbrace{\sum_{k=1}^{n-1}(n-k)(n-k+1)}_{\text{Additions}} \quad (4.5)$$

Letting $p = n - k$, Equation 4.5 is rewritten as (verify)

$$N_e = \sum_{p=1}^{n-1} p + 2\sum_{p=1}^{n-1} p(p+1) = 3\sum_{p=1}^{n-1} p + 2\sum_{p=1}^{n-1} p^2 \quad (4.6)$$

Using the well-known identities

$$\sum_{p=1}^{M} p = \frac{M(M+1)}{2} \quad \text{and} \quad \sum_{p=1}^{M} p^2 = \frac{M(M+1)(2M+1)}{6}$$

in Equation 4.6, the total number of operations in the elimination process is given by

$$N_e = 3\frac{(n-1)n}{2} + 2\frac{(n-1)n(2n-1)}{6} \overset{\text{For large } n}{\cong} \frac{2}{3}n^3 \quad (4.7)$$

where we have neglected lower powers of n. The approximation is particularly useful for a large system. With the above information, we can show, for example, that the total number of multiplications is roughly $\frac{1}{3}n^3$.

4.3.3.2 Back Substitution

When back substitution is used to determine x_k, one performs

$n - k$ multiplications
$n - k$ subtractions
1 division

In Example 4.1, for instance, $n = 4$, and solving for x_2 (so that $k = 2$) requires two multiplications ($n - k = 2$), two subtractions, and one division. So, the total number of operations N_s for the back substitution process is

$$N_s = \underbrace{\sum_{k=1}^{n} 1}_{\text{Divisions}} + \underbrace{\sum_{k=1}^{n} (n-k)}_{\text{Multiplications}} + \underbrace{\sum_{k=1}^{n} (n-k)}_{\text{Subtractions}}$$

$$= n + \frac{n(n-1)}{2} + \frac{n(n-1)}{2} \overset{\text{For large } n}{\cong} n^2 \quad (4.8)$$

If n is large, N_e dominates N_s, and the total number of operations in Gauss elimination (for a large system) is

$$N_o = N_e + N_s \cong \frac{2}{3}n^3$$

4.3.4 Tridiagonal Systems

Tridiagonal systems often arise in engineering applications and appear in the special form

$$\begin{bmatrix} d_1 & u_1 & 0 & 0 & \dots & 0 \\ l_2 & d_2 & u_2 & 0 & \dots & 0 \\ 0 & l_3 & d_3 & u_3 & \dots & 0 \\ \dots & \dots & \dots & \dots & \dots & \dots \\ \dots & \dots & 0 & l_{n-1} & d_{n-1} & u_{n-1} \\ 0 & 0 & \dots & 0 & l_n & d_n \end{bmatrix} \begin{Bmatrix} x_1 \\ x_2 \\ x_3 \\ \dots \\ x_{n-1} \\ x_n \end{Bmatrix} = \begin{Bmatrix} b_1 \\ b_2 \\ b_3 \\ \dots \\ b_{n-1} \\ b_n \end{Bmatrix} \tag{4.9}$$

where d_i ($i = 1, 2, \dots, n$) are the diagonal entries, l_i ($i = 2, \dots, n$) the lower diagonal entries, and u_i ($i = 1, 2, \dots, n-1$) the upper diagonal entries of the coefficient matrix. Gauss elimination can be used for solving such systems, but is not recommended. This is because Gauss elimination is not designed to accommodate the very special structure of a tridiagonal coefficient matrix, and as a result will perform unnecessary operations to find the solution. Instead, we use an efficient technique known as the Thomas method, which takes advantage of the fact that the coefficient matrix has several zero entries. The Thomas method uses Gauss elimination with the diagonal entry scaled to 1 in each step.

4.3.4.1 *Thomas Method*

Writing out the equations in Equation 4.9, we have

$$d_1x_1 + u_1x_2 = b_1$$

$$l_2x_1 + d_2x_2 + u_2x_3 = b_2$$

$$\dots$$

$$l_{n-1}x_{n-2} + d_{n-1}x_{n-1} + u_{n-1}x_n = b_{n-1}$$

$$l_nx_{n-1} + d_nx_n = b_n$$

In the first equation, the diagonal entry is scaled to 1, that is, multiply the equation by $1/a_{11}$. Therefore, in the first equation, the modified elements are

$$u_1 = \frac{u_1}{d_1}, \quad b_1 = \frac{b_1}{d_1}$$

All remaining equations, except the very last one, involve three terms. In these equations, the modified elements are

$$u_i = \frac{u_i}{d_i - u_{i-1}l_i}, \quad b_i = \frac{b_i - b_{i-1}l_i}{d_i - u_{i-1}l_i}, \quad i = 2, 3, \ldots, n-1$$

Note that in every stage, the (latest) modified values for all elements must be used. In the last equation

$$b_n = \frac{b_n - b_{n-1}l_n}{d_n - u_{n-1}l_n}$$

Finally, use back substitution to solve the system:

$$x_n = b_n$$

$$x_i = b_i - u_i x_{i+1}, \quad i = n-1, n-2, \ldots, 2, 1$$

Example 4.4: Thomas Method

Solve the following tridiagonal system using the Thomas method:

$$\begin{bmatrix} -2 & 1 & 0 \\ 3 & 2 & 1 \\ 0 & -1 & 3 \end{bmatrix} \begin{Bmatrix} x_1 \\ x_2 \\ x_3 \end{Bmatrix} = \begin{Bmatrix} 3 \\ 1 \\ 5 \end{Bmatrix}$$

SOLUTION

We first identify all elements:

$$d_1 = -2, d_2 = 2, d_3 = 3, l_2 = 3, l_3 = -1, u_1 = 1, u_2 = 1, b_1 = 3, b_2 = 1, b_3 = 5$$

In the first equation

$$u_1 = \frac{u_1}{d_1} = \frac{1}{-2} = -\frac{1}{2}, \quad b_1 = \frac{b_1}{d_1} = \frac{3}{-2} = -\frac{3}{2}$$

In the second equation

$$u_2 = \frac{u_2}{d_2 - u_1 l_2} = \frac{1}{2 - (-\frac{1}{2})(3)} = \frac{2}{7}, \quad b_2 = \frac{b_2 - b_1 l_2}{d_2 - u_1 l_2} = \frac{1 - (-\frac{3}{2})(3)}{2 - (-\frac{1}{2})(3)} = \frac{11}{7}$$

In the last equation

$$b_3 = \frac{b_3 - b_2 l_3}{d_3 - u_2 l_3} = \frac{5 - (\frac{11}{7})(-1)}{3 - (\frac{2}{7})(-1)} = 2$$

Back substitution yields

$$x_3 = b_3 = 2$$
$$x_2 = b_2 - u_2 x_3 = \frac{11}{7} - \left(\frac{2}{7}\right)(2) = 1$$
$$x_1 = b_1 - u_1 x_2 = -\frac{3}{2} - \left(-\frac{1}{2}\right)(1) = -1$$

The user-defined function ThomasMethod uses the Thomas method to solve an $n \times n$ tridiagonal system $\mathbf{Ax} = \mathbf{b}$. The inputs are $\mathbf{A}$ and $\mathbf{b}$. From $\mathbf{A}$, three $n \times 1$ vectors will be constructed:

$$\mathbf{d} = \begin{bmatrix} a_{11} & a_{22} & a_{33} & \dots & a_{nn} \end{bmatrix}^T$$
$$\boldsymbol{l} = \begin{bmatrix} 0 & a_{21} & a_{32} & \dots & a_{n,n-1} \end{bmatrix}^T$$
$$\mathbf{u} = \begin{bmatrix} a_{12} & a_{23} & \dots & a_{n-1,n} & 0 \end{bmatrix}^T$$

These are subsequently used in the procedure outlined above to determine the solution vector $\mathbf{x}$.

```
function x = ThomasMethod(A,b)
%
% ThomasMethod uses Thomas method to find the solution
% vector x of a tridiagonal system Ax = b.
%
%   x = ThomasMethod(A,b) where
%
%     A is a tridiagonal n-by-n coefficient matrix,
%     b is the n-by-1 vector of the right-hand sides,
%
%     x is the n-by-1 solution vector.
%
n = size(A,1);
d = diag(A);          % Vector of diagonal entries of A
l = [0;diag(A, -1)];        % Vector of lower diagonal elements
u = [diag(A,1);0];          % Vector of upper diagonal elements

u(1) = u(1)/d(1); b(1) = b(1)/d(1);    % First equation

for  i = 2:n-1,     % The next n-2 equations
     den = d(i) - u(i-1)*l(i);
     if den == 0
          x = 'failure, division by zero';
          return
     end
```

```
  u(i)=u(i)/den; b(i) = (b(i)-b(i-1)*l(i))/den;
end

b(n) = (b(n)-b(n-1)*l(n))/(d(n)-u(n-1)*l(n));
% Last equation
x(n) =b(n);
for i=n-1:-1:1,
      x(i)=b(i) - u(i)*x(i+1);
end
x=x';
```

The result obtained in Example 4.4 can be verified through the execution of this function as

```
>> A=[-2 1 0;3 2 1;0 -1 3];
>> b=[3;1;5];
>> x=ThomasMethod(A,b)

x=
    -1
     1
     2
```

4.3.4.2 MATLAB® Built-In Function "\"

The built-in function in MATLAB® for solving a linear system **Ax** = **b** is the backslash (\), and the solution vector is obtained via **x** = **A\b**. It is important to note that **x** = **A\b** computes the solution vector by Gauss elimination and not by $\mathbf{x} = \mathbf{A}^{-1}\mathbf{b}$.

For the linear system in Example 4.4, this yields

```
>> x=A\b

   x=
      -1
       1
       2
```

4.4 LU Factorization Methods

In the last section, we learned that solving a large $n \times n$ system **Ax** = **b** using Gauss elimination requires approximately $\frac{2}{3}n^3$ operations. There are other direct methods that require fewer operations than Gauss elimination. These methods make use of the LU factorization of the coefficient matrix **A**.

LU factorization of a matrix **A** means expressing it in the form **A** = **LU**, where **L** is a lower-triangular matrix and **U** is upper-triangular. There are different ways to accomplish this, depending on the specific restrictions imposed on **L** or **U**. For example, Crout factorization (see Problem Set at the end of this chapter) requires the diagonal entries of **U** be 1s, while **L** is a general lower-triangular matrix. Another technique, known as Doolittle factorization, uses the results from different steps of Gauss elimination. These two approaches have similar performances, but we will present Doolittle factorization here.

4.4.1 Doolittle Factorization

Doolittle factorization of **A** is **A** = **LU**, where **L** is lower-triangular consisting of 1s along the diagonal and negatives of the multipliers (from Gauss elimination) below the main diagonal, while **U** is the upper-triangular form of the coefficient matrix in the final step of Gauss elimination.

Example 4.5: Doolittle Factorization

Find the Doolittle factorization of

$$\mathbf{A} = \begin{bmatrix} 1 & 3 & 6 \\ 2 & -1 & 1 \\ 4 & -2 & 3 \end{bmatrix}$$

SOLUTION

Imagine **A** as the coefficient matrix in a linear system, which is being solved by Gauss elimination as shown in Figure 4.5.

The final upper-triangular form is **U**. Three multipliers, –2, –4, and –2, have been used to create zeros in the (2,1), (3,1), and (3,2) positions. Therefore, 2, 4, and 2 will occupy the respective slots in matrix **L**. As a result

$$\mathbf{L} = \begin{bmatrix} 1 & 0 & 0 \\ 2 & 1 & 0 \\ 4 & 2 & 1 \end{bmatrix}, \quad \mathbf{U} = \begin{bmatrix} 1 & 3 & 6 \\ 0 & -7 & -11 \\ 0 & 0 & 1 \end{bmatrix}$$

A more efficient way to find **L** and **U** is a direct approach as demonstrated in the following example.

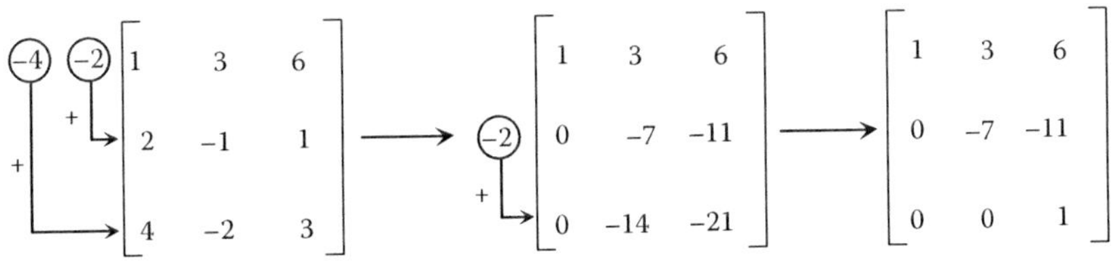

$$\begin{bmatrix} 1 & 3 & 6 \\ 2 & -1 & 1 \\ 4 & -2 & 3 \end{bmatrix} \longrightarrow \begin{bmatrix} 1 & 3 & 6 \\ 0 & -7 & -11 \\ 0 & -14 & -21 \end{bmatrix} \longrightarrow \begin{bmatrix} 1 & 3 & 6 \\ 0 & -7 & -11 \\ 0 & 0 & 1 \end{bmatrix}$$

FIGURE 4.5
Reduction to an upper-triangular matrix.

Example 4.6: Direct Calculation of L and U

Consider the matrix in Example 4.5.

$$\mathbf{A} = \begin{bmatrix} 1 & 3 & 6 \\ 2 & -1 & 1 \\ 4 & -2 & 3 \end{bmatrix}$$

Based on the structures of **L** and **U** in Doolittle factorization, we write

$$\mathbf{L} = \begin{bmatrix} 1 & 0 & 0 \\ l_{21} & 1 & 0 \\ l_{31} & l_{32} & 1 \end{bmatrix}, \quad \mathbf{U} = \begin{bmatrix} u_{11} & u_{12} & u_{13} \\ 0 & u_{22} & u_{23} \\ 0 & 0 & u_{33} \end{bmatrix}$$

Setting **A** = **LU**, we find

$$\begin{bmatrix} a_{11} & a_{12} & a_{13} \\ a_{21} & a_{22} & a_{23} \\ a_{31} & a_{32} & a_{33} \end{bmatrix} = \begin{bmatrix} u_{11} & u_{12} & u_{13} \\ l_{21}u_{11} & l_{21}u_{12} + u_{22} & l_{21}u_{13} + u_{23} \\ l_{31}u_{11} & l_{31}u_{12} + l_{32}u_{22} & l_{31}u_{13} + l_{32}u_{23} + u_{33} \end{bmatrix}$$

Each entry on the left must be equal to the corresponding entry on the right. This generates nine equations in nine unknowns. The entries in the first row of **U** are found immediately as

$$u_{11} = a_{11}, \quad u_{12} = a_{12}, \quad u_{13} = a_{13}$$

The elements in the first column of **L** are found as

$$l_{21} = \frac{a_{21}}{u_{11}}, \quad l_{31} = \frac{a_{31}}{u_{11}}$$

The entries in the second row of **U** are calculated via

$$u_{22} = a_{22} - l_{21}u_{12}, \quad u_{23} = a_{23} - l_{21}u_{13}$$

The element in the second column of **L** is found as

$$l_{32} = \frac{a_{32} - l_{31}u_{12}}{u_{22}}$$

Finally, the entry in the third row of **U** is given by

$$u_{33} = a_{33} - l_{31}u_{13} - l_{32}u_{23}$$

Using the entries of matrix **A** and solving the nine equations just listed, we find

$$\mathbf{L} = \begin{bmatrix} 1 & 0 & 0 \\ 2 & 1 & 0 \\ 4 & 2 & 1 \end{bmatrix}, \quad \mathbf{U} = \begin{bmatrix} 1 & 3 & 6 \\ 0 & -7 & -11 \\ 0 & 0 & 1 \end{bmatrix}$$

This clearly agrees with the outcome of Example 4.5.

The direct calculation of the entries of **L** and **U** in Doolittle factorization can be performed systematically for an $n \times n$ matrix **A** using the steps outlined in Example 4.6. The user-defined function `DoolittleFactor` performs all the operations in the order suggested in Example 4.6 and returns the appropriate **L** and **U** matrices.

```
function [L U] = DoolittleFactor(A)
%
% DoolittleFactor returns the Doolittle factorization of
% matrix A.
%
%   [L U] = DoolittleFactor(A) where
%
%     A is an n-by-n matrix,
%
%     L is the lower triangular matrix with 1's along the
%     diagonal,
%     U is an upper triangular matrix.
%
n = size(A,1);
L = eye(n); U = zeros(n,n);          % Initialize
for i = 1:n,
    U(i,i) = A(i,i)-L(i,1:i-1)*U(1:i-1,i);
    for j = i+1:n,
        U(i,j) = A(i,j)-L(i,1:i-1)*U(1:i-1,j);
        L(j,i) = (A(j,i)-L(j,1:i-1)*U(1:i-1,i))/U(i,i);
    end
end
```

The findings of the last example can easily be confirmed using this function.

```
>> A = [1 3 6;2 -1 1;4 -2 3];
>> [L U] = DoolittleFactor(A)

L =

   1   0   0
   2   1   0
   4   2   1
```

```
U =

     1     3     6
     0    -7   -11
     0     0     1
```

4.4.1.1 Doolittle's Method to Solve a Linear System

Doolittle's method uses Doolittle factorization to solve $\mathbf{Ax} = \mathbf{b}$. Substitution of $\mathbf{A} = \mathbf{LU}$ into the system yields

$$[\mathbf{LU}]\mathbf{x} = \mathbf{b} \quad \Rightarrow \quad \mathbf{L}[\mathbf{Ux}] = \mathbf{b}$$

which will be solved in two steps:

$$\begin{cases} \mathbf{Ly} = \mathbf{b} \\ \mathbf{Ux} = \mathbf{y} \end{cases} \tag{4.10}$$

Note that each of the two systems is triangular, and hence only requires forward and back substitution to solve. The user-defined function `DoolittleMethod` uses Doolittle factorization of the coefficient matrix, and subsequently solves the two triangular systems in Equation 4.10 to find the solution vector $\mathbf{x}$.

```
function x = DoolittleMethod(A,b)
%
% DoolittleMethod uses the Doolittle factorization of
% matrix A and solves the ensuing triangular systems to
% find the solution vector x.
%
%   x = DoolittleMethod(A,b) where
%
%     A is the n-by-n coefficient matrix,
%     b is the n-by-1 vector of the right-hand sides,
%
%     x is the n-by-1 solution vector.
%
[L U] = DoolittleFactor(A);
% Find Doolittle factorization of A
n = size(A,1);

% Solve the lower triangular system Ly = b (forward
% substitution)
y = zeros(n,1);
y(1) = b(1);
```

```
for i=2:n,
    y(i)=b(i)-L(i,1:i-1)*y(1:i-1);
end

% Solve the upper triangular system Ux=y (back
% substitution)
x=zeros(n,1);
x(n)=y(n)/U(n,n);
    for i=n-1:-1:1,
        x(i)=(y(i)-U(i,i+1:n)*x(i+1:n))/U(i,i);
    end
end
```

Example 4.7: Doolittle's Method to Solve a Linear System

Using Doolittle's method, solve $\mathbf{Ax}=\mathbf{b}$, where

$$\mathbf{A}=\begin{bmatrix}1 & 3 & 6\\ 2 & -1 & 1\\ 4 & -2 & 3\end{bmatrix},\quad \mathbf{b}=\begin{Bmatrix}19\\ -2\\ -1\end{Bmatrix},\quad \mathbf{x}=\begin{Bmatrix}x_1\\ x_2\\ x_3\end{Bmatrix}$$

SOLUTION

Doolittle factorization of **A** was done in Examples 4.5 and 4.6. Using **L** and **U** in Equation 4.10

$$\mathbf{Ly}=\mathbf{b}:\begin{bmatrix}1 & 0 & 0\\ 2 & 1 & 0\\ 4 & 2 & 1\end{bmatrix}\begin{Bmatrix}y_1\\ y_2\\ y_3\end{Bmatrix}=\begin{Bmatrix}19\\ -2\\ -1\end{Bmatrix}\Rightarrow \mathbf{y}=\begin{Bmatrix}19\\ -40\\ 3\end{Bmatrix}$$

$$\mathbf{Ux}=\mathbf{y}:\begin{bmatrix}1 & 3 & 6\\ 0 & -7 & -11\\ 0 & 0 & 1\end{bmatrix}\begin{Bmatrix}x_1\\ x_2\\ x_3\end{Bmatrix}=\begin{Bmatrix}19\\ -40\\ 3\end{Bmatrix}\Rightarrow \mathbf{x}=\begin{Bmatrix}-2\\ 1\\ 3\end{Bmatrix}$$

The result can be verified by executing the user-defined function `DoolittleMethod`.

```
>> A=[1 3 6;2 -1 1;4 -2 3];
>> b=[19;-2; -1];
>> x=DoolittleMethod(A,b)

x=

   -2
    1
    3
```

4.4.1.2 Operations Count

Doolittle's method comprises two parts: LU factorization of the coefficient matrix and back/forward substitution to solve the two triangular systems. For a large system $\mathbf{Ax} = \mathbf{b}$, the LU factorization of $\mathbf{A}$ requires roughly $\frac{1}{3}n^3$ operations. The ensuing triangular systems are simply solved by back and forward substitution, each of which requires n^2 operations; see Section 4.3. Therefore, the total number of operations is roughly $\frac{1}{3}n^3 + n^2$, which is approximately $\frac{1}{3}n^3$ since n is large. This implies that Doolittle's method requires half as many operations as the Gauss elimination method.

4.4.2 Cholesky Factorization

A very special matrix encountered in many engineering applications is a symmetric, positive definite matrix. An $n \times n$ matrix $\mathbf{A} = [a_{ij}]$ is positive definite if all of the following determinants are positive:

$$D_1 = a_{11} > 0, \quad D_2 = \begin{vmatrix} a_{11} & a_{12} \\ a_{21} & a_{22} \end{vmatrix} > 0, \quad D_3 = \begin{vmatrix} a_{11} & a_{12} & a_{13} \\ a_{21} & a_{22} & a_{23} \\ a_{31} & a_{32} & a_{33} \end{vmatrix} > 0, \ldots,$$

$$D_n = |\mathbf{A}| > 0$$

Of course, $\mathbf{A}$ is symmetric if $\mathbf{A} = \mathbf{A}^T$. For a symmetric, positive definite matrix, there is a very special form of LU factorization, where the upper-triangular matrix is the transpose of the lower-triangular matrix. This is known as Cholesky factorization

$$\mathbf{A} = \mathbf{LL}^T$$

For instance, in the case of a 3×3 matrix

$$\begin{bmatrix} a_{11} & a_{12} & a_{13} \\ a_{12} & a_{22} & a_{23} \\ a_{13} & a_{23} & a_{33} \end{bmatrix} = \begin{bmatrix} l_{11} & 0 & 0 \\ l_{21} & l_{22} & 0 \\ l_{31} & l_{32} & l_{33} \end{bmatrix} \begin{bmatrix} l_{11} & l_{21} & l_{31} \\ 0 & l_{22} & l_{32} \\ 0 & 0 & l_{33} \end{bmatrix}$$

$$= \begin{bmatrix} l_{11}^2 & l_{11}l_{21} & l_{11}l_{31} \\ l_{21}l_{11} & l_{21}^2 + l_{22}^2 & l_{21}l_{31} + l_{22}l_{32} \\ l_{31}l_{11} & l_{21}l_{31} + l_{22}l_{32} & l_{31}^2 + l_{32}^2 + l_{33}^2 \end{bmatrix} \quad (4.11)$$

Owing to symmetry, only six equations—as opposed to nine for Doolittle—need to be solved. The user-defined function `CholeskyFactor` performs all the operations and returns the appropriate $\mathbf{L}$ and $\mathbf{U} = \mathbf{L}^T$ matrices.

```
function [L U] = CholeskyFactor(A)
%
% CholeskyFactor returns the Cholesky factorization of
% matrix A.
%
%   [L U] = CholeskyFactor(A) where
%
%     A is a symmetric, positive definite n-by-n matrix,
%
%     L is a lower triangular matrix,
%     U = L' is an upper triangular matrix.
%
n = size(A,1);
L = zeros(n,n);      % Initialize
for i = 1:n,
    L(i,i) = sqrt(A(i,i)-L(i,1:i-1)*L(i,1:i-1)');
    for j = i+1:n,
        L(j,i) = (A(j,i)-L(j,1:i-1)*L(i,1:i-1)')/L(i,i);
    end
end
U = L';
```

Example 4.8: Cholesky Factorization

Find the Cholesky factorization of

$$\mathbf{A} = \begin{bmatrix} 9 & 6 & -3 \\ 6 & 13 & -5 \\ -3 & -5 & 18 \end{bmatrix}$$

SOLUTION

The elements listed in Equation 4.11 can be directly used to determine the six entries of **L**. For instance

$$l_{11}^2 = a_{11} \quad \Rightarrow \quad l_{11}^2 = 9 \quad \Rightarrow \quad l_{11} = 3$$

$$l_{11}l_{21} = a_{12} \quad \Rightarrow \quad l_{21} = \frac{a_{12}}{l_{11}} = \frac{6}{3} \quad \Rightarrow \quad l_{21} = 2$$

and so on. The user-defined function `CholeskyFactor` will confirm the results.

```
>> A = [9 6 -3;6 13 -5; -3 -5 18];
>> [L U] = CholeskyFactor(A)

L =

     3     0     0
     2     3     0
    -1    -1     4
```

```
U =
     3     2    -1
     0     3    -1
     0     0     4
```

4.4.2.1 Cholesky's Method to Solve a Linear System

Cholesky's method uses Cholesky factorization to solve $\mathbf{Ax} = \mathbf{b}$. Substitution of $\mathbf{A} = \mathbf{LL}^T$ into the system yields

$$\left[\mathbf{LL}^T\right]\mathbf{x} = \mathbf{b} \quad \Rightarrow \quad \mathbf{L}\left[\mathbf{L}^T\mathbf{x}\right] = \mathbf{b}$$

which will be solved in two steps:

$$\begin{cases} \mathbf{Ly} = \mathbf{b} \\ \mathbf{L}^T\mathbf{x} = \mathbf{y} \end{cases} \tag{4.12}$$

Both systems are triangular, for which the solutions are found by forward and back substitution. The user-defined function `CholeskyMethod` uses Cholesky factorization of the coefficient matrix, and subsequently solves the two triangular systems in Equation 4.12 to find the solution vector $\mathbf{x}$.

```
function x = CholeskyMethod(A,b)
%
% CholeskyMethod uses the Cholesky factorization of
% matrix A and solves the ensuing triangular systems to
% find the solution vector x.
%
%   x = CholeskyMethod(A,b) where
%
%     A is a symmetric, positive definite n-by-n
%     coefficient matrix,
%     b is the n-by-1 vector of the right-hand sides,
%
%     x is the n-by-1 solution vector.
%
[L U] = CholeskyFactor(A);
% Find Cholesky factorization of A
n = size(A,1);

% Solve the lower triangular system Ly = b (forward
% substitution)
y = zeros(n,1);
y(1) = b(1)/L(1,1);
```

```
for  i=2:n,
     y(i) = (b(i)-L(i,1:i-1)*y(1:i-1))/L(i,i);
end

% Solve the upper triangular system L'x=y (back
% substitution)
x=zeros(n,1);
x(n)=y(n)/U(n,n);
    for i=n-1:-1:1,
        x(i) = (y(i)-U(i,i+1:n)*x(i+1:n))/U(i,i);
    end
end
```

Example 4.9: Cholesky's Method to Solve a Linear System

Using Cholesky's method, solve $\mathbf{Ax} = \mathbf{b}$, where

$$\mathbf{A} = \begin{bmatrix} 9 & 6 & -3 \\ 6 & 13 & -5 \\ -3 & -5 & 18 \end{bmatrix}, \quad \mathbf{b} = \begin{Bmatrix} -15 \\ 17 \\ -52 \end{Bmatrix}, \quad \mathbf{x} = \begin{Bmatrix} x_1 \\ x_2 \\ x_3 \end{Bmatrix}$$

SOLUTION

The coefficient matrix **A** is symmetric, positive definite. Cholesky factorization of **A** was done in Example 4.8. Using **L** and **U** in Equation 4.12

$$\mathbf{Ly} = \mathbf{b} : \begin{bmatrix} 3 & 0 & 0 \\ 2 & 3 & 0 \\ -1 & -1 & 4 \end{bmatrix} \begin{Bmatrix} y_1 \\ y_2 \\ y_3 \end{Bmatrix} = \begin{Bmatrix} -15 \\ 17 \\ -52 \end{Bmatrix} \Rightarrow \mathbf{y} = \begin{Bmatrix} -5 \\ 9 \\ -12 \end{Bmatrix}$$

$$\mathbf{L}^T\mathbf{x} = \mathbf{y} : \begin{bmatrix} 3 & 2 & -1 \\ 0 & 3 & -1 \\ 0 & 0 & 4 \end{bmatrix} \begin{Bmatrix} x_1 \\ x_2 \\ x_3 \end{Bmatrix} = \begin{Bmatrix} -5 \\ 9 \\ -12 \end{Bmatrix} \Rightarrow \mathbf{x} = \begin{Bmatrix} -4 \\ 2 \\ -3 \end{Bmatrix}$$

The result can be verified by executing the user-defined function `CholeskyMethod`.

```
>> A=[9 6 -3;6 13 -5; -3 -5 18];
>> b=[-15;17;-52];
>> x=CholeskyMethod(A,b)

x=
   -4
    2
   -3
```

4.4.2.2 Operations Count

Cholesky's method comprises two parts: LU factorization of the coefficient matrix and back/forward substitution to solve the two triangular systems. For a large system **Ax** = **b**, the LU factorization of **A** requires roughly $\frac{1}{3}n^3$ operations. The ensuing triangular systems are solved by back and forward substitution, each requiring n^2 operations. Therefore, the total number of operations is roughly $\frac{1}{3}n^3 + n^2$, which is approximately $\frac{1}{3}n^3$ since n is large. This implies that Cholesky's method requires half as many operations as the Gauss elimination method.

4.4.2.3 MATLAB® Built-In Functions `lu` and `chol`

MATLAB has built-in functions to perform LU factorization of a square matrix: `lu` for general square matrices and `chol` for symmetric, positive definite matrices. There are different ways of calling the function `lu`. For example, the outputs in `[L U] = lu(A)` are **U**, which is upper-triangular and **L**, which is the product of a lower-triangular matrix and permutation matrices such that **LU** = **A**. On the other hand, `[L U P] = lu(A)` returns a lower-triangular **L**, an upper-triangular **U**, and permutation matrices **P**, such that **LU** = **PA**. Other options in `lu` allow for the control of pivoting when working with sparse matrices.

For a symmetric, positive definite matrix **A**, the function call `U = chol(A)` returns an upper-triangular matrix **U** such that $\mathbf{U}^T\mathbf{U} = \mathbf{A}$. If the matrix is not positive definite, `chol` returns an error message.

4.5 Iterative Solution of Linear Systems

In Sections 4.3 and 4.4, we introduced direct methods for solving **Ax** = **b**, which included the Gauss elimination and methods based on LU factorization of the coefficient matrix **A**. We now turn our attention to indirect, or iterative, methods. In principle, a successful iteration process starts with an initial vector and generates successive approximations that eventually converge to the solution vector **x**.

Unlike direct methods, where the total number of operations is identified in advance, the number of operations required by an iterative method depends on how many iteration steps must be performed for satisfactory convergence, as well as the nature of the system at hand. What is meant by convergence is that the iteration must be terminated as soon as two successive vectors are close to one another. A measure of the proximity of two vectors is provided by a vector norm.

4.5.1 Vector Norms

The norm of a vector $\mathbf{v}_{n\times 1}$, denoted by $\|\mathbf{v}\|$, provides a measure of how small or large $\mathbf{v}$ is, and has the following properties:

$\|\mathbf{v}\| \geq 0$ for all $\mathbf{v}$, and $\|\mathbf{v}\| = 0$ if and only if $\mathbf{v} = \mathbf{0}_{n\times 1}$
$\|\alpha\, \mathbf{v}\| = |\alpha|\, \|\mathbf{v}\|$, $\alpha =$ scalar
$\|\mathbf{v} + \mathbf{w}\| \leq \|\mathbf{v}\| + \|\mathbf{w}\|$ for all vectors $\mathbf{v}$ and $\mathbf{w}$

There are three commonly used vector norms as listed below. In all cases, vector $\mathbf{v}$ is assumed in the form

$$\mathbf{v} = \begin{Bmatrix} v_1 \\ v_2 \\ \dots \\ v_n \end{Bmatrix}$$

l_1-norm, denoted by $\|\mathbf{v}\|_1$, is the sum of the absolute values of all components of $\mathbf{v}$:

$$\|\mathbf{v}\|_1 = |v_1| + |v_2| + \cdots + |v_n| \tag{4.13}$$

l_∞-norm, denoted by $\|\mathbf{v}\|_\infty$, is the largest (in absolute value) of all components of $\mathbf{v}$:

$$\|\mathbf{v}\|_\infty = \max\{|v_1|, |v_2|, \dots, |v_n|\} \tag{4.14}$$

l_2-norm, denoted by $\|\mathbf{v}\|_2$, is the square root of the sum of the squares of all components of $\mathbf{v}$:

$$\|\mathbf{v}\|_2 = \left[v_1^2 + v_2^2 + \cdots + v_n^2\right]^{1/2} \tag{4.15}$$

Example 4.10: Vector Norms

Find the three norms of

$$\mathbf{v} = \begin{Bmatrix} -2 \\ 0 \\ 1.3 \\ 0.5 \end{Bmatrix}$$

a. Using Equations 4.13 through 4.15
b. Using the MATLAB built-in function `norm`

SOLUTION

a. By Equations 4.13 through 4.15

$$\|\mathbf{v}\|_1 = |-2| + |1.3| + |0.5| = 3.8$$

$$\|\mathbf{v}\|_\infty = \max\{|-2|,|1.3|,|0.5|\} = 2$$

$$\|\mathbf{v}\|_2 = \sqrt{(-2)^2 + 1.3^2 + 0.5^2} = 2.4372$$

Note that all three norms return values that are of the same order of magnitude, as is always the case. If a certain norm of a vector happens to be small, the other norms will also be somewhat small, and so on.

b. MATLAB built-in function `norm` calculates vector and matrix norms.

```
>> v = [-2;0;1.3;0.5];
>> [norm(v,1) norm(v,inf) norm(v,2)]

ans =

    3.8000    2.0000    2.4372
```

4.5.2 Matrix Norms

The norm of a matrix $\mathbf{A}_{n\times n}$, denoted by $\|\mathbf{A}\|$, is a nonnegative real number that provides a measure of how small or large $\mathbf{A}$ is, and has the following properties:

$\|\mathbf{A}\| \ge 0$ for all $\mathbf{A}$, and $\|\mathbf{A}\| = 0$ if and only if $\mathbf{A} = \mathbf{0}_{n\times n}$

$\|\alpha\,\mathbf{A}\| = |\alpha|\;\|\mathbf{A}\|$, α = scalar

$\|\mathbf{A}+\mathbf{B}\| \le \|\mathbf{A}\| + \|\mathbf{B}\|$ for all $n \times n$ matrices $\mathbf{A}$ and $\mathbf{B}$

$\|\mathbf{AB}\| \le \|\mathbf{A}\|\;\|\mathbf{B}\|$ for all $n \times n$ matrices $\mathbf{A}$ and $\mathbf{B}$

There are three commonly used matrix norms as listed below. In all cases, matrix $\mathbf{A}$ is in the form $\mathbf{A} = [a_{ij}]_{n\times n}$.

1-norm (column-sum norm), denoted by $\|\mathbf{A}\|_1$, is defined as

$$\|\mathbf{A}\|_1 = \max_{1\le j\le n}\left\{\sum_{i=1}^{n} |a_{ij}|\right\} \tag{4.16}$$

The sum of the absolute values of entries in each column of **A** is calculated, and the largest is selected.

Infinite-norm (row-sum norm), denoted by $\|\mathbf{A}\|_\infty$, is defined as

$$\|\mathbf{A}\|_\infty = \max_{1\le i\le n}\left\{\sum_{j=1}^{n}|a_{ij}|\right\} \tag{4.17}$$

The sum of the absolute values of entries in each row of **A** is calculated, and the largest is selected.

Euclidean norm (2-norm, Frobenius norm), denoted by $\|\mathbf{A}\|_E$, is defined as

$$\|\mathbf{A}\|_E = \left[\sum_{i=1}^{n}\sum_{j=1}^{n} {a_{ij}}^2\right]^{1/2} \tag{4.18}$$

Example 4.11: Matrix Norms

Find the three norms of

$$\mathbf{A} = \begin{bmatrix} -1 & 2 & -3 & 0 \\ 0 & -5 & 0 & 2 \\ 2 & 2 & 0 & -4 \\ 3 & 6 & 2 & 5 \end{bmatrix}$$

a. Using Equations 4.16 through 4.18
b. Using the MATLAB built-in function `norm`

SOLUTION

a. By Equations 4.16 through 4.18

$$\|\mathbf{A}\|_1 = \max_{1\le j\le 4}\left\{\sum_{i=1}^{4}|a_{ij}|\right\} = \max\{6,15,5,11\} = 15$$

$$\|\mathbf{A}\|_\infty = \max_{1\le i\le 4}\left\{\sum_{j=1}^{4}|a_{ij}|\right\} = \max\{6,7,8,16\} = 16$$

$$\|\mathbf{A}\|_E = \left\{\sum_{i=1}^{4}\sum_{j=1}^{4} a_{ij}^2\right\}^{1/2} = 11.8743$$

As it was the case with vector norms, the values returned by all three matrix norms are of the same order of magnitude.

b.

```
>> A = [-1 2 -3 0;0 -5 0 2;2 2 0 -4;3 6 2 5];
>> [norm(A,1) norm(A,inf) norm(A,'fro')]

ans =

   15.0000    16.0000    11.8743
```

4.5.2.1 Compatibility of Vector and Matrix Norms

The three matrix norms above are compatible with the three vector norms introduced earlier, in the exact order they were defined. More specifically, the compatibility relations are

$$\begin{aligned} \|\mathbf{Av}\|_1 &\le \|\mathbf{A}\|_1 \|\mathbf{v}\|_1 \\ \|\mathbf{Av}\|_\infty &\le \|\mathbf{A}\|_\infty \|\mathbf{v}\|_\infty \\ \|\mathbf{Av}\|_2 &\le \|\mathbf{A}\|_E \|\mathbf{v}\|_2 \end{aligned} \tag{4.19}$$

Example 4.12: Compatibility Relations

The relations in Equation 4.19 can be verified for the vector and the matrix used in Examples 4.10 and 4.11 as follows:

$$\mathbf{Av} = \begin{Bmatrix} -1.9 \\ 1 \\ -6 \\ -0.9 \end{Bmatrix} \overset{\text{Calculate vector norms}}{\Rightarrow} \|\mathbf{Av}\|_1 = 9.8, \quad \|\mathbf{Av}\|_\infty = 6, \quad \|\mathbf{Av}\|_2 = 6.4358$$

Then

$$9.8 \le (15)(3.8), \quad 6 \le (16)(2), \quad 6.4358 \le (11.8743)(2.4372)$$

4.5.3 General Iterative Method

The idea behind the general iterative method to solve $\mathbf{Ax} = \mathbf{b}$ is outlined as follows: Split the coefficient matrix as $\mathbf{A} = \mathbf{Q} - \mathbf{P}$ and substitute into $\mathbf{Ax} = \mathbf{b}$ to obtain

$$[\mathbf{Q} - \mathbf{P}]\mathbf{x} = \mathbf{b} \Rightarrow \mathbf{Qx} = \mathbf{Px} + \mathbf{b}$$

This system cannot be solved in its present form, as the solution vector $\mathbf{x}$ appears on both sides. Instead, it will be solved by iterations. Choose an initial vector $\mathbf{x}^{(0)}$ and generate a sequence of vectors via

$$\mathbf{Qx}^{(k+1)} = \mathbf{Px}^{(k)} + \mathbf{b}, \quad k = 0, 1, 2, \ldots \tag{4.20}$$

Assuming $\mathbf{Q}$ is nonsingular, this is easily solved at each step for the updated vector $\mathbf{x}^{(k+1)}$ as

$$\mathbf{x}^{(k+1)} = \mathbf{Q}^{-1}\mathbf{P}\mathbf{x}^{(k)} + \mathbf{Q}^{-1}\,\mathbf{b}, \quad k = 0, 1, 2, \ldots \tag{4.21}$$

In the general procedure, splitting of $\mathbf{A}$ is arbitrary, except that $\mathbf{Q}$ must be nonsingular. This arbitrary nature of the split causes the procedure to be generally inefficient. In specific iterative methods, matrices $\mathbf{P}$ and $\mathbf{Q}$ obey very specific formats.

4.5.3.1 Convergence of the General Iterative Method

The sequence of vectors obtained via Equation 4.21 converges if the sequence of error vectors associated with each iteration step approaches the zero vector. The error vector at iteration k is defined as

$$\mathbf{e}^{(k)} = \mathbf{x}^{(k)} - \mathbf{x}_a$$

Note that the actual solution $\mathbf{x}_a$ is unknown, and is being used in the analysis merely for the development of the important theoretical results. That said, since $\mathbf{x}_a$ is the actual solution of $\mathbf{A}\mathbf{x} = \mathbf{b}$, then

$$\mathbf{A}\mathbf{x}_a = \mathbf{b} \;\Rightarrow\; [\mathbf{Q} - \mathbf{P}]\mathbf{x}_a = \mathbf{b}$$

Inserting this into Equation 4.20 yields

$$\mathbf{Q}\mathbf{x}^{(k+1)} = \mathbf{P}\mathbf{x}^{(k)} + [\mathbf{Q} - \mathbf{P}]\mathbf{x}_a \;\Rightarrow\; \mathbf{Q}[\mathbf{x}^{(k+1)} - \mathbf{x}_a] = \mathbf{P}[\mathbf{x}^{(k)} - \mathbf{x}_a] \;\Rightarrow\; \mathbf{Q}\mathbf{e}^{(k+1)} = \mathbf{P}\mathbf{e}^{(k)}$$

Premultiplication of this last equation by $\mathbf{Q}^{-1}$, and letting $\mathbf{M} = \mathbf{Q}^{-1}\mathbf{P}$, results in

$$\mathbf{e}^{(k+1)} = \mathbf{Q}^{-1}\,\mathbf{P}\mathbf{e}^{(k)} = \mathbf{M}\mathbf{e}^{(k)}, \quad k = 0, 1, 2, \ldots$$

so that

$$\mathbf{e}^{(1)} = \mathbf{M}\mathbf{e}^{(0)}, \quad \mathbf{e}^{(2)} = \mathbf{M}\mathbf{e}^{(1)} = \mathbf{M}^2\,\mathbf{e}^{(0)}, \ldots, \mathbf{e}^{(k)} = \mathbf{M}^k\,\mathbf{e}^{(0)}$$

Taking the infinite-norm of both sides of the last equation and k applications of the second compatibility relation in Equation 4.19, we find

$$\left\|\mathbf{e}^{(k)}\right\|_\infty \le \left\|\mathbf{M}\right\|_\infty^k \left\|\mathbf{e}^{(0)}\right\|_\infty$$

Thus, a sufficient condition for $\|\mathbf{e}^{(k)}\|_\infty \to 0$ as $k \to \infty$ is that $\|\mathbf{M}\|_\infty^k \to 0$ as $k \to \infty$, which is met if $\|\mathbf{M}\|_\infty < 1$. The matrix $\mathbf{M} = \mathbf{Q}^{-1}\mathbf{P}$ plays a key role in the convergence of iterative schemes. The above analysis suggests that in splitting matrix $\mathbf{A}$, matrices $\mathbf{Q}$ and $\mathbf{P}$ must be chosen so that the norm of $\mathbf{M} = \mathbf{Q}^{-1}\mathbf{P}$

is small. We note that $\|\mathbf{M}\|_\infty < 1$ is only a sufficient condition and not necessary. This means that if it holds, the iteration converges, but if it does not hold, convergence is not automatically ruled out.

4.5.4 Jacobi Iteration Method

Let **D**, **L**, and **U** be the diagonal, lower-, and upper-triangular portions of matrix $\mathbf{A} = [a_{ij}]_{n\times n}$, respectively, that is

$$\mathbf{D} = \begin{bmatrix} a_{11} & & & \\ & a_{22} & & \\ & & \ldots & \\ & & & a_{nn} \end{bmatrix}, \quad \mathbf{L} = \begin{bmatrix} 0 & 0 & \ldots & 0 \\ a_{21} & 0 & & 0 \\ \ldots & \ldots & \ldots & \ldots \\ a_{n1} & a_{n2} & \ldots & 0 \end{bmatrix}, \quad \mathbf{U} = \begin{bmatrix} 0 & a_{12} & \ldots & a_{1n} \\ 0 & 0 & & a_{2n} \\ \ldots & & \ldots & \ldots \\ 0 & 0 & \ldots & 0 \end{bmatrix}$$

In the Jacobi method, **A** is split as

$$\begin{aligned} \mathbf{A} &= \mathbf{Q} - \mathbf{P} \\ &= \mathbf{D} + [\mathbf{L} + \mathbf{U}] \end{aligned} \quad \text{so that} \quad \begin{aligned} \mathbf{Q} &= \mathbf{D} \\ \mathbf{P} &= -[\mathbf{L} + \mathbf{U}] \end{aligned}$$

Subsequently, Equation 4.20 takes the specific form

$$\mathbf{D}\mathbf{x}^{(k+1)} = -[\mathbf{L} + \mathbf{U}]\mathbf{x}^{(k)} + \mathbf{b}, \quad k = 0, 1, 2, \ldots \tag{4.22}$$

For $\mathbf{D}^{-1}$ to exist, the diagonal entries of **D**, and hence of **A**, must all be nonzero. If a zero entry appears in a diagonal slot, the equations in the original system must be rearranged in such a way that no zero entry appears along the diagonal in the resulting matrix. Then, premultiplication of Equation 4.22 by $\mathbf{D}^{-1}$ yields

$$\mathbf{x}^{(k+1)} = \mathbf{D}^{-1}\{-[\mathbf{L} + \mathbf{U}]\mathbf{x}^{(k)} + \mathbf{b}\}, \quad k = 0, 1, 2, \ldots \tag{4.23}$$

known as the Jacobi method. Note that $\mathbf{L} + \mathbf{U}$ is precisely **A**, with zero diagonal entries, and that the diagonal elements of $\mathbf{D}^{-1}$ are $1/a_{ii}$ for $i = 1, 2, \ldots, n$. Denoting the vector generated at the kth iteration by $\mathbf{x}^{(k)} = [x_1^{(k)} \ \ldots \ x_n^{(k)}]^T$, Equation 4.23 can be expressed component-wise as

$$x_i^{(k+1)} = \frac{1}{a_{ii}}\left\{-\sum_{\substack{j=1 \\ j\neq i}}^{n} a_{ij}x_j^{(k)} + b_i\right\}, \quad i = 1, 2, \ldots, n \tag{4.24}$$

The very important matrix $\mathbf{M} = \mathbf{Q}^{-1}\mathbf{P}$ takes the special form

$$\mathbf{M}_J = -\mathbf{D}^{-1}[\mathbf{L} + \mathbf{U}]$$

and is called the Jacobi iteration matrix. A sufficient condition for Jacobi iteration to converge is that $\|\mathbf{M}_J\|_\infty < 1$.

4.5.4.1 Convergence of the Jacobi Iteration Method

Convergence of the Jacobi method relies on a special class of matrices known as diagonally dominant. An $n \times n$ matrix $\mathbf{A}$ is diagonally dominant if in each row, the absolute value of the diagonal entry is greater than the sum of the absolute values of all the off-diagonal entries, that is

$$|a_{ii}| > \sum_{\substack{j=1 \\ j \neq i}}^{n} |a_{ij}|, \quad i = 1, 2, \ldots, n \tag{4.25}$$

or equivalently

$$\sum_{\substack{j=1 \\ j \neq i}}^{n} \frac{|a_{ij}|}{|a_{ii}|} < 1, \quad i = 1, 2, \ldots, n \tag{4.26}$$

Theorem 4.1

Suppose $\mathbf{A}$ is diagonally dominant. Then, the linear system $\mathbf{Ax} = \mathbf{b}$ has a unique solution $\mathbf{x}_a$, and the sequence of vectors generated by Jacobi iteration, Equation 4.23, converges to $\mathbf{x}_a$ regardless of the initial vector $\mathbf{x}^{(0)}$.

Proof

Since $\mathbf{A}$ is diagonally dominant, Equation 4.26 holds. The Jacobi iteration matrix is formed as

$$\mathbf{M}_J = -\mathbf{D}^{-1}[\mathbf{L} + \mathbf{U}] = \begin{bmatrix} 0 & a_{12}/a_{11} & \cdots & \cdots & a_{1n}/a_{11} \\ a_{21}/a_{22} & 0 & \cdots & \cdots & a_{2n}/a_{22} \\ \cdots & \cdots & & & \cdots \\ \cdots & \cdots & & & a_{n-1,n}/a_{n-1,n-1} \\ a_{n1}/a_{nn} & a_{n2}/a_{nn} & \cdots & \cdots & 0 \end{bmatrix}$$

It is clear that every row of the above matrix satisfies Equation 4.26 so that every row-sum is less than 1. This means the row-sum norm of $\mathbf{M}_J$ is less than 1, that is, $\|\mathbf{M}_J\|_\infty < 1$. Since this is a sufficient condition for convergence of the Jacobi method, the proof is complete.

The user-defined function `Jacobi` uses the Jacobi iteration method to solve the linear system $\mathbf{Ax} = \mathbf{b}$, and returns the approximate solution vector, the number of iterations needed for convergence, and $\|\mathbf{M}_J\|_\infty$. The terminating condition is $\|\mathbf{x}^{(k+1)} - \mathbf{x}^{(k)}\| < \varepsilon$ for a prescribed tolerance ε.

```
function [x, k, MJnorm] = Jacobi(A, b, x0, tol, kmax)
%
% Jacobi uses the Jacobi iteration method to approximate
% the solution of Ax = b.
%
%  [x, k, MJnorm] = Jacobi(A, b, x0, tol, kmax) where
%
%    A is the n-by-n coefficient matrix,
%    b is the n-by-1 right-hand side vector,
%    x0 is the n-by-1 initial vector (default zeros),
%    tol is the scalar tolerance for convergence (default 1e-4),
%    kmax is the maximum number of iterations (default 100),
%
%    x is the n-by-1 solution vector,
%    k is the number of iterations required for convergence,
%    MJnorm is the infinite norm of the Jacobi iteration
%    matrix.
if nargin < 3 || isempty(x0)
   x0 = zeros(size(b));
end
if nargin < 4 || isempty(tol)
   tol = 1e-4;
end
if nargin < 5 || isempty(kmax)
   kmax = 100;
end
x(:, 1) = x0;

D = diag(diag(A)); At = A - D;
L = tril(At);
U = triu(At);

% Norm of Jacobi iteration matrix
M = -D\(L + U); MJnorm = norm(M, inf);
B = D\b;

% Perform iterations up to kmax
for k = 1:kmax,
   x(:, k+1) = M*x(:, k) + B;    % Compute next approximation
   if norm(x(:, k+1) - x(:, k)) < tol, break; end
   % Check convergence
end
x = x(:, end);
```

Example 4.13: Jacobi Iteration

Consider the linear system

$$\begin{bmatrix} 4 & 1 & -1 \\ -2 & 5 & 0 \\ 2 & 1 & 6 \end{bmatrix} \mathbf{x} = \begin{Bmatrix} 1 \\ -7 \\ 13 \end{Bmatrix}, \quad \underbrace{\mathbf{x}^{(0)}}_{\text{initial vector}} = \begin{Bmatrix} 0 \\ 1 \\ 1 \end{Bmatrix}$$

where the coefficient matrix is diagonally dominant because

$$4 > 1 + |-1|, \quad 5 > |-2|, \quad 6 > 2 + 1$$

Theorem 4.1 guarantees convergence of the sequence of vectors generated by the Jacobi iteration method to the actual solution. Starting with the given initial vector $\mathbf{x}^{(0)}$, we will find the components of the next vector $\mathbf{x}^{(1)}$ generated by Jacobi iteration using Equation 4.24

$$x_i^{(1)} = \frac{1}{a_{ii}} \left\{ -\sum_{\substack{j=1 \\ j \neq i}}^{3} a_{ij} x_j^{(0)} + b_i \right\}, \quad i = 1,2,3$$

Specifically

$$x_1^{(1)} = \frac{1}{a_{11}} \left\{ -\left[a_{12} x_2^{(0)} + a_{13} x_3^{(0)} \right] + b_1 \right\} = \frac{1}{4} \left\{ -[(1)(1) + (-1)(1)] + 1 \right\} = 0.25$$

$$x_2^{(1)} = \frac{1}{a_{22}} \left\{ -\left[a_{21} x_1^{(0)} + a_{23} x_3^{(0)} \right] + b_2 \right\} = \frac{1}{5} \left\{ -[(-2)(0) + (0)(1)] + (-7) \right\} = -1.4$$

$$x_3^{(1)} = \frac{1}{a_{33}} \left\{ -\left[a_{31} x_1^{(0)} + a_{32} x_2^{(0)} \right] + b_3 \right\} = \frac{1}{6} \left\{ -[(2)(0) + (1)(1)] + 13 \right\} = 2$$

Therefore

$$\mathbf{x}^{(1)} = \begin{Bmatrix} 0.25 \\ -1.4 \\ 2 \end{Bmatrix}$$

Subsequent vectors may be found in a similar manner. This vector can be verified by executing the user-defined function `Jacobi` with `kmax = 1` so that only one iteration is performed.

```
>> A = [4 1 -1; -2 5 0;2 1 6]; b = [1; -7;13]; x0 = [0;1;1];
>> [x, k, MJnorm] = Jacobi(A, b, x0, 1e-4, 1)
```

```
x =

      0.2500
     -1.4000
      2.0000      % Agrees with earlier result

k =

      1

MJnorm =

      0.5000
```

Finally, we will solve the system using Jacobi iteration with initial vector $\mathbf{x}^{(0)}$ as given above and terminating condition $\|\mathbf{x}^{(k+1)} - \mathbf{x}^{(k)}\| < 10^{-4}$.

```
>> [x, k, MJnorm] = Jacobi(A, b, x0)
% Default values for tolerance and kmax

x =
      1.0000
     -1.0000
      2.0000

k =
      13

MJnorm =

      0.5000
```

4.5.5 Gauss–Seidel Iteration Method

Based on Equations 4.23 and 4.24, every component of $\mathbf{x}^{(k+1)}$ is calculated entirely from $\mathbf{x}^{(k)}$ of the previous iteration. In other words, to have access to $\mathbf{x}^{(k+1)}$, the kth iteration has to be completed so that $\mathbf{x}^{(k)}$ is available. The performance of Jacobi iteration can be improved if the most updated components of a vector are utilized, as soon as they are available, to compute the subsequent components of the same vector. Consider two successive vectors, as well as the actual solution

$$\mathbf{x}^{(k)} = \begin{Bmatrix} x_1^{(k)} \\ \cdots \\ x_p^{(k)} \\ x_{p+1}^{(k)} \\ \cdots \\ x_n^{(k)} \end{Bmatrix}, \quad \mathbf{x}^{(k+1)} = \begin{Bmatrix} x_1^{(k+1)} \\ \cdots \\ x_p^{(k+1)} \\ x_{p+1}^{(k+1)} \\ \cdots \\ x_n^{(k+1)} \end{Bmatrix}, \quad \mathbf{x}_a = \begin{Bmatrix} x_1 \\ \cdots \\ x_p \\ x_{p+1} \\ \cdots \\ x_n \end{Bmatrix}$$

Generally speaking, $x_p^{(k+1)}$ is a better estimate of x_p than $x_p^{(k)}$. Then, using $x_p^{(k+1)}$ instead of $x_p^{(k)}$ should lead to a better approximation of the next component, $x_{p+1}^{(k+1)}$, in the current vector. This is the logic behind the Gauss–Seidel iteration method, which is considered a refinement of the Jacobi method. To comply with this logic, the coefficient matrix **A** is split as

$$\begin{aligned}\mathbf{A} &= \mathbf{Q} - \mathbf{P} \\ &= [\mathbf{D} + \mathbf{L}] + \mathbf{U}\end{aligned} \quad \text{so that} \quad \begin{aligned}\mathbf{Q} &= \mathbf{D} + \mathbf{L} \\ \mathbf{P} &= -\mathbf{U}\end{aligned}$$

As a result, Equation 4.20 takes the specific form

$$[\mathbf{D} + \mathbf{L}]\mathbf{x}^{(k+1)} = -\mathbf{U}\mathbf{x}^{(k)} + \mathbf{b}, \quad k = 0, 1, 2, \ldots \tag{4.27}$$

But $\mathbf{D} + \mathbf{L}$ is a lower-triangular matrix whose diagonal entries are those of **A**. Thus, $[\mathbf{D} + \mathbf{L}]^{-1}$ exists if **A** has nonzero diagonal entries to begin with. Subsequently, premultiplication of Equation 4.27 by $[\mathbf{D} + \mathbf{L}]^{-1}$ yields

$$\mathbf{x}^{(k+1)} = -[\mathbf{D} + \mathbf{L}]^{-1}\,\mathbf{U}\mathbf{x}^{(k)} + [\mathbf{D} + \mathbf{L}]^{-1}\,\mathbf{b}, \quad k = 0, 1, 2, \ldots \tag{4.28}$$

known as the Gauss–Seidel iteration method. Denoting the vector at the kth iteration by $\mathbf{x}^{(k)} = [x_1^{(k)} \;\ldots\; x_n^{(k)}]^T$, Equation 4.28 can be expressed componentwise as

$$x_i^{(k+1)} = \frac{1}{a_{ii}}\left\{-\sum_{j=1}^{i-1} a_{ij}x_j^{(k+1)} - \sum_{j=i+1}^{n} a_{ij}x_j^{(k)} + b_i\right\}, \quad i = 1, 2, \ldots, n \tag{4.29}$$

where the first sum on the right side is considered zero when $i = 1$. The very important matrix $\mathbf{M} = \mathbf{Q}^{-1}\,\mathbf{P}$ now takes the special form

$$\mathbf{M}_{GS} = -[\mathbf{D} + \mathbf{L}]^{-1}\,\mathbf{U}$$

known as the Gauss–Seidel iteration matrix. A sufficient condition for the Gauss–Seidel iteration to converge is that $\|\mathbf{M}_{GS}\|_\infty < 1$.

4.5.5.1 Convergence of the Gauss–Seidel Iteration Method

Since the Gauss–Seidel method is a refinement of the Jacobi method, it converges whenever the Jacobi method does, and usually faster. Recall that if **A** is diagonally dominant, Jacobi iteration is guaranteed to converge to the solution vector. This implies that if **A** is diagonally dominant, Gauss–Seidel iteration is also guaranteed to converge, and faster than Jacobi.

If **A** is not diagonally dominant, the convergence of the Gauss–Seidel method relies on another special class of matrices known as symmetric, positive definite (Section 4.4).

Theorem 4.2

Suppose $\mathbf{A}$ is symmetric, positive definite. Then, the linear system $\mathbf{Ax} = \mathbf{b}$ has a unique solution $\mathbf{x}_a$, and the sequence of vectors generated by Gauss–Seidel iteration, Equation 4.28, converges to $\mathbf{x}_a$ regardless of the initial vector $\mathbf{x}^{(0)}$.

The user-defined function GaussSeidel uses the Gauss–Seidel iteration method to solve the linear system $\mathbf{Ax} = \mathbf{b}$, and returns the approximate solution vector, the number of iterations needed for convergence, and $\|\mathbf{M}_{GS}\|_\infty$. The terminating condition is $\|\mathbf{x}^{(k+1)} - \mathbf{x}^{(k)}\| < \varepsilon$ for a prescribed tolerance ε.

```
function [x, k, MGSnorm] = GaussSeidel(A, b, x0, tol, kmax)
%
% GaussSeidel uses the Gauss-Seidel iteration method to
% approximate the solution of Ax=b.
%
% [x, k, MGSnorm] = GaussSeidel(A, b, x0, tol, kmax) where
%
%   A is the n-by-n coefficient matrix,
%   b is the n-by-1 right-hand side vector,
%   x0 is the n-by-1 initial vector (default zeros),
%   tol is the scalar tolerance for convergence
%   (default 1e-4),
%   kmax is the maximum number of iterations (default 100),
%
%   x is the n-by-1 solution vector,
%   k is the number of iterations required for convergence,
%   MGSnorm is the infinite norm of the Gauss-Seidel
%   iteration matrix.
if nargin<3 || isempty(x0)
    x0 = zeros(size(b));
end
if nargin<4 || isempty(tol)
    tol = 1e-4;
end
if nargin<5 || isempty(kmax)
    kmax = 100;
end
x(:, 1) = x0;

D = diag(diag(A)); At = A - D;
L = tril(At);
U = At - L;
```

```
% Norm of Gauss-Seidel iteration matrix
M=-(D+L)\U; MGSnorm=norm(M, inf);
B=(D+L)\b;

% Perform iterations up to kmax
for k=1:kmax,
   x(:, k+1)=M*x(:, k)+B;
   if norm(x(:,k+1)-x(:, k))<tol,
      break
   end
end
x=x(:, end);
```

Example 4.14: Gauss–Seidel Iteration

Consider the linear system of Example 4.13, where the coefficient matrix was diagonally dominant:

$$\begin{bmatrix} 4 & 1 & -1 \\ -2 & 5 & 0 \\ 2 & 1 & 6 \end{bmatrix} \mathbf{x} = \begin{Bmatrix} 1 \\ -7 \\ 13 \end{Bmatrix}, \quad \underset{\text{initial vector}}{\mathbf{x}^{(0)}} = \begin{Bmatrix} 0 \\ 1 \\ 1 \end{Bmatrix}$$

Gauss–Seidel iteration is guaranteed to converge because Jacobi is guaranteed to converge. Starting with the given initial vector $\mathbf{x}^{(0)}$, we will find the components of the next vector $\mathbf{x}^{(1)}$ generated by Gauss–Seidel iteration using Equation 4.29

$$x_i^{(k+1)} = \frac{1}{a_{ii}} \left\{ -\sum_{j=1}^{i-1} a_{ij} x_j^{(k+1)} - \sum_{j=i+1}^{3} a_{ij} x_j^{(k)} + b_i \right\}, \quad i = 1,2,3$$

As previously mentioned, the first sum on the right side is considered zero when $i = 1$. For the problem at hand

$$x_1^{(1)} = \frac{1}{a_{11}} \left[-a_{12} x_2^{(0)} - a_{13} x_3^{(0)} + b_1 \right] = \frac{1}{4} \left[-(1)(1) - (-1)(1) + 1 \right] = 0.25$$

$$x_2^{(1)} = \frac{1}{a_{22}} \left[-a_{21} x_1^{(1)} - a_{23} x_3^{(0)} + b_2 \right] = \frac{1}{5} \left[-(-2)(0.25) - (0)(1) - 7 \right] = -1.3$$

$$x_3^{(1)} = \frac{1}{a_{33}} \left[-a_{31} x_1^{(1)} - a_{32} x_2^{(1)} + b_3 \right] = \frac{1}{6} \left[-(2)(0.25) - (1)(-1.3) + 13 \right] = 2.3$$

Therefore

$$\mathbf{x}^{(1)} = \begin{Bmatrix} 0.25 \\ -1.3 \\ 2.3 \end{Bmatrix}$$

Subsequent vectors may be calculated in a similar manner. This vector can be verified by executing the user-defined function GaussSeidel with kmax = 1 so that only one iteration is performed.

```
>> A = [4 1 -1; -2 5 0;2 1 6]; b = [1; -7;13]; x0 = [0;1;1];
>> [x, k, MGSnorm] = GaussSeidel(A, b, x0, 1e-4, 1)

x =
    0.2500
   -1.3000
    2.3000       % Agrees with the above result

k =
   1

MGSnorm =

   0.5000
```

Using Gauss–Seidel iteration with the given $\mathbf{x}^{(0)}$ and terminating condition $\|\mathbf{x}^{(k+1)} - \mathbf{x}^{(k)}\| < 10^{-4}$, we find

```
>> [x, k, MGSnorm] = GaussSeidel(A, b, x0)

x =
    1.0000
   -1.0000
    2.0000

k =
   8

MGSnorm =

   0.5000
```

As expected, Gauss–Seidel exhibits a faster convergence (8 iterations) than Jacobi (13 iterations).

Example 4.15: Gauss–Seidel Iteration

Consider

$$\begin{bmatrix} 1 & 1 & -2 \\ 1 & 10 & 4 \\ -2 & 4 & 24 \end{bmatrix} \mathbf{x} = \begin{Bmatrix} -6 \\ 15 \\ 58 \end{Bmatrix}, \quad \underbrace{\mathbf{x}^{(0)}}_{\text{initial vector}} = \begin{Bmatrix} 0 \\ 0 \\ 0 \end{Bmatrix}$$

The coefficient matrix is symmetric, positive definite; thus Gauss–Seidel iteration will converge to the solution vector for any initial vector.

Executing the user-defined function GaussSeidel with default values for tol and kmax, we find

```
>> A = [1 1 -2;1 10 4;-2 4 24]; b = [-6;15;58];
>> [x, k, MGSnorm] = GaussSeidel(A, b)

x =

   -3.0000
    1.0000
    2.0000

k =

    15

MGSnorm =

    3
```

The input argument x0 was left out because the initial vector here happens to be the zero vector, which agrees with the default. Also note that $\|\mathbf{M}_{GS}\|_\infty = 3 > 1$ even though iterations did converge. This is because the condition $\|\mathbf{M}_{GS}\|_\infty < 1$ is only sufficient and not necessary. Also note that unlike the fact that a diagonally dominant coefficient matrix guarantees $\|\mathbf{M}_J\|_\infty < 1$, a symmetric, positive definite coefficient matrix does not guarantee $\|\mathbf{M}_{GS}\|_\infty < 1$, but does guarantee convergence for the Gauss–Seidel method.

4.5.6 Indirect Methods versus Direct Methods for Large Systems

Indirect methods such as Gauss–Seidel are mostly used when the linear system $\mathbf{Ax} = \mathbf{b}$ is large. Suppose a large system is being solved by the general iterative method, Equation 4.21

$$\mathbf{x}^{(k+1)} = \mathbf{Q}^{-1}\,\mathbf{P}\mathbf{x}^{(k)} + \mathbf{Q}^{-1}\,\mathbf{b}, \quad k = 0, 1, 2, \ldots$$

and that convergence is observed after m iterations. Since each iteration requires roughly n^2 multiplications, a total of n^2m multiplications are performed by the time convergence is achieved. On the other hand, a direct method such as Gauss elimination requires $\frac{1}{3}n^3$ multiplications to find the solution. Therefore, an indirect method is superior to a direct method as long as

$$n^2m < 1/3n^3 \quad \Rightarrow \quad m < 1/3n$$

For example, for a 100×100 system, this yields $m < \frac{1}{3}(100)$ so that an iterative method is preferred as long as it converges within 33 iterations. In many physical applications, not only is the size of the coefficient matrix $\mathbf{A}$ large, it is also sparse, that is, it contains a large number of zero entries. As one example, consider the numerical solution of partial differential equations using the finite-differences method (Chapter 10). In these cases, we encounter a large, sparse system where the coefficient matrix has at most five nonzero entries in each row. Therefore, based on Equations 4.24 and/or 4.29, six multiplications must be performed to find each component $x_i^{(k+1)}$ of the generated vector. But each vector has n components; thus, a total of $6n$ multiplications per iteration are performed. If it takes m iterations for convergence, then a total of $6nm$ multiplications are required for the indirect method. Therefore, the indirect method is computationally more efficient than a direct method as long as

$$6nm < 1/3n^3 \quad \Rightarrow \quad m < 1/18n^2$$

For a 1000×1000 system with a sparse coefficient matrix, this translates to $m < \frac{1}{18}(1000)^2$ so that an iterative method such as Gauss–Seidel is superior to Gauss elimination if it converges within 55,556 iterations, which is quite likely.

4.6 Ill-Conditioning and Error Analysis

So far, this chapter has focused on methods to find the solution vector for linear systems in the form $\mathbf{Ax} = \mathbf{b}$. In this section, we study the conditioning of a linear system and how it may impact the error associated with a computed solution.

4.6.1 Condition Number

The condition number of a nonsingular matrix $\mathbf{A}_{n\times n}$ is defined as

$$\kappa(\mathbf{A}) = \|\mathbf{A}\|\,\|\mathbf{A}^{-1}\| \tag{4.30}$$

where the same matrix norm is used for both $\mathbf{A}$ and $\mathbf{A}^{-1}$. It can be shown that for any $\mathbf{A}_{n\times n}$

$$\kappa(\mathbf{A}) \geq 1$$

We will learn that the smaller the condition number of a matrix, the better the condition of the matrix. A useful measure of the condition of a matrix is provided by the ratio of the largest (in magnitude) to the smallest (in magnitude) eigenvalue of the matrix.

Example 4.16: Condition Number

Calculate the condition number of the following matrix using all three norms, and verify the results using the MATLAB built-in command cond:

$$\mathbf{A} = \begin{bmatrix} 6 & 4 & 3 \\ 4 & 3 & 2 \\ 3 & 4 & 2 \end{bmatrix}$$

SOLUTION

The inverse is found as

$$\mathbf{A}^{-1} = \begin{bmatrix} -2 & 4 & -1 \\ -2 & 3 & 0 \\ 7 & -12 & 2 \end{bmatrix}$$

Then

$$\|\mathbf{A}\|_1 = 13, \quad \|\mathbf{A}^{-1}\|_1 = 19 \quad \Rightarrow \quad \kappa(\mathbf{A}) = 247$$

$$\|\mathbf{A}\|_\infty = 13, \quad \|\mathbf{A}^{-1}\|_\infty = 21 \quad \Rightarrow \quad \kappa(\mathbf{A}) = 273$$

$$\|\mathbf{A}\|_E = 10.9087, \quad \|\mathbf{A}^{-1}\|_E = 15.1987 \quad \Rightarrow \quad \kappa(\mathbf{A}) = 165.7982$$

In MATLAB, `cond(A,P)` returns the condition number of matrix **A** in P-norm.

```
>>A= [6 4 3;4 3 2;3 4 2];
>> [cond(A,1) cond(A,inf) cond(A,'fro')]
% Using three different matrix norms

ans =

   247.0000      273.0000      165.7981
```

Note that all three returned values are of the same order of magnitude, regardless of the choice of norm used.

4.6.2 Ill-Conditioning

The system $\mathbf{Ax} = \mathbf{b}$ is said to be well-conditioned if small errors generated during the solution process, or small changes in the coefficients, have small effects on the solution. For instance, if the diagonal entries of **A** are much larger in magnitude than the off-diagonal ones, the system is well-conditioned. If small errors and changes during the solution process have large impacts on the solution, the system is ill-conditioned. Ill-conditioned systems often arise in areas such as statistical analysis and least-squares fits.

Example 4.17: Ill-Conditioning

Investigate the ill-conditioning of

$$\mathbf{Ax} = \mathbf{b}, \quad \mathbf{A} = \begin{bmatrix} 1 & -2 \\ 1.0001 & -1.9998 \end{bmatrix}, \quad \mathbf{b} = \begin{Bmatrix} 2 \\ 2 \end{Bmatrix}$$

SOLUTION

The actual solution of this system can be easily verified to be

$$\mathbf{x}_a = \begin{Bmatrix} 1 \\ -0.5 \end{Bmatrix}$$

Suppose the first component of vector **b** is slightly perturbed by a very small $\varepsilon > 0$ so that the new vector is

$$\tilde{\mathbf{b}} = \begin{Bmatrix} 2+\varepsilon \\ 2 \end{Bmatrix}$$

The ensuing system $\mathbf{A}\tilde{\mathbf{x}} = \tilde{\mathbf{b}}$ is then solved via Gauss elimination as

$$\left[\begin{array}{cc|c} 1 & -2 & 2+\varepsilon \\ 1.0001 & -1.9998 & 2 \end{array}\right] \xrightarrow{-1.0001(\text{row}_1)+\text{row}_2} \left[\begin{array}{cc|c} 1 & -2 & 2+\varepsilon \\ 0 & 0.0004 & -0.0002-1.0001\varepsilon \end{array}\right]$$

$$\to \left[\begin{array}{cc|c} 1 & -2 & 2+\varepsilon \\ 0 & 1 & -0.5-2500.25\varepsilon \end{array}\right]$$

so that

$$\tilde{\mathbf{x}} = \begin{Bmatrix} 1-4999.50\varepsilon \\ -0.5-2500.25\varepsilon \end{Bmatrix} = \begin{Bmatrix} 1 \\ -0.5 \end{Bmatrix} - \begin{Bmatrix} 4999.5 \\ 2500.25 \end{Bmatrix}\varepsilon = \mathbf{x}_a - \begin{Bmatrix} 4999.5 \\ 2500.25 \end{Bmatrix}\varepsilon$$

Therefore, even though one of the components of **b** was subjected to a very small change of ε, the resulting solution vector shows very large relative changes in its components. This indicates that the system is ill-conditioned.

4.6.2.1 Indicators of Ill-Conditioning

There are essentially three indicators of ill-conditioning for a system $\mathbf{Ax} = \mathbf{b}$:

1. det (**A**) is very small in absolute value relative to the largest entries of **A** and **b** in absolute value.

2. The entries of $\mathbf{A}^{-1}$ are large in absolute value relative to the components of the solution vector.
3. $\kappa(\mathbf{A})$ is very large.

Example 4.18: Indicators of Ill-Conditioning

Consider the system in Example 4.17:

$$\mathbf{Ax} = \mathbf{b}, \quad \mathbf{A} = \begin{bmatrix} 1 & -2 \\ 1.0001 & -1.9998 \end{bmatrix}, \quad \mathbf{b} = \begin{Bmatrix} 2 \\ 2 \end{Bmatrix}$$

We will see that this system is ill-conditioned by verifying all three indicators listed above.

1. $\det(\mathbf{A}) = 0.0004$, which is considerably smaller than the absolute values of entries of $\mathbf{A}$ and $\mathbf{b}$.
2. The inverse of $\mathbf{A}$ is found as

$$\mathbf{A}^{-1} = \begin{bmatrix} -4999.50 & 5000 \\ -2500.25 & 2500 \end{bmatrix}$$

The entries are very large in magnitude relative to the components of the solution vector

$$\mathbf{x}_a = \begin{Bmatrix} 1 \\ -0.5 \end{Bmatrix}$$

3. Using the 1-norm, we find the condition number of $\mathbf{A}$ as

$$\|\mathbf{A}\|_1 = 3.9998, \quad \|\mathbf{A}^{-1}\|_1 = 7500 \Rightarrow \kappa(\mathbf{A}) = 29{,}998.5$$

which is quite large.

4.6.3 Computational Error

Suppose $\mathbf{x}_c$ is the computed solution of a linear system $\mathbf{Ax} = \mathbf{b}$, while $\mathbf{x}_a$ is the actual solution. The corresponding residual vector is defined as

$$\mathbf{r} = \mathbf{Ax}_c - \mathbf{b}$$

The norm of the residual vector $\|\mathbf{r}\|$ gives a measure of the accuracy of the computed solution, and so does the absolute error $\|\mathbf{x}_c - \mathbf{x}_a\|$. The most commonly used measure is the relative error

$$\frac{\|\mathbf{x}_c - \mathbf{x}_a\|}{\|\mathbf{x}_a\|}$$

Note that in applications the actual solution vector $\mathbf{x}_a$ is not available and the notation is being used here merely to develop some important results.

Theorem 4.3

Let $\mathbf{x}_a$ and $\mathbf{x}_c$ be the actual and computed solutions of $\mathbf{Ax} = \mathbf{b}$, respectively. If $\mathbf{r} = \mathbf{Ax}_c - \mathbf{b}$ is the residual vector and $\kappa(\mathbf{A})$ is the condition number of $\mathbf{A}$, then

$$\frac{1}{\kappa(\mathbf{A})}\frac{\|\mathbf{r}\|}{\|\mathbf{b}\|} \le \frac{\|\mathbf{x}_c - \mathbf{x}_a\|}{\|\mathbf{x}_a\|} \le \kappa(\mathbf{A})\frac{\|\mathbf{r}\|}{\|\mathbf{b}\|} \tag{4.31}$$

A selected matrix norm and its compatible vector norm must be used throughout.

Proof

We first write

$$\mathbf{r} = \mathbf{Ax}_c - \mathbf{b} = \mathbf{Ax}_c - \mathbf{Ax}_a = \mathbf{A}(\mathbf{x}_c - \mathbf{x}_a) \quad \Rightarrow \quad \mathbf{x}_c - \mathbf{x}_a = \mathbf{A}^{-1}\,\mathbf{r}$$

so that

$$\|\mathbf{x}_c - \mathbf{x}_a\| = \|\mathbf{A}^{-1}\mathbf{r}\| \le \|\mathbf{A}^{-1}\|\|\mathbf{r}\| \underset{\text{by } \|\mathbf{x}_a\|}{\overset{\text{Divide both sides}}{\Rightarrow}} \frac{\|\mathbf{x}_c - \mathbf{x}_a\|}{\|\mathbf{x}_a\|} \le \|\mathbf{A}^{-1}\|\frac{\|\mathbf{r}\|}{\|\mathbf{x}_a\|} \tag{4.32}$$

But by Equation 4.30

$$\|\mathbf{A}^{-1}\| = \frac{\kappa(\mathbf{A})}{\|\mathbf{A}\|}$$

and

$$\mathbf{b} = \mathbf{Ax}_a \quad \Rightarrow \quad \|\mathbf{b}\| \le \|\mathbf{A}\|\|\mathbf{x}_a\| \quad \Rightarrow \quad \frac{1}{\|\mathbf{x}_a\|} \le \frac{\|\mathbf{A}\|}{\|\mathbf{b}\|}$$

Inserting these into Equation 4.32 yields

$$\frac{\|\mathbf{x}_c - \mathbf{x}_a\|}{\|\mathbf{x}_a\|} \le \kappa(\mathbf{A})\frac{\|\mathbf{r}\|}{\|\mathbf{b}\|}$$

which establishes the upper bound for relative error. To derive the lower bound, we first note that

$$\mathbf{r} = \mathbf{A}\mathbf{x}_c - \mathbf{A}\mathbf{x}_a \;\Rightarrow\; \|\mathbf{r}\| \le \|\mathbf{A}\|\|\mathbf{x}_c - \mathbf{x}_a\| \;\underset{\|\mathbf{A}\|>0 \text{ for any nonzero } \mathbf{A}}{\overset{\text{Divide by } \|\mathbf{A}\|}{\Longrightarrow}}\; \|\mathbf{x}_c - \mathbf{x}_a\| \ge \frac{\|\mathbf{r}\|}{\|\mathbf{A}\|}$$

Also

$$\mathbf{x}_a = \mathbf{A}^{-1}\mathbf{b} \;\Rightarrow\; \|\mathbf{x}_a\| \le \|\mathbf{A}^{-1}\|\|\mathbf{b}\| \;\Rightarrow\; \frac{1}{\|\mathbf{x}_a\|} \ge \frac{1}{\|\mathbf{A}^{-1}\|\|\mathbf{b}\|}$$

Multiplication of the last two inequalities results in

$$\frac{\|\mathbf{x}_c - \mathbf{x}_a\|}{\|\mathbf{x}_a\|} \ge \frac{1}{\kappa(\mathbf{A})}\frac{\|\mathbf{r}\|}{\|\mathbf{b}\|}$$

This completes the proof.

4.6.3.1 Consequences of Ill-Conditioning

Ill-conditioning has an immediate impact on the accuracy of the computed solution. Consider the relative error bounds given in Equation 4.31. For a computed solution, it is safe to assume that the norm of the residual vector $\|\mathbf{r}\|$ is relatively small compared to $\|\mathbf{b}\|$. A small condition number for **A** raises the lower bound while lowering the upper bound, thus narrowing the interval for relative error. A large condition number for **A**, on the other hand, lowers the lower bound and raises the upper bound, widening the interval and allowing for a large relative error associated with the computed solution.

Another consequence of ill-conditioning is less conspicuous in the sense that a poor approximation of the actual solution vector may come with a very small residual vector norm. Once again, refer to the system in Examples 4.17 and 4.18, and consider

$$\hat{\mathbf{x}} = \begin{Bmatrix} 2 \\ 0.0002 \end{Bmatrix}$$

which is clearly a poor approximation of the actual solution

$$\mathbf{x}_a = \begin{Bmatrix} 1 \\ -0.5 \end{Bmatrix}$$

The corresponding residual vector is

$$\mathbf{r} = \mathbf{A}\hat{\mathbf{x}} - \mathbf{b} = \begin{bmatrix} 1 & -2 \\ 1.0001 & -1.9998 \end{bmatrix} \begin{Bmatrix} 2 \\ 0.0002 \end{Bmatrix} - \begin{Bmatrix} 2 \\ 2 \end{Bmatrix} = \begin{Bmatrix} -0.0004 \\ -0.0002 \end{Bmatrix}$$

Any one of the three vector norms returns a very small value for $\|\mathbf{r}\|$, suggesting that $\hat{\mathbf{x}}$ may be a valid solution.

4.6.4 Effects of Parameter Changes on the Solution

The following theorem illustrates how changes in the entries of **A** or components of **b** affect the resulting solution vector **x**, as well as the role of condition number of **A**.

Theorem 4.4

Consider the linear system $\mathbf{Ax} = \mathbf{b}$. Let $\Delta\mathbf{A}$, $\Delta\mathbf{b}$, and $\Delta\mathbf{x}$ reflect the changes in the entries or components of **A**, **b**, and **x**, respectively. Then

$$\frac{\|\Delta\mathbf{x}\|}{\|\mathbf{x}\|} \le \frac{\|\Delta\mathbf{A}\|}{\|\mathbf{A}\|}\kappa(\mathbf{A}) \tag{4.33}$$

and

$$\frac{\|\Delta\mathbf{x}\|}{\|\mathbf{x}\|} \le \frac{\|\Delta\mathbf{b}\|}{\|\mathbf{b}\|}\kappa(\mathbf{A}) \tag{4.34}$$

A selected matrix norm and its compatible vector norm must be used throughout.

Proof

Suppose that entries of **A** have been changed and these changes are recorded in $\Delta\mathbf{A}$. As a result, the solution **x** will also change, say, by $\Delta\mathbf{x}$. In $\mathbf{Ax} = \mathbf{b}$, insert $\mathbf{A} + \Delta\mathbf{A}$ for **A** and $\mathbf{x} + \Delta\mathbf{x}$ for **x** to obtain

$$(\mathbf{A} + \Delta\mathbf{A})(\mathbf{x} + \Delta\mathbf{x}) = \mathbf{b} \quad \overset{\text{Expand}}{\Rightarrow} \quad \mathbf{Ax} + \mathbf{A}(\Delta\mathbf{x}) + [\Delta\mathbf{A}]\mathbf{x} + [\Delta\mathbf{A}](\Delta\mathbf{x}) = \mathbf{b}$$

$$\underset{\text{from both sides}}{\overset{\text{Cancel } \mathbf{Ax}=\mathbf{b}}{\Rightarrow}} \quad \mathbf{A}(\Delta\mathbf{x}) + [\Delta\mathbf{A}]\mathbf{x} + [\Delta\mathbf{A}](\Delta\mathbf{x}) = \mathbf{0}$$

Solving for $\Delta\mathbf{x}$

$$\mathbf{A}(\Delta\mathbf{x}) = -[\Delta\mathbf{A}](\mathbf{x}+\Delta\mathbf{x}) \quad \Rightarrow \quad \Delta\mathbf{x} = -\mathbf{A}^{-1}[\Delta\mathbf{A}](\mathbf{x}+\Delta\mathbf{x})$$

Taking the (vector) norm of both sides, and applying compatibility relations, Equation 4.19, twice, we have

$$\|\Delta\mathbf{x}\| = \|-\mathbf{A}^{-1}[\Delta\mathbf{A}](\mathbf{x}+\Delta\mathbf{x})\| \le \|\mathbf{A}^{-1}\|\,\|\Delta\mathbf{A}\|\,\|\mathbf{x}+\Delta\mathbf{x}\|$$

Inserting $\|\mathbf{A}^{-1}\| = \kappa(\mathbf{A})/\|\mathbf{A}\|$ in this last equation, and dividing both sides by $\|\mathbf{x}+\Delta\mathbf{x}\|$, yields

$$\frac{\|\Delta\mathbf{x}\|}{\|\mathbf{x}+\Delta\mathbf{x}\|} \le \frac{\|\Delta\mathbf{A}\|}{\|\mathbf{A}\|}\kappa(\mathbf{A})$$

Since $\Delta\mathbf{x}$ represents small changes in $\mathbf{x}$, then

$$\|\mathbf{x}+\Delta\mathbf{x}\| \cong \|\mathbf{x}\| \quad \Rightarrow \quad \frac{\|\Delta\mathbf{x}\|}{\|\mathbf{x}+\Delta\mathbf{x}\|} \cong \frac{\|\Delta\mathbf{x}\|}{\|\mathbf{x}\|}$$

Using this in the previous equation establishes Equation 4.33. To verify Equation 4.34, insert $\mathbf{b}+\Delta\mathbf{b}$ for $\mathbf{b}$, and $\mathbf{x}+\Delta\mathbf{x}$ for $\mathbf{x}$ in $\mathbf{Ax}=\mathbf{b}$ and proceed as before. This completes the proof.

Equations 4.33 and 4.34 assert that if $\kappa(\mathbf{A})$ is small, then small percent changes in $\mathbf{A}$ or $\mathbf{b}$ will result in small percent changes in $\mathbf{x}$. This, of course, is in line with the previous findings in this section. Furthermore, Equations 4.33 and 4.34 only provide upper bounds and not estimates of the percent change in solution.

Example 4.19: Percent Change

Consider the linear system

$$\mathbf{Ax}=\mathbf{b}, \quad \mathbf{A} = \begin{bmatrix} 1 & 2 \\ 2 & 4.0001 \end{bmatrix}, \quad \mathbf{b} = \begin{Bmatrix} -1 \\ -2.0002 \end{Bmatrix} \quad \Rightarrow \quad \underset{\text{solution}}{\mathbf{x}} = \begin{Bmatrix} 3 \\ -2 \end{Bmatrix}$$

Suppose the (1,2) entry of $\mathbf{A}$ is reduced by 0.01 while its (2,1) entry is increased by 0.01 so that

$$\Delta\mathbf{A} = \begin{bmatrix} 0 & -0.01 \\ 0.01 & 0 \end{bmatrix} \quad \overset{\text{New coefficient matrix}}{\Rightarrow} \quad \bar{\mathbf{A}} = \mathbf{A}+\Delta\mathbf{A} = \begin{bmatrix} 1 & 1.99 \\ 2.01 & 4.0001 \end{bmatrix}$$

Solving the new system yields

$$\bar{\mathbf{A}}\bar{\mathbf{x}} = \mathbf{b} \quad \Rightarrow \quad \bar{\mathbf{x}} = \begin{Bmatrix} -98.51 \\ 49 \end{Bmatrix} \quad \text{so that} \quad \Delta\mathbf{x} = \mathbf{x} - \bar{\mathbf{x}} = \begin{Bmatrix} 101.51 \\ -51 \end{Bmatrix}$$

From this point forward, the 1-norm will be used for all vectors and matrices. The condition number of $\mathbf{A}$ is calculated as $\kappa(\mathbf{A}) = 3.6001 \times 10^5$ indicating ill-conditioning. The upper bound for the percent change in solution can be found as

$$\frac{\|\Delta\mathbf{x}\|_1}{\|\mathbf{x}\|_1} \le \frac{\|\Delta\mathbf{A}\|_1}{\|\mathbf{A}\|_1}\kappa(\mathbf{A}) = \frac{0.01}{6.0001}(3.6001 \times 10^5) = 600.0100$$

The upper bound is rather large as a consequence of ill-conditioning. The actual percent change is calculated as

$$\frac{\|\Delta\mathbf{x}\|_1}{\|\mathbf{x}\|_1} = \frac{152.51}{5} = 30.5020$$

which is much smaller than the upper bound offered by Equation 4.33, thus asserting that the upper bound is in no way an estimate of the actual percent change.

4.7 Systems of Nonlinear Equations

Systems of nonlinear equations can be solved numerically by either using Newton's method (for small systems) or the fixed-point iteration method* (for large systems).

4.7.1 Newton's Method for a System of Nonlinear Equations

Newton's method for solving a single nonlinear equation was discussed previously. An extension of that technique can be used for solving a system of nonlinear equations. We will first present the idea and the details as pertained to a system of two nonlinear equations, followed by a general system of n nonlinear equations.

4.7.1.1 Method for Solving a System of Two Nonlinear Equations

A system of two (nonlinear) equations in two unknowns can generally be expressed as

* Refer to Newton's method and fixed-point iteration method for a single nonlinear equation, Chapter 3.

$$
\begin{aligned}
f_1(x,y) &= 0 \\
f_2(x,y) &= 0
\end{aligned}
\tag{4.35}
$$

We begin by selecting (x_1, y_1) as an initial estimate of the solution. Suppose (x_2, y_2) denotes the actual solution so that $f_1(x_2, y_2) = 0$ and $f_2(x_2, y_2) = 0$. If x_1 is sufficiently close to x_2, and y_1 to y_2, then by Taylor series expansion

$$
\begin{aligned}
f_1(x_2,y_2) &= f_1(x_1,y_1) + \left.\frac{\partial f_1}{\partial x}\right|_{(x_1,y_1)}(x_2 - x_1) + \left.\frac{\partial f_1}{\partial y}\right|_{(x_1,y_1)}(y_2 - y_1) + \cdots \\
f_2(x_2,y_2) &= f_2(x_1,y_1) + \left.\frac{\partial f_2}{\partial x}\right|_{(x_1,y_1)}(x_2 - x_1) + \left.\frac{\partial f_2}{\partial y}\right|_{(x_1,y_1)}(y_2 - y_1) + \cdots
\end{aligned}
$$

where the terms involving higher powers of small quantities $x_2 - x_1$ and $y_2 - y_1$ have been neglected. Let $\Delta x = x_2 - x_1$ and $\Delta y = y_2 - y_1$ and recall that $f_1(x_2, y_2) = 0$ and $f_2(x_2, y_2) = 0$ to rewrite the above equations as

$$
\begin{aligned}
\left.\frac{\partial f_1}{\partial x}\right|_{(x_1,y_1)}\Delta x + \left.\frac{\partial f_1}{\partial y}\right|_{(x_1,y_1)}\Delta y + \cdots &= -f_1(x_1,y_1) \\
\left.\frac{\partial f_2}{\partial x}\right|_{(x_1,y_1)}\Delta x + \left.\frac{\partial f_2}{\partial y}\right|_{(x_1,y_1)}\Delta y + \cdots &= -f_2(x_1,y_1)
\end{aligned}
$$

which can be expressed as

$$
\begin{bmatrix}
\dfrac{\partial f_1}{\partial x} & \dfrac{\partial f_1}{\partial y} \\
\dfrac{\partial f_2}{\partial x} & \dfrac{\partial f_2}{\partial y}
\end{bmatrix}_{(x_1,y_1)}
\begin{Bmatrix} \Delta x \\ \Delta y \end{Bmatrix}
=
\begin{Bmatrix} -f_1 \\ -f_2 \end{Bmatrix}_{(x_1,y_1)}
\tag{4.36}
$$

Equation 4.36 is a linear system that can be solved for $\Delta x, \Delta y$ as long as the coefficient matrix is nonsingular. The quantity

$$
J(f_1, f_2) = \begin{vmatrix}
\dfrac{\partial f_1}{\partial x} & \dfrac{\partial f_1}{\partial y} \\
\dfrac{\partial f_2}{\partial x} & \dfrac{\partial f_2}{\partial y}
\end{vmatrix}
$$

is called the Jacobian of f_1 and f_2. Therefore, for Equation 4.36 to have a nontrivial solution, we must have

$$\det\left\{\begin{bmatrix} \frac{\partial f_1}{\partial x} & \frac{\partial f_1}{\partial y} \\ \frac{\partial f_2}{\partial x} & \frac{\partial f_2}{\partial y} \end{bmatrix}_{(x_1,y_1)}\right\} \neq 0 \quad \Rightarrow \quad J(f_1,f_2)\Big|_{(x_1,y_1)} \neq 0$$

This means that the values of $x_2 = x_1 + \Delta x$ and $y_2 = y_1 + \Delta y$ are now known. And they clearly do not describe the actual solution because higher-order terms in Taylor series expansions were neglected earlier. Since (x_2, y_2) is closer to the actual solution than (x_1, y_1), we use (x_2, y_2) as the new estimate to the solution and solve Equation 4.36, with (x_2, y_2) replacing (x_1, y_1), and continue the process until values generated at successive iterations meet a prescribed tolerance condition. A reasonable terminating condition is

$$\left\| \begin{Bmatrix} x_{k+1} \\ y_{k+1} \end{Bmatrix} - \begin{Bmatrix} x_k \\ y_k \end{Bmatrix} \right\|_2 \leq \varepsilon \tag{4.37}$$

where ε is a specified tolerance. Keep in mind that in each iteration step, the Jacobian must be nonzero.

Example 4.20: Newton's Method

Solve the nonlinear system below using Newton's method, terminating condition as in Equation 4.37 with tolerance $\varepsilon = 10^{-3}$ and maximum number of iterations set to 20:

$$\begin{cases} 3.2x^3 + 1.8y^2 + 24.43 = 0 \\ -2x^2 + 3y^3 = 5.92 \end{cases}$$

SOLUTION

The original system is written in the form of Equation 4.35 as

$$\begin{cases} f_1(x,y) = 3.2x^3 + 1.8y^2 + 24.43 = 0 \\ f_2(x,y) = -2x^2 + 3y^3 - 5.92 = 0 \end{cases}$$

First, we need to find approximate locations of the roots using a graphical approach.

```
>> f1 = inline('3.2*x^3+1.8*y^2+24.43','x','y');
>> f2 = inline('-2*x^2+3*y^3-5.92','x','y');
>> ezplot(f1)
>> hold on
>> ezplot(f2)      % Figure 4.6
```

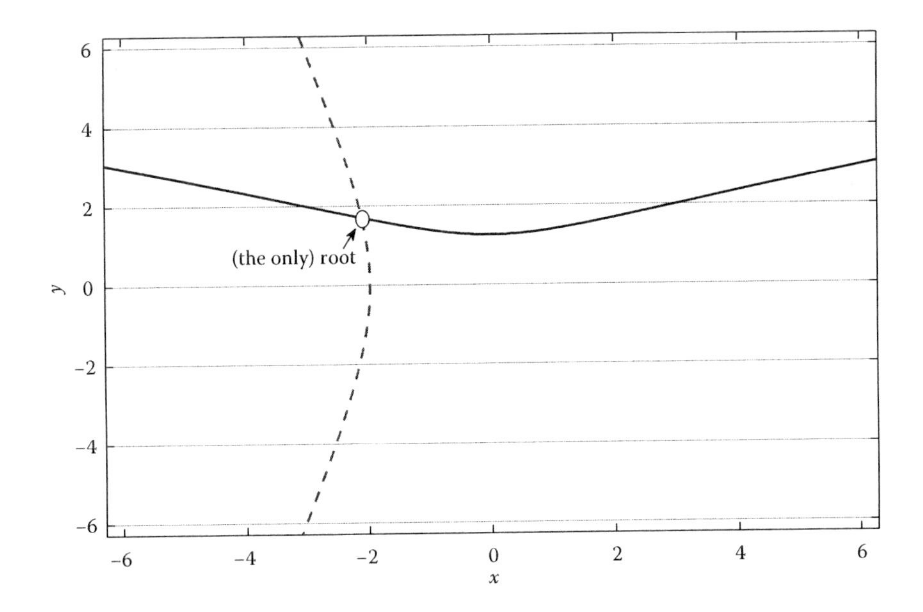

FIGURE 4.6
Graph of the nonlinear system in Example 4.20.

Based on Figure 4.6, there is only one solution, and a logical initial estimate for this solution is (−2, 2). Performing partial differentiations in Equation 4.36, we find

$$\begin{bmatrix} 9.6x^2 & 3.6y \\ -4x & 9y^2 \end{bmatrix}_{(-2,2)} \begin{Bmatrix} \Delta x \\ \Delta y \end{Bmatrix} = \begin{Bmatrix} -f_1 \\ -f_2 \end{Bmatrix}_{(-2,2)} \overset{\text{Solve}}{\Rightarrow} \begin{Bmatrix} \Delta x \\ \Delta y \end{Bmatrix}$$

The next solution estimate is then found as

$$\begin{Bmatrix} x_2 \\ y_2 \end{Bmatrix} = \begin{Bmatrix} x_1 \\ y_1 \end{Bmatrix} + \begin{Bmatrix} \Delta x \\ \Delta y \end{Bmatrix}$$

The following MATLAB code will perform these tasks. In addition to the terminating condition in Equation 4.37, the code includes a section that checks to see if $|f_1(x, y)| < \varepsilon$ and $|f_2(x, y)| < \varepsilon$ after each iteration. This is because sometimes an acceptable estimate of a root may have been found, but because of the nature of f_1 and/or f_2, the current vector and the subsequent vector do not meet the terminating condition in Equation 4.37.

```
% Define the entries of the coefficient matrix and
% the right-hand side vector as inline functions
```

```
A11=inline('9.6*x^2','x','y'); A12=inline('3.6*y','x','y');
A21=inline('-4*x','x','y'); A22=inline('9*y^2','x','y');

f1=inline('3.2*x^3+1.8*y^2+24.43','x','y');
f2=inline('-2*x^2+3*y^3-5.92','x','y');

tol=1e-3; kmax=20;
% Set tolerance and maximum number of
% iterations allowed
v(:,1) = [-2;2]; % Initial estimate

for k=1:kmax,
    A=[A11(v(1,k),v(2,k)) A12(v(1,k),v(2,k));
        A21(v(1,k),v(2,k)) A22(v(1,k),v(2,k))];
    b=[-f1(v(1,k),v(2,k));-f2(v(1,k),v(2,k))];

% Check to see if a root has been found, while two
% successive vectors do not satisfy terminating
% condition due to the nature of function(s) near
% a root

  if abs(b(1))<tol && abs(b(2))<tol,
        root=v(:,k);
    return
  end

    delv=A\b;
    v(:,k+1)=v(:,k)+delv;       % Update solution
    if norm(v(:,k+1)-v(:,k))<tol,
    % Check terminating condition
        root=v(:,k+1);
         break
    end
end
```

The execution of this code results in

```
>> v

v =

   -2.0000    -2.1091    -2.1001    -2.0999
    2.0000     1.7442     1.7012     1.7000
```

Therefore, it took three iterations for the tolerance to be met. The computed solution is (–2.0999, 1.7000).

4.7.1.2 Method for Solving a System of *n* Nonlinear Equations

A system of n (nonlinear) equations in n unknowns can in general be expressed as

$$
\begin{aligned}
f_1(x_1, x_2, \ldots, x_n) &= 0 \\
f_2(x_1, x_2, \ldots, x_n) &= 0 \\
&\ldots \\
f_n(x_1, x_2, \ldots, x_n) &= 0
\end{aligned} \tag{4.38}
$$

Choose $(x_{1,1}, x_{2,1}, \ldots, x_{n,1})$ as the initial estimate and follow the steps that led to Equation 4.36 to arrive at

$$
\begin{bmatrix}
\frac{\partial f_1}{\partial x_1} & \frac{\partial f_1}{\partial x_2} & \ldots & \frac{\partial f_1}{\partial x_n} \\
\frac{\partial f_2}{\partial x_1} & \frac{\partial f_2}{\partial x_2} & \ldots & \frac{\partial f_2}{\partial x_n} \\
\ldots & \ldots & \ldots & \ldots \\
\frac{\partial f_n}{\partial x_1} & \frac{\partial f_n}{\partial x_2} & \ldots & \frac{\partial f_n}{\partial x_n}
\end{bmatrix}_{(x_{1,1}, x_{2,1}, \ldots, x_{n,1})}
\begin{Bmatrix} \Delta x_1 \\ \Delta x_2 \\ \ldots \\ \Delta x_n \end{Bmatrix}
=
\begin{Bmatrix} -f_1 \\ -f_2 \\ \ldots \\ -f_n \end{Bmatrix}_{(x_{1,1}, x_{2,1}, \ldots, x_{n,1})} \tag{4.39}
$$

The partial derivatives in Equation 4.39 can be found either analytically or numerically.* Solve this system to obtain the vector composed of increments $\Delta x_1, \ldots, \Delta x_n$. Then, update the solution estimate

$$
\begin{Bmatrix} x_{1,2} \\ x_{2,2} \\ \ldots \\ x_{n,2} \end{Bmatrix}
=
\begin{Bmatrix} x_{1,1} \\ x_{2,1} \\ \ldots \\ x_{n,1} \end{Bmatrix}
+
\begin{Bmatrix} \Delta x_1 \\ \Delta x_2 \\ \ldots \\ \Delta x_n \end{Bmatrix}
$$

If a specified terminating condition is not met, solve Equation 4.39 with $(x_{1,2}, x_{2,2}, \ldots, x_{n,2})$ replacing $(x_{1,1}, x_{2,1}, \ldots, x_{n,1})$ and continue the process until the terminating condition is satisfied.

4.7.1.3 Convergence of Newton's Method

Convergence of Newton's method is not guaranteed, but it is expected if these conditions hold:

- $f_1, f_2, \ldots, f_n$ and their partial derivatives are continuous and bounded near the actual solution.
- The Jacobian is nonzero, $J(f_1, f_2, \ldots, f_n) \neq 0$, near the solution.
- The initial solution estimate is sufficiently close to the actual solution.

* See Chapter 10.

As it was the case with a single nonlinear equation, if Newton's method does not exhibit convergence, it is usually because the initial solution estimate is not sufficiently close to the actual solution.

4.7.2 Fixed-Point Iteration Method for a System of Nonlinear Equations

The fixed-point iteration* to solve a single nonlinear equation can be extended to handle systems of nonlinear equations in the form of Equation 4.38. The idea is to find suitable auxiliary functions $g_i(x_1, x_2, \ldots, x_n)$, $i = 1, 2, \ldots, n$, and rewrite Equation 4.38 as

$$\begin{aligned} x_1 &= g_1(x_1, x_2, \ldots, x_n) \\ x_2 &= g_2(x_1, x_2, \ldots, x_n) \\ &\ldots \\ x_n &= g_n(x_1, x_2, \ldots, x_n) \end{aligned} \tag{4.40}$$

or in vector form

$$\mathbf{x} = \mathbf{g}(\mathbf{x}), \quad \mathbf{x} = \begin{Bmatrix} x_1 \\ x_2 \\ \ldots \\ x_n \end{Bmatrix}, \quad \mathbf{g} = \begin{Bmatrix} g_1(\mathbf{x}) \\ g_2(\mathbf{x}) \\ \ldots \\ g_n(\mathbf{x}) \end{Bmatrix} \tag{4.41}$$

Choose $(x_{1,1}, x_{2,1}, \ldots, x_{n,1})$ as the initial estimate and substitute into the right sides of the equations in Equation 4.40. The updated estimates are calculated as

$$\begin{aligned} x_{1,2} &= g_1(x_{1,1}, x_{2,1}, \ldots, x_{n,1}) \\ x_{2,2} &= g_2(x_{1,1}, x_{2,1}, \ldots, x_{n,1}) \\ &\ldots \\ x_{n,2} &= g_n(x_{1,1}, x_{2,1}, \ldots, x_{n,1}) \end{aligned}$$

These new values are then inserted in the right sides of Equation 4.40 to generate the new updates, and so on. The process continues until convergence is observed.

4.7.2.1 *Convergence of the Fixed-Point Iteration Method*

The conditions for convergence of the fixed-point iteration

$$\mathbf{x}^{(k+1)} = \mathbf{g}(\mathbf{x}^{(k)}), \quad k = 0, 1, 2, \ldots \tag{4.42}$$

* See Section 3.4.

are similar to those for the case of a function of one variable. Let R be an n-dimensional rectangular region composed of points $x_1, x_2, \ldots, x_n$ such that $a_i \le x_i \le b_i$ $(i = 1,2, \ldots, n)$ for constants $a_1, a_2, \ldots, a_n$ and $b_1, b_2, \ldots, b_n$. Suppose $\mathbf{g}(\mathbf{x})$ is defined on R. Then the sufficient conditions for convergence of the fixed-point iteration method, Equation 4.42, are*

- Auxiliary functions $g_1, g_2, \ldots, g_n$ and their partial derivatives with respect to $x_1, x_2, \ldots, x_n$ are continuous near the actual solution.
- There exists a constant $K < 1$ such that for each $\mathbf{x} \in R$

$$\left|\frac{\partial g_j(\mathbf{x})}{\partial x_i}\right| \le \frac{K}{n}, \quad j = 1, 2, \ldots, n, \quad i = 1, 2, \ldots, n \tag{4.43}$$

which may also be interpreted as

$$\begin{gathered}
\left|\frac{\partial g_1}{\partial x_1}\right| + \left|\frac{\partial g_1}{\partial x_2}\right| + \cdots + \left|\frac{\partial g_1}{\partial x_n}\right| \le 1 \\
\left|\frac{\partial g_2}{\partial x_1}\right| + \left|\frac{\partial g_2}{\partial x_2}\right| + \cdots + \left|\frac{\partial g_2}{\partial x_n}\right| \le 1 \\
\cdots \\
\left|\frac{\partial g_n}{\partial x_1}\right| + \left|\frac{\partial g_n}{\partial x_2}\right| + \cdots + \left|\frac{\partial g_n}{\partial x_n}\right| \le 1
\end{gathered} \tag{4.44}$$

- The initial estimate $(x_{1,1}, x_{2,1}, \ldots, x_{n,1})$ is sufficiently close to the actual solution.

Example 4.21: Fixed-Point Iteration

Using the fixed-point iteration method, solve the nonlinear system in Example 4.20:

$$\begin{cases} 3.2x^3 + 1.8y^2 + 24.43 = 0 \\ -2x^2 + 3y^3 = 5.92 \end{cases}$$

Use the same initial estimate and terminating condition as before.

SOLUTION

We first need to rewrite the given equations in the form of Equation 4.40 by selecting suitable auxiliary functions. These auxiliary functions are not unique, and one way to rewrite the original system is

$$x = g_1(x, y) = -\left(\frac{1.8y^2 + 24.43}{3.2}\right)^{1/3}$$

* Refer to Atkinson, K.E., *An Introduction to Numerical Analysis*, 2nd ed., John Wiley, NY, 1989.

$$y = g_2(x,y) = \left(\frac{2x^2 + 5.92}{3}\right)^{1/3}$$

Based on Figure 4.6, a reasonable rectangular region R is chosen as $-4 \le x \le -2$, $0 \le y \le 2$. We next examine the conditions listed in Equation 4.44 in relation to our choices of g_1 and g_2. Noting that in this example, $n = 2$, the four conditions to be met are

$$\left|\frac{\partial g_1}{\partial x}\right| < \frac{1}{2}, \quad \left|\frac{\partial g_1}{\partial y}\right| < \frac{1}{2}, \quad \left|\frac{\partial g_2}{\partial x}\right| < \frac{1}{2}, \quad \left|\frac{\partial g_2}{\partial y}\right| < \frac{1}{2}$$

Of course, $|\partial g_1/\partial x| = 0 < \frac{1}{2}$ and $|\partial g_2/\partial y| = 0 < \frac{1}{2}$ satisfy two of the above. The other two may be inspected with the aid of MATLAB as follows:

```
>> g1 = '-((1.8*y^2+24.43)/3.2)^(1/3)';
>> g2 = '((2*x^2+5.92)/3)^(1/3)';
>> ezplot(abs(diff(g1,'y')),[0 2])
% First plot in Figure 4.7
>> ezplot(abs(diff(g2,'x')),[-4 -2])
% Complete Figure 4.7
```

The two plots in Figure 4.7 clearly indicate that the two remaining partial derivatives satisfy their respective conditions as well. This means that the vector function

$$\mathbf{g} = \begin{Bmatrix} g_1 \\ g_2 \end{Bmatrix}$$

has a fixed point in region R, and the fixed-point iteration in Equation 4.42 is guaranteed to converge to this fixed point.

The following code will use the fixed-point iteration to generate a sequence of values for x and y and terminates the iterations as soon as the tolerance is met. For simplicity, we define a vector

$$\mathbf{v} = \begin{Bmatrix} x \\ y \end{Bmatrix}$$

and subsequently define g_1 and g_2 as functions of the components of vector $\mathbf{v}$.

```
% Define the auxiliary functions g1 and g2
g1 = inline('-((24.43+1.8*v(2,1)^2)/3.2)^(1/3)','v');
g2 = inline('((5.92+2*v(1,1)^2)/3)^(1/3)','v');

tol = 1e-3; kmax = 10;
v(:,1) = [-1;-2];    % Initial estimate
for k = 1:kmax,
```

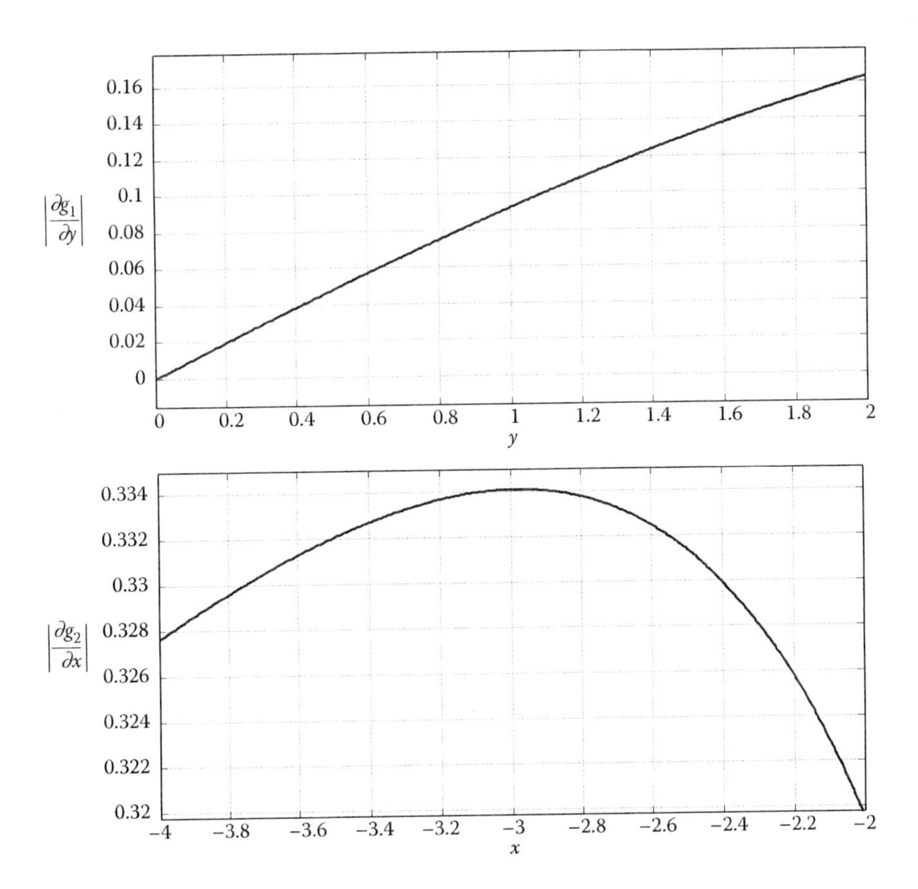

FIGURE 4.7
Graphical inspection of upper bounds for $|\partial g_1/\partial y|$ and $|\partial g_2/\partial x|$.

```
        v(:,k+1) = [g1(v(:,k));g2(v(:,k))];
        % Fixed-point iteration
        if norm(v(:,k+1)-v(:,k))<tol,
            break
        end
    end
```

Execution of this code results in

```
>> v

v =
   -1.0000   -2.1461   -2.0574   -2.1021   -2.0980   -2.1000   -2.0998
   -2.0000    1.3821    1.7150    1.6863    1.7007    1.6994    1.7000
```

Convergence to the true solution is observed after six iterations. The results agree with those of Example 4.20.

PROBLEM SET

Gauss Elimination Method (Section 4.3)

✍ In Problems 1 through 12, solve the linear system using basic Gauss elimination with partial pivoting, if necessary.

1. $$\begin{cases} -2x_1 - x_2 = 2 \\ 4x_1 + 3x_2 = 0 \end{cases}$$

2. $$\begin{cases} 3x_1 - 4x_2 = 4 \\ 2x_1 + 6x_2 = 7 \end{cases}$$

3. $$\begin{cases} 0.6x_1 + 1.3x_2 = 0.2 \\ 2.1x_1 - 3.2x_2 = 3.8 \end{cases}$$

4. $$\begin{cases} x_2 + 3x_3 = -11 \\ 2x_1 - 3x_2 + 2x_3 = 1 \\ -6x_1 + 9x_2 - 7x_3 = 1 \end{cases}$$

5. $$\begin{cases} -2x_1 + x_2 + 2x_3 = 8 \\ x_1 + 3x_2 - 2x_3 = -7 \\ 3x_1 + 8x_2 + x_3 = 0 \end{cases}$$

6. $$\begin{cases} 3x_1 + 5x_3 = -1 \\ -x_1 + 5x_2 - 2x_3 = 12 \\ 2x_1 - 7x_2 + 4x_3 = -18 \end{cases}$$

7. $$\begin{bmatrix} 4 & -1 & 0 \\ -1 & 3 & -5 \\ 2 & 5 & 6 \end{bmatrix} \begin{Bmatrix} x_1 \\ x_2 \\ x_3 \end{Bmatrix} = \begin{Bmatrix} -9 \\ -15 \\ 25 \end{Bmatrix}$$

8. $$\begin{bmatrix} 6 & -29 & -1 \\ 0 & 7 & 3 \\ 3 & -4 & -1 \end{bmatrix} \begin{Bmatrix} x_1 \\ x_2 \\ x_3 \end{Bmatrix} = \begin{Bmatrix} -47 \\ -1 \\ 0 \end{Bmatrix}$$

9. $$\begin{bmatrix} -3 & -3 & 2 \\ 1 & 4 & -1 \\ 0 & 9 & 2 \end{bmatrix} \begin{Bmatrix} x_1 \\ x_2 \\ x_3 \end{Bmatrix} = \begin{Bmatrix} -12 \\ 5 \\ -6 \end{Bmatrix}$$

10. $$\begin{bmatrix} 1.5 & 1.0 & -2.5 \\ 3.0 & 2.0 & -6.2 \\ -2.4 & 0.4 & 5.0 \end{bmatrix} \begin{Bmatrix} x_1 \\ x_2 \\ x_3 \end{Bmatrix} = \begin{Bmatrix} 0.5 \\ -0.2 \\ -1.3 \end{Bmatrix}$$

11. $\begin{bmatrix} -2.0 & 2.5 & -8.0 \\ 1.0 & 0 & 6.0 \\ 3.0 & 5.0 & 15.2 \end{bmatrix} \begin{Bmatrix} x_1 \\ x_2 \\ x_3 \end{Bmatrix} = \begin{Bmatrix} 6.0 \\ 0 \\ 9.3 \end{Bmatrix}$

12. $\begin{bmatrix} -5 & 1 & 16 & -12 \\ 1 & 0 & -4 & 3 \\ 0 & -3 & 10 & -5 \\ 4 & 8 & -24 & -3 \end{bmatrix} \begin{Bmatrix} x_1 \\ x_2 \\ x_3 \\ x_4 \end{Bmatrix} = \begin{Bmatrix} -28 \\ 6 \\ -2 \\ 1 \end{Bmatrix}$

Small Pivot

13. ✍ Consider a linear 2×2 system (with unknowns x_1 and x_2) described by its augmented matrix

$$\left[\begin{array}{cc|c} \varepsilon & 1 & 2 \\ 1 & 1 & 1 \end{array}\right]$$

where $\varepsilon > 0$ is a very small constant.

a. Solve by Gauss elimination without partial pivoting.

b. Solve by Gauss elimination with partial pivoting. Compare the results and discuss their validity.

In Problems 14 through 18, a linear system is given.

a. ✍ Solve using Gauss elimination with partial pivoting and row scaling.

b. ▲ Solve by executing the user-defined function `GaussPivotScale` (Section 4.3).

14. $\begin{cases} 2x_1 + x_2 - 6x_3 = 8 \\ 3x_1 - 2x_2 + 2x_3 = 1 \\ x_1 + 5x_2 = 1 \end{cases}$

15. $\begin{cases} -3x_1 + 3x_2 + 5x_3 = -14 \\ 3x_1 - 2x_2 - 2x_3 = 3 \\ 9x_1 + 10x_2 - x_3 = 5 \end{cases}$

16. $\begin{cases} 5x_1 + 8x_2 + 12x_3 = 15 \\ -3x_1 + 5x_2 - 8x_3 = 11 \\ 2x_1 - x_2 + 4x_3 = -2 \end{cases}$

17. $$\begin{bmatrix} 2 & -1 & 3 \\ 5 & 6 & 4 \\ 1 & 0 & 1 \end{bmatrix} \begin{Bmatrix} x_1 \\ x_2 \\ x_3 \end{Bmatrix} = \begin{Bmatrix} 17 \\ 3 \\ 5 \end{Bmatrix}$$

18. $$\begin{bmatrix} 3 & 2 & 0 \\ -3 & 5 & -1 \\ 6 & 5 & 12 \end{bmatrix} \begin{Bmatrix} x_1 \\ x_2 \\ x_3 \end{Bmatrix} = \begin{Bmatrix} -7 \\ 9 \\ 47 \end{Bmatrix}$$

Tridiagonal Systems

In Problems 19 through 24, a tridiagonal system is given.

a. ✍ Solve using the Thomas method.
b. Solve by executing the user-defined function `ThomasMethod` (Section 4.3).

19. $$\begin{bmatrix} 2 & 1 & 0 \\ 1 & 2 & 1 \\ 0 & 1 & 2 \end{bmatrix} \begin{Bmatrix} x_1 \\ x_2 \\ x_3 \end{Bmatrix} = \begin{Bmatrix} -1 \\ 5 \\ 5 \end{Bmatrix}$$

20. $$\begin{bmatrix} 3 & -1 & 0 \\ 1 & 2 & 1 \\ 0 & -1 & -3 \end{bmatrix} \begin{Bmatrix} x_1 \\ x_2 \\ x_3 \end{Bmatrix} = \begin{Bmatrix} 6 \\ -4 \\ 0 \end{Bmatrix}$$

21. $$\begin{bmatrix} 1 & -2 & 0 & 0 \\ 2 & 3 & 1 & 0 \\ 0 & 1 & -2 & 1 \\ 0 & 0 & -1 & 4 \end{bmatrix} \begin{Bmatrix} x_1 \\ x_2 \\ x_3 \\ x_4 \end{Bmatrix} = \begin{Bmatrix} -4 \\ 5 \\ 7 \\ 13 \end{Bmatrix}$$

22. $$\begin{bmatrix} 0.1 & 0.09 & 0 & 0 \\ 0.12 & 1.2 & 0.8 & 0 \\ 0 & 1.1 & 0.9 & 0.6 \\ 0 & 0 & -1.3 & 0.9 \end{bmatrix} \begin{Bmatrix} x_1 \\ x_2 \\ x_3 \\ x_4 \end{Bmatrix} = \begin{Bmatrix} -0.01 \\ 0.28 \\ 1.4 \\ 3.1 \end{Bmatrix}$$

23. $$\begin{bmatrix} -1 & 1 & 0 & 0 & 0 \\ 1 & -3 & -1 & 0 & 0 \\ 0 & 0 & 1 & 2 & 0 \\ 0 & 0 & -2 & 3 & 4 \\ 0 & 0 & 0 & 0 & 2 \end{bmatrix} \begin{Bmatrix} x_1 \\ x_2 \\ x_3 \\ x_4 \\ x_5 \end{Bmatrix} = \begin{Bmatrix} -1 \\ 1 \\ -4 \\ 9 \\ 4 \end{Bmatrix}$$

24. $$\begin{bmatrix} -1 & 1 & 0 & 0 & 0 \\ 1 & 2 & 0 & 0 & 0 \\ 0 & -2 & 3 & 1 & 0 \\ 0 & 0 & 1 & -4 & 0 \\ 0 & 0 & 0 & 3 & 1 \end{bmatrix} \begin{Bmatrix} x_1 \\ x_2 \\ x_3 \\ x_4 \\ x_5 \end{Bmatrix} = \begin{Bmatrix} 2 \\ 7 \\ -8 \\ 8 \\ -7 \end{Bmatrix}$$

25. Alternating direct implicit (ADI) methods are used to numerically solve a certain type of partial differential equation in a rectangular region; Chapter 10. These methods are specifically designed to generate tridiagonal systems in their solution process. In one such application, the following tridiagonal system has been created:

$$\begin{bmatrix} -4 & 1 & 0 & 0 & 0 & 0 \\ 1 & -4 & 1 & 0 & 0 & 0 \\ 0 & 1 & -4 & 0 & 0 & 0 \\ 0 & 0 & 0 & -4 & 1 & 0 \\ 0 & 0 & 0 & 1 & -4 & 1 \\ 0 & 0 & 0 & 0 & 1 & -4 \end{bmatrix} \begin{Bmatrix} x_1 \\ x_2 \\ x_3 \\ x_4 \\ x_5 \\ x_6 \end{Bmatrix} = \begin{Bmatrix} -0.6056 \\ -1.0321 \\ -1.1389 \\ -0.2563 \\ -0.4195 \\ -0.3896 \end{Bmatrix}$$

a. ✍ Solve using the Thomas method. Use four-digit rounding up.

b. Solve by executing the user-defined function `ThomasMethod`. Compare with the results in (a).

26. Finite difference methods are used to numerically solve boundary-value problems; Chapter 8. These methods are designed so that tridiagonal systems are generated in their solution process. In one such application, the following tridiagonal system has been created:

$$\begin{bmatrix} -6 & 3.5 & 0 & 0 & 0 \\ 3.5 & -8 & 4.5 & 0 & 0 \\ 0 & 4.5 & -10 & 5.5 & 0 \\ 0 & 0 & 5.5 & -12 & 6.5 \\ 0 & 0 & 0 & 4 & -4 \end{bmatrix} \begin{Bmatrix} w_1 \\ w_2 \\ w_3 \\ w_4 \\ w_5 \end{Bmatrix} = \begin{Bmatrix} -4 \\ -2 \\ -2.5 \\ -3 \\ -1 \end{Bmatrix}$$

a. ✍ Solve using the Thomas method. Use four-digit rounding up.

b. Solve by executing the user-defined function `ThomasMethod`. Compare with the results in (a).

LU Factorization Methods (Section 4.4)

Doolittle Factorization

✍ In Problems 27 through 32, find the Doolittle factorization of each matrix using the Gauss elimination method.

27. $\mathbf{A} = \begin{bmatrix} -2 & -1 \\ 4 & 3 \end{bmatrix}$

28. $\mathbf{A} = \begin{bmatrix} 2 & -3 & 2 \\ 0 & 1 & 3 \\ -6 & 9 & -7 \end{bmatrix}$

29. $\mathbf{A} = \begin{bmatrix} -1 & 3 & -5 \\ 4 & -1 & 0 \\ 2 & 5 & 6 \end{bmatrix}$

30. $\mathbf{A} = \begin{bmatrix} -1 & 2 & 2 \\ 3 & -4 & -5 \\ -2 & 6 & 3 \end{bmatrix}$

31. $\mathbf{A} = \begin{bmatrix} 3 & 6 & 3 & 9 \\ 1 & 5 & 4 & 7 \\ -2 & -1 & 2 & -3 \\ 3 & 0 & 1 & 7 \end{bmatrix}$

32. $\mathbf{A} = \begin{bmatrix} 2 & 4 & -2 & 6 \\ 1 & 3 & 2 & 5 \\ 4 & 7 & -3 & 10 \\ 3 & 5 & 1 & 11 \end{bmatrix}$

In Problems 33 through 38, find the Doolittle factorization of each matrix by

a. ✍ Direct calculation of **L** and **U**

b. Executing `DoolittleFactor` (Section 4.4)

33. $\mathbf{A} = \begin{bmatrix} 3 & 1 & 1 \\ -3 & -3 & 1 \\ 3 & -3 & 6 \end{bmatrix}$

34. $\mathbf{A} = \begin{bmatrix} 2 & 2 & 1 \\ 1 & -1 & \frac{13}{2} \\ -2 & -\frac{10}{3} & 6 \end{bmatrix}$

35. $\mathbf{A} = \begin{bmatrix} 1 & 3 & -3 \\ \frac{1}{3} & -5 & 2 \\ \frac{2}{3} & 14 & -3 \end{bmatrix}$

36. $\mathbf{A} = \begin{bmatrix} 4 & -2 & 8 \\ -4 & 5 & -13 \\ 1 & -\frac{19}{2} & 19 \end{bmatrix}$

37. $\mathbf{A} = \begin{bmatrix} 3 & 0 & -1 & 2 \\ -3 & 2 & 2 & 1 \\ 0 & -2 & -5 & -2 \\ 6 & 6 & -7 & 20 \end{bmatrix}$

38. $\mathbf{A} = \begin{bmatrix} -2 & 1 & 3 & -1 \\ -4 & 5 & 6 & 0 \\ 4 & -2 & -1 & -1 \\ -2 & 13 & -12 & 18 \end{bmatrix}$

Doolittle's Method

In Problems 39 through 44

a. ✍ Using Doolittle's method, solve each linear system $\mathbf{Ax} = \mathbf{b}$.
b. Confirm the results by executing the user-defined function `DoolittleMethod` (Section 4.4).

39. $\mathbf{A} = \begin{bmatrix} -1 & 3 & -5 \\ 4 & -1 & 0 \\ 2 & 5 & 6 \end{bmatrix}, \mathbf{b} = \begin{Bmatrix} -14 \\ 5 \\ 9 \end{Bmatrix}$

40. $\mathbf{A} = \begin{bmatrix} -1 & 2 & 2 \\ 3 & -4 & -5 \\ -2 & 6 & 3 \end{bmatrix}, \mathbf{b} = \begin{Bmatrix} -8 \\ 21 \\ -13 \end{Bmatrix}$

41. $\mathbf{A} = \begin{bmatrix} 3 & 1 & 1 \\ -3 & -3 & 1 \\ 3 & -3 & 6 \end{bmatrix}, \quad \mathbf{b} = \begin{Bmatrix} 2 \\ -4 \\ 0 \end{Bmatrix}$

42. $\mathbf{A} = \begin{bmatrix} 2 & 2 & 1 \\ 1 & -1 & \frac{13}{2} \\ -2 & -\frac{10}{3} & 6 \end{bmatrix}, \quad \mathbf{b} = \begin{Bmatrix} 6 \\ -15 \\ -24 \end{Bmatrix}$

43. $\mathbf{A} = \begin{bmatrix} 3 & 0 & -1 & 2 \\ -3 & 2 & 2 & 1 \\ 0 & -2 & -5 & -2 \\ 6 & 6 & -7 & 20 \end{bmatrix}, \quad \mathbf{b} = \begin{Bmatrix} 4 \\ -1 \\ 6 \\ 40 \end{Bmatrix}$

44. $\mathbf{A} = \begin{bmatrix} -2 & 1 & 3 & -1 \\ -4 & 5 & 6 & 0 \\ 4 & -2 & -1 & -1 \\ -2 & 13 & -12 & 18 \end{bmatrix}, \quad \mathbf{b} = \begin{Bmatrix} 0 \\ 7 \\ -11 \\ 65 \end{Bmatrix}$

Cholesky Factorization

In Problems 45 through 50, find the Cholesky factorization of each matrix by

a. Direct calculation of $\mathbf{L}$ and $\mathbf{L}^T$

b. Executing `CholeskyFactor` (Section 4.4)

45. $\mathbf{A} = \begin{bmatrix} 1 & 1 & -2 \\ 1 & 10 & 4 \\ -2 & 4 & 24 \end{bmatrix}$

46. $\mathbf{A} = \begin{bmatrix} 9 & -6 & 3 \\ -6 & 13 & 1 \\ 3 & 1 & 6 \end{bmatrix}$

47. $\mathbf{A} = \begin{bmatrix} 4 & 2 & -6 \\ 2 & 17 & 5 \\ -6 & 5 & 17 \end{bmatrix}$

48. $\mathbf{A} = \begin{bmatrix} 1 & -2 & -3 \\ -2 & 5 & 7 \\ -3 & 7 & 26 \end{bmatrix}$

49. $\mathbf{A} = \begin{bmatrix} 4 & 2 & 6 & -4 \\ 2 & 2 & 2 & -6 \\ 6 & 2 & 11 & -3 \\ -4 & -6 & -3 & 25 \end{bmatrix}$

50. $\mathbf{A} = \begin{bmatrix} 9 & 6 & 3 & 6 \\ 6 & 5 & 6 & 7 \\ 3 & 6 & 21 & 18 \\ 6 & 7 & 18 & 18 \end{bmatrix}$

Cholesky's Method

In Problems 51 through 56

a. ✍ Using Cholesky's method solve each linear system $\mathbf{Ax} = \mathbf{b}$.

b. Confirm the results by executing the user-defined function `CholeskyMethod` (Section 4.4).

51. $\mathbf{A} = \begin{bmatrix} 1 & 1 & -2 \\ 1 & 10 & 4 \\ -2 & 4 & 24 \end{bmatrix}, \quad \mathbf{b} = \begin{Bmatrix} -3 \\ 33 \\ 78 \end{Bmatrix}$

52. $\mathbf{A} = \begin{bmatrix} 9 & -6 & 3 \\ -6 & 13 & 1 \\ 3 & 1 & 6 \end{bmatrix}, \quad \mathbf{b} = \begin{Bmatrix} -30 \\ 53 \\ 9 \end{Bmatrix}$

53. $\mathbf{A} = \begin{bmatrix} 4 & 2 & -6 \\ 2 & 17 & 5 \\ -6 & 5 & 17 \end{bmatrix}, \quad \mathbf{b} = \begin{Bmatrix} -14 \\ 17 \\ 45 \end{Bmatrix}$

54. $\mathbf{A} = \begin{bmatrix} 1 & -2 & -3 \\ -2 & 5 & 7 \\ -3 & 7 & 26 \end{bmatrix}, \quad \mathbf{b} = \begin{Bmatrix} -6 \\ 13 \\ 83 \end{Bmatrix}$

55. $\mathbf{A} = \begin{bmatrix} 4 & 2 & 6 & -4 \\ 2 & 2 & 2 & -6 \\ 6 & 2 & 11 & -3 \\ -4 & -6 & -3 & 25 \end{bmatrix}, \quad \mathbf{b} = \begin{Bmatrix} -16 \\ -26 \\ -12 \\ 114 \end{Bmatrix}$

56. $\mathbf{A} = \begin{bmatrix} 9 & 6 & 3 & 6 \\ 6 & 5 & 6 & 7 \\ 3 & 6 & 21 & 18 \\ 6 & 7 & 18 & 18 \end{bmatrix}, \quad \mathbf{b} = \begin{Bmatrix} 3 \\ 1 \\ -3 \\ 1 \end{Bmatrix}$

Crout Factorization

57. Crout LU factorization requires the diagonal entries of **U** be 1's, while **L** is a general lower-triangular matrix. Perform direct calculation of the entries of **L** and **U** for the case of a 3×3 matrix, similar to that in Example 4.6. Based on the findings, write a user-defined function with function call `[L U] = Crout_Factor(A)` that returns the desired lower- and upper-triangular matrices for any $n \times n$ matrix. Apply `Crout_Factor` to

$$\mathbf{A} = \begin{bmatrix} 2 & 2 & 6 \\ 1 & 3 & -1 \\ -3 & -2 & -7 \end{bmatrix}$$

Crout's Method

58. Crout's method uses Crout factorization (see Problem 57) of the coefficient matrix **A** of the linear system $\mathbf{Ax} = \mathbf{b}$ and generates two triangular systems, which can be solved by back and forward substitution. Write a user-defined function with function call `x = Crout_Method(A,b)`. Apply `Crout_Method` to

$$\begin{bmatrix} 2 & 2 & 6 \\ 1 & 3 & -1 \\ -3 & -2 & -7 \end{bmatrix} \mathbf{x} = \begin{Bmatrix} -6 \\ 5 \\ 5 \end{Bmatrix}$$

Iterative Solution of Linear Systems (Section 4.5)

Vector/Matrix Norms

In Problems 59 through 70

a. Calculate the three norms of each vector or matrix.

b. Verify the results by using the MATLAB built-in function `norm`.

59. $\mathbf{v} = \dfrac{1}{\sqrt{2}} \begin{Bmatrix} 1 \\ 1 \end{Bmatrix}$

60. $\mathbf{v} = \begin{Bmatrix} -1 \\ 1.25 \\ 0.95 \end{Bmatrix}$

61. $\mathbf{v} = \begin{Bmatrix} \frac{1}{3} \\ \frac{2}{3} \\ \frac{1}{4} \end{Bmatrix}$

62. $\mathbf{v} = \sqrt{3}\begin{Bmatrix} 2 \\ -3 \\ 1 \end{Bmatrix}$

63. $\mathbf{v} = \begin{Bmatrix} -1 \\ 0 \\ 3 \\ -4 \end{Bmatrix}$

64. $\mathbf{v} = \begin{Bmatrix} 1 \\ -1 \\ -2 \\ 5 \end{Bmatrix}$

65. $\mathbf{A} = \begin{bmatrix} \dfrac{1}{\sqrt{2}} & 1 & 0 \\ -1 & \dfrac{1}{\sqrt{2}} & 1 \\ 0 & -2 & \dfrac{1}{\sqrt{2}} \end{bmatrix}$

66. $\mathbf{A} = \begin{bmatrix} -4 & 3 & 1 \\ 2 & 1 & -5 \\ 0 & -1 & 6 \end{bmatrix}$

67. $\mathbf{A} = \begin{bmatrix} 10 & 0.3 & -0.7 \\ 0.2 & 7 & 1.2 \\ 0.9 & -1.1 & 5 \end{bmatrix}$

68. $\mathbf{A} = \begin{bmatrix} -2 & 1.4 & -0.8 \\ 1.4 & 3 & 1.5 \\ -0.8 & 1.5 & -4 \end{bmatrix}$

69. $\mathbf{A} = \begin{bmatrix} -\frac{1}{5} & \frac{1}{2} & \frac{1}{3} & 0 \\ \frac{2}{3} & 1 & -\frac{1}{5} & \frac{1}{3} \\ \frac{1}{3} & -\frac{1}{5} & 1 & \frac{2}{3} \\ 0 & \frac{2}{5} & \frac{1}{3} & -\frac{1}{5} \end{bmatrix}$

70. $\mathbf{A} = \begin{bmatrix} 2 & -\frac{1}{2} & \frac{1}{3} & 0 \\ \frac{1}{2} & 1 & -\frac{1}{5} & \frac{1}{3} \\ -\frac{1}{3} & \frac{1}{5} & 3 & -\frac{2}{3} \\ 0 & -\frac{1}{3} & \frac{2}{3} & 4 \end{bmatrix}$

71. Write a user-defined function with function call `[x,k,Mnorm] = GenIter_1(A,b,x0,tol,kmax)` to solve $\mathbf{Ax} = \mathbf{b}$ using the general iterative method as follows: The coefficient matrix is split as $\mathbf{A} = \mathbf{Q} - \mathbf{P}$, where $\mathbf{Q}$ has the same diagonal and upper diagonal (one level higher than the diagonal) entries as $\mathbf{A}$ with all other entries zero. The input/output arguments, as well as the terminating condition are as in functions `Jacobi` and `GaussSeidel` (Section 4.5) with the same default values. The output `Mnorm` is the infinite norm of the corresponding iteration matrix. Apply `GenIter_1` to the linear system

$$\begin{bmatrix} 7 & 1 & -2 & 0 & 1 \\ -1 & 6 & 1 & -2 & 1 \\ 0 & 2 & 8 & 3 & -2 \\ 2 & -1 & 4 & 10 & 2 \\ 1 & 3 & -1 & 5 & 12 \end{bmatrix} \mathbf{x} = \begin{Bmatrix} 18 \\ 0 \\ -2 \\ 21 \\ 28 \end{Bmatrix}, \quad \mathbf{x}^{(0)} = \begin{Bmatrix} 0 \\ 0 \\ 0 \\ 1 \\ 0 \end{Bmatrix}, \quad \varepsilon = 10^{-6}$$

72. Write a user-defined function with function call `[x,k,Mnorm] = GenIter_2(A,b,x0,tol,kmax)` to solve $\mathbf{Ax} = \mathbf{b}$ using the general iterative method as follows: The coefficient matrix must be split as $\mathbf{A} = \mathbf{Q} - \mathbf{P}$, where $\mathbf{Q}$ has the same diagonal, upper diagonal (one level higher than the diagonal), and lower diagonal (one level lower than the diagonal) entries as $\mathbf{A}$ with all other entries zero. The input/output arguments, as well as the terminating condition are as in functions `Jacobi` and `GaussSeidel` (Section 4.5) with the same default values. The output `Mnorm` is the infinite norm of the corresponding iteration matrix. Apply `GenIter_2` to the linear system

$$\begin{bmatrix} -8 & 0 & 1 & -1 & 3 \\ 0 & 6 & 0 & 1 & 4 \\ 2 & -1 & -5 & 0 & -1 \\ 3 & 2 & 1 & -7 & 0 \\ -1 & 3 & 4 & -1 & 11 \end{bmatrix} \mathbf{x} = \begin{Bmatrix} 4 \\ 16 \\ -9 \\ -10 \\ 1 \end{Bmatrix}, \quad \mathbf{x}^{(0)} = \begin{Bmatrix} 0 \\ 0 \\ 0 \\ 1 \\ 1 \end{Bmatrix}, \quad \varepsilon = 10^{-6}$$

Jacobi Iteration Method

In Problems 73 through 76

a. ✍ For each linear system, find the components of the first vector generated by the Jacobi method.

b. ▲ Find the solution vector by executing the user-defined function `Jacobi` with default values for `tol` and `kmax`.

73. $\begin{bmatrix} 3 & 0 & 1 \\ -1 & 2 & 0 \\ 2 & -1 & 4 \end{bmatrix}\mathbf{x} = \begin{Bmatrix} 4 \\ -5 \\ 8 \end{Bmatrix}, \quad \mathbf{x}^{(0)} = \mathbf{0}_{3\times 1}$

74. $\begin{bmatrix} 1.9 & -0.7 & 0.9 \\ 0.6 & 2.3 & 1.2 \\ -0.8 & 1.3 & 3.2 \end{bmatrix}\mathbf{x} = \begin{Bmatrix} 1.5 \\ 0.7 \\ 9.1 \end{Bmatrix}, \quad \mathbf{x}^{(0)} = \begin{Bmatrix} 0 \\ 0 \\ 1 \end{Bmatrix}$

75. $\begin{bmatrix} 3 & 0 & 1 & -1 \\ 0 & -4 & 2 & 1 \\ 1 & -2 & 5 & 0 \\ -1 & 3 & 2 & 6 \end{bmatrix}\mathbf{x} = \begin{Bmatrix} 5 \\ -3 \\ -4 \\ 16 \end{Bmatrix}, \quad \mathbf{x}^{(0)} = \begin{Bmatrix} 1 \\ 0 \\ 0 \\ 1 \end{Bmatrix}$

76. $\begin{bmatrix} 6 & 2 & 1 & -2 \\ 2 & 5 & -1 & 0 \\ -1 & 3 & 7 & 1 \\ -2 & 1 & 4 & -8 \end{bmatrix}\mathbf{x} = \begin{Bmatrix} 5 \\ 7 \\ 28 \\ 6 \end{Bmatrix}, \quad \mathbf{x}^{(0)} = \begin{Bmatrix} 1 \\ 1 \\ 0 \\ 0 \end{Bmatrix}$

77. ✍ Calculate the components of the first two vectors generated by the Jacobi method when applied to

$$\begin{bmatrix} -3 & 1 & 2 \\ 2 & 4 & -1 \\ 1 & -2 & 4 \end{bmatrix}\mathbf{x} = \begin{Bmatrix} 24 \\ -5 \\ 12 \end{Bmatrix}, \quad \mathbf{x}^{(0)} = \begin{Bmatrix} 1 \\ 1 \\ 1 \end{Bmatrix}$$

78. ✍ Calculate the components of the first two vectors generated by the Jacobi method when applied to

$$\begin{bmatrix} -5 & 4 & 0 \\ 2 & 6 & -3 \\ -1 & 2 & 3 \end{bmatrix}\mathbf{x} = \begin{Bmatrix} 18 \\ 11 \\ 3 \end{Bmatrix}, \quad \mathbf{x}^{(0)} = \begin{Bmatrix} 1 \\ 0 \\ 1 \end{Bmatrix}$$

Gauss–Seidel Iteration Method

In Problems 79 through 82

a. ✍ For each linear system find the components of the first vector generated by Gauss–Seidel method.

b. ▲ Find the solution vector by executing the user-defined function `GaussSeidel` with `tol` and `kmax` set to default values. Discuss convergence!

79. $\begin{bmatrix} 4 & 2 & -6 \\ 2 & 17 & 5 \\ -6 & 5 & 17 \end{bmatrix}\mathbf{x} = \begin{Bmatrix} 2 \\ 25 \\ 5 \end{Bmatrix}, \quad \mathbf{x}^{(0)} = \begin{Bmatrix} 1 \\ 1 \\ 0 \end{Bmatrix}$

80. $\begin{bmatrix} 1 & 1 & -2 \\ 1 & 10 & 4 \\ -2 & 4 & 24 \end{bmatrix}\mathbf{x} = \begin{Bmatrix} -6 \\ 15 \\ 58 \end{Bmatrix}, \quad \mathbf{x}^{(0)} = \begin{Bmatrix} 0 \\ 1 \\ 1 \end{Bmatrix}$

81. $\begin{bmatrix} 6 & 3 & -2 & 0 \\ 3 & 7 & 1 & -2 \\ -2 & 1 & 8 & 3 \\ 0 & -2 & 3 & 9 \end{bmatrix}\mathbf{x} = \begin{Bmatrix} 14 \\ 3 \\ -9 \\ 6 \end{Bmatrix}, \quad \mathbf{x}^{(0)} = \begin{Bmatrix} 1 \\ 0 \\ 1 \\ 0 \end{Bmatrix}$

82. $\begin{bmatrix} 5 & 7 & 6 & 5 \\ 7 & 10 & 8 & 7 \\ 6 & 8 & 10 & 9 \\ 5 & 7 & 9 & 10 \end{bmatrix}\mathbf{x} = \begin{Bmatrix} 23 \\ 32 \\ 33 \\ 31 \end{Bmatrix}, \quad \mathbf{x}^{(0)} = \begin{Bmatrix} 0 \\ 0 \\ 1 \\ 1 \end{Bmatrix}$

83. ✍ Calculate the components of the first two vectors generated by Gauss–Seidel method when applied to

$$\begin{bmatrix} 1 & 1 & -3 \\ 1 & 10 & 4 \\ -3 & 4 & 24 \end{bmatrix}\mathbf{x} = \begin{Bmatrix} -5 \\ 0 \\ 38 \end{Bmatrix}, \quad \mathbf{x}^{(0)} = \begin{Bmatrix} 1 \\ 0 \\ 1 \end{Bmatrix}$$

84. ✍ Calculate the components of the first two vectors generated by Gauss–Seidel method when applied to

$$\begin{bmatrix} 4 & 2 & 6 \\ 2 & 17 & 5 \\ 6 & 5 & 17 \end{bmatrix}\mathbf{x} = \begin{Bmatrix} -10 \\ 21 \\ -35 \end{Bmatrix}, \quad \mathbf{x}^{(0)} = \begin{Bmatrix} 0 \\ 1 \\ 0 \end{Bmatrix}$$

85. Solve the linear system below by executing user-defined functions `Jacobi` and `GaussSeidel` with the initial vector and tolerance as indicated, and default `kmax`. Compare the results and discuss convergence.

$$\begin{bmatrix} -8 & 2 & 0 & 1 & 3 \\ 1 & 6 & -2 & 0 & 1 \\ 2 & -2 & 7 & 1 & -1 \\ 0 & 4 & -1 & 9 & 2 \\ -3 & 0 & 1 & 3 & 10 \end{bmatrix}\mathbf{x} = \begin{Bmatrix} -2 \\ 18 \\ -18 \\ 3 \\ 2 \end{Bmatrix}, \quad \mathbf{x}^{(0)} = \begin{Bmatrix} 1 \\ 0 \\ 0 \\ 1 \\ 0 \end{Bmatrix}, \quad \varepsilon = 10^{-6}$$

86. Solve the linear system below by executing user-defined functions `Jacobi` and `GaussSeidel` with the initial vector and tolerance as indicated, and default `kmax`. Compare the results and discuss convergence.

$$\begin{bmatrix} 8 & 1 & 0 & 1 & 3 \\ 1 & 6 & -1 & 0 & 1 \\ 0 & -1 & 10 & -1 & 2 \\ 1 & 0 & -1 & 9 & 1 \\ 3 & 1 & 2 & 1 & 10 \end{bmatrix} \mathbf{x} = \begin{Bmatrix} 41 \\ -6 \\ 16 \\ 24 \\ 44 \end{Bmatrix}, \quad \mathbf{x}^{(0)} = \begin{Bmatrix} 1 \\ 0 \\ 0 \\ 1 \\ 0 \end{Bmatrix}, \quad \varepsilon = 10^{-6}$$

Ill-Conditioning and Error Analysis (Section 4.6)

Condition Number

In Problems 87 through 92

a. Calculate the condition number of each matrix using all three matrix norms.

b. Verify the results using the MATLAB built-in function `cond`:

87. $\mathbf{A} = \begin{bmatrix} 1 & 0.3 \\ 0.3 & 0.1 \end{bmatrix}$

88. $\mathbf{A} = \begin{bmatrix} 3 & 1 & 0 \\ 1 & -1 & 2 \\ 1 & 1 & 1 \end{bmatrix}$

89. $\mathbf{A} = \begin{bmatrix} 5 & 3 & 0 \\ 3 & 2 & 0 \\ 0 & 0 & 1 \end{bmatrix}$

90. $\mathbf{A} = \begin{bmatrix} 2 & 7 & 4 \\ 2 & 1 & 2 \\ 5 & -1 & 2 \end{bmatrix}$

91. $\mathbf{A} = \begin{bmatrix} 1 & -2 & 1 & 3 \\ -1 & 1 & 1 & -6 \\ 1 & -3 & 4 & 4 \\ 2 & -4 & 2 & 5 \end{bmatrix}$

92. $\mathbf{A} = \begin{bmatrix} 1 & \frac{1}{2} & \frac{1}{3} & \frac{1}{4} \\ \frac{1}{2} & \frac{1}{3} & \frac{1}{4} & \frac{1}{5} \\ \frac{1}{3} & \frac{1}{4} & \frac{1}{5} & \frac{1}{6} \\ \frac{1}{4} & \frac{1}{5} & \frac{1}{6} & \frac{1}{7} \end{bmatrix}_{4\times 4 \text{ Hilbert matrix}}$

Ill-Conditioning

In Problems 93 through 96, a linear system $\mathbf{Ax} = \mathbf{b}$, its actual solution $\mathbf{x}_a$, and a poor approximation $\hat{\mathbf{x}}$ of the solution are given. Perform all of the following to inspect the ill-conditioning or well-conditioning of the system:

a. ✍ Perturb the second component of **b** by a small $\varepsilon > 0$ and find the solution of the ensuing system.

b. ▲ Find the condition number of **A** using the 1-norm.

c. ✍ Calculate the 1-norm of the residual vector corresponding to the poor approximation $\hat{\mathbf{x}}$.

93. $\begin{bmatrix} 2 & 2 \\ 1.0002 & 0.9998 \end{bmatrix} \mathbf{x} = \begin{Bmatrix} 4 \\ 2.0012 \end{Bmatrix}, \quad \mathbf{x}_a = \begin{Bmatrix} 4 \\ -2 \end{Bmatrix}, \quad \hat{\mathbf{x}} = \begin{Bmatrix} 2 \\ 0 \end{Bmatrix}$

94. $\begin{bmatrix} 1 & 2 \\ 2 & 4.0001 \end{bmatrix} \mathbf{x} = \begin{Bmatrix} -1 \\ -2.0002 \end{Bmatrix}, \quad \mathbf{x}_a = \begin{Bmatrix} 3 \\ -2 \end{Bmatrix}, \quad \hat{\mathbf{x}} = \begin{Bmatrix} -1 \\ 0 \end{Bmatrix}$

95. $\begin{bmatrix} 5 & 9 \\ 6 & 11 \end{bmatrix} \mathbf{x} = \begin{Bmatrix} -1 \\ -1 \end{Bmatrix}, \quad \mathbf{x}_a = \begin{Bmatrix} -2 \\ 1 \end{Bmatrix}, \quad \hat{\mathbf{x}} = \begin{Bmatrix} 7.2 \\ -4.1 \end{Bmatrix}$

96. $\begin{bmatrix} 13 & 14 & 14 \\ 11 & 12 & 13 \\ 12 & 13 & 14 \end{bmatrix} \mathbf{x} = \begin{Bmatrix} 55 \\ 48 \\ 52 \end{Bmatrix}, \quad \mathbf{x}_a = \begin{Bmatrix} 1 \\ 2 \\ 1 \end{Bmatrix}, \quad \hat{\mathbf{x}} = \begin{Bmatrix} -0.51 \\ 3.61 \\ 0.79 \end{Bmatrix}$

Percent Change

97. ✍ Consider the system in Example 4.19 and suppose the second component of vector **b** is increased by 0.0001, while the coefficient matrix is unchanged from its original form. Find the upper bound for, and the actual value of, the percent change in solution. Use the vector and matrix 1-norm.

98. ✍ Consider

$$\mathbf{Ax} = \mathbf{b}, \quad \mathbf{A} = \begin{bmatrix} 5 & 9 \\ 6 & 11 \end{bmatrix}, \quad \mathbf{b} = \begin{Bmatrix} -1 \\ -1 \end{Bmatrix}$$

a. Increase each of the second-column entries of **A** by 0.01, solve the ensuing system, and calculate the actual percent change in solution. Also find an upper bound for the percent change. Use matrix and vector 1-norm.

b. In the original system, increase each of the components of **b** by 0.01, and repeat (a).

Systems of Nonlinear Equations (Section 4.7)

99. Consider

$$\begin{cases} (x-1)^3 y + 1 = 0 \\ xy^2 - \sin x = 0 \end{cases}$$

First locate the roots graphically. Then find an approximate value for one of the roots using Newton's method with initial estimate (4, 0), a terminating condition with $\varepsilon = 10^{-4}$, and allow a maximum of 20 iterations.

100. Consider the nonlinear system

$$\begin{cases} y = x^2 - 4 \\ x^2 + y^2 = 10 \end{cases}$$

a. Solve using Newton's method with initial estimate (2, 0), tolerance $\varepsilon = 10^{-3}$, and a maximum of 10 iterations.

b. Repeat (a) but with an initial estimate (2, –2).

c. Use a graphical approach to validate the results in (a) and (b).

101. Consider the nonlinear system

$$\begin{cases} 2e^x + y = 0 \\ 3x^2 + 4y^2 = 8 \end{cases}$$

a. Solve using Newton's method with initial estimate (–1, –2), tolerance $\varepsilon = 10^{-4}$, and a maximum of 10 iterations.

b. Repeat (a) but with an initial estimate (–2, 0).

c. Use a graphical approach to validate the results in (a) and (b).

102. Consider the nonlinear system

$$\begin{cases} x^2 + y^2 = 1.2 \\ 2x^2 + 3y^2 = 3 \end{cases}$$

a. Solve using Newton's method with initial estimate (0.25,1), tolerance $\varepsilon = 10^{-4}$, and a maximum of 10 iterations.

b. Graph the two nonlinear equations to roughly locate their points of intersection. Then using Newton's method with an appropriate initial estimate, approximate the root directly below the one obtained in (a).

103. Solve the following system of three nonlinear equations in three unknowns:

$$\begin{cases} x^2 + y^2 = 2z \\ x^2 + z^2 = \dfrac{1}{3} \\ x^2 + y^2 + z^2 = 1 \end{cases}$$

using Newton's method with initial estimate (1,1,0.1), tolerance $\varepsilon = 10^{-4}$, and a maximum of 10 iterations.

104. A planar, two-link robot arm is shown in Figure 4.8. The coordinate system xy is the tool frame and is attached to the end-effector. The coordinates of the end-effector relative to the base frame are expressed as

$$\begin{cases} x = L_1 \cos\theta_1 + L_2 \cos(\theta_1 + \theta_2) \\ y = L_1 \sin\theta_1 + L_2 \sin(\theta_1 + \theta_2) \end{cases}$$

Suppose the lengths, in consistent physical units, of the two links are $L_1 = 1$ and $L_2 = 2$, and that $x = 2.5$, $y = 1.4$. Find the joint angles θ_1 and θ_2 (in radians) using Newton's method with an initial estimate of (0.8, 0.9), tolerance $\varepsilon = 10^{-4}$, and maximum number of iterations set to 10.

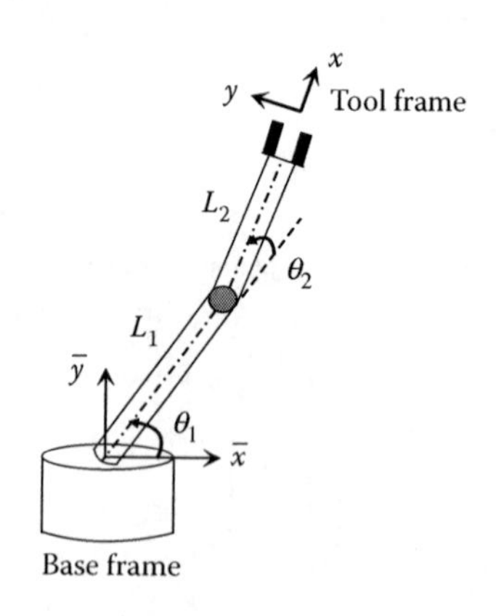

FIGURE 4.8
A two-link arm in plane motion.

105. Solve (with tolerance $\varepsilon = 10^{-4}$) the following nonlinear system using Newton's method:

$$\begin{cases} 2x^2 + 2\sqrt{2}xy + y^2 = 14 \\ x^2 - \sqrt{2}xy + 3y^2 = 9 \end{cases}$$

106. Solve the following nonlinear system using Newton's method, with initial estimate (0, 1, 1) and tolerance $\varepsilon = 10^{-3}$.

$$\begin{cases} xy - \cos x + z^2 = 3.6 \\ x^2 - 2y^2 + z = 2.8 \\ 3x + y \sin z = 2.8 \end{cases}$$

107. Consider

$$\begin{cases} \sin(\alpha + \beta) - \cos\beta = 0.17 \\ \cos(\alpha - \beta) + \sin\alpha = 1.8 \end{cases}$$

a. Locate the roots graphically for the ranges $0 \le \alpha \le 2$ and $-6 \le \beta \le -4$.

b. Using Newton's method with $\varepsilon = 10^{-4}$ and initial estimate (1.4, −5.4), find one root. Find the second root using initial estimate (1.2, −4.4).

108. Consider

$$\begin{cases} x^2 + y - 2x = 0.4 \\ 2y + 3xy - 2x^3 = 0 \end{cases}$$

a. Locate the roots graphically.

b. Using Newton's method with initial estimate (1,1), tolerance $\varepsilon = 10^{-4}$, and a maximum of 20 iterations, find one of the roots.

c. Find the second root using initial estimate (−0.5, −0.5).

d. Find the third root with initial estimate (−0.25, 0).

109. Consider the nonlinear system in Problem 99:

$$\begin{cases} xy^2 - \sin x = 0 \\ (x-1)^3 y + 1 = 0 \end{cases}$$

With auxiliary functions

$$g_1(x, y) = \frac{\sin x}{y^2}, \quad g_2(x, y) = \frac{-1}{(x-1)^3}$$

use the fixed-point iteration method with initial estimate (2, −2), tolerance $\varepsilon = 10^{-3}$, and maximum 20 iterations to estimate one solution of the system.

110. Reconsider the system in Problem 100:

$$\begin{cases} y = x^2 - 4 \\ x^2 + y^2 = 10 \end{cases}$$

a. Locate the roots graphically.

b. Find the root in the first quadrant by using the fixed-point iteration with auxiliary functions

$$g_1(x,y) = \sqrt{y+4}, \quad g_2(x,y) = \sqrt{10 - x^2}$$

with (2, 0) as initial estimate, $\varepsilon = 10^{-3}$, and maximum number of iterations set to 20.

c. There are three other roots, one in each quadrant. Find these roots by using combinations of the same g_1 and g_2 with positive and negative square roots.

111. Recall the nonlinear system in Problem 101:

$$\begin{cases} 2e^x + y = 0 \\ 3x^2 + 4y^2 = 8 \end{cases}$$

a. Locate the roots graphically.

b. Based on the location of the roots, select suitable auxiliary functions g_1 and g_2 and apply the fixed-point iteration method with (–1, –2) as the initial estimate, $\varepsilon = 10^{-4}$, and number of iterations not to exceed 20, to find one of the two roots.

c. To find the other root, write the original equations in reverse order, suitably select g_1 and g_2, and apply fixed-point iteration with all information as in (b).

112. Consider the nonlinear system in Problem 102.

a. Locate the roots using a graphical approach.

b. Select auxiliary functions

$$g_1(x,y) = \sqrt{1.2 - y^2}, \quad g_2(x,y) = \sqrt{\frac{3 - 2x^2}{3}}$$

and apply the fixed-point iteration method with initial estimate (0.25, 1) and tolerance $\varepsilon = 10^{-4}$ to find a root. Decide the maximum number of iterations to be performed.

c. There are three other roots, one in each of the remaining quadrants. Find these roots by using combinations of the same g_1 and g_2 with positive and negative square roots.

5

Curve Fitting (Approximation) and Interpolation

A set of data may come from various sources. In many scientific and engineering applications, the data comes from conducting experiments where physical quantities are measured; for instance, measuring the displacement of a coiled spring when subjected to tensile or compressive force. Other times, the data may be generated by other numerical methods; for instance, numerical solution of differential equations (Chapters 7, 8, and 10).

An available set of data can be used for different purposes. In some situations, the data is represented by a function, which in turn can be used for numerical differentiation or integration. Such function is obtained through curve fitting, or approximation, of the data. Curve fitting is a procedure where a function is used to fit a given set of data in the "best" possible way without having to match the data exactly. As a result, while the function does not necessarily yield the exact value at any of the data points, overall it fits the set of data well. Several types of functions and polynomials of different orders can be used for curve fitting purposes. Curve fitting is normally used when the data has substantial inherent error, such as data gathered from experimental measurements.

In many other cases, it may be desired to find estimates of values at intermediate points, that is, at the points between the given data points. This is done through interpolation, a procedure that first determines a polynomial that agrees exactly with the data points, and then uses the polynomial to find estimates of values at intermediate points. For a small set of data, interpolation over the entire data may be adequately accomplished using a single polynomial. For large sets of data, however, different polynomials are used in different intervals of the whole data. This is referred to as spline interpolation.

5.1 Least-Squares Regression

As mentioned above, a single polynomial may be sufficient for the interpolation of a small set of data. However, when the data has substantial error, even if the size of data is small, this may no longer be appropriate. Consider Figure 5.1, which shows a set of seven data points collected from

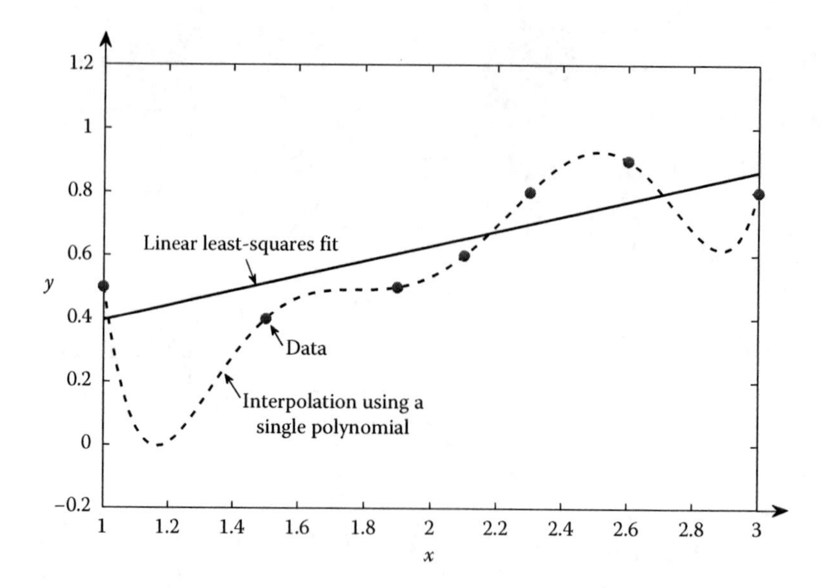

FIGURE 5.1
Interpolation by a single polynomial, and least-squares fit of a set of data.

an experiment. The nature of the data suggests that for the most part, the y values increase with the x values. A single interpolating polynomial goes through all of the data points, but displays large oscillations in some of the intervals. As a result, the interpolated values near $x = 1.2$ and $x = 2.9$ will be well outside of the range of the original data.

In these types of situations, it makes more sense to find a function that does not necessarily go through all of the data points, but fits the data well overall. One option, for example, is to fit the "best" straight line into the data. This line is not random and can be generated systematically via least-squares regression.

5.2 Linear Regression

The simplest case of a least-squares regression involves finding a straight line (linear function) in the form

$$y = a_1 x + a_0 \tag{5.1}$$

that best fits a set of n data points $(x_1, y_1), \ldots, (x_n, y_n)$. Of course, the data first needs to be plotted to see whether the independent and dependent variables have a somewhat linear relationship. If this is the case, then the coefficients a_1 and a_0 are determined such that the error associated with the line

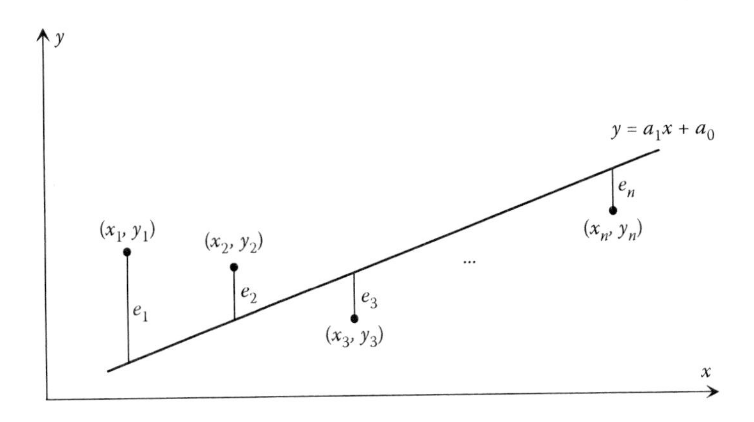

FIGURE 5.2
A linear fit of data, and individual errors.

is minimized. As shown in Figure 5.2, at each data point (x_i, y_i) the error e_i is defined as the difference between the true value y_i and the approximate value $a_1x_i + a_0$

$$e_i = y_i - (a_1x_i + a_0) \tag{5.2}$$

These individual errors will be used to calculate a total error associated with the line $y = a_1x + a_0$.

5.2.1 Deciding a "Best" Fit Criterion

Different strategies can be considered for determining the best linear fit of a set of n data points $(x_1, y_1), \ldots, (x_n, y_n)$. One such strategy is to minimize the sum of all the individual errors

$$E = \sum_{i=1}^{n} e_i = \sum_{i=1}^{n} [y_i - (a_1x_i + a_0)] \tag{5.3}$$

This criterion, however, does not offer a good measure of how well the line fits the data because, as shown in Figure 5.3, it allows for positive and negative individual errors—even very large errors—to cancel out and yield a zero sum.

Another strategy is to minimize the sum of the absolute values of the individual errors

$$E = \sum_{i=1}^{n} |e_i| = \sum_{i=1}^{n} |y_i - (a_1x_i + a_0)| \tag{5.4}$$

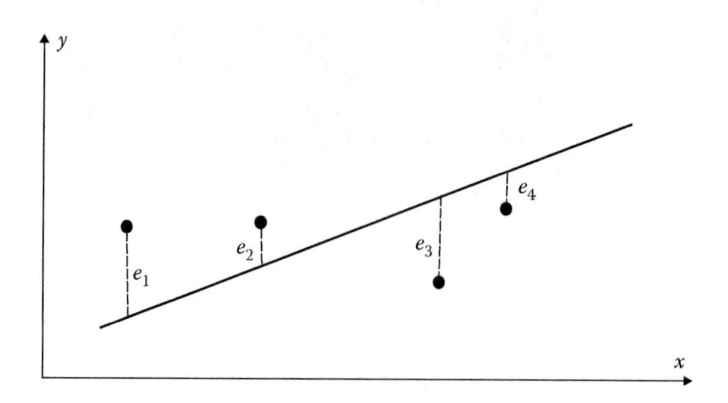

FIGURE 5.3
Zero total error based on the criterion defined by Equation 5.3.

As a result, the individual errors can no longer cancel out and the total error is always positive. This criterion, however, is not able to uniquely determine the coefficients that describe the best line fit because for a given set of data, several lines can have the same total error. Figure 5.4 shows a set of four data points with two line fits that have the same total error.

The third strategy is to minimize the sum of the squares of the individual errors

$$E = \sum_{i=1}^{n} e_i^2 = \sum_{i=1}^{n} [y_i - (a_1 x_i + a_0)]^2 \tag{5.5}$$

This criterion uniquely determines the coefficients that describe the best line fit for a given set of data. As in the second strategy, individual errors cannot cancel each other and the total error is always positive. Also note that small errors get smaller and large errors get larger. This means that larger

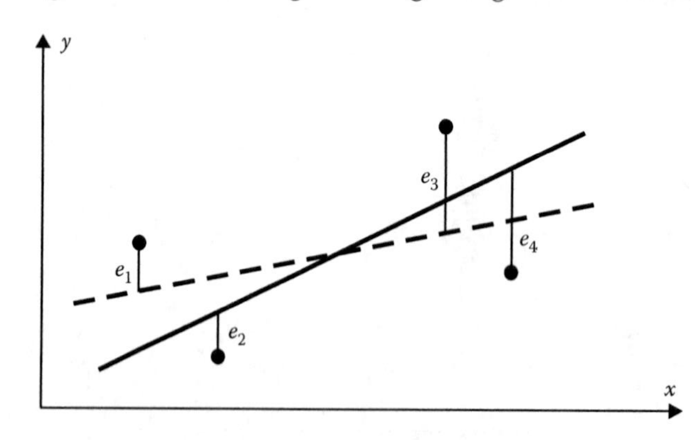

FIGURE 5.4
Two linear fits with the same total error calculated by Equation 5.4.

individual errors have larger contributions to the total error being minimized so that this strategy essentially minimizes the maximum distance that an individual data point is located relative to the line.

5.2.2 Linear Least-Squares Regression

As decided above, the criterion to find the line $y = a_1 x + a_0$ that best fits the data $(x_1, y_1), \ldots, (x_n, y_n)$ is to determine the coefficients a_1 and a_0 that minimize

$$E = \sum_{i=1}^{n} [y_i - (a_1 x_i + a_0)]^2 \tag{5.6}$$

Noting that E is a (nonlinear) function of a_0 and a_1, it attains its minimum where $\partial E/\partial a_0$ and $\partial E/\partial a_1$ vanish, that is,

$$\frac{\partial E}{\partial a_0} = -2\sum_{i=1}^{n} [y_i - (a_1 x_i + a_0)] = 0 \quad \Rightarrow \quad \sum_{i=1}^{n} [y_i - (a_1 x_i + a_0)] = 0$$

$$\frac{\partial E}{\partial a_1} = -2\sum_{i=1}^{n} x_i[y_i - (a_1 x_i + a_0)] = 0 \quad \Rightarrow \quad \sum_{i=1}^{n} \{x_i[y_i - (a_1 x_i + a_0)]\} = 0$$

Expanding and rearranging the above equations yields a system of two linear equations to be solved for a_0 and a_1:

$$n a_0 + \left(\sum_{i=1}^{n} x_i\right) a_1 = \sum_{i=1}^{n} y_i$$

$$\left(\sum_{i=1}^{n} x_i\right) a_0 + \left(\sum_{i=1}^{n} x_i^2\right) a_1 = \sum_{i=1}^{n} x_i y_i$$

By Cramer's rule, the solutions are found as

$$a_1 = \frac{n\left(\sum_{i=1}^{n} x_i y_i\right) - \left(\sum_{i=1}^{n} x_i\right)\left(\sum_{i=1}^{n} y_i\right)}{n\left(\sum_{i=1}^{n} x_i^2\right) - \left(\sum_{i=1}^{n} x_i\right)^2},$$

$$a_0 = \frac{\left(\sum_{i=1}^{n} x_i^2\right)\left(\sum_{i=1}^{n} y_i\right) - \left(\sum_{i=1}^{n} x_i\right)\left(\sum_{i=1}^{n} x_i y_i\right)}{n\left(\sum_{i=1}^{n} x_i^2\right) - \left(\sum_{i=1}^{n} x_i\right)^2} \tag{5.7}$$

The user-defined function `LinearRegression` uses the linear least-squares regression approach to find the straight line that best fits a set of data. The function plots this line, as well as the original data.

```
function [a1 a0] = LinearRegression(x,y)
%
% LinearRegression uses linear least-squares
% approximation to fit a data by a line in the form
% y = a1*x + a0.
%
% [a1 a0] = LinearRegression(x,y) where
%
%   x, y are n-dimensional row or column vectors of data,
%
%   a1 and a0 are the coefficients that describe the
%   linear fit.
%
n = length(x);
Sumx = sum(x); Sumy = sum(y); Sumxx = sum(x.*x);
Sumxy = sum(x.*y);
den = n*Sumxx - Sumx^2;
a1 = (n*Sumxy - Sumx*Sumy)/den;
a0 = (Sumxx*Sumy - Sumxy*Sumx)/den;
% Plot the data and the line fit
l = zeros(n,1);      % Pre-allocate
for i = 1:n,
    l(i) = a1*x(i) + a0;        % Calculate n points on the line
end
plot(x,y,'o')
hold on
plot(x,l)
end
```

Example 5.1: Linear Least-Squares Regression

Consider the data in Table 5.1.

TABLE 5.1

Data in Example 5.1

x_i	y_i
0.2	8.0
0.4	8.4
0.6	8.8
0.8	8.6
1.0	8.5
1.2	8.7

a. Using least-squares regression, find a straight line that best fits the data.
b. Confirm the results by executing the user-defined function `LinearRegression`.

SOLUTION

a. Noting $n = 6$, we first calculate all the essential sums involved in Equation 5.7:

$$\sum_{i=1}^{6} x_i = 0.2 + 0.4 + \cdots + 1.2 = 4.2,$$

$$\sum_{i=1}^{6} y_i = 8.0 + 8.4 + \cdots + 8.7 = 51$$

$$\sum_{i=1}^{6} x_i^2 = (0.2)^2 + (0.4)^2 + \cdots + (1.2)^2 = 3.64$$

$$\sum_{i=1}^{6} x_i y_i = (0.2)(8.0) + (0.4)(8.4) + \cdots + (1.2)(8.7) = 36.06$$

Then, following Equation 5.7, the coefficients are found as

$$a_1 = \frac{(6)(36.06) - (4.2)(51)}{(6)(3.64) - (4.2)^2} = 0.5143,$$

$$a_0 = \frac{(3.64)(51) - (4.2)(36.06)}{(6)(3.64) - (4.2)^2} = 8.1400$$

Therefore, the line that best fits the data is described by

$$y = 0.5143x + 8.1400$$

b. Execution of `LinearRegression` yields the coefficients a_1 and a_0, which describe the best line fit, as well as the plot of the line and the original set of data; Figure 5.5.

```
>> x = 0.2:0.2:1.2;
>> y = [8 8.4 8.8 8.6 8.5 8.7];
>> [a1 a0] = LinearRegression(x,y)

a1 =
     0.5143

a0 =
     8.1400
```

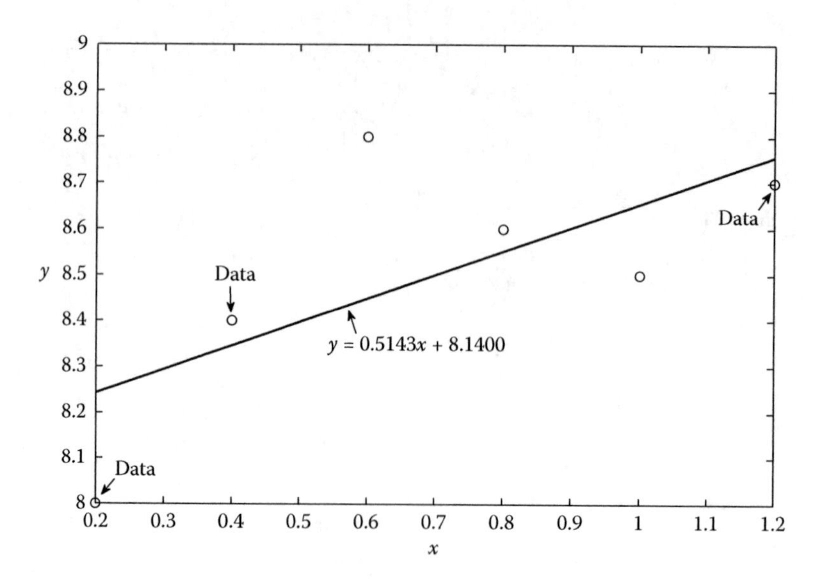

FIGURE 5.5
The data and the best line fit in Example 5.1.

5.3 Linearization of Nonlinear Data

If the relationship between the independent and dependent variables is nonlinear, curve-fitting techniques other than linear regression must be used. One such method is polynomial regression, to be presented in Section 5.4. Others involve conversion of the data into a form that could be handled by linear regression. Three examples of nonlinear functions that are used for curve fitting are as follows.

5.3.1 Exponential Function

One is the exponential function

$$y = ae^{bx} \quad (a,b = \text{constant}) \tag{5.8}$$

Because differentiation of the exponential function returns a constant multiple of the exponential function, this technique applies to situations where the rate of change of a quantity is directly proportional to the quantity itself; for instance, radioactive decay. Conversion into linear form is made by taking the natural logarithm of Equation 5.8 to obtain

$$\ln y = bx + \ln a \tag{5.9}$$

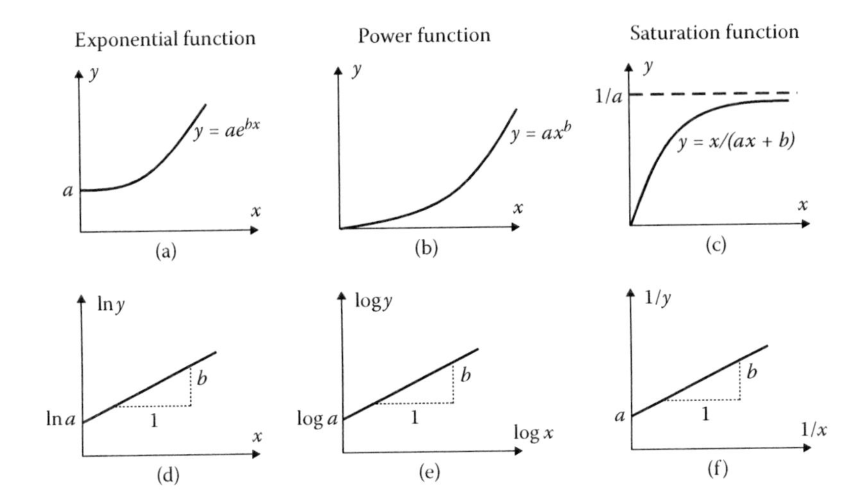

FIGURE 5.6
Linearization of three nonlinear functions for curve fitting: (a,d) exponential, (b,e) power, and (c,f) saturation.

Therefore, the plot of ln y versus x is a straight line with slope b and intercept ln a; see Figure 5.6a and d.

5.3.2 Power Function

Another example of a nonlinear function is the power function

$$y = ax^b \quad (a,b = \text{constant}) \tag{5.10}$$

Linearization is achieved by taking the standard (base 10) logarithm of Equation 5.10

$$\log y = b\log x + \log a \tag{5.11}$$

so that the plot of log y versus log x is a straight line with slope b and intercept log a; see Figure 5.6b and e.

5.3.3 Saturation Function

The saturation function is in the form

$$y = \frac{x}{ax + b} \quad (a,b = \text{constant}) \tag{5.12}$$

Inverting Equation 5.12 yields

$$\frac{1}{y} = b\left(\frac{1}{x}\right) + a \tag{5.13}$$

so that the plot of $1/y$ versus $1/x$ is a straight line with slope b and intercept a; see Figure 5.6c and f.

Example 5.2: Curve Fit: Saturation Function

Consider the data in Table 5.2. Plot of the data reveals that the saturation function is suitable for a curve fit; Figure 5.7.

TABLE 5.2
Data in Example 5.2

x	y
10	1.9
20	3.0
30	3.2
40	3.9
50	3.7
60	4.2
70	4.1
80	4.4
90	4.5
100	4.4

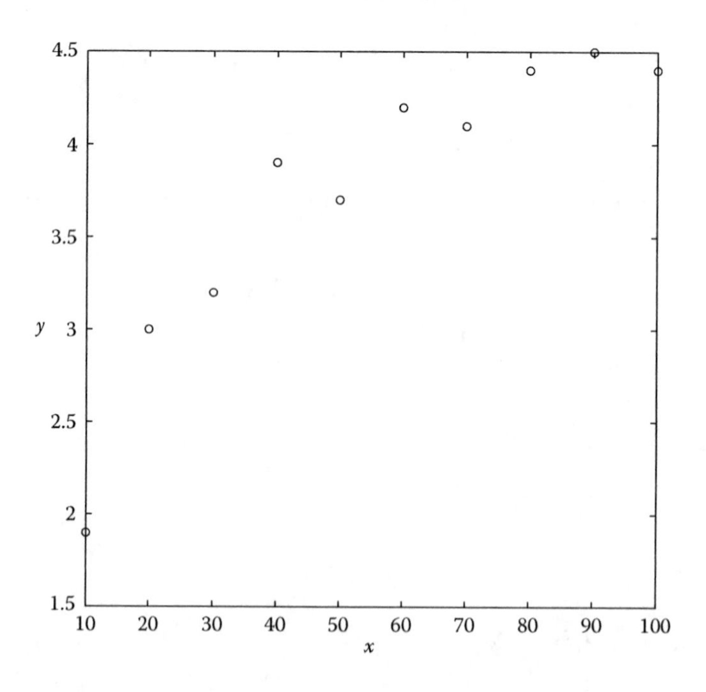

FIGURE 5.7
Plot of the data in Table 5.2.

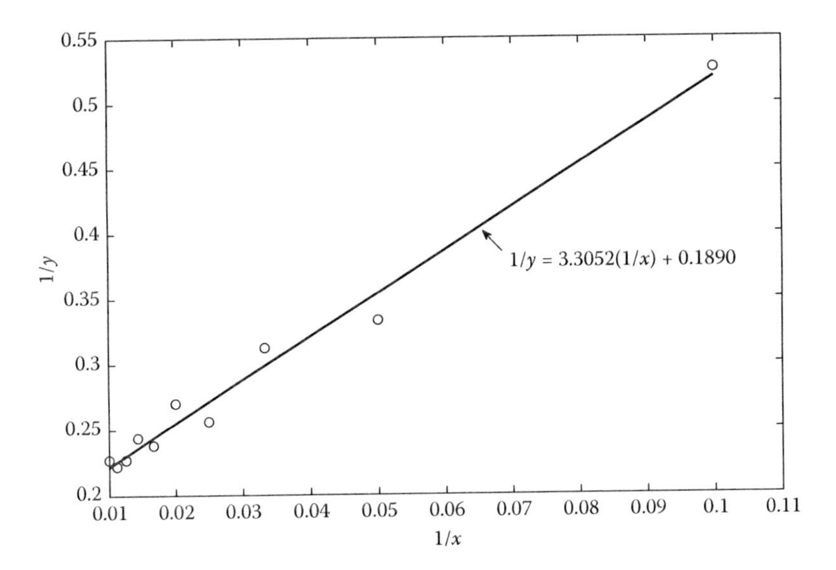

FIGURE 5.8
Linear fit of the converted data in Example 5.2.

```
>> x = 10:10:100; y = [1.9 3.0 3.2 3.9 3.7 4.2 4.1 4.4 4.5 4.4];
>> plot(x,y,'o')        % Figure 5.7
```

Therefore, the plot of $1/y$ versus $1/x$ should be somewhat linear. We will apply linear regression to this converted data to find the slope and the intercept of the line fit. Execution of the user-defined function `LinearRegression` (Section 5.2) provides this information and also plots the result; Figure 5.8.

```
>> xx = 1./x; yy = 1./y;     % Element-by-element reciprocals
>> [a1 a0] = LinearRegression(xx,yy)

a1 =
     3.3052

a0 =
     0.1890
```

Based on the form in Equation 5.13, we have $b = 3.3052$ (slope) and $a = 0.1890$ (intercept). Consequently, the saturation function of interest is

$$y = \frac{x}{ax + b} = \frac{x}{0.1890x + 3.3052}$$

Figure 5.9 shows the plot of this function together with the original data.

Example 5.3: Curve Fit: Power Function

Consider the data in Table 5.3. Plot of the data reveals that the power function is suitable for a curve fit; Figure 5.10.

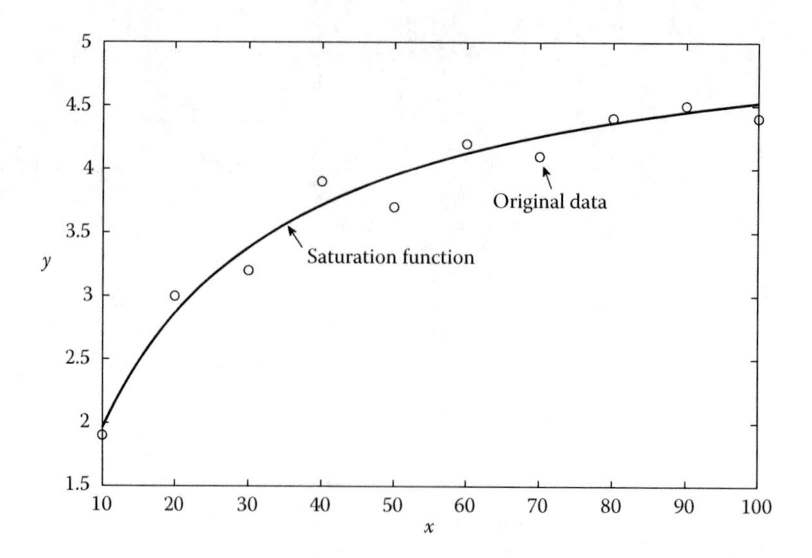

FIGURE 5.9
Curve fit using the saturation function; Example 5.2.

TABLE 5.3
Data in Example 5.3

x	y
0.1	0.02
0.2	0.10
0.3	0.20
0.4	0.35
0.5	0.56
0.6	0.75
0.7	1.04
0.8	1.30
0.9	1.70
1.0	2.09

```
>> x = 0.1:0.1:1;
>> y = [0.02 0.1 0.2 0.35 0.56 0.75 1.04 1.3 1.7 2.09];
>> plot(x,y,'o')      % Figure 5.10
```

If the power function is appropriate, then $\log y$ and $\log x$ should have a linear relationship. Applying linear regression to the converted data, we find the slope and the intercept of the line fit. Execution of the user-defined function `LinearRegression` provides this information as well as the plot of the result; Figure 5.11.

```
>> xx = log10(x); yy = log10(y);
>> [a1 a0]=LinearRegression(xx,yy)
```

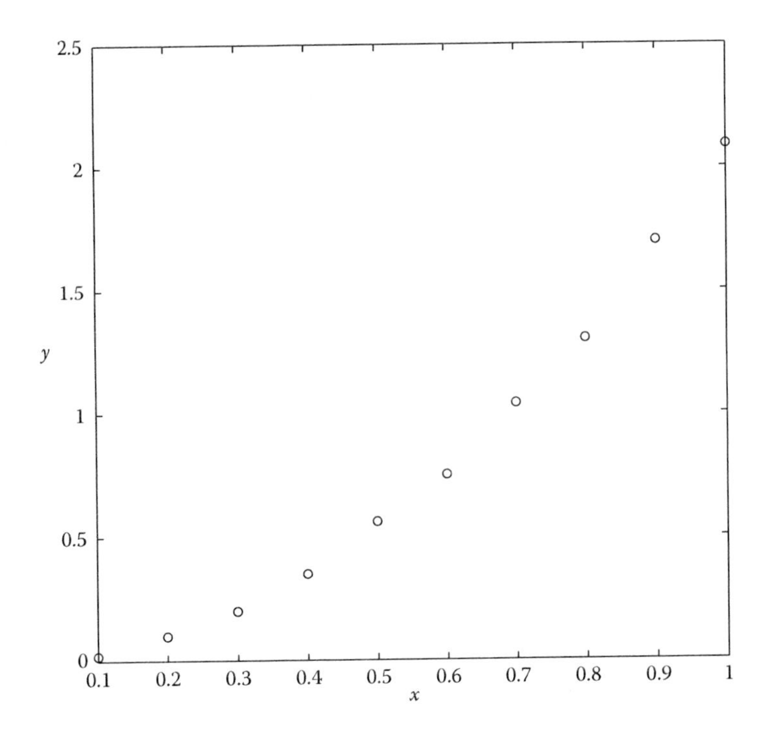

FIGURE 5.10
Plot of the data in Table 5.3.

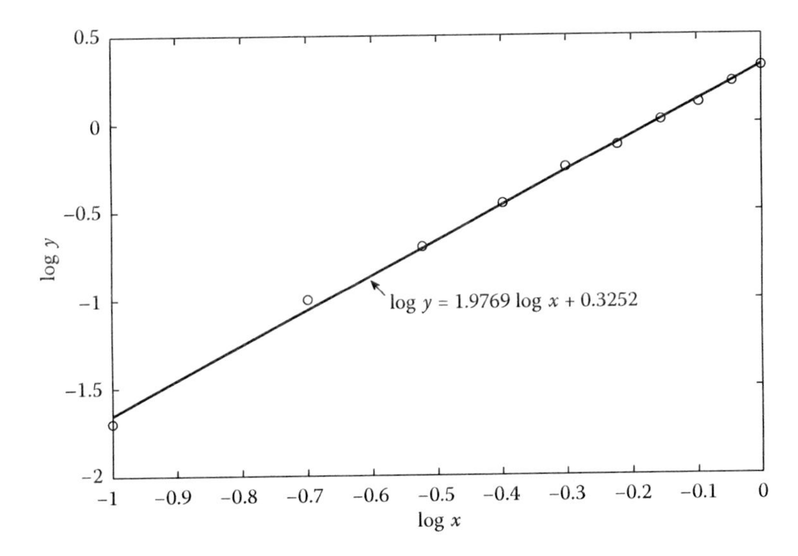

FIGURE 5.11
Linear fit of the converted data in Example 5.3.

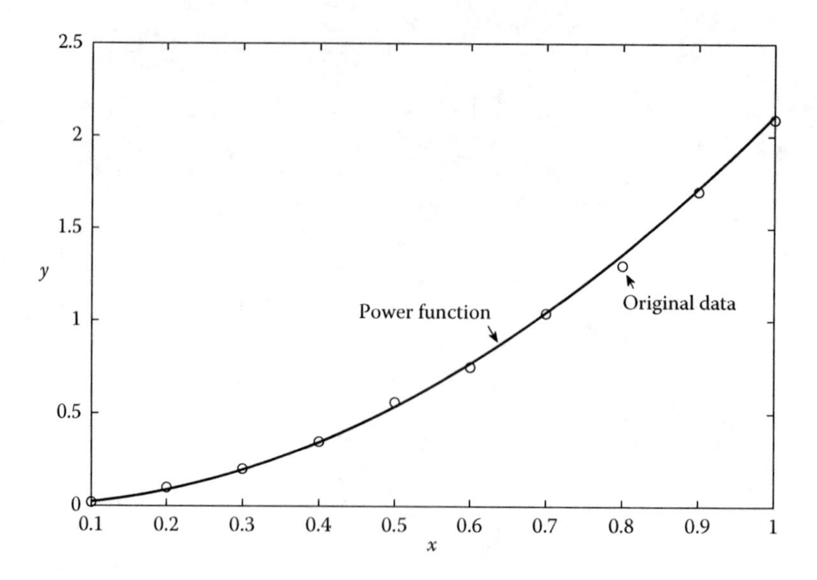

FIGURE 5.12
Curve fit using the power function; Example 5.3.

```
a1 =
     1.9769

a0 =
     0.3252
```

Based on Equation 5.11, we have $b = 1.9769$ (slope) and $\log a = 0.3252$ (intercept) so that $a = 2.1145$. The corresponding power function is

$$y = ax^b = 2.1145x^{1.9769}$$

Figure 5.12 shows the plot of this function together with the original data.

5.4 Polynomial Regression

In the previous section, we learned that a curve can fit into nonlinear data by transforming the data into a form that can be handled by linear regression. Another method is to fit polynomials of different orders to the data by means of polynomial regression.

The linear least-squares regression of Section 5.2 can be extended to fit a set of n data points $(x_1, y_1), \ldots, (x_n, y_n)$ with an mth-degree polynomial in the form

$$y = a_m x^m + a_{m-1} x^{m-1} + \cdots + a_2 x^2 + a_1 x + a_0$$

with a total error

$$E = \sum_{i=1}^{n} [y_i - (a_m x_i^m + \cdots + a_2 x_i^2 + a_1 x_i + a_0)]^2 \tag{5.14}$$

The coefficients $a_m, \ldots, a_2, a_1, a_0$ are to be determined such that E is minimized. A necessary condition for E to attain a minimum is that its partial derivative with respect to each of these coefficients vanishes, that is

$$\frac{\partial E}{\partial a_0} = -2\sum_{i=1}^{n} \left[y_i - \left(a_m x_i^m + \cdots + a_2 x_i^2 + a_1 x_i + a_0 \right) \right] = 0$$

$$\frac{\partial E}{\partial a_1} = -2\sum_{i=1}^{n} \left\{ x_i \left[y_i - \left(a_m x_i^m + \cdots + a_2 x_i^2 + a_1 x_i + a_0 \right) \right] \right\} = 0$$

$$\frac{\partial E}{\partial a_2} = -2\sum_{i=1}^{n} \left\{ x_i^2 \left[y_i - \left(a_m x_i^m + \cdots + a_2 x_i^2 + a_1 x_i + a_0 \right) \right] \right\} = 0$$

$$\cdots \qquad \cdots \qquad \cdots$$

$$\frac{\partial E}{\partial a_m} = -2\sum_{i=1}^{n} \left\{ x_i^m \left[y_i - \left(a_m x_i^m + \cdots + a_2 x_i^2 + a_1 x_i + a_0 \right) \right] \right\} = 0$$

Manipulation of these equations yields a system of $m + 1$ linear equations to be solved for $a_m, \ldots, a_2, a_1, a_0$:

$$
\begin{aligned}
n a_0 + \left(\sum_{i=1}^{n} x_i \right) a_1 + \left(\sum_{i=1}^{n} x_i^2 \right) a_2 + \cdots + \left(\sum_{i=1}^{n} x_i^m \right) a_m &= \sum_{i=1}^{n} y_i \\
\left(\sum_{i=1}^{n} x_i \right) a_0 + \left(\sum_{i=1}^{n} x_i^2 \right) a_1 + \left(\sum_{i=1}^{n} x_i^3 \right) a_2 + \cdots + \left(\sum_{i=1}^{n} x_i^{m+1} \right) a_m &= \sum_{i=1}^{n} x_i y_i \\
\left(\sum_{i=1}^{n} x_i^2 \right) a_0 + \left(\sum_{i=1}^{n} x_i^3 \right) a_1 + \left(\sum_{i=1}^{n} x_i^4 \right) a_2 + \cdots + \left(\sum_{i=1}^{n} x_i^{m+2} \right) a_m &= \sum_{i=1}^{n} x_i^2 y_i \\
\cdots \qquad \cdots \qquad \cdots & \\
\left(\sum_{i=1}^{n} x_i^m \right) a_0 + \left(\sum_{i=1}^{n} x_i^{m+1} \right) a_1 + \left(\sum_{i=1}^{n} x_i^{m+2} \right) a_2 + \cdots + \left(\sum_{i=1}^{n} x_i^{2m} \right) a_m &= \sum_{i=1}^{n} x_i^m y_i
\end{aligned}
\tag{5.15}
$$

5.4.1 Quadratic Least-Squares Regression

The objective is to fit a set of n data points $(x_1, y_1), \ldots, (x_n, y_n)$ with a second-degree polynomial

$$y = a_2x^2 + a_1x + a_0$$

such that the total error

$$E = \sum_{i=1}^{n}\left[y_i - \left(a_2x_i^2 + a_1x_i + a_0\right)\right]^2$$

is minimized. Following the procedure outlined above, which ultimately led to Equation 5.15, the coefficients a_2, a_1, a_0 are determined by solving a system of three linear equations:

$$\begin{aligned}
na_0 + \left(\sum_{i=1}^{n} x_i\right)a_1 + \left(\sum_{i=1}^{n} x_i^2\right)a_2 &= \sum_{i=1}^{n} y_i \\
\left(\sum_{i=1}^{n} x_i\right)a_0 + \left(\sum_{i=1}^{n} x_i^2\right)a_1 + \left(\sum_{i=1}^{n} x_i^3\right)a_2 &= \sum_{i=1}^{n} x_i y_i \\
\left(\sum_{i=1}^{n} x_i^2\right)a_0 + \left(\sum_{i=1}^{n} x_i^3\right)a_1 + \left(\sum_{i=1}^{n} x_i^4\right)a_2 &= \sum_{i=1}^{n} x_i^2 y_i
\end{aligned} \tag{5.16}$$

The user-defined function `QuadraticRegression` uses the quadratic least-squares regression approach to find the second-degree polynomial that best fits a set of data. The coefficients a_2, a_1, a_0 are found by writing Equation 5.16 in matrix form and applying the built-in backslash "\" operator in MATLAB®. The function also returns the plot of the data and the best quadratic polynomial fit.

```
function [a2 a1 a0] = QuadraticRegression(x,y)
%
% QuadraticRegression uses quadratic least-squares
% approximation to fit a data by a 2nd-degree polynomial
% in the form y = a2*x^2 + a1*x + a0.
%
%   [a2 a1 a0] = QuadraticRegression(x,y) where
%
%     x, y are n-dimensional row or column vectors of data,
%
%     a2, a1 and a0 are the coefficients that describe
%     the quadratic fit.
%
```

```
n = length(x);
Sumx = sum(x); Sumy = sum(y);
Sumx2 = sum(x.^2); Sumx3 = sum(x.^3); Sumx4 = sum(x.^4);
Sumxy = sum(x.*y); Sumx2y = sum(x.*x.*y);

% Form the coefficient matrix and the vector of right-hand
% sides
A = [n Sumx Sumx2;Sumx Sumx2 Sumx3;Sumx2 Sumx3 Sumx4];
b = [Sumy;Sumxy;Sumx2y];
w = A\b;  % Solve
a2 = w(3); a1 = w(2); a0 = w(1);
% Plot the data and the quadratic fit
xx = linspace(x(1),x(end));
% Generate 100 points for plotting purposes
p = zeros(100,1);  % Pre-allocate
for i = 1:100,
    p(i) = a2*xx(i)^2 + a1*xx(i) + a0; % Calculate 100 points
end
plot(x,y,'o')
hold on
plot(xx,p)
end
```

Example 5.4: Curve Fit: Quadratic Regression

Using quadratic least-squares regression, find the second-degree polynomial that best fits the data in Table 5.4.

SOLUTION

```
>> x = 0:0.4:1.6;
>> y = [2.90 3.10 3.56 4.60 6.70];
>> [a2 a1 a0] = QuadraticRegression(x,y)

a2 =
     1.9554

a1 =
    -0.8536

a0 =
     2.9777
```

The results are shown in Figure 5.13.

TABLE 5.4

Data in Example 5.4

x	y
0	2.90
0.4	3.10
0.8	3.56
1.2	4.60
1.6	6.70

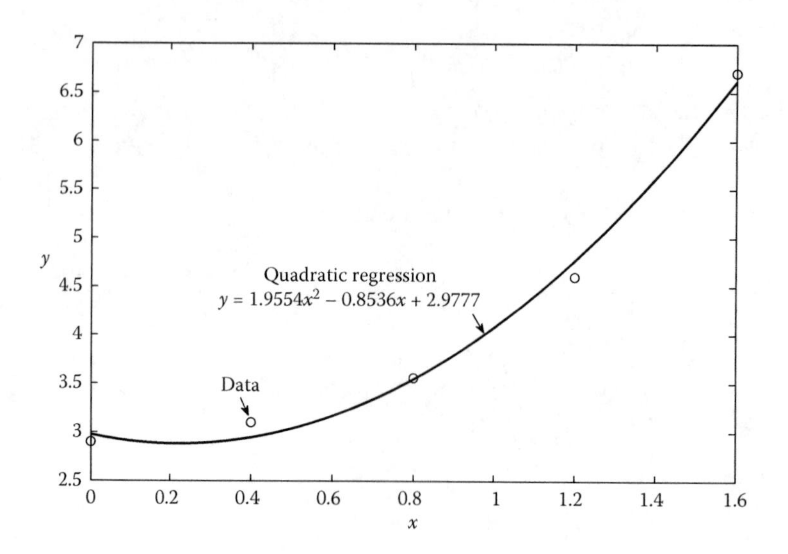

FIGURE 5.13
Quadratic polynomial fit in Example 5.4.

5.4.2 Cubic Least-Squares Regression

The objective is to fit a set of n data points $(x_1, y_1), \ldots, (x_n, y_n)$ with a third-degree polynomial

$$y = a_3x^3 + a_2x^2 + a_1x + a_0$$

such that the total error

$$E = \sum_{i=1}^{n}\left[y_i - \left(a_3x_i^3 + a_2x_i^2 + a_1x_i + a_0\right)\right]^2$$

is minimized. Proceeding as before, a_3, a_2, a_1, a_0 are determined by solving a system of four linear equations:

$$\begin{aligned}
na_0 + \left(\sum_{i=1}^{n} x_i\right)a_1 + \left(\sum_{i=1}^{n} x_i^2\right)a_2 + \left(\sum_{i=1}^{n} x_i^3\right)a_3 &= \sum_{i=1}^{n} y_i \\
\left(\sum_{i=1}^{n} x_i\right)a_0 + \left(\sum_{i=1}^{n} x_i^2\right)a_1 + \left(\sum_{i=1}^{n} x_i^3\right)a_2 + \left(\sum_{i=1}^{n} x_i^4\right)a_3 &= \sum_{i=1}^{n} x_iy_i \\
\left(\sum_{i=1}^{n} x_i^2\right)a_0 + \left(\sum_{i=1}^{n} x_i^3\right)a_1 + \left(\sum_{i=1}^{n} x_i^4\right)a_2 + \left(\sum_{i=1}^{n} x_i^5\right)a_3 &= \sum_{i=1}^{n} x_i^2y_i \\
\left(\sum_{i=1}^{n} x_i^3\right)a_0 + \left(\sum_{i=1}^{n} x_i^4\right)a_1 + \left(\sum_{i=1}^{n} x_i^5\right)a_2 + \left(\sum_{i=1}^{n} x_i^6\right)a_3 &= \sum_{i=1}^{n} x_i^3y_i
\end{aligned} \tag{5.17}$$

The user-defined function `CubicRegression` uses the cubic least-squares regression approach to find the third-degree polynomial that best fits a set of data. The coefficients a_3, a_2, a_1, a_0 are found by writing Equation 5.17 in matrix form and solving by "\" in MATLAB. The function also returns the plot of the data and the best cubic polynomial fit.

```
function [a3 a2 a1 a0]= CubicRegression(x,y)
%
% CubicRegression uses cubic least-squares approximation
% to fit a data by a 3rd-degree polynomial
% y = a3*x^3 + a2*x^2 + a1*x + a0.
%
%   [a3 a2 a1 a0]= CubicRegression(x,y) where
%
%      x, y are n-dimensional row or column vectors of
%      data,
%
%      a3, a2, a1 and a0 are the coefficients that
%      describe the cubic fit.
%

n = length(x);
Sumx = sum(x); Sumy = sum(y);
Sumx2 = sum(x.^2); Sumx3 = sum(x.^3); Sumx4 = sum(x.^4);
Sumx5 = sum(x.^5); Sumx6 = sum(x.^6);
Sumxy = sum(x.*y); Sumx2y = sum(y.*x.^2);
Sumx3y = sum(y.*x.^3);

% Form the coefficient matrix and the vector of right-
% hand sides
A = [n Sumx Sumx2 Sumx3;Sumx Sumx2 Sumx3 Sumx4;Sumx2 Sumx3
   Sumx4 Sumx5; Sumx3 Sumx4 Sumx5 Sumx6];
b = [Sumy;Sumxy;Sumx2y;Sumx3y];
w = A\b;    % Solve
a3 = w(4); a2 = w(3); a1 = w(2); a0 = w(1);
% Plot the data and the cubic fit
xx = linspace(x(1),x(end));
p = zeros(100,1);    % Pre-allocate
for i = 1:100,
    p(i)= a3*xx(i)^3 + a2*xx(i)^2 + a1*xx(i) + a0;
    % Calculate 100 points
end
plot(x,y,'o')
hold on
plot(xx,p)
end
```

Example 5.5: Curve Fit: Cubic Regression

Find the cubic polynomial that best fits the data in Example 5.4. Plot the quadratic and cubic polynomial fits in one graph and compare.

SOLUTION

```
>> x=0:0.4:1.6;
>> y=[2.90 3.10 3.56 4.60 6.70];
>> [a3 a2 a1 a0]=CubicRegression(x,y)

a3 =
     1.0417

a2 =
    -0.5446

a1 =
     0.5798

a0 =
     2.8977

>> [a2 a1 a0]=QuadraticRegression(x,y)
% Previously done in Example 5.4

a2 =
     1.9554

a1 =
    -0.8536

a0 =
     2.9777
```

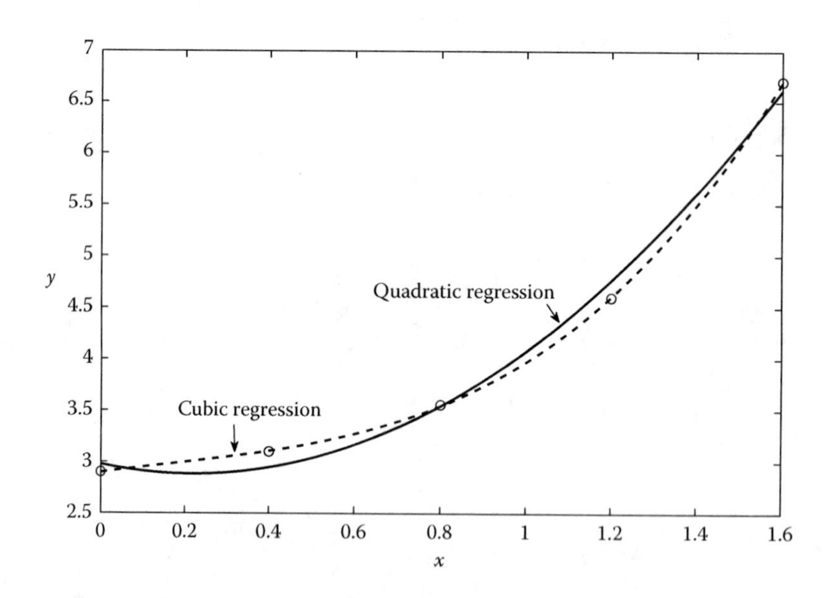

FIGURE 5.14
Quadratic and cubic polynomial fits; Example 5.5.

Figure 5.14 clearly shows that the cubic polynomial fit is superior to the quadratic one. The cubic polynomial fit almost goes through all five data points, but not exactly. This is because a third-degree polynomial has four coefficients but the data contains five points. A fourth-degree polynomial, which has five coefficients, would exactly agree with the data in this example. In general, if a set of n data points is given, then the $(n-1)$th-degree polynomial that best fits the data will agree exactly with the data points. This is the main idea behind interpolation, and such polynomial is called an interpolating polynomial. We will discuss this topic in Section 5.5.

5.4.3 MATLAB® Built-In Functions `polyfit` and `polyval`

A brief description of the MATLAB built-in function `polyfit` is given as

```
POLYFIT Fit polynomial to data.

   P = POLYFIT(X,Y,N) finds the coefficients of a polynomial
   P(X) of degree N that fits the data Y best in a least-
   squares sense. P is a row vector of length N+1 containing
   the polynomial coefficients in descending powers, P(1)*
   X^N + P(2)*X^(N-1) + ··· + P(N)*X + P(N+1).
```

This polynomial can then be evaluated at any x using the built-in function `polyval`.

Example 5.6: Curve Fit: `polyfit` Function

Using the `polyfit` function, find and plot the third-degree polynomial that best fits the data in Table 5.5. Apply the user-defined function `CubicRegression` and compare the results.

SOLUTION

```
>> x = 0:0.3:1.2;
>> y = [3.6 4.8 5.9 7.6 10.9];
>> P = polyfit(x,y,3)

P =
     5.2469     -5.6349      5.2897      3.5957
     % Coefficients of the 3rd-deg polynomial fit
```

TABLE 5.5

Data in Example 5.6

x	y
0	3.6
0.3	4.8
0.6	5.9
0.9	7.6
1.2	10.9

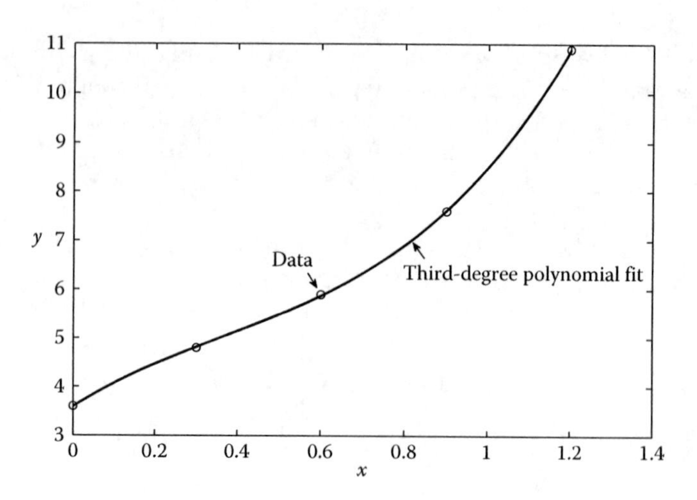

FIGURE 5.15
Cubic polynomial fit using `polyfit`.

```
>> xi = linspace(0,1.2);
% Generate 100 points for plotting purposes
>> yi = polyval(P,xi);
% Evaluate the polynomial at these points
>> plot(xi,yi)    % Figure 5.15
>> hold on
>> plot(x,y,'o')
```

Execution of `CubicRegression` yields:

```
>> [a3 a2 a1 a0] = CubicRegression(x,y)

a3 =

    5.2469

a2 =

   -5.6349

a1 =

    5.2897

a0 =

    3.5957
```

The coefficients of the third-degree polynomial are precisely those returned by `polyfit`.

5.5 Polynomial Interpolation

Given a set of $n+1$ data points $(x_1, y_1), \ldots, (x_{n+1}, y_{n+1})$, there is only one polynomial of degree at most n in the form

$$p(x) = a_{n+1}x^n + a_n x^{n-1} + \cdots + a_3 x^2 + a_2 x + a_1$$

that goes through all the points. Although this polynomial is unique, it can be expressed in different forms. The two most commonly used forms are provided by Lagrange interpolating polynomials and Newton interpolating polynomials, which are presented in this section.

5.5.1 Lagrange Interpolating Polynomials

The first-degree Lagrange interpolating polynomial that goes through the two points (x_1, y_1) and (x_2, y_2) is in the form

$$p_1(x) = L_1(x)y_1 + L_2(x)y_2$$

where $L_1(x)$ and $L_2(x)$ are the Lagrange coefficient functions and described by

$$L_1(x) = \frac{x - x_2}{x_1 - x_2}, \quad L_2(x) = \frac{x - x_1}{x_2 - x_1}$$

Then, $L_1(x_1) = 1$ and $L_1(x_2) = 0$, while $L_2(x_1) = 0$ and $L_2(x_2) = 1$. As a result, $p_1(x_1) = y_1$ and $p_1(x_2) = y_2$, which means the polynomial, in this case a straight line, passes through the two points; see Figure 5.16a.

The second-degree Lagrange interpolating polynomial that goes through the three points (x_1, y_1), (x_2, y_2), and (x_3, y_3) is in the form

$$p_2(x) = L_1(x)y_1 + L_2(x)y_2 + L_3(x)y_3$$

where

$$L_1(x) = \frac{(x - x_2)(x - x_3)}{(x_1 - x_2)(x_1 - x_3)}, \quad L_2(x) = \frac{(x - x_1)(x - x_3)}{(x_2 - x_1)(x_2 - x_3)},$$

$$L_3(x) = \frac{(x - x_1)(x - x_2)}{(x_3 - x_1)(x_3 - x_2)}$$

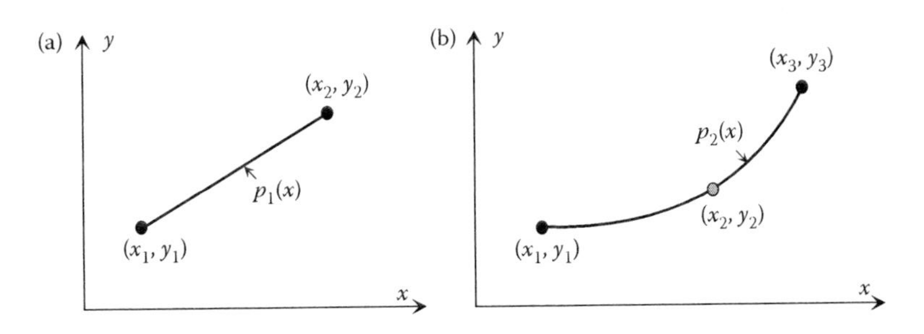

FIGURE 5.16
(a) First-degree and (b) second-degree Lagrange interpolating polynomials.

Then, $L_1(x_1) = 1 = L_2(x_2) = L_3(x_3)$, while all other $L_i(x_j) = 0$ for $i \neq j$. This guarantees $p_2(x_1) = y_1$, $p_2(x_2) = y_2$, and $p_2(x_3) = y_3$, so that the polynomial goes through the three points; see Figure 5.16b.

In general, the nth-degree Lagrange interpolating polynomial that goes through $n+1$ points $(x_1, y_1), \ldots, (x_{n+1}, y_{n+1})$ is formed as

$$p_n(x) = L_1(x)y_1 + \cdots + L_{n+1}(x)y_{n+1} = \sum_{i=1}^{n+1} L_i(x)y_i \tag{5.18}$$

where each $L_i(x)$ is defined as

$$L_i(x) = \prod_{\substack{j=1 \\ j \neq i}}^{n+1} \frac{x - x_j}{x_i - x_j} \tag{5.19}$$

and "Π" denotes the product of terms.

The user-defined function `LagrangeInterp` finds the Lagrange interpolating polynomial that fits a set of data and uses this polynomial to calculate the interpolated value at a specified point.

```
function yi = LagrangeInterp(x,y,xi)
%
% LagrangeInterp finds the Lagrange interpolating
% polynomial that fits the data (x,y) and uses it to find the
% interpolated value at xi.
%
%   yi = LagrangeInterp(x,y,xi) where
%
%    x, y are n-dimensional row or column vectors of data,
%    xi is a specified point,
%
%    yi is the interpolated value at xi.
%
n = length(x);
L = zeros(1,n);     % Pre-allocate

for i = 1:n,
    L(i) = 1;
    for j = 1:n,
        if j ~= i,
            L(i) = L(i)* (xi - x(j))/(x(i) - x(j));
        end
    end
end
yi = sum(y.* L);
```

TABLE 5.6

Data in Example 5.7

i	x_i	y_i
1	0.1	0.12
2	0.5	0.47
3	0.9	0.65

Example 5.7: Lagrange Interpolation

Find the second-degree Lagrange interpolating polynomial for the data in Table 5.6. Use this polynomial to find the interpolated value at $x = 0.3$. Confirm the result by executing the user-defined function `LagrangeInterp`.

SOLUTION

The three Lagrange coefficient functions are first found as

$$L_1(x) = \frac{(x - x_2)(x - x_3)}{(x_1 - x_2)(x_1 - x_3)} = \frac{(x - 0.5)(x - 0.9)}{(0.1 - 0.5)(0.1 - 0.9)}$$

$$L_2(x) = \frac{(x - x_1)(x - x_3)}{(x_2 - x_1)(x_2 - x_3)} = \frac{(x - 0.1)(x - 0.9)}{(0.5 - 0.1)(0.5 - 0.9)}$$

$$L_3(x) = \frac{(x - x_1)(x - x_2)}{(x_3 - x_1)(x_3 - x_2)} = \frac{(x - 0.1)(x - 0.5)}{(0.9 - 0.1)(0.9 - 0.5)}$$

The second-degree polynomial is then formed as

$$\begin{aligned} p_2(x) &= L_1(x)y_1 + L_2(x)y_2 + L_3(x)y_3 \\ &= \frac{(x - 0.5)(x - 0.9)}{(0.1 - 0.5)(0.1 - 0.9)}(0.12) + \frac{(x - 0.1)(x - 0.9)}{(0.5 - 0.1)(0.5 - 0.9)}(0.47) \\ &\quad + \frac{(x - 0.1)(x - 0.5)}{(0.9 - 0.1)(0.9 - 0.5)}(0.65) \\ &\overset{\text{simplify}}{=} -0.5312x^2 + 1.1937x + 0.0059 \end{aligned}$$

Using this polynomial, we can interpolate at $x = 0.3$ to find

$$p_2\,(0.3) = 0.3162$$

The result can be readily verified in MATLAB as follows:

```
>> x= [0.1 0.5 0.9];
>> y= [0.12 0.47 0.65];
>> yi =LagrangeInterp(x,y,0.3)

yi =
    0.3162
```

5.5.2 Drawbacks of Lagrange Interpolation

With Lagrange interpolating polynomials, we learned that going from one degree to the next, no information is stored from the lower-degree polynomial and used in the construction of the new higher-degree polynomial. This is particularly inconvenient in two situations: (1) when the exact degree of the interpolating polynomial is not known in advance; for instance, it might be better to use a portion of the set of data, or (2) when additional points are added to the data. In these cases, a more suitable form is provided by the Newton interpolating polynomials.

5.5.3 Newton Divided-Difference Interpolating Polynomials

The Newton interpolating polynomials are recursively constructed as

$$
\begin{aligned}
p_1(x) &= a_1 + a_2(x - x_1) \\
p_2(x) &= a_1 + a_2(x - x_1) + a_3(x - x_1)(x - x_2) = p_1(x) + a_3(x - x_1)(x - x_2) \\
p_3(x) &= p_2(x) + a_4(x - x_1)(x - x_2)(x - x_3) \\
&\dots \\
p_n(x) &= P_{n-1}(x) + a_{n+1}(x - x_1)(x - x_2)\dots(x - x_n)
\end{aligned}
\tag{5.20}
$$

where the coefficients $a_1, a_2, \dots, a_{n+1}$ are determined inductively as follows: Since $p_1(x)$ must agree with the data at (x_1, y_1) and (x_2, y_2), we have

$$p_1(x_1) = y_1, \quad p_1(x_2) = y_2$$

so that

$$
\begin{aligned}
a_1 + a_2(x_1 - x_1) &= y_1 \\
a_1 + a_2(x_2 - x_1) &= y_2
\end{aligned}
\quad \Rightarrow \quad a_1 = y_1, \quad a_2 = \frac{y_2 - y_1}{x_2 - x_1}
$$

Similarly, $p_2(x)$ must agree with the data at (x_1, y_1), (x_2, y_2), and (x_3, y_3); hence

$$p_2(x_1) = y_1, \quad p_2(x_2) = y_2, \quad p_2(x_3) = y_3$$

This gives a_1 and a_2 as above, and

$$a_3 = \frac{(y_3 - y_2)/(x_3 - x_2) - (y_2 - y_1)/(x_2 - x_1)}{x_3 - x_1}$$

Continuing this process yields all remaining coefficients. These coefficients follow a systematic pattern that is best understood through Newton's divided differences. For two points (x_1, y_1) and (x_2, y_2), the first divided difference is the slope of the line connecting them and denoted by

$$f[x_2, x_1] = \frac{y_2 - y_1}{x_2 - x_1} = a_2$$

For three points (x_1, y_1), (x_2, y_2), and (x_3, y_3), the second divided difference is defined as

$$f[x_3, x_2, x_1] = \frac{f[x_3, x_2] - f[x_2, x_1]}{x_3 - x_1} = \frac{(y_3 - y_2)/(x_3 - x_2) - (y_2 - y_1)/(x_2 - x_1)}{x_3 - x_1} = a_3$$

In general, the kth divided difference is described by

$$f[x_{k+1}, x_k, \ldots, x_2, x_1] = \frac{f[x_{k+1}, x_k, \ldots, x_3, x_2] - f[x_k, x_{k-1}, \ldots, x_2, x_1]}{x_{k+1} - x_1} = a_{k+1} \quad (5.21)$$

Therefore, the nth-degree Newton divided-difference interpolating polynomial for the data $(x_1, y_1), \ldots, (x_{n+1}, y_{n+1})$ is formed as

$$p_n(x) = \underset{\substack{\downarrow \\ y_1}}{a_1} + \underset{\substack{\downarrow \\ f[x_2, x_1]}}{a_2}(x - x_1) + \underset{\substack{\downarrow \\ f[x_3, x_2, x_1]}}{a_3}(x - x_1)(x - x_2) + \cdots$$

$$+ \underset{\substack{\downarrow \\ f[x_{n+1}, \ldots, x_1]}}{a_{n+1}}(x - x_1)(x - x_2)\cdots(x - x_n) \quad (5.22)$$

where the coefficients $a_1, \ldots, a_{n+1}$ are best calculated with the aid of a divided-differences table. A typical such table is presented in Table 5.7 corresponding to a set of five data points.

The user-defined function `NewtonInterp` finds the Newton divided-difference interpolating polynomial that fits a set of data and uses this polynomial to calculate the interpolated value at a specific point.

```
function yi = NewtonInterp(x,y,xi)
%
% NewtonInterp finds the Newton divided-difference
% interpolating polynomial that fits the data (x,y) and
% uses it to find the interpolated value at xi.
%
%   yi = NewtonInterp(x,y,xi) where
%
%     x, y are n-dimensional row or column vectors of data,
%     xi is a specified point,
```

```
%
%   yi is the interpolated value at xi.
%
n = length(x);
a = zeros(1,n);     % Pre-allocate
a(1) = y(1);

DivDiff = zeros(1,n-1);     % Pre-allocate

for i = 1:n-1,
    DivDiff(i,1) = (y(i+1) - y(i))/(x(i+1) - x(i));
end
for j = 2:n-1,
    for i = 1:n-j,
        DivDiff(i,j) = (DivDiff(i+1,j-1) - DivDiff(i,j-1))/
        (x(j+i) - x(i));
    end
end
for k = 2:n,
    a(k) = DivDiff(1,k-1);
end
yi = a(1);
xprod = 1;
for m = 2:n,
    xprod = xprod*(xi - x(m-1));
    yi = yi + a(m)*xprod;
end
```

TABLE 5.7

Divided Differences Table

x_i	y_i	First Divided Diff.	Second Divided Diff.	Third Divided Diff.	Fourth Divided Diff.
x_1	$y_1 = \boxed{a_1}$				
		$f[x_2,x_1] = \boxed{a_2}$			
x_2	y_2		$f[x_3,x_2,x_1] = \boxed{a_3}$		
		$f[x_3, x_2]$		$f[x_4,x_3,x_2,x_1] = \boxed{a_4}$	
x_3	y_3		$f[x_4, x_3, x_2]$		$f[x_5,x_4,x_3,x_2,x_1] = \boxed{a_5}$
		$f[x_4, x_3]$		$f[x_5, x_4, x_3, x_2]$	
x_4	y_4		$f[x_5, x_4, x_3]$		
		$f[x_5, x_4]$			
x_5	y_5				

TABLE 5.8
Data in Example 5.8

x_i	y_i
0	0
0.1	0.1210
0.2	0.2258
0.5	0.4650
0.8	0.6249

Example 5.8: Newton Interpolation, Divided Differences

Find the fourth-degree Newton interpolating polynomial for the data in Table 5.8. Use this polynomial to find the interpolated value at $x = 0.7$. Confirm the result by executing the user-defined function `NewtonInterp`.

SOLUTION

The divided differences are calculated according to Equation 5.21 and recorded in Table 5.9.

The fourth-degree Newton interpolating polynomial is then formed as

$$\begin{aligned} p_4(x) &= a_1 + a_2(x - x_1) + a_3(x - x_1)(x - x_2) + a_4(x - x_1)(x - x_2)(x - x_3) \\ &\quad + a_5(x - x_1)(x - x_2)(x - x_3)(x - x_4) \\ &= 0 + 1.2100(x - 0) - 0.8100(x - 0)(x - 0.1) + 0.3664(x - 0) \\ &\quad \times (x - 0.1)(x - 0.2) - 0.1254(x - 0)(x - 0.1)(x - 0.2)(x - 0.5) \\ &= -0.1254x^4 + 0.4667x^3 - 0.9412x^2 + 1.2996x \end{aligned}$$

Using this polynomial, we can find the interpolated value at $x = 0.7$, as

$$p_4\,(0.7) = 0.5784$$

The result can be readily verified in MATLAB as follows:

```
>> x = [0 0.1 0.2 0.5 0.8];
>> y = [0 0.1210 0.2258 0.4650 0.6249];
>> yi = NewtonInterp(x,y,0.7)

yi =
      0.5785
```

5.5.4 Special Case: Equally Spaced Data

In the derivation of Newton divided difference interpolating polynomial, Equation 5.22, no restriction was placed on how the data was spaced. In

TABLE 5.9

Divided Differences Table for Example 5.8

x_i	y_i	First Divided Diff.	Second Divided Diff.	Third Divided Diff.	Fourth Divided Diff.
0	$\overset{a_1}{\boxed{0}}$				
		$\frac{0.1210-0}{0.1-0}=\overset{a_2}{\boxed{1.2100}}$			
0.1	0.1210		$\frac{1.0480-1.2100}{0.2-0}=\overset{a_3}{\boxed{-0.8100}}$		
		$\frac{0.2258-0.1210}{0.2-0.1}=1.0480$		$\frac{-0.6268-(-0.8100)}{0.5-0}=\overset{a_4}{\boxed{0.3664}}$	
0.2	0.2258		$\frac{0.7973-1.0480}{0.5-0.1}=-0.6268$		$\frac{0.2661-0.3664}{0.8-0}=\overset{a_5}{\boxed{-0.1254}}$
		$\frac{0.4650-0.2258}{0.5-0.2}=0.7973$		$\frac{-0.4405-(-0.6268)}{0.8-0.1}=0.2661$	
0.5	0.4650		$\frac{0.5330-0.7973}{0.8-0.2}=-0.4405$		
		$\frac{0.6249-0.4650}{0.8-0.5}=0.5330$			
0.8	0.6249				

the event that the data is equally spaced, as it is often the case in practice, the divided differences reduce to simpler forms. Let every two successive x values in the data $(x_1, y_1), \ldots, (x_{n+1}, y_{n+1})$ be separated by distance h so that

$$x_{i+1} - x_i = h, \quad i = 1, 2, \ldots, n$$

Consequently,

$$x_2 = x_1 + h, \quad x_3 = x_1 + 2h, \ldots, \quad x_{n+1} = x_1 + nh$$

The first forward difference at x_i is defined as

$$\Delta y_i = y_{i+1} - y_i, \quad i = 1, 2, \ldots, n$$

The second forward difference at x_i is defined as

$$\Delta^2 y_i = \Delta y_{i+1} - \Delta y_i, \quad i = 1, 2, \ldots, n$$

In general, the kth forward difference is described as

$$\Delta^k y_i = \Delta^{k-1} y_{i+1} - \Delta^{k-1} y_i, \quad k = 1, 2, \ldots, n \tag{5.23}$$

We next find out how the divided differences and forward differences are related. For (x_1, y_1) and (x_2, y_2), the first divided difference can be written as

$$f[x_2, x_1] = \frac{y_2 - y_1}{x_2 - x_1} = \frac{\Delta y_1}{h}$$

For (x_1, y_1), (x_2, y_2), and (x_3, y_3), the second divided difference is

$$f[x_3, x_2, x_1] = \frac{(y_3 - y_2)/(x_3 - x_2) - (y_2 - y_1)/(x_2 - x_1)}{x_3 - x_1} = \frac{(\Delta y_2 - \Delta y_1)/h}{2h}$$

$$= \frac{\Delta y_2 - \Delta y_1}{2h^2} = \frac{\Delta^2 y_1}{2h^2}$$

In general, for $(x_1, y_1), \ldots, (x_{k+1}, y_{k+1})$, the kth divided difference can be expressed as

$$f[x_{k+1}, x_k, \ldots, x_2, x_1] = \frac{\Delta^k y_1}{k!h^k}$$

5.5.5 Newton Forward-Difference Interpolating Polynomials

Any arbitrary x between x_1 and x_{n+1} can be expressed as $x = x_1 + mh$ for a suitable real value m. Then

$$x - x_2 = (x_1 + mh) - x_2 = mh - (x_2 - x_1) = mh - h = (m-1)h$$

$$x - x_3 = (m-2)h$$

$$\ldots$$

$$x - x_n = (m - (n-1))h$$

Substitution of these, together with the relations between divided differences and forward differences established above, into Equation 5.22, yields the Newton forward-difference interpolating polynomial as

$$\begin{aligned} p_n(x) &= y_1 + \frac{\Delta y_1}{h}(mh) + \frac{\Delta^2 y_1}{2h^2}(mh)((m-1)h) + \cdots \\ &\quad + \frac{\Delta^n y_1}{n!h^n}(mh)((m-1)h)\cdots(m-(n-1))h \\ &= y_1 + m\Delta y_1 + \frac{m(m-1)}{2!}\Delta^2 y_1 + \cdots + \frac{m(m-1)\cdots(m-(n-1))}{n!}\Delta^n y_1 \\ &\quad \text{where } m = \frac{x - x_1}{h} \end{aligned} \tag{5.24}$$

This polynomial is best formed with the aid of a forward-differences table, as in Table 5.10.

TABLE 5.10

Forward Differences Table

x_i	y_i	First Forward Diff.	Second Forward Diff.	Third Forward Diff.
x_1	$\boxed{y_1}$			
		$\boxed{\Delta y_1}$		
x_2	y_2		$\boxed{\Delta^2 y_1}$	
		Δy_2		$\boxed{\Delta^3 y_1}$
x_3	y_3		$\Delta^2 y_2$	
		Δy_3		
x_4	y_4			

Example 5.9: Newton Interpolation, Forward Differences

For the data in Table 5.11, interpolate at $x = 0.52$ using the third-degree Newton forward-difference interpolating polynomial. Confirm the result by executing the user-defined function `NewtonInterp`.

SOLUTION

The forward differences are calculated according to Equation 5.23 and recorded in Table 5.12. Since we are interpolating at $x = 0.52$, we have

$$m = \frac{x - x_1}{h} = \frac{0.52 - 0.4}{0.1} = 1.2$$

Inserting this and the proper (boxed) entries of Table 5.12 into Equation 5.24, we find the interpolated value as

$$p_3(0.52) = 0.921061 + 1.2(-0.043478) + \frac{(1.2)(0.2)}{2}(-0.008769)$$
$$+ \frac{(1.2)(0.2)(-0.8)}{6}(0.000522) = 0.867818$$

The actual value is $\cos(0.52) = 0.867819$, indicating a 5-decimal place accuracy. The result can be verified in MATLAB using the user-defined function `NewtonInterp`.

TABLE 5.11

Data in Example 5.9

x_i	$y_i = \cos(x_i)$
0.4	0.921061
0.5	0.877583
0.6	0.825336
0.7	0.764842

TABLE 5.12

Forward Differences Table for Example 5.9

x_i	y_i	First Forward Diff.	Second Forward Diff.	Third Forward Diff.
0.4	[0.921061]			
		[−0.043478]		
0.5	0.877583		[−0.008769]	
		−0.052247		[0.000522]
0.6	0.825336		−0.008247	
		−0.060494		
0.7	0.764842			

```
>> x = [0.4 0.5 0.6 0.7];
>> y = [0.921061 0.877583 0.825336 0.764842];
>> format long
>> yi = NewtonInterp(x,y,0.52)

yi =
    0.867818416000000
```

5.6 Spline Interpolation

In Section 5.5, we used nth-degree polynomials to interpolate $n+1$ data points. For example, we learned that a set of 11 data points can be interpolated by a single 10th-degree polynomial. When there are a small number of points in the data, the degree of the interpolating polynomial will also be small, and the interpolated values are generally accurate. However, when a high-degree polynomial is used to interpolate a large number of points, large errors in interpolation are possible, as shown in Figure 5.17. The main contributing factor is the large number of peaks and valleys that accompany a high-degree polynomial. These problems can be avoided by using several low-degree polynomials, each of which is valid in one interval between one or more data points. These low-degree polynomials are known as splines. The term "spline" originated from a thin, flexible strip, known as a spline,

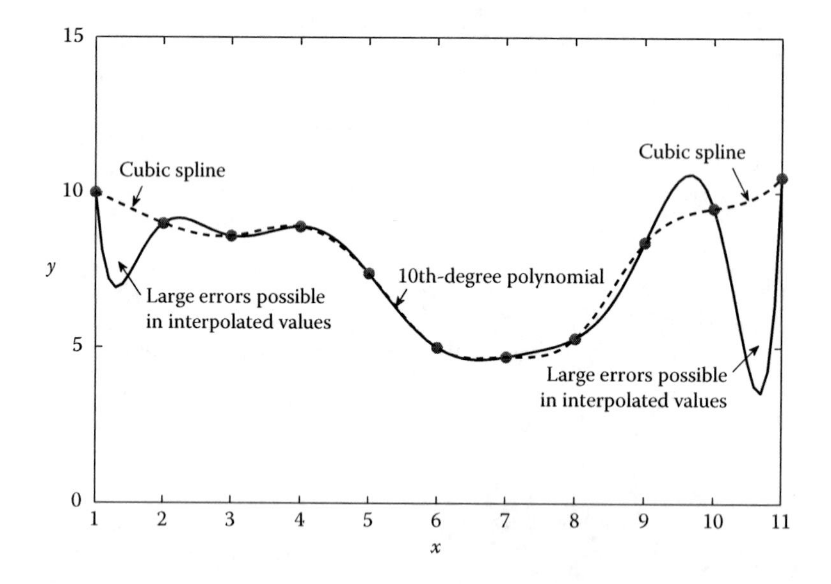

FIGURE 5.17
A 10th-degree polynomial and cubic splines for interpolation of 11 data points.

used by draftsmen to draw smooth curves over a set of points marked by pegs or nails. The data points at which two splines meet are called knots.

The most commonly used splines are cubic splines, which produce very smooth connections over adjacent intervals. Figure 5.17 shows the clear advantage of using several cubic splines as opposed to one single high-degree polynomial for interpolation of a large set of data.

5.6.1 Linear Splines

With linear splines, straight lines (linear functions) are used for interpolation between the data points. Figure 5.18 shows the linear splines used for a set of four data points, as well as the third-degree interpolating polynomial. If the data points are labeled (x_1, y_1), (x_2, y_2), (x_3, y_3), and (x_4, y_4), then, using the Lagrange form, the linear splines are simply three linear functions defined as

$$S_1(x) = \frac{x - x_2}{x_1 - x_2} y_1 + \frac{x - x_1}{x_2 - x_1} y_2, \quad x_1 \le x \le x_2$$

$$S_2(x) = \frac{x - x_3}{x_2 - x_3} y_2 + \frac{x - x_2}{x_3 - x_2} y_3, \quad x_2 \le x \le x_3$$

$$S_3(x) = \frac{x - x_4}{x_3 - x_4} y_3 + \frac{x - x_3}{x_4 - x_3} y_4, \quad x_3 \le x \le x_4$$

This is clearly the same as linear interpolation as discussed in Section 5.5. The obvious drawback of linear splines is that they are not smooth so that

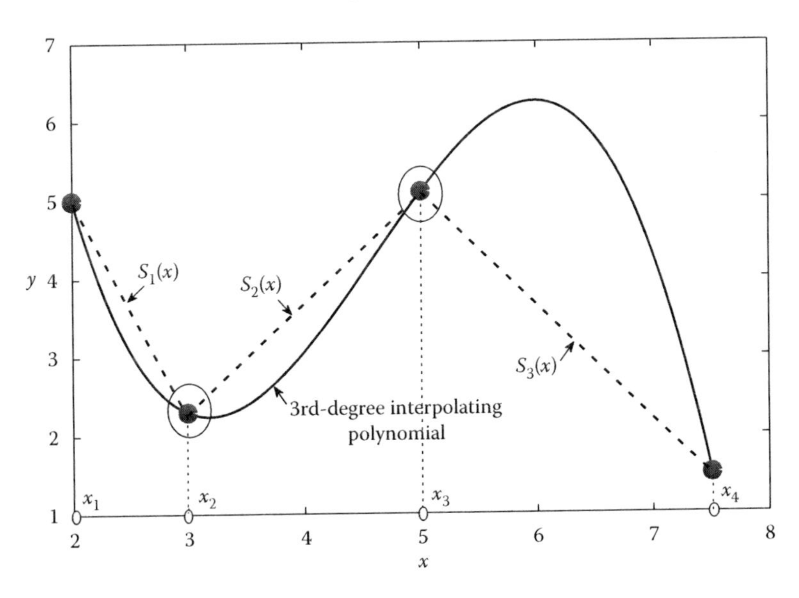

FIGURE 5.18
Linear splines.

the slope experiences sudden changes at the knots. This is because the first derivatives of neighboring linear functions do not agree. To circumvent this problem, higher-degree polynomial splines are used such that the derivatives of every two successive splines agree at the point (knot) they meet. Quadratic splines ensure continuous first derivatives at the knots, but not the second derivatives. Cubic splines ensure continuity of both first and second derivatives at the knots, and are most commonly used in practice.

5.6.2 Quadratic Splines

The idea behind quadratic splines is to use second-degree polynomials to interpolate over each interval between data points. Suppose there are $n + 1$ data points $(x_1, y_1), \ldots, (x_{n+1}, y_{n+1})$ so that there are n intervals and thus n quadratic polynomials; see Figure 5.19. Each quadratic polynomial is in the form

$$S_i(x) = a_i\, x^2 + b_i\, x + c_i, \quad i = 1, 2, \ldots, n \tag{5.25}$$

where a_i, b_i, c_i $(i = 1, 2, \ldots, n)$ are unknown constants to be determined. Since there are n such polynomials, and each has three unknown constants, there are a total of $3n$ unknown constants. Therefore, exactly $3n$ equations are needed to determine all the unknowns. These equations are generated as follows:

5.6.2.1 Function Values at the Endpoints

The first polynomial $S_1(x)$ must go through (x_1, y_1), and the last polynomial $S_n(x)$ must go through (x_{n+1}, y_{n+1}):

$$\left.\begin{array}{l} S_1(x_1) = y_1 \\ S_n(x_{n+1}) = y_{n+1} \end{array}\right\} \text{2equations}$$

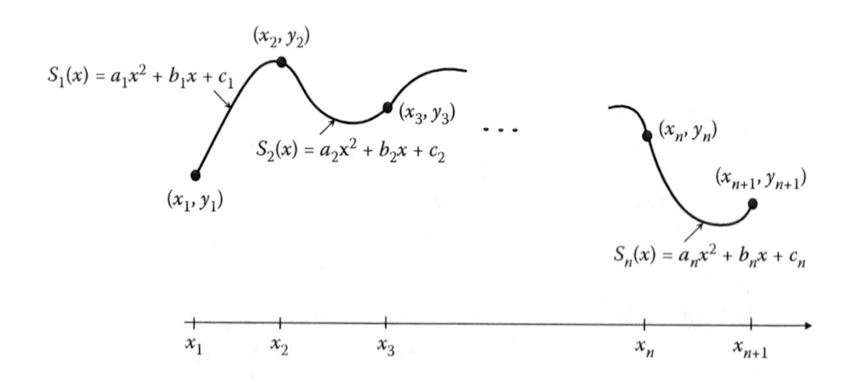

FIGURE 5.19
Quadratic splines.

More specifically,

$$\begin{aligned} a_1x_1^2 + b_1x_1 + c_1 &= y_1 \\ a_nx_{n+1}^2 + b_nx_{n+1} + c_n &= y_{n+1} \end{aligned} \tag{5.26}$$

5.6.2.2 Function Values at the Interior Knots

At the interior knots, two conditions must hold: (1) polynomials must go through the data points, and (2) adjacent polynomials must agree at the data points:

$$\left.\begin{aligned} &S_1(x_2) = y_2,\quad S_2(x_3) = y_3, \ldots, S_{n-1}(x_n) = y_n \quad n-1 \text{ equations} \\ &S_2(x_2) = y_2,\quad S_3(x_3) = y_3, \ldots, S_n(x_n) = y_n \quad n-1 \text{ equations} \end{aligned}\right\}$$

More specifically,

$$\begin{aligned} S_i(x_{i+1}) &= y_{i+1}, \quad i = 1, 2, \ldots, n-1 \\ S_i(x_i) &= y_i, \quad i = 2, 3, \ldots, n \end{aligned}$$

so that

$$\begin{aligned} a_ix_{i+1}^2 + b_ix_{i+1} + c_i &= y_{i+1}, \quad i = 1, 2, \ldots, n-1 \\ a_ix_i^2 + b_ix_i + c_i &= y_i, \quad i = 2, 3, \ldots, n \end{aligned} \tag{5.27}$$

5.6.2.3 First Derivatives at the Interior Knots

At the interior knots, the first derivatives of adjacent quadratic polynomials must agree:

$$S_1'(x_2) = S_2'(x_2),\quad S_2'(x_3) = S_3'(x_3), \ldots,\quad \left.S_{n-1}'(x_n) = S_n'(x_n)\right\} \, n-1 \text{ equations}$$

More specifically,

$$S_i'(x_{i+1}) = S_{i+1}'(x_{i+1}), \quad i = 1, 2, \ldots, n-1$$

Noting that $S_i'(x) = 2a_ix + b_i$, the above yields

$$2a_i\, x_{i+1} + b_i = 2a_{i+1}\, x_{i+1} + b_{i+1}, \quad i = 1, 2, \ldots, n-1 \tag{5.28}$$

So far, we have managed to generate $2 + (n-1) + (n-1) + (n-1) = 3n - 1$ equations. One more equation is needed to complete the task. Among several available choices, we select the following:

5.6.2.4 Second Derivative at the Left Endpoint is Zero

$$S_1''(x_1) = 0\}_{1\text{ equation}}$$

Noting that $S_1''(x_1) = 2a_1$, this yields

$$a_1 = 0 \tag{5.29}$$

A total of $3n$ equations have therefore been generated.

In summary, one equation simply gives $a_1 = 0$ and the remaining $3n - 1$ unknowns are found by solving the $3n - 1$ equations provided by Equations 5.26 through 5.28.

Example 5.10: Quadratic Splines

Construct the quadratic splines for the data in Table 5.13.

SOLUTION

Since there are 4 data points, we have $n = 3$ so that there are 3 quadratic splines with a total of 9 unknown constants. Of these, one is given by $a_1 = 0$. The remaining 8 equations to solve are provided by Equations 5.26 through 5.28 as follows.

Equation 5.26 yields

$$\begin{aligned} a_1(2)^2 + b_1(2) + c_1 &= 5 \\ a_3(7.5)^2 + b_3(7.5) + c_3 &= 1.5 \end{aligned}$$

Equation 5.27 yields

$$\begin{aligned} a_1(3)^2 + b_1(3) + c_1 &= 2.3 \\ a_2(5)^2 + b_2(5) + c_2 &= 5.1 \\ a_2(3)^2 + b_2(3) + c_2 &= 2.3 \\ a_3(5)^2 + b_3(5) + c_3 &= 5.1 \end{aligned}$$

TABLE 5.13

Data in Example 5.10

x_i	y_i
2	5
3	2.3
5	5.1
7.5	1.5

Finally, Equation 5.28 gives

$$\begin{aligned} 2a_1(3) + b_1 &= 2a_2(3) + b_2 \\ 2a_2(5) + b_2 &= 2a_3(5) + b_3 \end{aligned}$$

Substituting $a_1 = 0$ and writing the above equations in matrix form, we arrive at

$$\begin{bmatrix} 2 & 1 & 0 & 0 & 0 & 0 & 0 & 0 \\ 0 & 0 & 0 & 0 & 0 & 56.25 & 7.5 & 1 \\ 3 & 1 & 0 & 0 & 0 & 0 & 0 & 0 \\ 0 & 0 & 25 & 5 & 1 & 0 & 0 & 0 \\ 0 & 0 & 9 & 3 & 1 & 0 & 0 & 0 \\ 0 & 0 & 0 & 0 & 0 & 25 & 5 & 1 \\ 1 & 0 & -6 & -1 & 0 & 0 & 0 & 0 \\ 0 & 0 & 10 & 1 & 0 & -10 & -1 & 0 \end{bmatrix} \begin{Bmatrix} b_1 \\ c_1 \\ a_2 \\ b_2 \\ c_2 \\ a_3 \\ b_3 \\ c_3 \end{Bmatrix} = \begin{Bmatrix} 5 \\ 1.5 \\ 2.3 \\ 5.1 \\ 2.3 \\ 5.1 \\ 0 \\ 0 \end{Bmatrix}$$

This system is subsequently solved to obtain

$$\begin{array}{lll} \boxed{a_1 = 0} & a_2 = 2.05 & a_3 = -2.776 \\ b_1 = -2.7 & b_2 = -15 & b_3 = 33.26 \\ c_1 = 10.4 & c_2 = 28.85 & c_3 = -91.8 \end{array}$$

Therefore, the quadratic splines are completely defined by the following three second-degree polynomials:

$$\begin{aligned} S_1(x) &= -2.7x + 10.4, \quad 2 \le x \le 3 \\ S_2(x) &= 2.05x^2 - 15x + 28.85, \quad 3 \le x \le 5 \\ S_3(x) &= -2.776x^2 + 33.26x - 91.8, \quad 5 \le x \le 7.5 \end{aligned}$$

Results are shown graphically in Figure 5.20. Note that the first spline, $S_1(x)$, describes a straight line since $a_1 = 0$.

5.6.3 Cubic Splines

In cubic splines, third-degree polynomials are used to interpolate over each interval between data points. Suppose there are $n + 1$ data points (x_1, y_1), …, (x_{n+1}, y_{n+1}) so that there are n intervals and thus n cubic polynomials. Each cubic polynomial is conveniently expressed in the form

$$S_i(x) = a_i(x - x_i)^3 + b_i(x - x_i)^2 + c_i(x - x_i) + d_i, \quad i = 1, 2, \dots, n \tag{5.30}$$

where a_i, b_i, c_i, d_i $(i = 1, 2, \dots, n)$ are unknown constants to be determined. Since there are n such polynomials, and each has 4 unknown constants, there

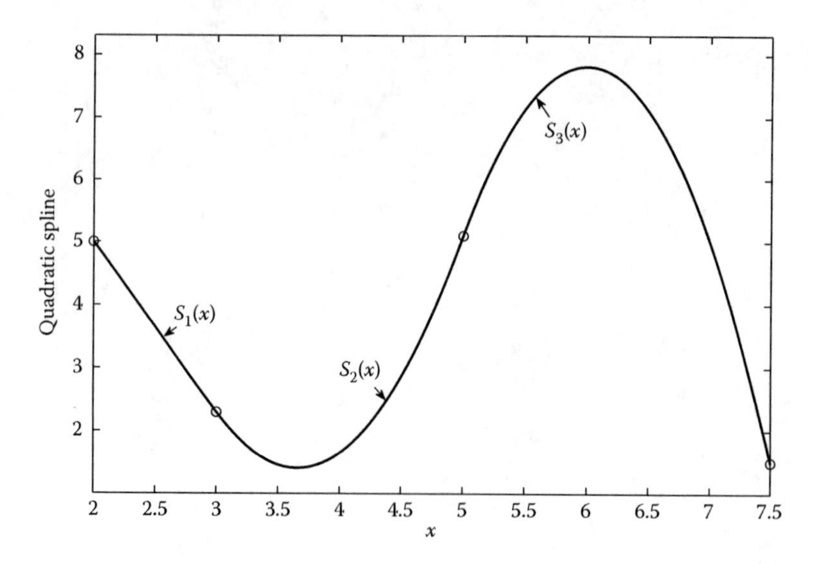

FIGURE 5.20
Quadratic splines in Example 5.10.

are a total of $4n$ unknown constants. Therefore, $4n$ equations are needed to determine all the unknowns. These equations are derived based on the same logic as quadratic splines, except that second derivatives of adjacent splines also agree at the interior knots and two boundary conditions are required.

Splines go through the endpoints and interior knots, and adjacent splines agree at the interior knots ($2n$ equations)

$$\begin{aligned} &S_1(x_1) = y_1, \quad S_n(x_{n+1}) = y_{n+1} \\ &S_{i+1}(x_{i+1}) = S_i(x_{i+1}), \quad i = 1, 2, \ldots, n-1 \\ &S_i(x_i) = y_i, \quad i = 2, 3, \ldots, n \end{aligned} \tag{5.31}$$

First derivatives of adjacent splines agree at the common interior knots ($n - 1$ equations)

$$S_i'(x_{i+1}) = S_{i+1}'(x_{i+1}), \quad i = 1, 2, \ldots, n-1 \tag{5.32}$$

Second derivatives of adjacent splines agree at the common interior knots ($n - 1$ equations)

$$S_i''(x_{i+1}) = S_{i+1}''(x_{i+1}), \quad i = 1, 2, \ldots, n-1 \tag{5.33}$$

A total of $4n - 2$ equations have been generated up to this point. The other two are provided by the boundary conditions. Boundary conditions indicate the manner in which the first spline departs from the first data point and the last spline arrives at the last data point. There are two sets of boundary conditions that are generally used for this purpose.

Clamped Boundary Conditions
The slopes with which S_1 departs from (x_1, y_1) and S_n arrives at (x_{n+1}, y_{n+1}) are specified:

$$S_1'(x_1) = p, \quad S_n'(x_{n+1}) = q \tag{5.34}$$

Free Boundary Conditions

$$S_1''(x_1) = 0, \quad S_n''(x_{n+1}) = 0 \tag{5.35}$$

The clamped boundary conditions generally yield more accurate approximations because they contain more specific information about the splines; see Example 5.12.

5.6.4 Construction of Cubic Splines: Clamped Boundary Conditions

The coefficients a_i, b_i, c_i, d_i ($i = 1, 2, \ldots, n$) will be determined by Equations 5.31 through 5.33, together with clamped boundary conditions given by Equation 5.34.

By Equation 5.30, $S_i(x_i) = d_i$ ($i = 1, 2, \ldots, n$). The first and last equations in Equation 5.31 yield $S_i(x_i) = y_i$ ($i = 1, 2, \ldots, n$). Therefore,

$$d_i = y_i, \quad i = 1, 2, \ldots, n \tag{5.36}$$

Let $h_i = x_{i+1} - x_i$ ($i = 1, 2, \ldots, n$) define the spacing between the data points. Using this in the second equation in Equation 5.31, while noting $S_{i+1}(x_{i+1}) = d_{i+1}$, we have

$$d_{i+1} = a_i h_i^3 + b_i h_i^2 + c_i h_i + d_i, \quad i = 1, 2, \ldots, n-1$$

If we define $d_{n+1} = y_{n+1}$, then the above equation will be valid for the range $i = 1, 2, \ldots, n$ since $S_n(x_{n+1}) = y_{n+1}$. Thus,

$$d_{i+1} = a_i h_i^3 + b_i h_i^2 + c_i h_i + d_i, \quad i = 1, 2, \ldots, n \tag{5.37}$$

Taking the first derivative of $S_i(x)$ and applying Equation 5.32, we find

$$c_{i+1} = 3a_i h_i^2 + 2b_i h_i + c_i, \quad i = 1, 2, \ldots, n-1$$

If we define $c_{n+1} = S'_n(x_{n+1})$, then the above equation will be valid for the range $i = 1, 2, \ldots, n$. Therefore,

$$c_{i+1} = 3a_i h_i^2 + 2b_i h_i + c_i, \quad i = 1, 2, \ldots, n \tag{5.38}$$

Taking the second derivative of $S_i(x)$ and applying Equation 5.33, yields

$$2b_{i+1} = 6a_i h_i + 2b_i, \quad i = 1, 2, \ldots, n-1$$

If we define $b_{n+1} = \frac{1}{2} S''_n(x_{n+1})$, then the above equation will be valid for the range $i = 1, 2, \ldots, n$. Therefore,

$$b_{i+1} = 3a_i h_i + b_i, \quad i = 1, 2, \ldots, n \tag{5.39}$$

The goal is to derive a system of equations for b_i ($i = 1, 2, \ldots, n+1$) only. Solve Equation 5.39 for $a_i = (b_{i+1} - b_i)/3h_i$ and substitute into Equations 5.37 and 5.38 to obtain

$$d_{i+1} = \frac{1}{3}(2b_i + b_{i+1})\, h_i^2 + c_i h_i + d_i, \quad i = 1, 2, \ldots, n \tag{5.40}$$

and

$$c_{i+1} = (b_i + b_{i+1})h_i + c_i, \quad i = 1, 2, \ldots, n \tag{5.41}$$

Solve Equation 5.40 for c_i:

$$c_i = \frac{d_{i+1} - d_i}{h_i} - \frac{1}{3}(2b_i + b_{i+1})h_i \tag{5.42}$$

Change i to $i-1$ and rewrite the above as

$$c_{i-1} = \frac{d_i - d_{i-1}}{h_{i-1}} - \frac{1}{3}(2b_{i-1} + b_i)\, h_{i-1} \tag{5.43}$$

Also change i to $i-1$ and rewrite Equation 5.41 as

$$c_i = (b_{i-1} + b_i)h_{i-1} + c_{i-1} \tag{5.44}$$

Finally, insert Equations 5.42 and 5.43 into Equation 5.44 to derive

$$b_{i-1}h_{i-1} + 2b_i(h_i + h_{i-1}) + b_{i+1}h_i = \frac{3(d_{i+1} - d_i)}{h_i} - \frac{3(d_i - d_{i-1})}{h_{i-1}}, \quad i = 2, 3, \ldots, n \tag{5.45}$$

This describes a system whose only unknowns are b_i $(i = 1, 2, \ldots, n+1)$ because d_i $(i = 1, 2, \ldots, n+1)$ are simply the values at the data points and h_i $(i = 1, 2, \ldots, n)$ define the spacing of the data. Equation 5.45, however, generates a system of $n - 1$ equations in $n + 1$ unknowns, which means two more equations are still needed. These come from the clamped boundary conditions, Equation 5.34. First, Equation 5.42 with $i = 1$ gives

$$c_1 = \frac{d_2 - d_1}{h_1} - \frac{1}{3}(2b_1 + b_2)\,h_1$$

But $c_1 = S_1'(x_1) = p$. Then, the above equation can be rewritten as

$$(2b_1 + b_2)\,h_1 = \frac{3(d_2 - d_1)}{h_1} - 3p \tag{5.46}$$

By Equation 5.41,

$$c_{n+1} = (b_n + b_{n+1})h_n + c_n$$

Knowing that $c_{n+1} = S_n'(x_{n+1}) = q$, we have

$$c_n = q - (b_n + b_{n+1})h_n \tag{5.47}$$

Equation 5.42 with $i = n$ gives

$$c_n = \frac{d_{n+1} - d_n}{h_n} - \frac{1}{3}(2b_n + b_{n+1})\,h_n$$

Substituting Equation 5.47 into the above, we have

$$(2b_{n+1} + b_n)\,h_n = -\frac{3(d_{n+1} - d_n)}{h_n} + 3q \tag{5.48}$$

Combining Equations 5.48, 5.46, and 5.45 yields a system of $n + 1$ equations in $n + 1$ unknowns b_i $(i = 1, 2, \ldots, n+1)$.

In summary, b_is are obtained by solving the system

$$\begin{aligned}
(2b_1 + b_2)\,h_1 &= \frac{3(d_2 - d_1)}{h_1} - 3p \\
b_{i-1}h_{i-1} + 2b_i\,(h_i + h_{i-1}) + b_{i+1}h_i &= \frac{3(d_{i+1} - d_i)}{h_i} - \frac{3(d_i - d_{i-1})}{h_{i-1}}, \quad i = 2, 3, \ldots, n \\
(2b_{n+1} + b_n)\,h_n &= -\frac{3(d_{n+1} - d_n)}{h_n} + 3q
\end{aligned} \tag{5.49}$$

Recall that d_i ($i = 1, 2, \ldots, n+1$) are the values at the data points and h_i ($i = 1, 2, \ldots, n$) define the spacing of the data. The system is tridiagonal (see Section 4.3) with a unique solution. Once b_is are known, Equation 5.42 is used to find c_is:

$$c_i = \frac{d_{i+1} - d_i}{h_i} - \frac{1}{3}(2b_i + b_{i+1})h_i, \quad i = 1, 2, \ldots, n \tag{5.50}$$

Finally, Equation 5.39 is used to determine a_is:

$$a_i = \frac{b_{i+1} - b_i}{3h_i}, \quad i = 1, 2, \ldots, n \tag{5.51}$$

Example 5.11: Cubic Splines, Clamped Boundary Conditions

For the data in Table 5.13 of Example 5.10 construct the cubic splines with clamped boundary conditions

$$p = -1, \quad q = 1$$

SOLUTION

Since there are 4 data points, we have $n = 3$ so that there are 3 cubic polynomials

$$S_i(x) = a_i(x - x_i)^3 + b_i(x - x_i)^2 + c_i(x - x_i) + d_i, \quad i = 1, 2, 3$$

Following the summarized procedure outlined above, we first find b_is by solving the system in Equation 5.49:

$$\begin{aligned}
(2b_1 + b_2)\,h_1 &= \frac{3(d_2 - d_1)}{h_1} - 3p \\
b_1h_1 + 2b_2\,(h_2 + h_1) + b_3h_2 &= \frac{3(d_3 - d_2)}{h_2} - \frac{3(d_2 - d_1)}{h_1} \\
b_2h_2 + 2b_3\,(h_3 + h_2) + b_4h_3 &= \frac{3(d_4 - d_3)}{h_3} - \frac{3(d_3 - d_2)}{h_2} \\
(2b_4 + b_3)\,h_3 &= -\frac{3(d_4 - d_3)}{h_3} + 3q
\end{aligned}$$

Note that d_is are simply the values at data points; hence

$$d_1 = 5, \quad d_2 = 2.3, \quad d_3 = 5.1, \quad d_4 = 1.5$$

Also, $h_1 = 1$, $h_2 = 2$, and $h_3 = 2.5$. Substituting these, together with $p = -1$ and $q = 1$, the system reduces to

$$\begin{bmatrix} 2 & 1 & 0 & 0 \\ 1 & 6 & 2 & 0 \\ 0 & 2 & 9 & 2.5 \\ 0 & 0 & 2.5 & 5 \end{bmatrix} \begin{Bmatrix} b_1 \\ b_2 \\ b_3 \\ b_4 \end{Bmatrix} = \begin{Bmatrix} -5.1 \\ 12.3 \\ -8.52 \\ 7.32 \end{Bmatrix}$$

which is tridiagonal, as asserted, and its solution is

$$b_1 = -4.3551, \quad b_2 = 3.6103, \quad b_3 = -2.5033, \quad b_4 = 2.7157$$

Next, c_is are found by solving Equation 5.50:

$$\begin{aligned} c_1 &= \frac{d_2 - d_1}{h_1} - \frac{1}{3}(2b_1 + b_2)\,h_1 = -1 \\ c_2 &= \frac{d_3 - d_2}{h_2} - \frac{1}{3}(2b_2 + b_3)\,h_2 = -1.7449 \\ c_3 &= \frac{d_4 - d_3}{h_3} - \frac{1}{3}(2b_3 + b_4)\,h_3 = 0.4691 \end{aligned}$$

Finally, a_is come from Equation 5.51:

$$\begin{aligned} a_1 &= \frac{b_2 - b_1}{3h_1} = 2.6551 \\ a_2 &= \frac{b_3 - b_2}{3h_2} = -1.0189 \\ a_3 &= \frac{b_4 - b_3}{3h_3} = 0.6959 \end{aligned}$$

Therefore, the three cubic splines are determined as

$$S_1(x) = 2.6551(x-2)^3 - 4.3551(x-2)^2 - (x-2) + 5, \quad 2 \le x \le 3$$

$$S_2(x) = -1.0189(x-3)^3 + 3.6103(x-3)^2 - 1.7449(x-3) + 2.3, \quad 3 \le x \le 5$$

$$S_3(x) = 0.6959(x-5)^3 - 2.5033(x-5)^2 + 0.4691(x-5) + 5.1, \quad 5 \le x \le 7.5$$

The results are illustrated in Figure 5.21, where it is clearly seen that cubic splines are much more desirable than the quadratic splines obtained for the same set of data.

5.6.5 Construction of Cubic Splines: Free Boundary Conditions

Recall that free boundary conditions are $S_1''(x_1) = 0$, $S_n''(x_{n+1}) = 0$. Knowing $S_1''(x) = 6a_1(x - x_1) + 2b_1$, the first condition yields $b_1 = 0$. From previous work, $b_{n+1} = \frac{1}{2}S_n''(x_{n+1})$ so that the second condition implies $b_{n+1} = 0$. Combining these with Equation 5.45 forms a system of $n + 1$ equations in $n + 1$ unknowns that can be solved for b_is:

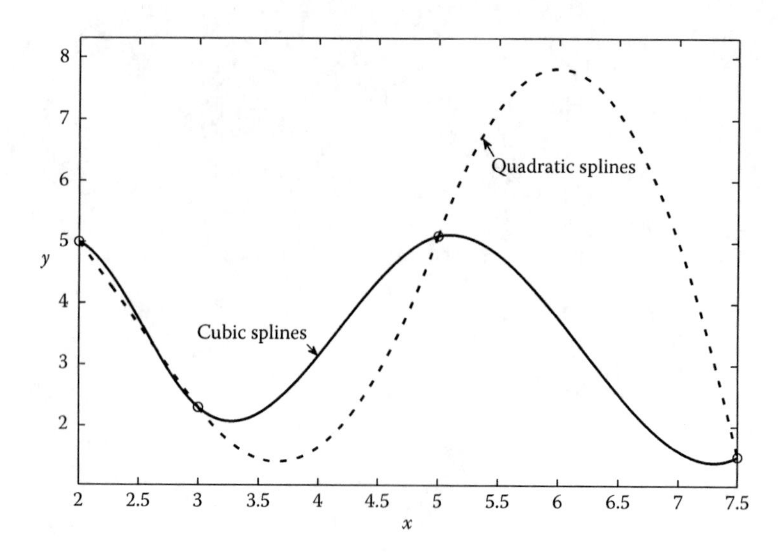

FIGURE 5.21
Cubic and quadratic splines for the same set of data.

$$\begin{aligned}
b_1 &= 0 \\
b_{i-1}h_{i-1} + 2b_i(h_i + h_{i-1}) + b_{i+1}h_i &= \frac{3(d_{i+1} - d_i)}{h_i} - \frac{3(d_i - d_{i-1})}{h_{i-1}}, \quad i = 2, 3, \ldots, n \\
b_{n+1} &= 0
\end{aligned} \tag{5.52}$$

Once b_is are available, all other unknown constants are determined as in the case of clamped boundary conditions. In summary, d_i $(i = 1, 2, \ldots, n+1)$ are the values at the data points, h_i $(i = 1, 2, \ldots, n)$ define the spacing of the data, b_is come from Equation 5.52, c_is from Equation 5.50, and a_is from Equation 5.51.

Example 5.12: Cubic Splines, Free Boundary Conditions

For the data in Table 5.13 of Examples 5.10 and 5.11, construct the cubic splines with free boundary conditions.

SOLUTION

The free boundary conditions imply that $b_1 = 0$, $b_4 = 0$. Consequently, the system in Equation 5.52 simplifies to

$$\begin{aligned}
b_1 &= 0 \\
6b_2 + 2b_3 &= 12.3 \\
2b_2 + 9b_3 &= -8.52 \\
b_4 &= 0
\end{aligned} \quad \Rightarrow \quad \begin{aligned}
b_2 &= 2.5548 \\
b_3 &= -1.5144
\end{aligned}$$

Next, c_is are found by solving Equation 5.50:

$$c_1 = \frac{d_2 - d_1}{h_1} - \frac{1}{3}(2b_1 + b_2)\,h_1 = -3.5516$$
$$c_2 = \frac{d_3 - d_2}{h_2} - \frac{1}{3}(2b_2 + b_3)\,h_2 = -0.9968$$
$$c_3 = \frac{d_4 - d_3}{h_3} - \frac{1}{3}(2b_3 + b_4)\,h_3 = 1.0840$$

Finally, a_is come from Equation 5.51:

$$a_1 = \frac{b_2 - b_1}{3h_1} = 0.8516$$
$$a_2 = \frac{b_3 - b_2}{3h_2} = -0.6782$$
$$a_3 = \frac{b_4 - b_3}{3h_3} = 0.2019$$

Therefore, the three cubic splines are determined as

$$S_1(x) = 0.8516(x-2)^3 - 3.5516(x-2) + 5, \quad 2 \le x \le 3$$
$$S_2(x) = -0.6782(x-3)^3 + 2.5548(x-3)^2 - 0.9968(x-3) + 2.3, \quad 3 \le x \le 5$$
$$S_3(x) = 0.2019(x-5)^3 - 1.5144(x-5)^2 + 1.0840(x-5) + 5.1, \quad 5 \le x \le 7.5$$

The graphical results are shown in Figure 5.22, where it is observed that the clamped boundary conditions lead to more accurate approximations, as stated earlier.

5.6.6 MATLAB® Built-In Functions `interp1` and `spline`

Brief descriptions of the MATLAB built-in functions `interp1` and `spline` are given as

```
YI = INTERP1(X,Y,XI,METHOD) specifies alternate methods.
    The default is linear interpolation. Use an empty matrix []
    to specify the default. Available methods are:

'nearest' - nearest neighbor interpolation
'linear'  - linear interpolation
'spline'  - piecewise cubic spline interpolation (SPLINE)
'pchip'   - shape-preserving piecewise cubic Hermite interpolation
```

Of the four methods, the nearest-neighbor interpolation is the fastest, and does not generate new points. It only returns values that already exist in the `Y` vector. The linear method is slightly slower than the nearest-neighbor

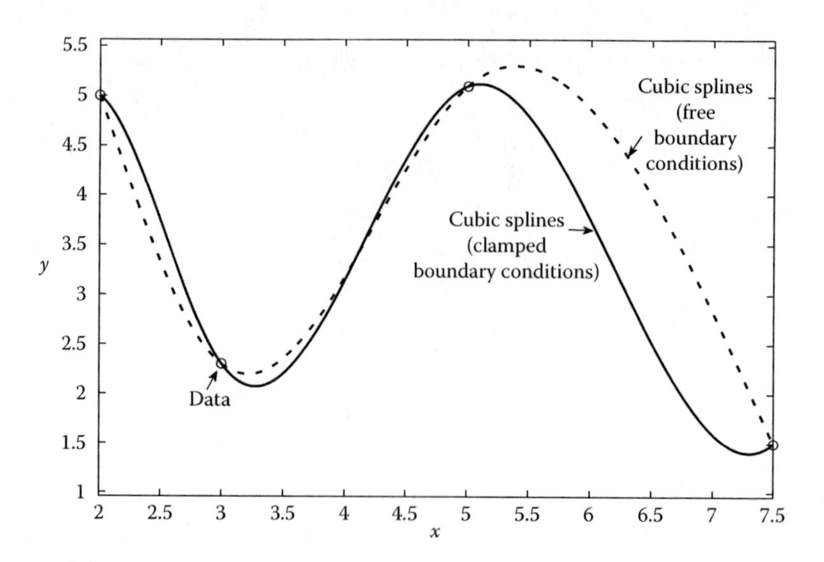

FIGURE 5.22
Cubic splines with clamped and free boundary conditions; Example 5.12.

method and returns values that approximate a continuous function. Each of the `pchip` and `spline` methods generates a different cubic polynomial between any two data points, and uses these points as two of the constraints when determining the polynomial. The difference between these two methods is that `pchip` seeks to match the first-order derivatives at these points with those of the intervals before and after, which is a characteristic of Hermite interpolation. The `spline` method tries to match the second-order derivatives at these points with those of the intervals before and after.

The `pchip` method produces a function whose minimums match the minimums of the data. Also, the function is monotonic over intervals where the data are monotonic. The `spline` method produces a smooth (twice-continuously differentiable) function, but will overshoot and undershoot the given data.

Example 5.13: MATLAB® Function `interp1`

Consider the data for `x = -2:0.5:2` generated by $y = \frac{1}{4}x^4 - \frac{1}{2}x^2$. Interpolate and plot using the four different methods listed in `interp1`.

SOLUTION

```
>> x = -2:0.5:2;
>> y = 1./4.*x.^4-1./2.*x.^2;
>> xi = linspace(-2, 2);
>> ynear = interp1(x, y, xi, 'nearest');
>> ylin = interp1(x, y, xi, 'linear');
>> ypc = interp1(x, y, xi, 'pchip');
```

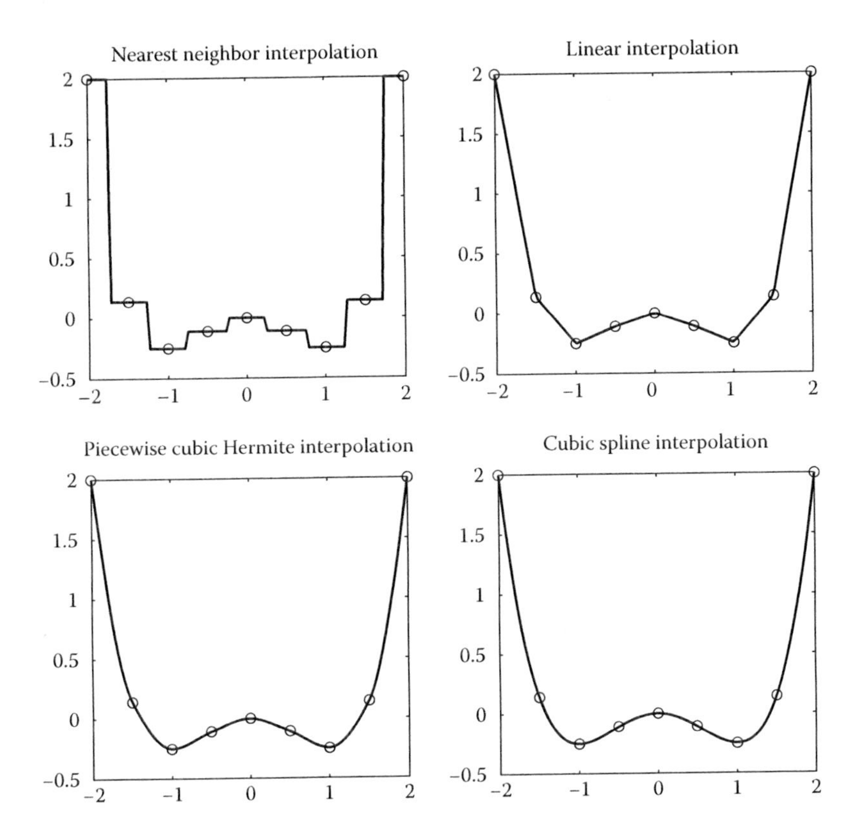

FIGURE 5.23
Interpolation by `interp1` using four methods; Example 5.13.

```
>> yspl = interp1(x, y, xi, 'spline');
% Start Figure 5.23
>> subplot(2,2,1), plot(xi,ynear,x,y,'o'), title('Nearest
   neighbor interpolation')
>> hold on
>> subplot(2,2,2), plot(xi,ylin,x,y,'o'), title('Linear
   interpolation')
>> subplot(2,2,3), plot(xi,ypc,x,y,'o'), title('Piecewise
   cubic Hermite interpolation')
>> subplot(2,2,4), plot(xi,yspl,x,y,'o'), title('Cubic
   spline interpolation')
```

5.6.7 Boundary Conditions

`YI = INTERP1(X,Y,XI,'spline') uses piecewise cubic splines interpolation. Note that the option 'spline' does not allow for specific boundary conditions.`

`PP = SPLINE(X,Y) provides the piecewise polynomial form of the cubic spline interpolant to the data values Y at the data sites`

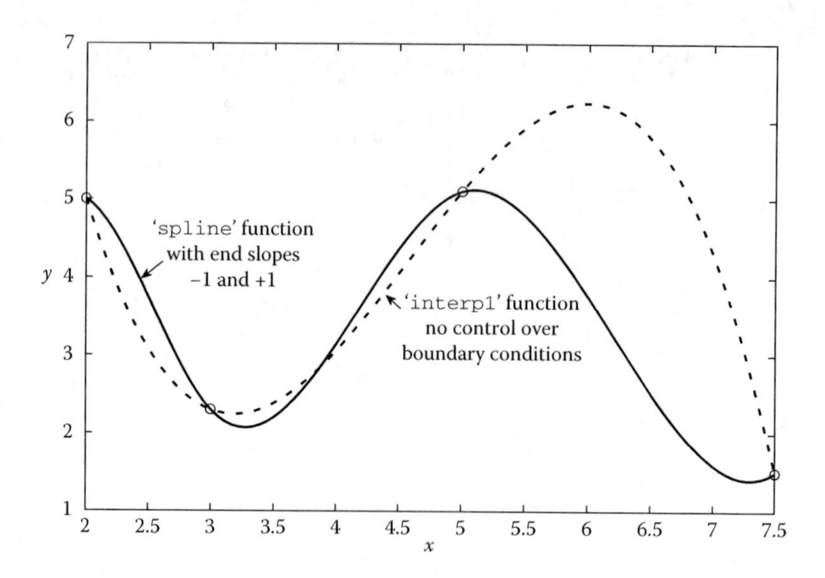

FIGURE 5.24
Cubic splines using MATLAB built-in functions.

```
X. Ordinarily, the not-a-knot end conditions are used. However,
if Y contains two more values than X has entries, then the
first and last value in Y are used as the end slopes for the
cubic spline.
```

We will apply these two functions to the set of data considered in Examples 5.10 through 5.12.

```
>> x = [2 3 5 7.5] y = [5 2.3 5.1 1.5];
>> xi = linspace(2,7.5);
>> yi = interp1(x,y,xi,'spline');
% No control over boundary conditions
>> plot(x,y,'o',xi,yi)
>> cs = spline(x,[-1 y 1]);
% Specify boundary conditions: end slopes of -1 and 1
>> hold on
>> plot(x,y,'o',xi,ppval(cs,xi),'-');     % Figure 5.24
```

5.6.8 Interactive Curve Fitting and Interpolation in MATLAB®

The basic fitting interface in MATLAB allows for interactive curve fitting of data. First plot the data. Then, under the "tools" menu choose "basic fitting." This opens a new window on the side with a list of fitting methods, including spline interpolation and different degree polynomial interpolation. By simply checking the box next to the desired method, the corresponding curve is generated and plotted. Figure 5.25 shows the spline and seventh-degree polynomial interpolations of a set of 21 data points.

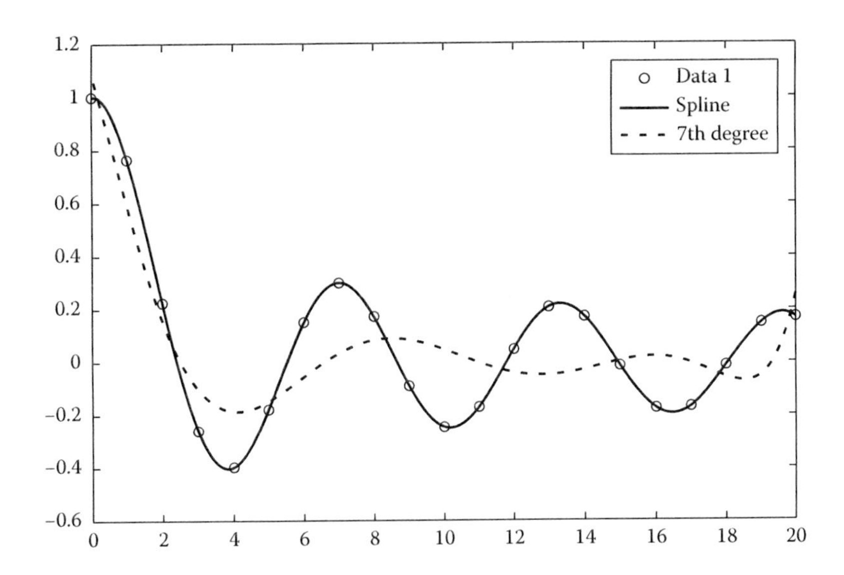

FIGURE 5.25
The basic fitting interface in MATLAB.

5.7 Fourier Approximation and Interpolation

So far in this chapter we have discussed curve fitting and interpolation of data using polynomials. But in engineering we often deal with systems that oscillate. To that end, trigonometric functions 1, cos t, cos $2t$, . . . , sin t, sin $2t$, . . . are used for modeling such systems. Fourier approximation outlines the systematic use of trigonometric series for this purpose.

5.7.1 Sinusoidal Curve Fitting

To present the idea, we first consider a very special set of equally spaced data and then use a linear transformation to apply the results to any given equally spaced data.

Consider N data points $(\sigma_1, x_1), (\sigma_2, x_2), \ldots, (\sigma_N, x_N)$, where σ_k $(k = 1, 2, \ldots, N)$ are assumed to be equally spaced along the interval $[0,2\pi)$, that is

$$\sigma_1 = 0,\ \sigma_2 = \frac{2\pi}{N},\ \sigma_3 = 2\left(\frac{2\pi}{N}\right), \ldots,\ \sigma_N = (N-1)\left(\frac{2\pi}{N}\right)$$

It is desired to interpolate or approximate this set of data by means of a function in the form

$$f(\sigma) = \frac{1}{2}a_0 + \sum_{j=1}^{m}\left[a_j \cos j\sigma + b_j \sin j\sigma\right] = \frac{1}{2}a_0 + a_1 \cos\sigma + \cdots +$$

$$a_m \cos m\sigma + b_1 \sin\sigma + \cdots + b_m \sin m\sigma \qquad (5.53)$$

where $f(\sigma)$ is a trigonometric polynomial of degree m if both a_m and b_m are not zero. Interpolation requires $f(\sigma)$ to go through the data points, while approximation (curve fit) is in the sense of least squares; Section 5.4. More specifically, the coefficients $a_0, a_1, \ldots, a_m, b_1, \ldots, b_m$ are determined so as to minimize

$$Q = \sum_{k=1}^{N}\left\{\left(\frac{1}{2}a_0 + \sum_{j=1}^{m}\left[a_j \cos j\sigma_k + b_j \sin j\sigma_k\right]\right) - x_k\right\}^2 \qquad (5.54)$$

A necessary condition for Q to attain a minimum is that its partial derivatives with respect to all coefficients vanish, that is

$$\frac{\partial Q}{\partial a_0} = 0, \quad \frac{\partial Q}{\partial a_j} = 0 \quad (j = 1, 2, \ldots, m), \quad \frac{\partial Q}{\partial b_j} = 0 \quad (j = 1, 2, \ldots, m)$$

The ensuing system of $2m + 1$ equations can then be solved to yield

$$\begin{aligned} a_j &= \frac{2}{N}\sum_{k=1}^{N} x_k \cos j\sigma_k, \quad j = 0, 1, 2, \ldots, m \\ b_j &= \frac{2}{N}\sum_{k=1}^{N} x_k \sin j\sigma_k, \quad j = 1, 2, \ldots, m \end{aligned} \qquad (5.55)$$

Fourier Approximation

If $2m + 1 < N$ (there are more data points than unknown coefficients), then the least-squares approximation of the data is described by Equation 5.53 with coefficients given by Equation 5.55.

Fourier Interpolation

For interpolation, the suitable form of the trigonometric polynomial depends on whether N is odd or even.

- **Case (1)** $N = odd = 2m + 1$

  In this case, the interpolating trigonometric polynomial is exactly in the form of the approximating polynomial.

- **Case (2)** $N = even = 2m$

  The interpolating polynomial is in the form

$$f(\sigma) = \frac{1}{2}a_0 + a_1 \cos\sigma + a_2 \cos 2\sigma + \cdots + a_{m-1} \cos(m-1)\sigma + \frac{1}{2}a_m \cos m\sigma$$
$$+ b_1 \sin\sigma + b_2 \sin 2\sigma + \cdots + b_{m-1} \sin(m-1)\sigma \qquad (5.56)$$

where the coefficients are once again given by Equation 5.55.

5.7.2 Linear Transformation of Data

Fourier approximation or interpolation of an equally spaced data $(t_1, x_1),(t_2, x_2), \ldots, (t_N, x_N)$ is handled as follows. First assume the data is $(\sigma_1, x_1),(\sigma_2, x_2), \ldots, (\sigma_N, x_N)$ equally spaced over $[0,2\pi)$ and apply the results presented above. Then transform the data back to its original form via (see Figure 5.26)

$$t = \frac{N(t_N - t_1)}{(N-1)2\pi}\sigma + t_1 \qquad (5.57)$$

The user-defined function `TrigPoly` finds the appropriate Fourier approximation or interpolation of an equally spaced data by first assuming the data is equally spaced over $[0,2\pi)$ and then transforming the data to agree with the range of the original set. The function also returns the plot of the interpolating/approximating trigonometric polynomial and the given set of data.

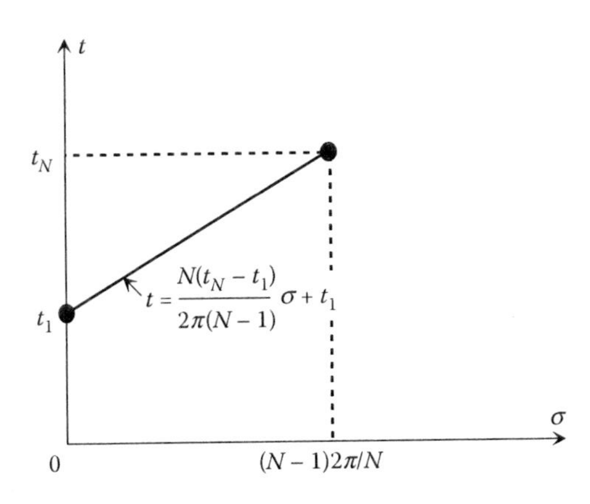

FIGURE 5.26
Linear transformation of data.

```
function [a, b] = TrigPoly(x, m, t1, tN)
%
% TrigPoly approximates or interpolates a set of equally
% spaced data (t1, x1),. . ., (tN, xN) by a trigonometric
% polynomial of degree m.

%     [a, b] = TrigPoly(x, m, t1, tN) where
%
%        x = [x1 x2. . . xN],
%        m is the degree of the trigonometric polynomial,
%        t1 and tN define the interval endpoints (interval
%        open at tN),
%
%        a and b are the vectors of coefficients of the
%        polynomial.
%
% Case(1) Approximation if 2*m+1<N,
% Case(2) Interpolation if 2*m+1=N or 2*m=N.

N = length(x);

% Consider an equally-spaced data from s=0 to s=2*pi
h = 2*pi/N; s = 0:h:2*pi-h; s = s';
a = zeros(m,1);      % Pre-allocate
b = a;

for i = 1:m,
    a(i) = x*cos(i*s);
    b(i) = x*sin(i*s);
end
a = 2*a/N; b = 2*b/N; a0 = sum(x)/N;
if N == 2*m,
    a(m) = a(m)/2;
end

ss = linspace(0,2*pi*(N-1)/N,500);
% 500 points for plotting
xx = a0 + a(1)*cos(ss) + b(1)*sin(ss);
for i = 2:m,
    xx = xx + a(i)*cos(i*ss) + b(i)*sin(i*ss);
end

% Transform from s to t
t = N*((tN-t1)/(2*pi*(N-1)))*s + t1;
tt = N*((tN-t1)/(2*pi*(N-1)))*ss + t1;

plot(tt,xx,t,x,'o')
a = [a0;a];
```

TABLE 5.14

Data in Example 5.14

t_i	x_i
0.5	6.8
0.7	3.2
0.9	−4.1
1.1	−3.9
1.3	3.3

Example 5.14: Fourier Approximation

Find the first-degree approximating or interpolating trigonometric polynomial for the data in Table 5.14. Confirm the results by executing the user-defined function `TrigPoly`.

SOLUTION

First treat the data as $(\sigma_1, x_1),(\sigma_2, x_2), \ldots, (\sigma_5, x_5)$, equally spaced over $[0,2\pi)$. That is

$$\sigma_1 = 0, \quad \sigma_2 = \frac{2\pi}{5}, \quad \sigma_3 = \frac{4\pi}{5}, \quad \sigma_4 = \frac{6\pi}{5}, \quad \sigma_5 = \frac{8\pi}{5}$$

Since $m = 1$ and $N = 5$, we have $2m + 1 < N$ so that the polynomial in Equation 5.53 is the suitable form for approximation; in particular, $f(\sigma) = \frac{1}{2}a_0 + a_1\cos\sigma + b_1\sin\sigma$. The coefficients are provided by Equation 5.55 as

$$a_0 = \frac{2}{5}\sum_{k=1}^{5} x_k = \frac{2}{5}(6.8 + 3.2 - 4.1 - 3.9 + 3.3) = 2.1200$$

$$a_1 = \frac{2}{5}\sum_{k=1}^{5} x_k\cos\sigma_k = \frac{2}{5}[6.8\cos(0) + 3.2\cos(2\pi/5)$$

$$- 4.1\cos(4\pi/5) - 3.9\cos(6\pi/5) + 3.3\cos(8\pi/5)] = 6.1123$$

$$b_1 = \frac{2}{5}\sum_{k=1}^{5} x_k\sin\sigma_k = \frac{2}{5}[6.8\sin(0) + 3.2\sin(2\pi/5) - 4.1\sin(4\pi/5)$$

$$- 3.9\sin(6\pi/5) + 3.3\sin(8\pi/5)] = -0.0851$$

Therefore, the least-squares approximating polynomial is $f(\sigma) = 1.06 + 6.1123\cos\sigma - 0.0851\sin\sigma$. But variables t and σ are related via Equation 5.57,

$$t = \frac{N(t_N - t_1)}{(N-1)2\pi}\sigma + t_1 = \frac{5(1.3 - 0.5)}{4(2\pi)}\sigma + 0.5 = 0.1592\sigma + 0.5$$

$$\Rightarrow \quad \sigma = 6.2832(t - 0.5)$$

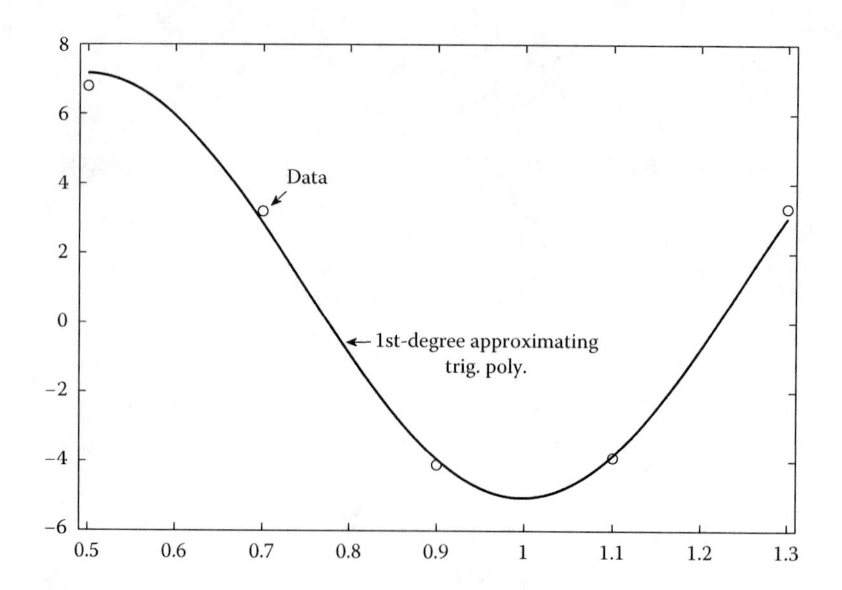

FIGURE 5.27
Fourier approximation of the data in Example 5.14.

The approximating trigonometric polynomial is then formed as

$$f(t) = 1.06 + 6.1123\cos(6.2832(t - 0.5)) - 0.0851\sin(6.2832(t - 0.5))$$

Executing the user-defined function `TrigPoly` yields the coefficients of the trigonometric polynomial as well as the plot of this polynomial and the original data. This is shown in Figure 5.27.

```
>> x = [6.8 3.2 -4.1 -3.9 3.3];
>> [a, b] = TrigPoly(x, 1, 0.5, 1.3)

a =
     1.0600
     6.1123

b =
    -0.0851
```

Example 5.15: Fourier Interpolation

Find the third-degree approximating or interpolating trigonometric polynomial for the data in Table 5.15. Confirm the results by executing the user-defined function `TrigPoly`. Find the interpolated value at $t = 0.63$.

SOLUTION

First treat the data as $(\sigma_1, x_1), (\sigma_2, x_2), \ldots, (\sigma_6, x_6)$, equally spaced over $[0,2\pi)$. That is

TABLE 5.15

Data in Example 5.15

t_i	x_i
0.10	0
0.25	0
0.40	0
0.55	1
0.70	1
0.85	1

$$\sigma_1 = 0, \quad \sigma_2 = \frac{\pi}{3}, \quad \sigma_3 = \frac{2\pi}{3}, \quad \sigma_4 = \pi, \quad \sigma_5 = \frac{4\pi}{3}, \quad \sigma_6 = \frac{5\pi}{3}$$

Since $m = 3$ and $N = 6$, we have $2m = N$ so that the trigonometric polynomial interpolates the data and is given by Equation 5.56, more specifically

$$f(\sigma) = \frac{1}{2}a_0 + a_1 \cos\sigma + a_2 \cos 2\sigma + \frac{1}{2}a_3 \cos 3\sigma + b_1 \sin\sigma + b_2 \sin 2\sigma$$

The coefficients are provided by Equation 5.55 as

$$a_0 = \frac{2}{6}\sum_{k=1}^{6} x_k = 1, \quad a_1 = \frac{2}{6}\sum_{k=1}^{6} x_k \cos\sigma_k = -0.3333,$$

$$a_2 = \frac{2}{6}\sum_{k=1}^{6} x_k \cos 2\sigma_k = 0, \quad a_3 = \frac{2}{6}\sum_{k=1}^{6} x_k \cos 3\sigma_k = -0.3333,$$

$$b_1 = \frac{2}{6}\sum_{k=1}^{6} x_k \sin\sigma_k = -0.5774, \quad b_2 = \frac{2}{6}\sum_{k=1}^{6} x_k \sin 2\sigma_k = 0$$

This yields $f(\sigma) = \frac{1}{2} - 0.3333\cos\sigma - 0.1667\cos 3\sigma - 0.5774\sin\sigma$. But variables t and σ are related via Equation 5.57

$$t = \frac{N(t_N - t_1)}{(N-1)2\pi}\sigma + t_1 = \frac{6(0.85 - 0.10)}{5(2\pi)}\sigma + 0.10 = 0.1432\sigma + 0.10$$

$$\Rightarrow \quad \sigma = 6.9813(t - 0.10)$$

Therefore, the approximating trigonometric polynomial is formed as

$$f(t) = \frac{1}{2} - 0.3333\cos(6.9813(t - 0.10)) - 0.1667$$
$$\times \cos(3 \times 6.9813(t - 0.10)) - 0.5774\sin(6.9813(t - 0.10))$$

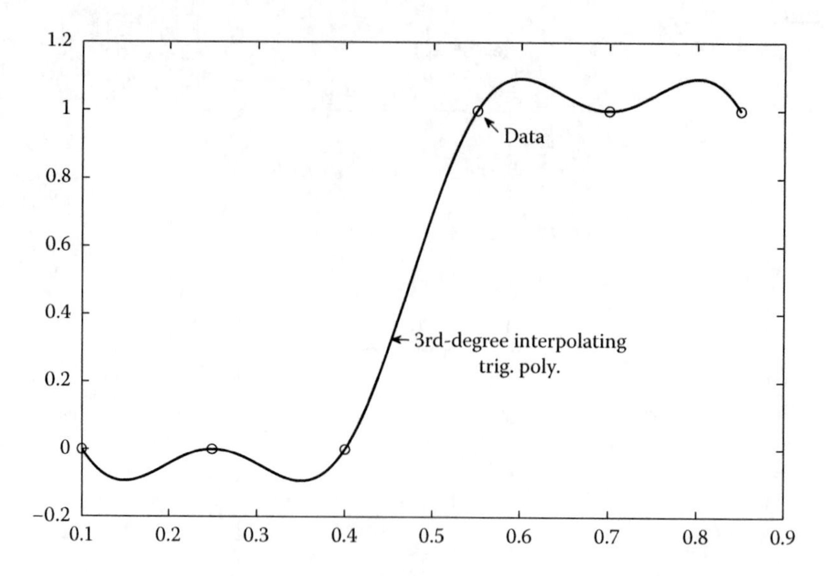

FIGURE 5.28
Fourier interpolation of the data in Example 5.15.

The interpolated value at $t = 0.63$ is

$$f(0.63) = 1.0712$$

Executing the user-defined function `TrigPoly` yields the coefficients of the trigonometric polynomial as well as the plot of this polynomial and the original data. This is shown in Figure 5.28.

```
>> x = [0 0 0 1 1 1];
>> [a, b] = TrigPoly(x, 3, 0.10, 0.85)

a =
     0.5000
    -0.3333
    -0.0000
    -0.1667

b =
    -0.5774
     0.0000
     0.0000
```

The numerical results agree with those obtained earlier.

5.7.3 Discrete Fourier Transform

Periodic functions can conveniently be represented by Fourier series. But there are many functions or signals that are not periodic; for example, an impulsive force applied to a mechanical system will normally have a relatively large magnitude and will be applied for a very short period of time. Such nonperiodic signals are best represented by the Fourier integral. The Fourier integral of a function can be obtained while taking the Fourier transform of that function. The Fourier transform pair for a continuous function $x(t)$ is defined as

$$\hat{x}(\omega) = \int_{-\infty}^{\infty} x(t)e^{-i\omega t}dt$$
$$x(t) = \frac{1}{2\pi}\int_{-\infty}^{\infty} \hat{x}(\omega)e^{i\omega t}d\omega \tag{5.58}$$

The first equation gives the Fourier transform $\hat{x}(\omega)$ of $x(t)$. The second equation uses $\hat{x}(\omega)$ to represent $x(t)$ as an integral known as the Fourier integral. In applications, data is often collected as a discrete set of values and hence $x(t)$ is not available in the form of a continuous function.

Divide an interval $[0,T]$ into N equally spaced subintervals, each of width $h = T/N$. Consider the set of data chosen as $(t_0, x_0), \ldots, (t_{N-1}, x_{N-1})$ such that $t_0 = 0, t_1 = t_0 + h, \ldots, t_{N-1} = t_0 + (N-1)h$. Note that the point $t_N = T$ is not included.* Equation 5.58 is then discretized as

$$\hat{x}_k = \sum_{n=0}^{N-1} x_n e^{-ik\omega n}, \quad k = 0, 1, \ldots, N-1, \quad \omega = \frac{2\pi}{N} \tag{5.59}$$

$$x_n = \frac{1}{N}\sum_{k=0}^{N-1} \hat{x}_k e^{ik\omega n}, \quad n = 0, 1, \ldots, N-1 \tag{5.60}$$

where $\hat{x}_k$ is known as the discrete Fourier transform (DFT). Equations 5.59 and 5.60 can be used to compute the Fourier and inverse Fourier transform for a set of discrete data. Calculation of the DFT in Equation 5.59 requires N^2 complex operations. Therefore, even for data of moderate size, such calculations can be quite time-consuming. To remedy that, the fast Fourier transform (FFT) is developed for efficient computation of the DFT. What makes FFT computationally attractive is that it reduces the number of operations by using the results of previous calculations.

5.7.4 Fast Fourier Transform

The FFT algorithm requires roughly $N \log_2 N$ operations as opposed to N^2 by the DFT, and it does so by using the fact that trigonometric functions

---

* Refer to R.W. Ramirez, *The FFT, Fundamentals and Concepts*, Prentice-Hall, 1985.

are periodic and symmetric. For instance, for $N = 100$, the FFT is roughly 15 times faster than the DFT. For $N = 500$, it is about 56 times faster. The first major contribution leading to an algorithm for computing the FFT was made by J. W. Cooley and J. W. Tukey in 1965, known as the Cooley–Tukey algorithm. Since then, a number of other methods have been developed that are essentially consequences of their approach.

The basic idea behind all of these techniques is to decompose, or decimate, a DFT of length N into successively smaller length DFTs. This can be achieved via decimation-in-time or decimation-in-frequency techniques. The Cooley–Tukey method, for example, is a decimation-in-time technique. Here, we will discuss an alternative approach, the Sande–Tukey algorithm, which is a decimation-in-frequency method. The two decimation techniques differ in how they are organized, but they both require $N \log_2 N$ operations. We will limit our presentation to the case $N = 2^p$ (integer p) for which the techniques work best, but analogous methods will clearly work for the general case $N = N_1 N_2 \ldots N_m$, where each N_i is an integer.

5.7.4.1 Sande–Tukey Algorithm ($N = 2^p$, p = integer)

We will present the simplified algorithm for the special case $N = 2^p$, where p is some integer. Recall from Equation 5.59 that the DFT is given by

$$\hat{x}_k = \sum_{n=0}^{N-1} x_n e^{-ik(2\pi/N)n}, \quad k = 0, 1, \ldots, N-1 \tag{5.61}$$

Define the weighting function $W = e^{-(2\pi/N)i}$ so that Equation 5.61 may also be written as

$$\hat{x}_k = \sum_{n=0}^{N-1} x_n W^{kn}, \quad k = 0, 1, \ldots, N-1 \tag{5.62}$$

We next divide the sample of length N in half, each half containing $N/2$ points, and write Equation 5.61 as

$$\hat{x}_k = \sum_{n=0}^{(N/2)-1} x_n e^{-ik(2\pi/N)n} + \sum_{n=N/2}^{N-1} x_n e^{-ik(2\pi/N)n}$$

Since summations can only be combined if their indices cover the same range, introduce a change of variables in the second summation and rewrite this last equation as

$$\hat{x}_k = \sum_{n=0}^{(N/2)-1} x_n e^{-ik(2\pi/N)n} + \sum_{n=0}^{(N/2)-1} x_{n+N/2}\, e^{-ik(2\pi/N)(n+N/2)}$$

$$\overset{\text{Combine}}{=} \sum_{n=0}^{(N/2)-1} [x_n + e^{-i\pi k} x_{n+N/2}] e^{-2\pi kni/N} \tag{5.63}$$

But

$$e^{-i\pi k} = \cos k\pi - i\sin k\pi = (-1)^k = \begin{cases} 1 & \text{if} \quad k = \text{even} \\ -1 & \text{if} \quad k = \text{odd} \end{cases}$$

Therefore, the expression for $\hat{x}_k$ will depend on whether the index is even or odd. For even index, Equation 5.63 yields

$$\hat{x}_{2k} \underset{2k \text{ for } k}{\overset{\text{Substitute}}{=}} \sum_{n=0}^{(N/2)-1} \left[x_n + x_{n+N/2}\right] e^{-2\pi(2k)ni/N} \overset{\text{Rewrite}}{=} \sum_{n=0}^{(N/2)-1} \left[x_n + x_{n+N/2}\right] e^{-[(2\pi/N)i]2kn}$$

For odd index

$$\hat{x}_{2k+1} \underset{2k+1 \text{ for } k}{\overset{\text{Substitute}}{=}} \sum_{n=0}^{(N/2)-1} \left[x_n - x_{n+N/2}\right] e^{-2\pi(2k+1)ni/N}$$

$$\overset{\text{Rewrite}}{=} \sum_{n=0}^{(N/2)-1} \left[x_n - x_{n+N/2}\right] e^{-[(2\pi/N)i]n} e^{-[(2\pi/N)i]2kn}$$

In terms of $W = e^{-(2\pi/N)i}$, defined earlier

$$\hat{x}_{2k} = \sum_{n=0}^{(N/2)-1} [x_n + x_{n+N/2}] W^{2kn} \tag{5.64}$$

$$\hat{x}_{2k+1} = \sum_{n=0}^{(N/2)-1} \{[x_n - x_{n+N/2}] W^n\} W^{2kn} \tag{5.65}$$

Next, define

$$\begin{aligned} y_n &= x_n + x_{n+N/2} \\ z_n &= \left[x_n - x_{n+N/2}\right] W^n \end{aligned}, \quad n = 0, 1, \ldots, (N/2) - 1 \tag{5.66}$$

Inspired by Equation 5.62, it is easy to see that the summations in Equations 5.64 and 5.65 simply represent the transforms of y_n and z_n. That is

$$\begin{aligned}\hat{x}_{2k} &= \hat{y}_k \\ \hat{x}_{2k+1} &= \hat{z}_k\end{aligned}, \quad k = 0, 1, \ldots, (N/2) - 1$$

Consequently, the original N-point computation has been replaced by two $(N/2)$-point computations, each requiring $(N/2)^2 = N^2/4$ operations for a total of $N^2/2$. Comparing with N^2 for the original data, the algorithm manages to reduce the number of operations by a factor of 2. The decomposition continues, with the number of sample points divided by two in each step, until $N/2$ two-point DFTs are computed. To better understand how the scheme works, we present the details involving an 8-point sample.

Case Study: $N = 2^3 = 8$

An 8-point DFT is to be decomposed successively using Sande–Tukey algorithm (decimation-in-frequency) into smaller DFTs. Figure 5.29 shows the details in the first stage where two 4-point DFTs are generated. The intersections that are accompanied by "+" and/or "−" signs act as summing junctions. For example, by Equation 5.66, we have

$$y_0 = x_0 + x_4, \quad z_0 = (x_0 - x_4)W^0$$

The operation $y_0 = x_0 + x_4$ is handled by a simple addition of two signals. To obtain z_0, we first perform $x_0 - x_4$, then send the outcome to a block of W^0. The same logic applies to the remainder of the sample.

Next, each of the 4-point DFTs will be decomposed into two 2-point DFTs, which will mark the end of the process for the case of $N = 8$; see Figure 5.30.

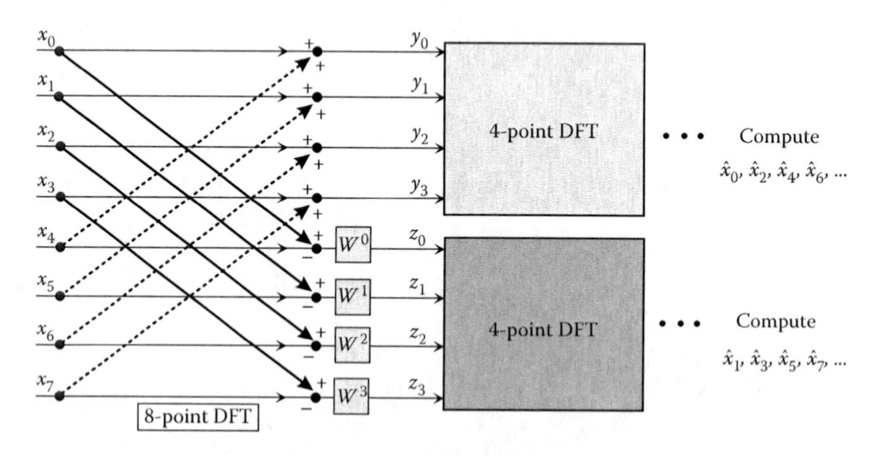

FIGURE 5.29
First stage of decomposition (decimation-in-frequency) of an 8-point DFT into two 4-point DFTs.

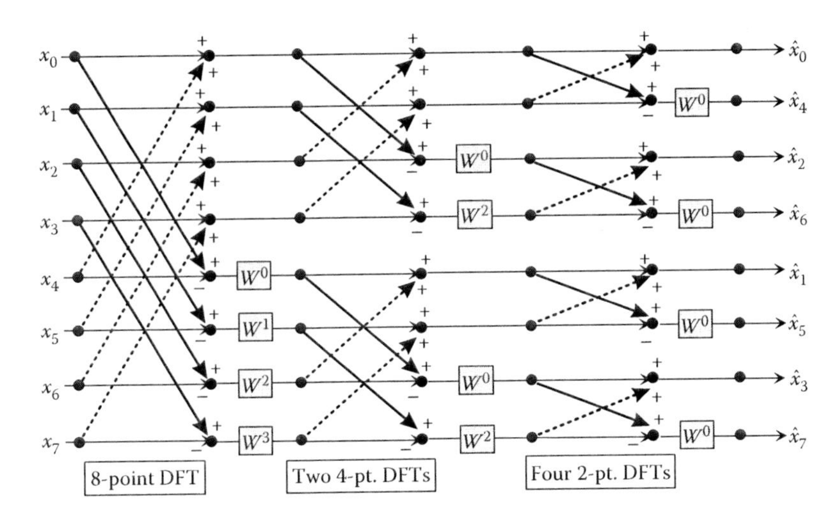

FIGURE 5.30
Complete decomposition (decimation-in-frequency) of an 8-point DFT.

TABLE 5.16
Bit Reversal Applied to the Scenario in Figure 5.30

Mixed Order	Binary Subscripts	Reverse Bits	Unscrambled Order
$\hat{x}_0$	$0 \to 000$	$000 \to 0$	$\hat{x}_0$
$\hat{x}_4$	$4 \to 100$	$001 \to 1$	$\hat{x}_1$
$\hat{x}_2$	$2 \to 010$	$010 \to 2$	$\hat{x}_2$
$\hat{x}_6$	$6 \to 110$	$011 \to 3$	$\hat{x}_3$
$\hat{x}_1$	$1 \to 001$	$100 \to 4$	$\hat{x}_4$
$\hat{x}_5$	$5 \to 101$	$101 \to 5$	$\hat{x}_5$
$\hat{x}_3$	$3 \to 011$	$110 \to 6$	$\hat{x}_6$
$\hat{x}_7$	$7 \to 111$	$111 \to 7$	$\hat{x}_7$

Also, $N = 2^p = 8$ implies $p = 3$, and there are exactly three stages involved in the process. Furthermore, since $N = 8$, we have $W = e^{-(2\pi/N)i} = e^{-(\pi/4)i}$, and

$$W^0 = 1,\ W^1 = e^{-(\pi/4)i} = \frac{\sqrt{2}}{2}(1-i),\quad W^2 = e^{-(\pi/2)i} = -i,\ W^3 = e^{-(3\pi/4)i} = \frac{-\sqrt{2}}{2}(1+i)$$

The computed Fourier coefficients appear in a mixed order but can be unscrambled using bit reversal as follows: (1) express the subscripts 0 through 7 in binary form,* (2) reverse the bits, and (3) express the reversed bits in decimal form. The details are depicted in Table 5.16.

* For example, $5 = \mathbf{1} \times 2^2 + \mathbf{0} \times 2^1 + \mathbf{1} \times 2^0 = (\mathbf{101})_2$.

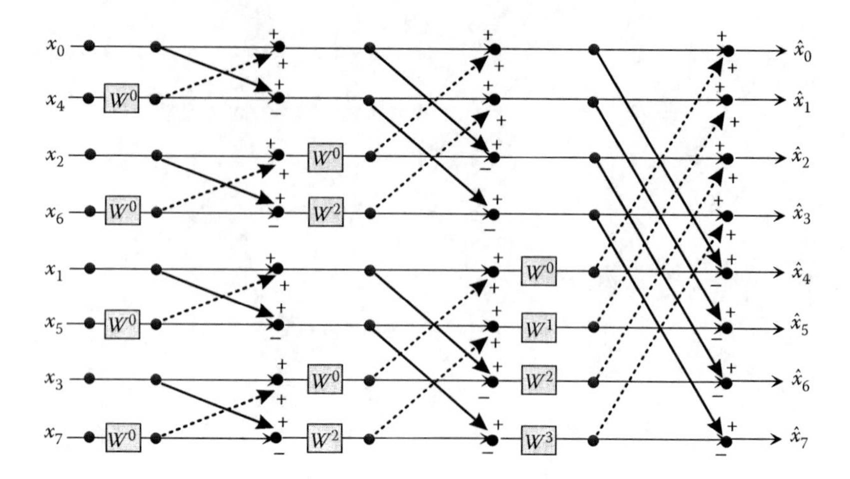

FIGURE 5.31
Complete decomposition (decimation-in-time) of an 8-point DFT.

5.7.4.2 Cooley–Tukey Algorithm ($N = 2^p$, p = integer)

The flow graph for the Cooley–Tukey algorithm is shown in Figure 5.31. The initial sample is divided into groups of even-indexed and odd-indexed data points, but the final outcome appears in correct order.

5.7.5 MATLAB® Built-In Function `fft`

The MATLAB built-in function `fft` computes the DFT of an N-dimensional vector using the efficient FFT method. The DFT of the evenly spaced data points $x(1)$, $x(2)$, … , $x(N)$ is another N-dimensional vector $X(1)$, $X(2)$, … , $X(N)$, where

$$X(k) = \sum_{n=1}^{N} x(n)e^{-2\pi(k-1)(n-1)i/N}, \quad k = 1, 2, \ldots, N$$

5.7.5.1 Interpolation Using `fft`

A set of equally spaced data (t_k, x_k), $k = 1, 2, \ldots, N$ is interpolated using `fft` as follows. The data is first treated as

$$(\sigma_1, x_1), (\sigma_2, x_2), \ldots, (\sigma_N, x_N)$$

where σ_k ($k = 1, 2, \ldots, N$) are equally spaced along the interval $[0,2\pi)$, that is

$$\sigma_1 = 0, \quad \sigma_2 = \frac{2\pi}{N}, \quad \sigma_3 = 2\left(\frac{2\pi}{N}\right), \ldots, \quad \sigma_N = (N-1)\left(\frac{2\pi}{N}\right)$$

The FFT of the vector $[x_1 \quad x_2 \quad \ldots \quad x_N]$ is computed using MATLAB built-in function `fft`. The resulting data is then used in Equation 5.60, with index ranging from 1 to N, to reconstruct x_k. Finally, the data is transformed to its original form and plotted.

Example 5.16: Interpolation Using `fft`

Table 5.17 contains the equally spaced data for one period of a periodic waveform. Construct and plot the interpolating function for this data using the MATLAB function `fft`.

SOLUTION

We will accomplish this in two steps: First, we compute the FFT of the given data, and then use the transformed data to find the interpolating function by essentially reconstructing the original waveform. Note that the reconstruction is done via Equation 5.60, rewritten as

$$x_n = \frac{1}{16}\sum_{k=1}^{16} \hat{x}_k e^{2\pi i(k-1)(n-1)/16}$$

where we will use $n = 1{:}200$ for plotting purposes. The 16 values of $\hat{x}_k$ are obtained as follows:

```
>> x = [2.95  2.01  0.33  .71  .11  .92  -.16  .68  -1.57  -1.12
-.58  -.69  -.21  -.54  -.63  -2.09];
>> Capx = fft(x)'

Capx =

   0.1200
   5.1408 + 6.1959i
   0.8295 + 1.9118i
   4.4021 + 5.0122i
   2.3200 + 2.6600i
   4.0157 + 3.7006i
   2.1305 + 0.8318i
```

TABLE 5.17

Data in Example 5.16

t_i	x_i	t_i	x_i
0.0	2.95	0.8	−1.57
0.1	2.01	0.9	−1.12
0.2	0.33	1.0	−0.58
0.3	0.71	1.1	−0.69
0.4	0.11	1.2	−0.21
0.5	0.92	1.3	−0.54
0.6	−0.16	1.4	−0.63
0.7	0.68	1.5	−2.09

```
4.5215 + 3.6043i
0.3600
4.5215 - 3.6043i
2.1305 - 0.8318i
4.0157 - 3.7006i
2.3200 - 2.6600i
4.4021 - 5.0122i
0.8295 - 1.9118i
5.1408 - 6.1959i
```

Let $\hat{x}_k = \alpha_k + i\beta_k$ so that

$$x_n = \frac{1}{16}\sum_{k=1}^{16} \hat{x}_k e^{2\pi i(k-1)(n-1)/16}$$

$$= \frac{1}{16}\sum_{k=1}^{16} [\alpha_k + i\beta_k]\left[\cos\left(2\pi(k-1)(n-1)/16\right) + i\sin\left(2\pi(k-1)(n-1)/16\right)\right]$$

Note that $\alpha_2 = \alpha_{16}, \ldots, \alpha_8 = \alpha_{10}$ and $\beta_2 = -\beta_{16}, \ldots, \beta_8 = -\beta_{10}$. Also, α_9 multiplies $\cos((n-1)\pi)$, which alternates between 1 and −1 so that over a range of 200 values for n will cancel out. Finally, α_1 multiplies $\cos 0 = 1$. Then, the above can be written as

$$x_n = \frac{1}{16}\alpha_1 + \frac{1}{8}\sum_{k=2}^{8} [\alpha_k \cos(2\pi(k-1)(n-1)/16) - \beta_k \sin(2\pi(k-1)(n-1)/16)]$$

The following MATLAB code will use this to reconstruct the original waveform:

```
x = [2.95  2.01  .33  .71  .11  .92  -.16  .68  -1.57  -1.12
-.58  -.69  -.21  -.54  -.63  -2.09];
N = length(x);
tN = 1.5; t1 = 0;
Capx = fft(x);    % Compute FFT of data
% Treat data as equally spaced on [0,2*pi)
h = 2*pi/N; s = 0:h:2*pi-h; s = s';
ss = linspace(0,2*pi*(N-1)/N,200);
% 200 points for plotting purposes
y = zeros(200,1);    % Pre-allocate
% Start reconstruction & interpolation
for i = 1:200,
    y(i) = Capx(1)/2;
    for k = 1:8,
      y(i) = y(i) + real(Capx(k+1))*cos(k*ss(i))
      - imag(Capx(k+1))*sin(k*ss(i));
    end
    y(i) = (1/8)*y(i);
end
% Transform data to original form
t = N*((tN-t1)/(2*pi*(N-1)))*s + t1;
tt = N*((tN-t1)/(2*pi*(N-1)))*ss + t1;

plot(tt,y,t,x,'o')    % Figure 5.32
```

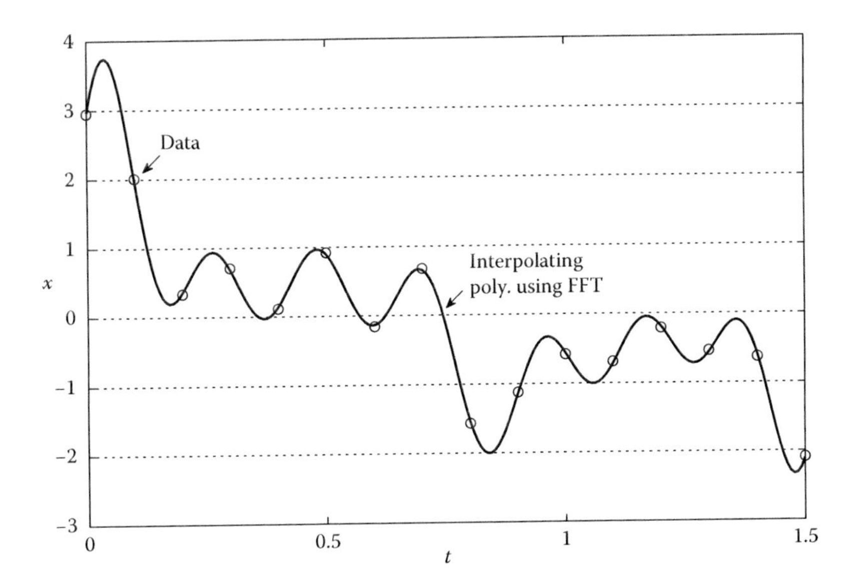

FIGURE 5.32
Interpolations using FFT in Example 5.16.

PROBLEM SET

Linear Regression (Section 5.2)

In Problems 1 through 9, for each set of data

a. ✍ Using least-squares regression, find a straight line that best fits the data.
b. Confirm the results by executing the user-defined function `LinearRegression`.

1. For this problem, use the following table.

x_i	y_i
0.7	0.12
0.8	0.32
0.9	0.58
1.0	0.79
1.1	1.05

2. For this problem, use the following table.

x_i	y_i
0	2.4
0.3	2.79
0.6	3.02
0.9	3.33
1.2	3.55
1.5	3.89

3. For this problem, use the following table.

x_i	y_i
−1	1.2
0	0.5
1	−0.2
2	−1.1
3	−1.8
4	−2.3

4. For this problem, use the following table.

x_i	y_i
−1	−0.42
−0.6	0.03
−0.2	0.51
0.2	1.04
0.6	1.60

5. For this problem, use the following table.

x_i	y_i
1	2.6
2	3.0
4	3.5
6	4.5
7	5.0
8	5.1
10	6.2

6. For this problem, use the following table.

x_i	y_i
4	6
10	20
13	27
25	47
29	59
40	81
45	87

7. For this problem, use the following table.

x_i	y_i
1	6.25
2	8.8
3	11.2
4	13.2
5	15.8
6	18
7	20.5
8	23
9	25
10	28

8. For this problem, use the following table.

x_i	y_i
−2	2.2
0	2.9
1	3.6
3	4.6
4	4.9
6	6.1
7	6.6
9	7.4
10	8
12	9.2

9. For this problem, use the following table.

x_i	y_i
−4.2	6
−3.1	4.6
−1.5	2.8
−0.2	1.2
0.9	−0.08
2.2	−1.7
3.2	−2.91
3.9	−3.75
4.5	−4.5
5.8	−6

10. Students' scores on the mathematics portion of the SAT exam and their GPA follow a linear probabilistic model. Data from 10 students have been collected and recorded in the following table.

   a. Execute the user-defined function `LinearRegression` to find and plot a straight line that best fits the data.

   b. Using the line fit, find an estimate for the GPA of a student whose test score was 400.

x_i **(Test Score)**	y_i **(GPA)**
340	1.60
380	1.80
460	2.25
480	2.36
500	2.40
520	2.50
590	2.80
610	3.00
640	3.15
720	3.60

11. The yield of a chemical reaction (%) at several temperatures (°C) is recorded in the following table.

   a. Execute the user-defined function `LinearRegression` to find and plot a straight line that best fits the data.

   b. Using the line fit, find an estimate for the yield at 240°C.

x_i (Temperature °C)	y_i (Yield%)
160	79.1
190	83.5
210	84.4
230	86.2
250	88.7
260	89.4
280	92.2
290	92.9
300	95.1
310	96.3

12. In a linear coiled spring, the relation between spring force (F) and displacement (x) is described by $F = kx$, where k is the coefficient of stiffness of the spring, which is constant. Testing on a certain spring has led to the data recorded in the following table. All parameter values are in consistent physical units.

    a. Execute the user-defined function `LinearRegression` to find and plot a straight line that best fits the data.

    b. Using (a), find the estimated value for the stiffness coefficient, and the displacement corresponding to $F = 250$.

x (Displacement)	F (Force)
0.1	19.8
0.2	40.9
0.4	79.8
0.5	99.3
0.7	140.2
0.8	160.4
1.0	199.5
1.2	239.8
1.3	260.6
1.5	300.7

Linearization of Nonlinear Data (Section 5.3)

13. In Example 5.3, it seems as if an exponential function may fit the data. Follow the procedure outlined in Section 5.3 to find the appropriate exponential function, including the necessary plots. Compare with the results of Example 5.3, which used a power function to fit the data.

14. Fit an exponential function to the data in the following table.

x_i	y_i
0	1.2
0.2	2.2
0.4	3.1
0.6	5.3
0.8	8.3
1.0	13
1.2	21.5
1.4	34
1.6	53.1
1.8	86

15. Fit a saturation function to the data in the following table.

x_i	y_i
1	0.30
2	0.49
3.5	0.61
5	0.72
6	0.80
8	0.81
9.5	0.89
10	0.88

16. For the data in the following table, decide whether an exponential or a power function provides an appropriate curve fit.

x_i	y_i
0.2	3.0
0.3	3.7
0.5	5.2
0.6	6.2
0.8	8.7
1.0	12.5
1.1	15.2
1.3	22.0

17. Fit a power function to the data in the following table.

x_i	y_i
1	3.2
1.6	3.5
2	3.6
2.5	3.9
3	4.1
3.4	4.5
4	4.6
4.6	4.7
5	4.9
5.8	5.1

18. Fit a saturation function to the data in the following table.

x_i	y_i
0.5	0.10
1	0.20
1.5	0.28
2	0.34
2.7	0.40
3.5	0.48
4.2	0.52
5	0.55
6	0.60
7	0.65

19. In many applications involving chemical processes, the experimental data follows an s-shaped curve as in Figure 5.33, where the data approaches a steady-state value of 1. The following table contains one such set of data. For these cases, curve fitting is done by approximating the s-shaped curve by $y = 1 + Ae^{-\alpha t}$, where $A < 0$ since $y < 1$ for all data, and $\alpha > 0$. Rearrange and take the natural logarithm to obtain

$$\ln|y - 1| = -\alpha t + \ln|A|$$

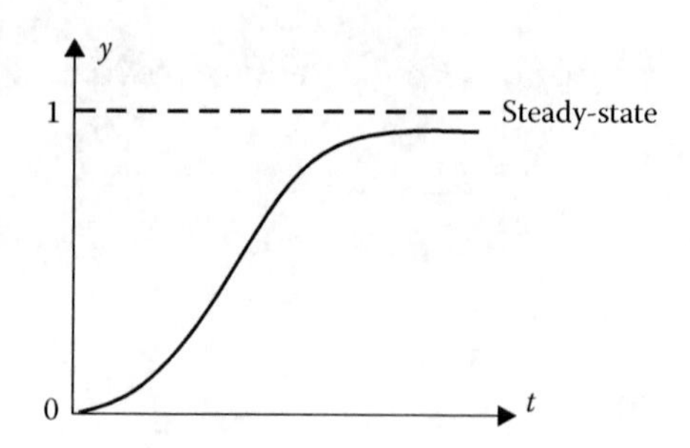

FIGURE 5.33
An s-shapted curve—chemical process.

so that ln $|y-1|$ versus t is linear with a slope of $-\alpha$ and an intercept of ln $|A|$. The slope and the intercept can be found by linear regression. Apply this procedure to the data in the following table to determine the parameters A and α. Plot the original data and the curve fit just obtained.

t_i	y_i
0.5	0.40
1	0.50
2	0.70
3	0.80
4	0.90
6	0.95
7	0.96
8	0.97
9	0.98
10	0.99

20. Repeat Problem 19 for the data in the following table.

t_i	y_i
2	0.78
4	0.87
6	0.89
8	0.94
10	0.95
12	0.97
14	0.98
16	0.98
18	0.99
20	0.99

Polynomial Regression (Section 5.4)

21. Using the user-defined functions of Section 5.4, find and plot the straight line, as well as the third-order polynomial that best fit the data in the following table.

x_i	y_i
1	5.0
3	5.8
4	6.8
5	6.5
7	7.3

22. Using the user-defined functions of Section 5.4, find and plot the straight line, as well as the second-order polynomial that best fit the data in the following table.

x_i	y_i
0	2.0
1	4.1
2	5.5
3	6.7
4	8.8
5	9.1

23. Repeat Problem 22 for the data in the following table.

x_i	y_i
0.1	0.8
0.3	1.0
0.5	1.5
0.7	2.2
0.9	3.3

24. Repeat Problem 21 for the data in the following table.

x_i	y_i
0.0	1.2
0.3	1.8
0.6	2.8
0.9	3.6
1.0	4.3

25. Using the `polyfit` function, find and plot the fourth-degree polynomial that best fits the data in the following table.

x_i	y_i
1	1
2	4
3	5
4	6
5	7
6	8

26. Using the `polyfit` function, find and plot the second-degree, as well as the fifth-degree polynomial that best fits the data in the following table.

x_i	y_i
0	3.4
1	4.9
2	6.2
3	7.3
4	9.2
5	10.2

27. During the free fall of a heavy object, the relationship between the velocity v of the object and the force r resisting its motion—both in consistent physical units—is described by the data in the following table.

   a. Using the `polyfit` function, find and plot the second-degree polynomial that best fits the data.

   b. Use the result of (a) to approximate the relationship between v and r, which is usually modeled as $r = bv^2$ for a suitable constant b.

Velocity (v)	**Resistance (r)**
0	0
0.4	0.1
0.8	0.5
1.2	1.0
1.6	1.8
2.0	2.7
2.4	4.0
2.8	5.5
3.2	7.2

28 Write a user-defined function `Deg_4_Regression`, which finds the fourth-degree polynomial that best fits a set of data (x,y). The function is called as

```
[a4 a3 a2 a1 a0] = Deg_4_Regression(x,y)
```

where the outputs are the coefficients of the desired polynomial. Apply the function to the data in the following table.

x	y
0.20	1.5
0.56	2.5
0.92	4.1
1.28	7.1
1.64	13
2.00	22

Polynomial Interpolation (Section 5.5)

Lagrange Interpolation

29. Given the data in the following table:
    a. Interpolate at $x = 0.7$ using the second-degree Lagrange interpolating polynomial.
    b. Confirm the results by executing the user-defined function `LagrangeInterp`.

x_i	y_i
0.3	0.53
0.6	0.22
0.8	0.10

30. Given the data in the following table:
    a. ✍ Interpolate at $x = 0.55$ using the second-degree Lagrange interpolating polynomial.
    b. Confirm the results by executing the user-defined function `LagrangeInterp`.

x_i	y_i
0.2	−0.16
0.4	−0.24
0.9	−0.09

31. ✍ Given the data in the following table:
    a. Interpolate at $x = 3$ with a first-degree Lagrange polynomial using two most suitable data points.
    b. Interpolate at $x = 3$ with a second-degree Lagrange polynomial using three most suitable data points.
    c. Compare the results of (a) and (b), and discuss.

x_i	$y_i = 2\sin(x_i/2)$
0	1
1	0.96
2	1.68
4	1.82

32. ✍ Given the data in the following table:
    a. Interpolate at $x = 2.5$ with a first-degree Lagrange polynomial using two most suitable data points.
    b. Interpolate at $x = 2.5$ with a second-degree Lagrange polynomial using three most suitable data points.
    c. Compare the results of (a) and (b), and discuss.

x_i	$y_i = \log_{10}(x_i)$
1	0
1.5	0.1761
2	0.3010
3	0.4771
5	0.6990

33. Using `format long` and the user-defined function `LagrangeInterp`, given the data in the following table:
    a. Interpolate at $x = 0.6$ with a second-degree Lagrange polynomial using three most suitable data points.
    b. Interpolate at $x = 0.6$ with a third-degree Lagrange polynomial using four most suitable data points.
    c. Compare the results of (a) and (b).

x_i	$y_i = e^{-x_i/2}$
0.2	0.9048
0.4	0.8187
0.5	0.7788
0.8	0.6703
1.1	0.5769
1.3	0.5220

34. Using `format long` and the user-defined function `LagrangeInterp`, given the data in the following table:
    a. Interpolate at $x = 0.6$ with a second-degree Lagrange polynomial using three most suitable data points.
    b. Interpolate at $x = 0.6$ with a third-degree Lagrange polynomial using four most suitable data points.
    c. Compare the results of (a) and (b).

x_i	$y_i = 2^{-x_i}$
0.1	0.9330
0.2	0.8706
0.4	0.7579
0.7	0.6156
0.9	0.5359
1.0	0.5000

35. Using the user-defined function `LagrangeInterp`, given the data in the following table:

a. Interpolate at $x = 1.7$ with a second-degree Lagrange polynomial using three most suitable data points.
b. Interpolate at $x = 9$ with a third-degree Lagrange polynomial using four most suitable data points.

x_i	$y_i = \sqrt[3]{x_i}$
0	0
1	1
3	1.44
7	1.91
12	2.29

36. Using the user-defined function `LagrangeInterp`, given the data in the following table:

a. Interpolate at $x = 1.5$ with a second-degree Lagrange polynomial using three most suitable data points.
b. Interpolate at $x = 3$ with a third-degree Lagrange polynomial using four most suitable data points.

x_i	$y_i = (x_i + 1)/(x_i^2 + 1)$
0	1
1	1
2	0.6
2.5	0.48
4	0.29

Newton Interpolation (Divided Differences)

37. For the data in the following table, construct a divided differences table and interpolate at $x = 0.25$ using Newton interpolating polynomials $p_1(x)$, $p_2(x)$, and $p_3(x)$.

x_i	y_i
0	1
0.5	0.9098
0.9	0.7725
1.2	0.6626

38. ✍ For the data in the following table, construct a divided differences table and interpolate at $x = 0.3$ using Newton interpolating polynomials $p_1(x)$, $p_2(x)$, and $p_3(x)$.

x_i	y_i
0	1
0.4	2.68
0.8	5.79
1	8.15

39. ✍ Consider the data in the following table.
   a. Construct a divided differences table and interpolate at $x = 1.75$ using the third-degree Newton interpolating polynomial $p_3(x)$.
   b. Suppose one more point $(x = 3,\ y = 9.11)$ is added to the data. Update the divided-differences table from (a) and interpolate at $x = 1.75$ using the fourth-degree Newton interpolating polynomial $p_4(x)$.

x_i	y_i
1	1.22
1.5	2.69
2	4.48
2.5	6.59

40. ✍ Consider the data in the following table.
   a. Construct a divided differences table and interpolate at $x = 4$ using the third-degree Newton interpolating polynomial $p_3(x)$.
   b. Suppose one more point $(x = 7,\ y = 0.18)$ is added to the data. Update the divided-difference table from (a) and interpolate at $x = 4$ using the fourth-degree Newton interpolating polynomial $p_4(x)$.

x_i	y_i
1	1
3	0.45
5	0.26
6	0.21

41. Given the data in the following table
    a. Construct a divided differences table and interpolate at $x = 2.4$ and $x = 4.2$ using the fourth-degree Newton interpolating polynomial $p_4(x)$.
    b. Confirm the results by executing the user-defined function `NewtonInterp`.

x_i	y_i
1	0.69
2	1.10
3	1.39
4	1.61
6	1.95

42. Given the data in the following table
    a. Construct a divided differences table and interpolate at $x = 2.5$ and $x = 5$ using the fourth-degree Newton interpolating polynomial $p_4(x)$.
    b. Confirm the results by executing the user-defined function `NewtonInterp`.

x_i	y_i
1	0.89
2	1.81
3	2.94
4	4.38
6	8.72

43. Using the user-defined function `NewtonInterp`, given the data in the following table, interpolate at $x = 4.5$ via Newton interpolating polynomials of all possible degrees, and comment on accuracy.

x_i	$y_i = \sqrt{x_i}$
1	1
2	1.4142
3.5	1.8708
5	2.2361
7	2.6458
9	3
10	3.1623

44. Using `format long` and the user-defined function `NewtonInterp`, given the data in the following table, interpolate at $x = 4.25$ via Newton interpolating polynomials of all possible degrees, and comment on accuracy.

x_i	$y_i = e^{x_i/3}$
1	1.3956
1.5	1.6487
2.5	2.3010
3	2.7183
4	3.7937
4.5	4.4817
5	5.2945
6	7.3891

Newton Interpolation (Forward Differences)

45. For the data in the following table, construct a forward-differences table and interpolate at $x = 3.3$ using Newton interpolating polynomial $p_4(x)$.

x_i	y_i
1	1.25
2	3.25
3	7.25
4	13.25
5	21.25

46. Consider the data in the following table.

a. Construct a forward-differences table and interpolate at $x = 1.25$ using Newton interpolating polynomials $p_3(x)$ and $p_4(x)$.

b. Suppose a new point ($x = 1.5$, $y = 4.27$) is added to the data. Interpolate at $x = 1.25$ using Newton interpolating polynomial $p_5(x)$.

x_i	y_i
1.0	1.30
1.1	1.75
1.2	2.27
1.3	2.86
1.4	3.52

47. For the data in the following table, construct a forward-differences table and interpolate at $x = 2.8$ using Newton interpolating polynomial $p_5(x)$.

x_i	y_i
1	0.92
1.5	0.80
2	0.64
2.5	0.46
3	0.29
3.5	0.14

48. Write a user-defined function with syntax `yi = Newton_FD (x,y,xi)` that finds the Newton forward-difference interpolating polynomial for the data `(x,y)` and uses this polynomial to interpolate at `xi` and returns the interpolated value in `yi`. For the data in the following table, find the interpolated value at $x = 6.5$ by executing `Newton_FD`.

x_i	y_i
1	1.0000
2	0.4444
3	0.2632
4	0.1818
5	0.1373
6	0.1096
7	0.0929
8	0.0775
9	0.0675
10	0.0597

49. Given the data in the following table, interpolate at $x = 5.5$ by executing the user-defined function `Newton_FD` (see Problem 48).

x_i	y_i
2	1.7100
4	2.0801
6	2.3513
8	2.5713
10	2.7589
12	2.9240
14	3.0723
16	3.2075
18	3.3322
20	3.4482

50. Given the data in the following table, interpolate at $x = 7.4$ by executing
    a. The user-defined function `Newton_FD` (see Problem 48).
    b. The user-defined function `NewtonInterp` (see Section 5.5). Compare the results.

x_i	y_i
1	1.4422
2	1.8171
3	2.2240
4	2.6207
5	3.0000
6	3.3620
7	3.7084
8	4.0412
9	4.3621
10	4.6723

Spline Interpolation (Section 5.6)

In Problems 51 through 54, find and plot the quadratic splines for the given data and interpolate at the specified point(s). Assume $S_1''(x_1) = 0$.

51. For this problem, use the following table ($x = 3.6$).

x_i	y_i
1	1.2
3	2.3
4	1.2
6	2.9

52. For this problem, use the following table ($x = 0.5$, $x = 2.2$).

x_i	y_i
0	1
1	2
2	5
3	9

53. For this problem, use the following table ($x = 5$, $x = 8.5$).

x_i	y_i
3	1
7	4
10	5
13	2

54. For this problem, use the following table ($x = 5$, $x = 8$).

x_i	y_i
1	10
4	8
6	12
9	14

55. ✍ Consider the data in the following table.
   a. Find and plot the quadratic splines, assuming $S_1''(x_1) = 0$.
   b. Find and plot the cubic splines with clamped boundary conditions $p = 1$, $q = -1$.
   c. Find the interpolated value at $x = 1.5$ using the splines of (a) and (b), and compare with the true value.

x_i	$y_i = 10 - e^{-x_i/2}$
0.1	9.0488
0.5	9.2212
1	9.3935
2	9.6321

56. ✍ Repeat Problem 55 but this time assume the cubic splines satisfy free boundary conditions.

57. ✍ For the data in the following table, construct and plot cubic splines that satisfy

   a. Clamped boundary conditions $p = -1$, $q = -0.5$

   b. Free boundary conditions

x_i	y_i
1	5
4	1
6	2
8	0.5

58. ✍ For the data in the following table construct and plot cubic splines that satisfy

   a. Clamped boundary conditions $p = 0$, $q = 0.3$

   b. Free boundary conditions

x_i	y_i
1	1
3	2
5	5
8	6

59. In an exercise session, maximum heart rates for eight individuals of different ages have been recorded as shown in the following table. Find the maximum heart rate of a 37-year-old individual by using

   a. `interp1`

   b. Clamped ($p = -1$, $q = -1$) cubic spline interpolation

Age x_i	**Maximum Heart Rate** y_i
12	204
20	198
25	192
30	185
35	180
40	175
45	168
50	162

60. The yield of a certain chemical reaction at various temperatures is recorded in the following table. Find the reaction yield at 260°C by using
    a. `interp1`.
    b. Clamped ($p = -1$, $q = 1$) cubic spline interpolation. Plot both splines and the original data.
    c. Clamped ($p = -0.5$, $q = 0.5$) cubic spline interpolation. Plot the new splines together with those of (a). Compare with (b) and discuss the results.

Temperature (°C) x_i	Reaction Yield (%) y_i
160	78.5
190	83.1
195	84.5
230	85.9
250	89.3
280	91.6
300	93.9

61. The data in the following table is generated by the function $f(x) = 1/(1 + 2x^2)$.
    a. Construct and plot the cubic splines with clamped boundary conditions $p = 0.1$, $q = -0.1$. Also plot the original function and the given data. Interpolate at $x = 1.5$ and compare with the true value at that point.
    b. Repeat (a) for boundary conditions $p = 0.2$, $q = -0.2$. Discuss the results.

x_i	y_i
−2	0.1111
−1	0.3333
0	1
1	0.3333
2	0.1111

62. For the data in the following table, construct and plot the cubic splines using `interp1` and find the interpolated value at $x = 2.5$. Repeat for cubic splines with clamped boundary conditions $p = 0$, $q = 0$, and compare the results.

x_i	y_i
–4	0
–3	0
–2	0
–1	1.6
0	2
1	1.6
2	0
3	0
4	0

63. Consider the data for `x = 0:20` generated by the Bessel function of the first kind of order zero $J_0(x)$, which in MATLAB is handled by `besselj(0,x)`. Construct and plot the cubic splines using `interp1`, interpolate at $x = 3.4$, and compare with the actual value at that point.

64. The data in Table 5.18 shows the (experimental) compressibility factor for air at several pressures when temperature is fixed at 180 K. Construct and plot the cubic splines using `interp1` and find the interpolated value at $x = 125$. Repeat for a third-degree polynomial using `polyfit` and compare the results.

TABLE 5.18

Experimental Compressibility Factor for Air at Several Pressures

Pressure (Bars) x_i	Comp. Factor y_i	Pressure (Bars) x_i	Comp. Factor y_i
1	0.9967	100	0.7084
5	0.9832	150	0.7180
10	0.9660	200	0.7986
20	0.9314	250	0.9000
40	0.8625	300	1.0068
60	0.7977	400	1.2232
80	0.7432	500	1.4361

Source: *Perry's Chemical Engineers' Handbook*. 6th edition, McGraw-Hill, 1984.

Fourier Approximation and Interpolation (Section 5.7)

In Problems 65 through 72, for each given set of data:

a. ✍ Find the approximating or interpolating trigonometric polynomial of the degree indicated.

b. Confirm the results of (a) by executing the user-defined function `TrigPoly`.

65. For this problem, use the following table ($m = 2$).

t_i	x_i
0.3	1
0.4	0.9
0.5	0
0.6	0.1
0.7	0.8
0.8	0.9

66. For this problem, use the following table ($m = 2$).

t_i	x_i
1	0.9
1.3	1
1.6	–1
1.9	–0.8
2.2	0.9
2.5	1

67. For this problem, use the following table ($m = 2$).

t_i	x_i
0.6	–0.60
0.8	0.52
1.0	0.98
1.2	0.75
1.4	1.03

68. For this problem, use the following table ($m = 3$).

t_i	x_i
2	1.40
2.3	1.06
2.6	0.77
2.9	0.15
3.2	−0.62
3.5	0.31

69. For this problem, use the following table ($m = 3$).

t_i	x_i
1.5	1.05
1.7	1.85
1.9	1.40
2.1	0.35
2.3	1.50
2.5	0.80

70. For this problem, use the following table ($m = 2$).

t_i	x_i
2.4	4.15
2.6	2.05
2.8	6.20
3.0	4.30
3.2	5.80

71. For this problem, use the following table ($m = 3$).

t_i	x_i	t_i	x_i
0.7	−0.20	1.9	1.02
1.0	−0.54	2.2	0.92
1.3	−0.12	2.5	0.56
1.6	0.38	2.8	0.19

72. For this problem, use the following table ($m = 4$).

t_i	x_i	t_i	x_i
1.0	1.00	3.0	0.95
1.5	0.82	3.5	1.16
2.0	0.13	4.0	0.85
2.5	0.74	4.5	−0.25

In Problems 73 through 76, for each given set of data, find the interpolating function using the MATLAB function `fft`.

73. For this problem, use the following table.

t_i	x_i	t_i	x_i
0.2	–1	1.4	–1
0.5	0	1.7	–2
0.8	1	2.0	–3
1.1	0	2.3	–2

74. For this problem, use the following table.

t_i	x_i	t_i	x_i
0	1	2	0
0.5	1	2.5	0
1	0	3	1
1.5	0	3.5	1

75. For this problem, use the following table.

t_i	x_i	t_i	x_i
0.0	4.001	0.8	0.102
0.1	3.902	0.9	0.251
0.2	1.163	1.0	0.229
0.3	0.997	1.1	0.143
0.4	0.654	1.2	0.054
0.5	0.803	1.3	0.001
0.6	0.407	1.4	–0.583
0.7	0.706	1.5	–0.817

76. For this problem, use the following table.

t_i	x_i	t_i	x_i
1.0	5.024	1.8	0.543
1.1	5.536	1.9	0.510
1.2	3.023	1.0	0.702
1.3	1.505	2.1	0.189
1.4	1.559	2.2	0.176
1.5	1.021	2.3	–0.096
1.6	0.965	2.4	–1.112
1.7	0.998	2.5	0.465

6

Numerical Differentiation and Integration

The numerical methods to find estimates for derivatives and definite integrals are presented and discussed in this chapter. Many applications in engineering involve rates of change of quantities with respect to variables such as time. For example, linear damping force is directly proportional to the rate of change of displacement with respect to time. In many other applications, definite integrals are involved. For example, the voltage across a capacitor at any specified time is proportional to the integral of the current taken from an initial time to that specified time.

6.1 Numerical Differentiation

There are many situations where numerical differentiation is needed. Sometimes the analytical expression of the function to be differentiated is known but analytical differentiation is either very difficult or impossible. In that case, the function is discretized to generate several points (values), which are subsequently used by a numerical method to approximate the derivative of the function at any of the generated points. Often, however, data is available in the form of a discrete set of points. These points may be recorded data from experimental measurements or generated as a result of some type of numerical computations. In these situations, the derivative can be numerically approximated in one of two ways. One way is to use finite differences, which utilize the data in the neighborhood of the point of interest. In Figure 6.1a, for instance, the derivative at the point x_i is approximated by the slope of the line connecting x_{i-1} and x_{i+1}. The other approach is to fit a suitable, easy-to-differentiate function into the data (Chapter 5) and then differentiate the analytical expression of the function and evaluate at the point of interest; see Figure 6.1b.

6.2 Finite-Difference Formulas for Numerical Differentiation

Finite-difference formulas are used to approximate the derivative at a point by making use of the values at the neighboring points. These formulas can be

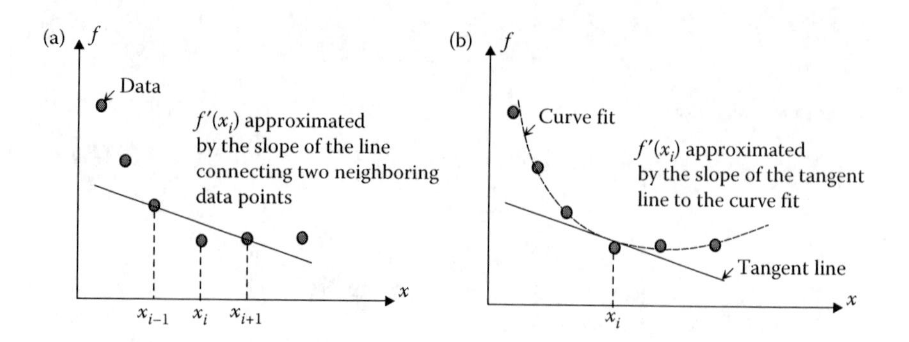

FIGURE 6.1
Approximating $f'(x_i)$ using (a) finite difference and (b) the curve fit.

derived to approximate derivatives of different orders at a specified point by using the Taylor series expansion. In this section, we present the derivation for finite-difference formulas to approximate first and second derivatives at a point, but those for the third and fourth derivatives will be provided without derivation.

6.2.1 Finite-Difference Formulas for the First Derivative

There are several methods for approximating the first derivative at a point and they use the values at two or more of its neighboring points. These points can be to the left, to the right, or on both sides of the point at which the first derivative is being approximated.

6.2.1.1 Two-Point Backward Difference Formula

The value of $f(x_{i-1})$ can be approximated by a Taylor series expansion at x_i. Letting $h = x_i - x_{i-1}$, this yields

$$f(x_{i-1}) = f(x_i) - hf'(x_i) + \frac{1}{2!}h^2 f''(x_i) - \frac{1}{3!}h^3 f'''(x_i) + \cdots$$

Retaining the linear terms only, we have

$$f(x_{i-1}) = f(x_i) - hf'(x_i) + \underbrace{\frac{1}{2!}h^2 f''(\xi)}_{\text{Remainder}}$$

where $x_{i-1} \le \xi \le x_i$. Solving for $f'(x_i)$, we find

$$f'(x_i) = \frac{f(x_i) - f(x_{i-1})}{h} + \underbrace{\frac{1}{2!}hf''(\xi)}_{\text{Truncation error}} \tag{6.1}$$

Approximating the first derivative can be done by neglecting the second term on the right side, which produces a truncation error. Since this is proportional to h, we say the truncation error is of the order of h and express it as $O(h)$:

$$f'(x_i) = \frac{f(x_i) - f(x_{i-1})}{h} + O(h) \tag{6.2}$$

The actual value of the truncation error is not available because the value of ξ is not exactly known. However, $O(h)$ means that the error gets smaller as h gets smaller.

6.2.1.2 Two-Point Forward Difference Formula

The value of $f(x_{i+1})$ can be approximated by a Taylor series expansion at x_i. Letting $h = x_{i+1} - x_i$

$$f(x_{i+1}) = f(x_i) + hf'(x_i) + \frac{1}{2!}h^2 f''(x_i) + \frac{1}{3!}h^3 f'''(x_i) + \cdots$$

Retaining the linear terms only

$$f(x_{i+1}) = f(x_i) + hf'(x_i) + \underbrace{\frac{1}{2!}h^2 f''(\xi)}_{\text{Remainder}}$$

where $x_i \le \xi \le x_{i+1}$. Solving for $f'(x_i)$, we find

$$f'(x_i) = \frac{f(x_{i+1}) - f(x_i)}{h} - \frac{1}{2!}hf''(\xi) \tag{6.3}$$

The first term on the right side of Equation 6.3 provides an approximation for the first derivative, while the neglected second term is of the order of h so that

$$f'(x_i) = \frac{f(x_{i+1}) - f(x_i)}{h} + O(h) \tag{6.4}$$

6.2.1.3 Two-Point Central Difference Formula

To derive the central difference formula, we retain up to the quadratic term in the Taylor series. Therefore

$$f(x_{i-1}) = f(x_i) - hf'(x_i) + \frac{1}{2!}h^2 f''(x_i) - \frac{1}{3!}h^3 f'''(\xi), \quad x_{i-1} \le \xi \le x_i$$

and

$$f(x_{i+1}) = f(x_i) + hf'(x_i) + \frac{1}{2!}h^2 f''(x_i) + \frac{1}{3!}h^3 f'''(\eta), \quad x_i \le \eta \le x_{i+1}$$

Subtracting the first equation from the second, we find

$$f(x_{i+1}) - f(x_{i-1}) = 2hf'(x_i) + \frac{1}{3!}h^3\left[f'''(\eta) + f'''(\xi)\right]$$

Solving for $f'(x_i)$ and proceeding as before

$$f'(x_i) = \frac{f(x_{i+1}) - f(x_{i-1})}{2h} + O(h^2) \tag{6.5}$$

Equation 6.5 reveals that the central difference formula provides a better accuracy than the backward and forward difference formulas. Figure 6.2 supports this observation.

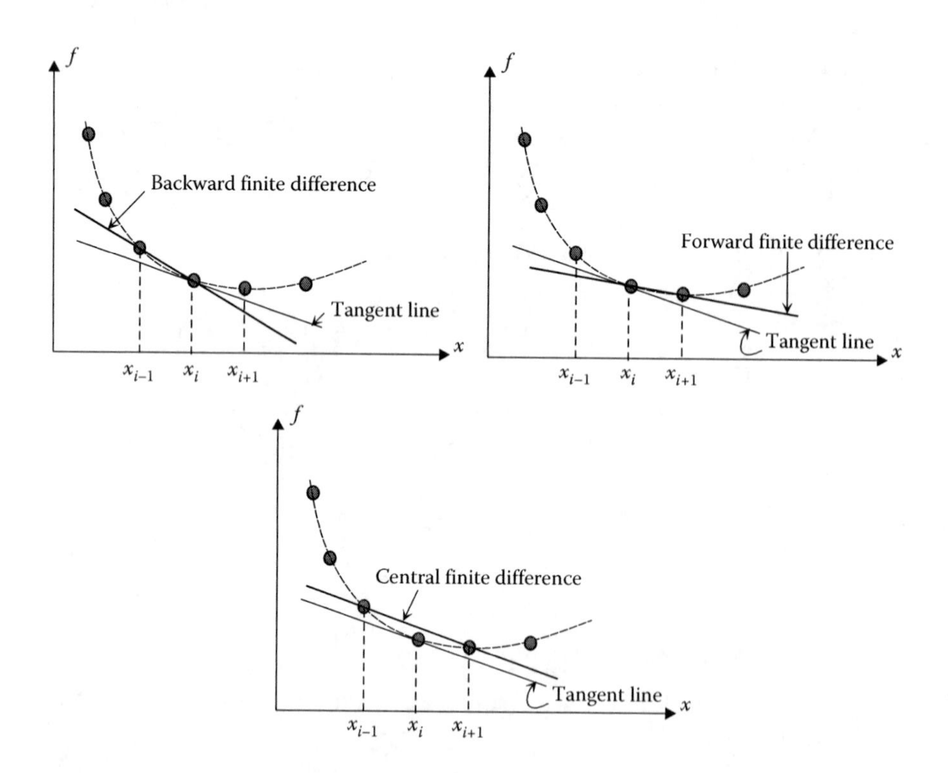

FIGURE 6.2
Two-point finite differences to approximate the first derivative.

Consider a set of data with $x_1, x_2, \dots, x_n$. Since the two-point backward difference formula uses x_i and the point to its left, x_{i-1}, it cannot be applied at the first data point x_1. But it can be used to approximate the first derivative at all interior points, as well as the last point x_n, with a truncation error of $O(h)$. The two-point forward difference formula cannot be applied at the last point x_n. But it can be used to approximate the first derivative at the first point x_1 and all the interior points with a truncation error of $O(h)$. The central difference formula approximates the first derivative at the interior points with a truncation error of $O(h^2)$ but cannot be applied at the first and last points. The central difference formula is therefore the preferred choice since it gives better accuracy, but cannot be used at the endpoints. This means the approximation of the first derivative at the interior points has an error of $O(h^2)$, while those at the endpoints come with $O(h)$. To have compatible accuracy, it is desired that the approximations at the endpoints also come with $O(h^2)$. These are provided by three-point difference formulas.

6.2.1.4 Three-Point Backward Difference Formula

We first approximate the value of $f(x_{i-1})$ by a Taylor series expansion at x_i

$$f(x_{i-1}) = f(x_i) - hf'(x_i) + \frac{1}{2!}h^2 f''(x_i) - \frac{1}{3!}h^3 f'''(\xi), \quad x_{i-1} \le \xi \le x_i$$

We also approximate the value of $f(x_{i-2})$ by a Taylor series expansion at x_i

$$f(x_{i-2}) = f(x_i) - (2h)f'(x_i) + \frac{1}{2!}(2h)^2 f''(x_i) - \frac{1}{3!}(2h)^3 f'''(\eta), \quad x_{i-2} \le \eta \le x_i$$

Multiplying the first equation by 4 and subtracting the result from the second equation yields

$$f(x_{i-2}) - 4f(x_{i-1}) = -3f(x_i) + 2hf'(x_i) + \frac{4}{3!}h^3 f'''(\xi) - \frac{8}{3!}h^3 f'''(\eta)$$

Solving for $f'(x_i)$, we arrive at

$$f'(x_i) = \frac{f(x_{i-2}) - 4f(x_{i-1}) + 3f(x_i)}{2h} - \frac{1}{3}h^2 f'''(\xi) + \frac{2}{3}h^2 f'''(\eta)$$

Then, $f'(x_i)$ can be approximated by neglecting the last two terms, which introduces a truncation error of the order of h^2, that is

$$f'(x_i) = \frac{f(x_{i-2}) - 4f(x_{i-1}) + 3f(x_i)}{2h} + O(h^2) \tag{6.6}$$

Therefore, the three-point backward difference formula approximates the first derivative at x_i by using the values at the points x_i, x_{i-1}, and x_{i-2}.

6.2.1.5 Three-Point Forward Difference Formula

The three-point forward difference formula approximates the first derivative at x_i by using the values at the points x_i, x_{i+1}, and x_{i+2}. The derivation is similar to that presented for the backward difference, except that the values of $f(x_{i+1})$ and $f(x_{i+2})$ are now considered as Taylor series expanded at x_i. This ultimately leads to

$$f'(x_i) = \frac{-3f(x_i) + 4f(x_{i+1}) - f(x_{i+2})}{2h} + O(h^2) \tag{6.7}$$

Example 6.1: Finite-Difference Formulas for the First Derivative

Consider the function $f(x) = e^{-x}\sin(x/2)$ with $x = 1.2, 1.4, 1.6, 1.8$. Approximate $f'(1.4)$ using

- Two-point backward difference formula
- Two-point forward difference formula
- Two-point central difference formula
- Three-point forward difference formula

Find the relative error in each case.

SOLUTION

Differentiation yields $f'(x) = e^{-x}\left[\frac{1}{2}\cos(x/2) - \sin(x/2)\right]$ so the actual value is $f'(1.4) = -0.0646$. The approximate first derivative is calculated via the four difference formulas listed above and are summarized in Table 6.1. As expected, the two-point central difference and three-point forward difference formulas provide better accuracy than the other two techniques.

TABLE 6.1

Summary of Calculations in Example 6.1

Difference Formula	Approximate $f'(1.4)$	Relative Error (%)
2-Point backward	$\frac{f(1.4) - f(1.2)}{0.2} = -0.0560$	13.22
2-Point forward	$\frac{f(1.6) - f(1.4)}{0.2} = -0.0702$	8.66
2-Point central	$\frac{f(1.6) - f(1.2)}{2(0.2)} = -0.0631$	2.28
3-Point forward	$\frac{-3f(1.4) + 4f(1.6) - f(1.8)}{2(0.2)} = -0.0669$	3.56

6.2.2 Finite-Difference Formulas for the Second Derivative

The second derivative at x_i can also be approximated by finite difference formulas. These formulas are derived in a similar manner as those for the first derivative. Below, we present three-point backward and forward difference, as well as three-point central difference formulas for approximating the second derivative.

6.2.2.1 Three-Point Backward Difference Formula

The values of $f(x_{i-1})$ and $f(x_{i-2})$ are first approximated by Taylor series expansions about x_i:

$$f(x_{i-1}) = f(x_i) - hf'(x_i) + \frac{1}{2!}h^2 f''(x_i) - \frac{1}{3!}h^3 f'''(\xi), \quad x_{i-1} \le \xi \le x_i$$

$$f(x_{i-2}) = f(x_i) - (2h)f'(x_i) + \frac{1}{2!}(2h)^2 f''(x_i) - \frac{1}{3!}(2h)^3 f'''(\eta), \quad x_{i-2} \le \eta \le x_i$$

Multiplying the first equation by 2 and subtracting from the second equation results in

$$f(x_{i-2}) - 2f(x_{i-1}) = -f(x_i) + h^2 f''(x_i) - \frac{4}{3}h^3 f'''(\eta) + \frac{1}{3}h^3 f'''(\xi)$$

Proceeding as before, we find

$$f''(x_i) = \frac{f(x_{i-2}) - 2f(x_{i-1}) + f(x_i)}{h^2} + O(h) \tag{6.8}$$

6.2.2.2 Three-Point Forward Difference Formula

The values of $f(x_{i+1})$ and $f(x_{i+2})$ are first approximated by Taylor series expansions about x_i:

$$f(x_{i+1}) = f(x_i) + hf'(x_i) + \frac{1}{2!}h^2 f''(x_i) + \frac{1}{3!}h^3 f'''(\xi), \quad x_i \le \xi \le x_{i+1}$$

$$f(x_{i+2}) = f(x_i) + (2h)f'(x_i) + \frac{1}{2!}(2h)^2 f''(x_i) + \frac{1}{3!}(2h)^3 f'''(\eta), \quad x_i \le \eta \le x_{i+2}$$

Multiplying the first equation by 2 and subtracting from the second equation results in

$$f(x_{i+2}) - 2f(x_{i+1}) = -f(x_i) + h^2 f''(x_i) + \frac{4}{3}h^3 f'''(\eta) - \frac{1}{3}h^3 f'''(\xi)$$

Therefore

$$f''(x_i) = \frac{f(x_{i+2}) - 2f(x_{i+1}) + f(x_i)}{h^2} + O(h) \tag{6.9}$$

6.2.2.3 Three-Point Central Difference Formula

Expanding $f(x_{i-1})$ and $f(x_{i+1})$ in Taylor series about x_i and retaining up to the third derivative terms, we find

$$f(x_{i-1}) = f(x_i) - hf'(x_i) + \frac{1}{2!}h^2 f''(x_i) - \frac{1}{3!}h^3 f'''(x_i) + \frac{1}{4!}h^4 f^{(4)}(\xi), \quad x_{i-1} \le \xi \le x_i$$

$$f(x_{i+1}) = f(x_i) + hf'(x_i) + \frac{1}{2!}h^2 f''(x_i) + \frac{1}{3!}h^3 f'''(x_i) + \frac{1}{4!}h^4 f^{(4)}(\eta), \quad x_i \le \eta \le x_{i+1}$$

Adding the two equations and proceeding as always, we have

$$f''(x_i) = \frac{f(x_{i-1}) - 2f(x_i) + f(x_{i+1})}{h^2} + O(h^2) \tag{6.10}$$

Therefore, in approximating the second derivative, the three-point central difference formula has a truncation error of $O(h^2)$ compared to $O(h)$ for the three-point backward and forward difference formulas.

Example 6.2: Finite-Difference Formulas for the Second Derivative

Consider $f(x) = e^{-x} \sin(x/2)$ of Example 6.1. Approximate $f''(1.4)$ using

- Three-point backward difference formula ($h = 0.2$)
- Three-point forward difference formula ($h = 0.2$)
- Three-point central difference formula ($h = 0.2$)
- Three-point central difference formula ($h = 0.1$)

Find the relative error in each case.

SOLUTION

The actual value is $f''(1.4) = -0.0695$. The numerical results are summarized in Table 6.2, where it is readily seen that the three-point central difference formula gives the most accurate estimate. It is also observed that reducing the spacing significantly improves the accuracy.

TABLE 6.2
Summary of Calculations in Example 6.2

Difference Formula	Approximate $f''(1.4)$	Relative Error (%)
3-Point backward $h = 0.2$	$\dfrac{f(1) - 2f(1.2) + f(1.4)}{(0.2)^2} = -0.1225$	76.4
3-Point forward $h = 0.2$	$\dfrac{f(1.4) - 2f(1.6) + f(1.8)}{(0.2)^2} = -0.0330$	52.6
3-Point central $h = 0.2$	$\dfrac{f(1.2) - 2f(1.4) + f(1.6)}{(0.2)^2} = -0.0706$	1.69
3-Point central $h = 0.1$	$\dfrac{f(1.3) - 2f(1.4) + f(1.5)}{(0.1)^2} = -0.0698$	0.42

6.2.2.4 *Summary of Finite-Difference Formulas for First to Fourth Derivatives*

Table 6.3 lists the difference formulas presented earlier, as well as additional formulas for the first and second derivatives. It also includes formulas that can similarly be derived for the third and fourth derivatives at a point x_i.

6.2.3 Estimate Improvement: Richardson's Extrapolation

Derivative estimates using finite differences can clearly be improved by either reducing the step size or using a higher-order difference formula that involves more points. A third method is to use Richardson's extrapolation, which combines two derivative approximations to obtain a more accurate estimate. The idea is best understood via a specific example.

Consider the approximation of the first derivative using the two-point central difference formula. We will repeat some of the analysis done earlier, but show more terms in Taylor series expansions for our purpose.

$$f(x_{i-1}) = f(x_i) - hf'(x_i) + \frac{1}{2!}h^2 f''(x_i) - \frac{1}{3!}h^3 f'''(x_i) + \frac{1}{4!}h^4 f^{(4)}(x_i) - \frac{1}{5!}h^5 f^{(5)}(\xi),$$

$$x_{i-1} \le \xi \le x_i$$

and

$$f(x_{i+1}) = f(x_i) + hf'(x_i) + \frac{1}{2!}h^2 f''(x_i) + \frac{1}{3!}h^3 f'''(x_i) + \frac{1}{4!}h^4 f^{(4)}(x_i) + \frac{1}{5!}h^5 f^{(5)}(\eta),$$

$$x_i \le \eta \le x_{i+1}$$

TABLE 6.3

Summary of Finite Difference Formulas for First, Second, Third, and Fourth Derivatives

Difference Formula	First Derivative	Truncation Error
2-Point backward	$f'(x_i) = \dfrac{f(x_i) - f(x_{i-1})}{h}$	$O(h)$
2-Point forward	$f'(x_i) = \dfrac{f(x_{i+1}) - f(x_i)}{h}$	$O(h)$
2-Point central	$f'(x_i) = \dfrac{f(x_{i+1}) - f(x_{i-1})}{2h}$	$O(h^2)$
3-Point backward	$f'(x_i) = \dfrac{f(x_{i-2}) - 4f(x_{i-1}) + 3f(x_i)}{2h}$	$O(h^2)$
3-Point forward	$f'(x_i) = \dfrac{-3f(x_i) + 4f(x_{i+1}) - f(x_{i+2})}{2h}$	$O(h^2)$
4-Point central	$f'(x_i) = \dfrac{f(x_{i-2}) - 8f(x_{i-1}) + 8f(x_{i+1}) - f(x_{i+2})}{12h}$	$O(h^4)$
Difference Formula	**Second Derivative**	**Truncation Error**
3-Point backward	$f''(x_i) = \dfrac{f(x_{i-2}) - 2f(x_{i-1}) + f(x_i)}{h^2}$	$O(h)$
3-Point forward	$f''(x_i) = \dfrac{f(x_{i+2}) - 2f(x_{i+1}) + f(x_i)}{h^2}$	$O(h)$

3-Point central	$f''(x_i) = \frac{f(x_{i-1}) - 2f(x_i) + f(x_{i+1})}{h^2}$	$O(h^2)$
4-Point backward	$f''(x_i) = \frac{-f(x_{i-3}) + 4f(x_{i-2}) - 5f(x_{i-1}) + 2f(x_i)}{h^2}$	$O(h^2)$
4-Point forward	$f''(x_i) = \frac{2f(x_i) - 5f(x_{i+1}) + 4f(x_{i+2}) - f(x_{i+3})}{h^2}$	$O(h^2)$
5-Point central	$f''(x_i) = \frac{-f(x_{i-2}) + 16f(x_{i-1}) - 30f(x_i) + 16f(x_{i+1}) - f(x_{i+2})}{12h^2}$	$O(h^4)$
Difference Formula	**Third Derivative**	**Truncation Error**
4-Point backward	$f'''(x_i) = \frac{-f(x_{i-3}) + 3f(x_{i-2}) - 3f(x_{i-1}) + f(x_i)}{h^3}$	$O(h)$
4-Point forward	$f'''(x_i) = \frac{-f(x_i) + 3f(x_{i+1}) - 3f(x_{i+2}) + f(x_{i+3})}{h^3}$	$O(h)$
4-Point central	$f'''(x_i) = \frac{-f(x_{i-2}) + 2f(x_{i-1}) - 2f(x_{i+1}) + f(x_{i+2})}{2h^3}$	$O(h^2)$
5-Point backward	$f'''(x_i) = \frac{3f(x_{i-4}) - 14f(x_{i-3}) + 24f(x_{i-2}) - 18f(x_{i-1}) + 5f(x_i)}{2h^3}$	$O(h^2)$
5-Point forward	$f'''(x_i) = \frac{-5f(x_i) + 18f(x_{i+1}) - 24f(x_{i+2}) + 14f(x_{i+3}) - 3f(x_{i+4})}{2h^3}$	$O(h^2)$
6-Point central	$f'''(x_i) = \frac{f(x_{i-3}) - 8f(x_{i-2}) + 13f(x_{i-1}) - 13f(x_{i+1}) + 8f(x_{i+2}) - f(x_{i+3})}{8h^3}$	$O(h^4)$

(continued)

TABLE 6.3 (continued)

Summary of Finite Difference Formulas for First, Second, Third, and Fourth Derivatives

Difference Formula	Fourth Derivative	Truncation Error
5-Point backward	$f^{(4)}(x_i) = \dfrac{f(x_{i-4}) - 4f(x_{i-3}) + 6f(x_{i-2}) - 4f(x_{i-1}) + f(x_i)}{h^4}$	$O(h)$
5-Point forward	$f^{(4)}(x_i) = \dfrac{f(x_i) - 4f(x_{i+1}) + 6f(x_{i+2}) - 4f(x_{i+3}) + f(x_{i+4})}{h^4}$	$O(h)$
5-Point central	$f^{(4)}(x_i) = \dfrac{f(x_{i-2}) - 4f(x_{i-1}) + 6f(x_i) - 4f(x_{i+1}) + f(x_{i+2})}{h^4}$	$O(h^2)$
6-Point backward	$f^{(4)}(x_i) = \dfrac{-2f(x_{i-5}) + 11f(x_{i-4}) - 24f(x_{i-3}) + 26f(x_{i-2}) - 14f(x_{i-1}) + 3f(x_i)}{h^4}$	$O(h^2)$
6-Point forward	$f^{(4)}(x_i) = \dfrac{3f(x_i) - 14f(x_{i+1}) + 26f(x_{i+2}) - 24f(x_{i+3}) + 11f(x_{i+4}) - 2f(x_{i+5})}{h^4}$	$O(h^2)$
7-Point central	$f^{(4)}(x_i) = \dfrac{f(x_{i-3}) + 12f(x_{i-2}) - 39f(x_{i-1}) + 56f(x_i) + 39f(x_{i+1}) + 12f(x_{i+2}) - f(x_{i+3})}{6h^4}$	$O(h^4)$

Subtracting the first equation from the second, and solving for $f'(x_i)$, yields

$$f'(x_i) = \frac{f(x_{i+1}) - f(x_{i-1})}{2h} - \frac{1}{3!}h^2 f'''(x_i) + O(h^4) \tag{6.11}$$

We next repeat this process with step size $\frac{1}{2}h$. In the meantime, let $f(x_{i-1/2}) = f(x_i - \frac{1}{2}h)$, $f(x_{i+1/2}) = f(x_i + \frac{1}{2}h)$. Then, it can be shown that

$$f'(x_i) = \frac{f(x_{i+1/2}) - f(x_{i-1/2})}{2(\frac{1}{2}h)} - \frac{1}{3!}\left(\frac{1}{2}h\right)^2 f'''(x_i) + O(h^4) \tag{6.12}$$

Multiply Equation 6.12 by 4 and subtract Equation 6.11 from the result to obtain

$$f'(x_i) = \frac{4}{3}\underbrace{\boxed{\frac{f(x_{i+1/2}) - f(x_{i-1/2})}{h}}}_{\substack{\text{2-pt central diff. formula} \\ \text{with } h/2 \text{ and error } O(h^2)}} - \frac{1}{3}\underbrace{\boxed{\frac{f(x_{i+1}) - f(x_{i-1})}{2h}}}_{\substack{\text{2-pt central diff. formula} \\ \text{with } h \text{ and error } O(h^2)}} + O(h^4) \tag{6.13}$$

Therefore, two approximations provided by the two-point central difference formula, one with spacing h and the other with $\frac{1}{2}h$, each with error $O(h^2)$, are combined to obtain a more accurate estimate of the first derivative with error $O(h^4)$.

Equation 6.13 can be expressed in a general form as

$$D = \frac{4}{3}D_{h/2} - \frac{1}{3}D_h + O(h^4) \tag{6.14}$$

where

D = value of the derivative

D_h = a function that approximates the derivative using h and has an error of $O(h^2)$

$D_{h/2}$ = a function that approximates the derivative using $\frac{1}{2}h$ and has an error of $O(h^2)$

Note that Equation 6.14 can be used in connection with any difference formula that has an error of $O(h^2)$. Also note that the coefficients in Equation 6.14 add up to 1, and hence act as weights attached to each estimate. With increasing accuracy, they place greater weight on the better estimate. For instance, using spacing size $\frac{1}{2}h$ generates a better estimate than the one using h, and consequently $D_{h/2}$ has a larger weight attached to it than D_h does.

Example 6.3: Richardson's Extrapolation

Consider $f(x) = e^{-x}\sin(x/2)$ studied in Example 6.2. We approximated $f''(1.4)$ using the three-point central difference formula, which has an error of $O(h^2)$. Using $h = 0.2$, we found the estimate to be −0.0706. Using $h = 0.1$, the estimate was −0.0698. Therefore, $D_h = -0.0706$ and $D_{h/2} = -0.0698$. By Equation 6.14

$$D \cong \frac{4}{3}(-0.0698) - \frac{1}{3}(-0.0706) = -0.0695$$

which agrees with the actual value to four decimal places and is a superior estimate to the first two.

Richardson's extrapolation can also be used in connection with estimates that have higher-order errors. In particular, it can combine two estimates, each with error $O(h^4)$, to compute a new, more accurate estimate with error $O(h^6)$:

$$D = \frac{16}{15}D_{h/2} - \frac{1}{15}D_h + O(h^6) \tag{6.15}$$

where

D = value of the derivative

D_h = a function that approximates the derivative using h and has an error of $O(h^4)$

$D_{h/2}$ = a function that approximates the derivative using $\frac{1}{2}h$ and has an error of $O(h^4)$

Once again, as mentioned before, the coefficients add up to 1 and act as weights attached to the two estimates, with greater weight placed on the better estimate.

6.2.4 Derivative Estimates for Nonevenly Spaced Data

The finite-difference formulas to approximate derivatives of various orders require that the data be equally spaced. Also, Richardson's extrapolation is applicable only to evenly spaced data and it computes better estimates by sequentially reducing the spacing by half. These techniques are appropriate if the data is equally spaced or if the data is generated by uniform discretization of a known function, such as that in Examples 6.1 and 6.2.

Empirical data—such as data resulting from experimental measurements—on the other hand, are often not evenly spaced. For these situations, one possible way to approximate the derivative is as follows: (1) consider a set of three consecutive data points that contains the point at which the derivative is to be estimated, (2) fit a second-degree Lagrange interpolating polynomial (Section 5.5) to the set, and (3) differentiate the polynomial and evaluate at the point of interest. The derivative estimate obtained in this manner has the same accuracy as that offered by the central difference formula, and exactly matches it for the case of equally spaced data.

TABLE 6.4

Data in Example 6.4

x_i	y_i
0	1
0.3	0.8228
0.8	0.4670
1.1	0.2617
1.3	0.1396

Example 6.4: Nonevenly Spaced Data

For the data in Table 6.4, approximate the first derivative at $x = 0.9$ using the data at 0.3, 0.8, and 1.1.

SOLUTION

The data is not evenly spaced. We will consider the set of three consecutive points 0.3, 0.8, and 1.1, which includes the point of interest $x = 0.9$, and fit a second-degree Lagrange interpolating polynomial to the set. Letting $x_1 = 0.3$, $x_2 = 0.8$, and $x_3 = 1.1$, we find

$$\begin{aligned} p_2(x) &= \frac{(x-x_2)(x-x_3)}{(x_1-x_2)(x_1-x_3)}(0.8228) + \frac{(x-x_1)(x-x_3)}{(x_2-x_1)(x_2-x_3)}(0.4670) \\ &\quad + \frac{(x-x_1)(x-x_2)}{(x_3-x_1)(x_3-x_2)}(0.2617) \\ &= 0.0341x^2 - 0.7491x + 1.0445 \end{aligned}$$

Differentiation yields $p_2'(x) = 0.0682x - 0.7491$ so that $p_2'(0.9) = -0.6877$.

6.2.5 MATLAB® Built-In Functions `diff` and `polyder`

The MATLAB® built-in function `diff` can be used to estimate derivatives for both cases of equally spaced and not equally spaced data. A brief description of `diff` is given as

```
diff(X) calculates differences between adjacent elements of X.

If X is a vector, then diff(X) returns a vector, one element
shorter than X, of differences between adjacent elements:

        [X(2) -X(1) X(3) -X(2)... X(n) -X(n-1)]

diff(X,n) applies diff recursively n times, resulting in the
nth difference. Thus, diff(X,2) is the same as diff(diff(X)).
```

Equally spaced data: Consider a set of equally spaced data $(x_1, y_1), \ldots, (x_n, y_n)$, where $x_{i+1} - x_i = h$ $(i = 1, \ldots, n-1)$. Then, by the description of `diff`, the command `diff(y)./h` returns the $(n-1)$-dimensional vector

$$\begin{bmatrix} \dfrac{y_2 - y_1}{h} & \cdots & \dfrac{y_n - y_{n-1}}{h} \end{bmatrix}$$

The first component is the first-derivative estimate at x_1 using the forward-difference formula; see Equation 6.4. Similarly, the second component is the derivative estimate at x_2. The last entry is the derivative estimate at x_{n-1}. As an example, consider $f(x) = e^{-x}\sin(x/2)$, $x = 1.2, 1.4, 1.6, 1.8$, of Example 6.1. We find an estimate for $f'(1.4)$ as follows:

```
>> h = 0.2;
>> x = 1.2:h:1.8;
>> y = [0.1701 0.1589 0.1448 0.1295];
%  Values of f at the discrete x values
>> y_prime = diff(y)./h

y_prime =

  -0.0560    -0.0702    -0.0767
```

Since 1.4 is the second point in the data, it is labeled x_2. This means an estimate for $f'(1.4)$ is provided by the second component of the output `y_prime`. That is, $f'(1.4) \cong -0.0702$. This agrees with the earlier numerical results in Table 6.1.

Nonequally spaced data: Consider a set of nonevenly spaced data $(x_1, y_1), \ldots, (x_n, y_n)$. Then, by the description of `diff`, the command `diff(y)./diff(x)` returns the $(n-1)$-dimensional vector

$$\begin{bmatrix} \dfrac{y_2 - y_1}{x_2 - x_1} & \cdots & \dfrac{y_n - y_{n-1}}{x_n - x_{n-1}} \end{bmatrix}$$

The first component is the first-derivative estimate at x_1 using the forward-difference formula, the second one is the derivative estimate at x_2, while the last entry is the derivative estimate at x_{n-1}.

As mentioned in the description of `diff` above, `diff(y,2)` is the same as `diff(diff(y))`. So, if $y = [y_1 \ldots y_n]$, then `diff(y)` returns

$$\begin{bmatrix} y_2 - y_1 & y_3 - y_2 & \cdots & y_n - y_{n-1} \end{bmatrix}_{(n-1)\text{ dim}}$$

and `diff(y,2)` returns

$$[(y_3 - y_2) - (y_2 - y_1) \quad (y_4 - y_3) - (y_3 - y_2) \quad \ldots \quad (y_n - y_{n-1}) - (y_{n-1} - y_{n-2})]_{(n-2)\text{ dim}}$$

which simplifies to

$$\begin{bmatrix} y_3 - 2y_2 + y_1 & y_4 - 2y_3 + y_2 & \cdots & y_n - 2y_{n-1} + y_{n-2} \end{bmatrix}$$

The first component is the numerator in the three-point forward difference formula for estimating the second derivative at x_1; see Equation 6.9. Similarly, the remaining components agree with the numerator of Equation 6.9 at $x_2, \ldots, x_{n-2}$. Therefore, for an equally spaced data $(x_1,y_1), \ldots, (x_n,y_n)$, an estimate of the second derivative at $x_1,x_2, \ldots, x_{n-2}$ is provided by

```
diff(y,2)./h^2
```

The MATLAB built-in function `polyder` finds the derivative of a polynomial:

```
polyder Differentiate polynomial.
  polyder(P) returns the derivative of the polynomial whose
  coefficients are the elements of vector P.
  polyder(A,B) returns the derivative of polynomial A*B.
  [Q,D] =polyder(B,A) returns the derivative of the polynomial
  ratio B/A, represented as Q/D.
```

For example, the derivative of a polynomial such as $2x^3 - x + 3$ is calculated as follows:

```
>> P = [2 0 -1 3];
>> polyder(P)
ans =
     6     0    -1
```

The output corresponds to $6x^2 - 1$.

6.3 Numerical Integration: Newton–Cotes Formulas

Definite integrals are encountered in a wide range of applications, generally in the form

$$\int_a^b f(x)\, dx$$

where $f(x)$ is the integrand and a and b are the limits of integration. The value of this definite integral is the area of the region between the graph of $f(x)$ and the x-axis, bounded by the lines $x = a$ and $x = b$. As an example of a definite integral, consider the relation between the bending moment M and shear force V along the longitudinal axis x of a beam, defined by

$$M_2 - M_1 = \int_{x_1}^{x_2} V dx$$

where M_2 is the bending moment at position x_2 and M_1 is the bending moment at x_1. In this case, the integrand is shear force $V(x)$ and the limits of integration are x_1 and x_2.

The integrand may be given analytically or as a set of discrete points. Numerical integration is used when the integrand is given as a set of data or, the integrand is an analytical function, but the antiderivative is not easily found. To carry out numerical integration, discrete values of the integrand are needed. This means that even if the integrand is an analytical function, it must be discretized and the discrete values will be used in the calculations.

6.3.1 Newton–Cotes Formulas

Newton–Cotes formulas provide the most commonly used integration techniques and are divided into two categories: closed form and open form. In closed form schemes, the data points at the endpoints of the interval are used in calculations; the trapezoidal and Simpson's rules are closed Newton–Cotes formulas. In open form methods, limits of integration extend beyond the range of the discrete data; the rectangular rule and the Gaussian quadrature (Section 6.4) are open Newton–Cotes formulas.

The main idea behind Newton–Cotes formulas is to replace the complicated integrand or data with an easy-to-integrate function, usually a polynomial. If the integrand is an analytical function, it is first discretized, and then the polynomial that interpolates this set is found and integrated. If the integrand is a set of data, the interpolating polynomial is found and integrated.

6.3.2 Rectangular Rule

In the rectangular rule, the definite integral $\int_a^b f(x)\,dx$ is approximated by the area of a rectangle. This rectangle may be built using the left endpoint, the right endpoint, or the midpoint of the interval $[a, b]$; Figure 6.3. The one that uses the midpoint is sometimes called the midpoint method. All three cases are Newton–Cotes formulas, where the integrand is replaced with a horizontal line (constant), that is, a zero-degree polynomial. But it is evident by Figure 6.3 that the error of approximation can be quite large depending

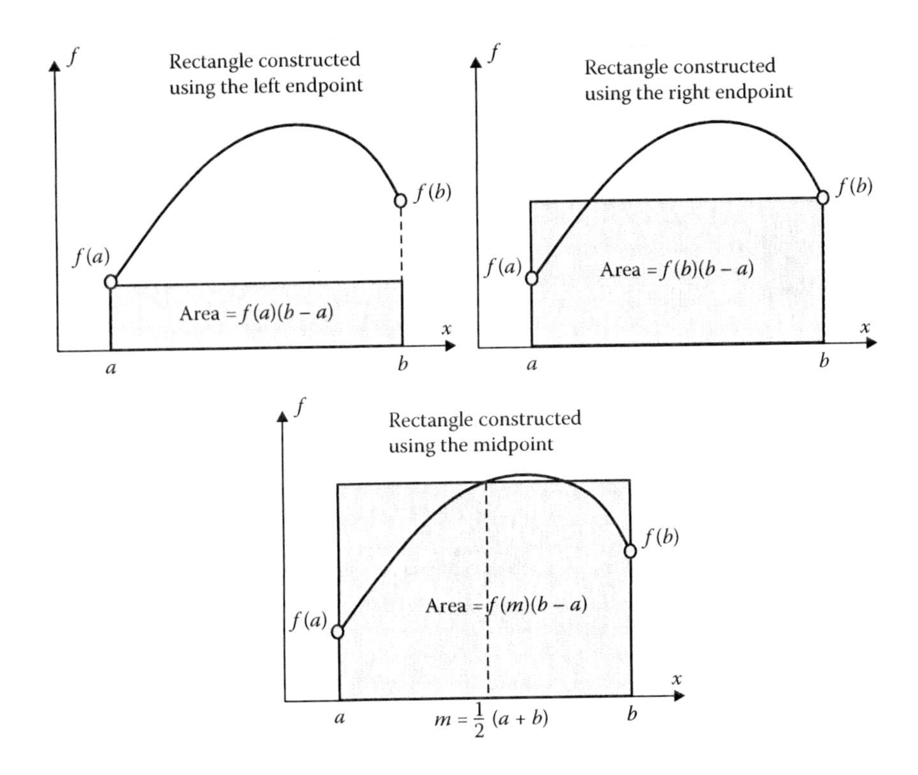

FIGURE 6.3
Rectangular rule.

on the nature of the integrand. The accuracy can be improved considerably by using the composite rectangular rule.

6.3.2.1 Composite Rectangular Rule

In applying the composite rectangular rule, the interval $[a, b]$ is divided into n subintervals defined by $n + 1$ points labeled $a = x_1, x_2, \ldots, x_n, x_{n+1} = b$. The subintervals can generally have different widths so that longer intervals may be chosen for regions where the integrand exhibits slow variations and shorter intervals where the integrand experiences rapid changes. In most of the results presented here, however, the data is assumed equally spaced. Over each subinterval $[x_i, x_{i+1}]$, the integral is approximated by the area of a rectangle. These rectangles are constructed using the left endpoint, the right endpoint, or the midpoint as described earlier; Figure 6.4. Adding the areas of rectangles yields the approximate value of the definite integral $\int_a^b f(x)dx$.

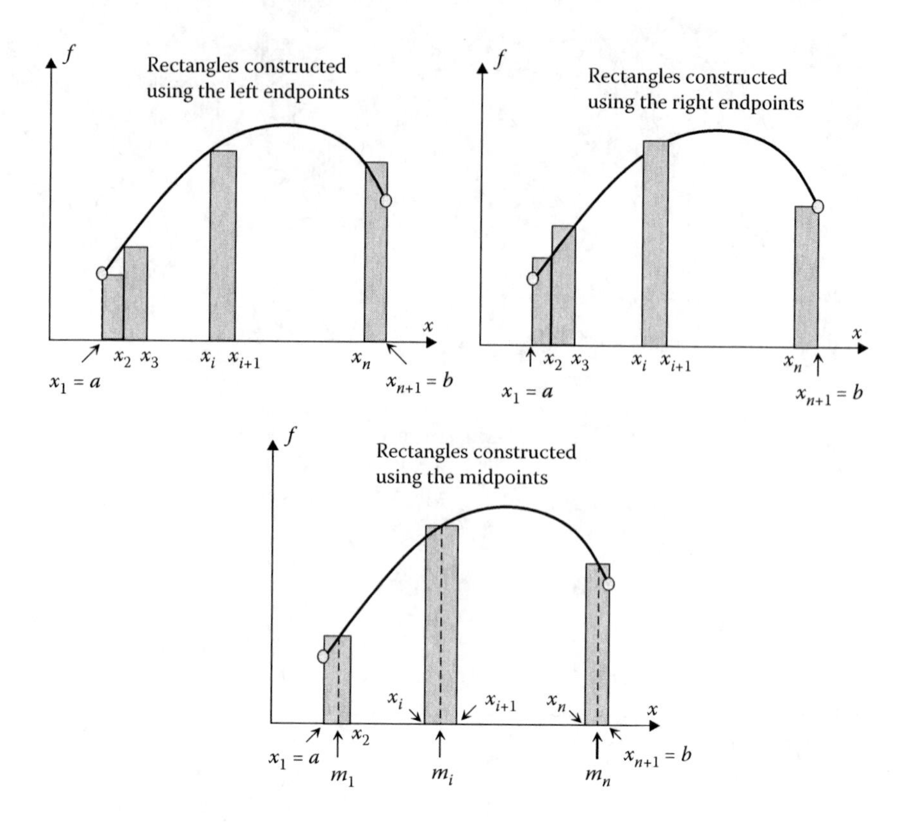

FIGURE 6.4
Composite rectangular rules.

Composite rectangular rule (using left endpoint):

$$\int_a^b f(x)\,dx \cong f(x_1)(x_2 - x_1) + f(x_2)(x_3 - x_2) + \cdots$$

$$+ f(x_n)(x_{n+1} - x_n) \underset{h=(b-a)/n}{\overset{\text{Equally spaced}}{=}} h\sum_{i=1}^{n} f(x_i) \tag{6.16}$$

Composite rectangular rule (using right endpoint):

$$\int_a^b f(x)\,dx \cong f(x_2)(x_2 - x_1) + f(x_3)(x_3 - x_2) + \cdots$$

$$+ f(x_{n+1})(x_{n+1} - x_n) \underset{h=(b-a)/n}{\overset{\text{Equally spaced}}{=}} h\sum_{i=2}^{n+1} f(x_i) \tag{6.17}$$

Composite rectangular rule (using midpoint):

$$\int_a^b f(x)\,dx \cong f(m_1)(x_2 - x_1) + \cdots$$

$$+ f(m_n)(x_{n+1} - x_n) \underset{h=(b-a)/n}{\overset{\text{Equally spaced}}{=}} h \sum_{i=1}^{n} f(m_i), \quad m_i = \frac{1}{2}(x_{i+1} + x_i) \quad (6.18)$$

6.3.3 Error Estimate for Composite Rectangular Rule

Consider $\int_a^b f(x)dx$, where $a = x_1, x_2, \ldots, x_n, x_{n+1} = b$ divide $[a, b]$ into n subintervals of equal length h. Assume that over each interval $[x_i, x_{i+1}]$ the rectangle is constructed using the left endpoint x_i so that it has an area of $hf(x_i)$. The error associated with the integral over each subinterval is

$$E_i = \underbrace{\int_{x_i}^{x_{i+1}} f(x)\,dx}_{\text{Actual value}} - \underbrace{hf(x_i)}_{\text{Estimate}}$$

By Taylor series expansion, we have

$$f(x) = f(x_i) + f'(\xi_i)(x - x_i), \quad x_i \le \xi_i \le x_{i+1}$$

Then

$$E_i = \int_{x_i}^{x_{i+1}} \left[f(x_i) + f'(\xi_i)(x - x_i)\right] dx - hf(x_i) \underset{\text{and simplify}}{\overset{\text{Evaluate}}{=}} \frac{1}{2}h^2 f'(\xi_i)$$

This indicates that each error E_i can be made very small by choosing a very small spacing size, that is, $h \ll 1$. The error associated with the entire interval $[a, b]$ is given by

$$E = \sum_{i=1}^{n} E_i = \sum_{i=1}^{n} \frac{1}{2}h^2 f'(\xi_i) = \frac{1}{2}h^2 \sum_{i=1}^{n} f'(\xi_i)$$

An average value for f' over $[a, b]$ may be estimated by

$$\bar{f'} \cong \frac{1}{n} \sum_{i=1}^{n} f'(\xi_i)$$

Consequently

$$E = \frac{1}{2}h^2 n\bar{f}' = \frac{1}{2}\left(\frac{b-a}{n}\right)hn\bar{f}' = \left[\frac{1}{2}(b-a)\bar{f}'\right]h$$

Since $\frac{1}{2}(b-a)\bar{f}' =$ constant, the error E is of the order of h, written $O(h)$. In summary

Composite rectangular rule (left endpoint):

$$E = \left[\frac{1}{2}(b-a)\bar{f}'\right]h = O(h) \tag{6.19}$$

Similarly, for the composite rectangular rule (using right endpoint), $E = O(h)$. Finally, we present without proof

Composite rectangular rule (midpoint):

$$E = \left[\frac{1}{24}(b-a)\bar{f}''\right]h^2 = O(h^2) \tag{6.20}$$

where $\bar{f}''$ is the estimated average value of f'' over $[a, b]$.

Example 6.5: Composite Rectangular Rule

Evaluate the following definite integral using all three composite rectangular rule strategies with $n = 8$:

$$\int_{-1}^{1} \frac{1}{x+2}\,dx$$

SOLUTION

With the limits of integration at $b = 1$, $a = -1$, we find the spacing size as $h = (b-a)/n = 2/8 = 0.25$. The nine nodes are thus defined as $x_1 = -1$, $-0.75, -0.5, \ldots, 0.75, 1 = x_9$. Letting $f(x) = 1/(x+2)$, the three integral estimates are found as follows:

Using left endpoint:

$$\int_{-1}^{1} f(x)\,dx \cong h\sum_{i=1}^{8} f(x_i) = 0.25[f(-1) + f(-0.75) + \cdots + f(0.75)]$$

$$= 1.1865$$

Using right endpoint:

$$\int_{-1}^{1} f(x)\,dx \cong h\sum_{i=2}^{9} f(x_i) = 0.25[f(-0.75) + f(-0.5) + \cdots$$
$$+ f(0.75) + f(1)] = 1.0199$$

Using midpoint:

$$\int_{-1}^{1} f(x)\,dx \cong h\sum_{i=1}^{8} f(m_i) = 0.25[f(-0.8750) + f(-0.6250) + \cdots$$
$$+ f(0.8750)] = 1.0963$$

Noting that the actual value of the integral is 1.0986, the above estimates come with relative errors of 8%, 7.17%, and 0.21%, respectively. As suggested by Equations 6.19 and 6.20, the midpoint method yields the best accuracy.

6.3.4 Trapezoidal Rule

The trapezoidal rule is a Newton–Cotes formula, where the integrand is replaced with a straight line (a first-degree polynomial) connecting the points $(a, f(a))$ and $(b, f(b))$ so that the definite integral $\int_a^b f(x)dx$ is approximated by the area of a trapezoid; Figure 6.5a. The equation of this connecting line is

$$p_1(x) = f(a) + \frac{f(b) - f(a)}{b - a}(x - a)$$

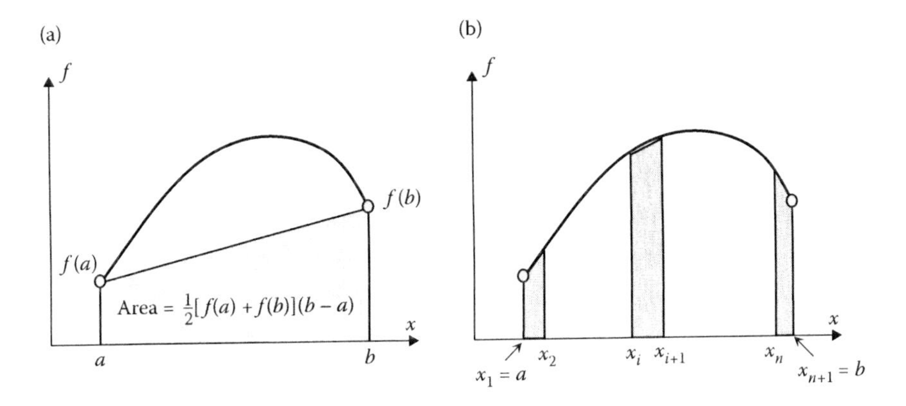

FIGURE 6.5
(a) Trapezoidal rule and (b) composite trapezoidal rule.

Therefore

$$\int_a^b f(x)\,dx \cong \int_a^b p_1(x)\,dx = \int_a^b \left\{ f(a) + \frac{f(b)-f(a)}{b-a}(x-a) \right\} dx$$

$$= \left[f(a)x + \frac{f(b)-f(a)}{b-a}\frac{(x-a)^2}{2} \right]_{x=a}^{b}$$

Evaluation of this last expression yields

$$\int_a^b f(x)\,dx \cong \frac{f(a)+f(b)}{2}(b-a) \tag{6.21}$$

The right side is indeed the area of the trapezoid as shown in Figure 6.5a. It is also evident by Figure 6.5a that the error of approximation can be quite large depending on the nature of the integrand. The accuracy of estimation can be improved significantly by using the composite trapezoidal rule; Figure 6.5b.

6.3.4.1 Composite Trapezoidal Rule

In the composite rectangular rule, the interval $[a, b]$ is divided into n subintervals defined by $n+1$ points labeled $a = x_1, x_2, \ldots, x_n, x_{n+1} = b$. As in the case of rectangular rule, the subintervals can have different widths so that longer intervals can be used for regions where the integrand shows slow variations and shorter intervals where the integrand shows rapid changes. In most of the results presented here, however, the data is assumed equally spaced. Over each subinterval $[x_i, x_{i+1}]$, the integral is approximated by the area of a trapezoid. Adding the areas of trapezoids yields the approximate value of the definite integral:

$$\int_a^b f(x)\,dx \cong \frac{f(x_1)+f(x_2)}{2}(x_2-x_1) + \frac{f(x_2)+f(x_3)}{2}(x_3-x_2) + \cdots$$

$$+ \frac{f(x_n)+f(x_{n+1})}{2}(x_{n+1}-x_n)$$

$$= \sum_{i=1}^{n} \frac{f(x_i)+f(x_{i+1})}{2}(x_{i+1}-x_i) \tag{6.22}$$

For the case of equally spaced data, $x_{i+1} - x_i = h$ ($i = 1, 2, \ldots, n$), Equation 6.22 simplifies to

$$\int_a^b f(x)\,dx \cong \frac{h}{2}\sum_{i=1}^{n}\left[f(x_i) + f(x_{i+1})\right] = \frac{h}{2}\left[f(a) + 2f(x_2) + 2f(x_3) + \cdots + 2f(x_n) + f(b)\right] \quad (6.23)$$

6.3.4.2 Error Estimate for Composite Trapezoidal Rule

The error for the composite trapezoidal rule can be shown to be

$$E = \left[-\frac{1}{12}(b-a)\bar{f}''\right]h^2 = O(h^2) \quad (6.24)$$

where $\bar{f}''$ is the estimated average value of f'' over $[a, b]$. Therefore, the error $O(h^2)$ is compatible with the midpoint method and superior to the rectangular rule using the endpoints whose error is $O(h)$.

The user-defined function TrapComp uses the composite trapezoidal rule to estimate the value of a definite integral.

```
function I = TrapComp(f,a,b,n)
%
% TrapComp estimates the value of the integral of f(x)
% from a to b by using the composite trapezoidal rule
% applied to n equal-length subintervals.
%
%   I = TrapComp(f,a,b,n) where
%
%     f is an inline function representing the integrand,
%     a and b are the limits of integration,
%     n is the number of equal-length subintervals in [a,b],
%
%     I is the integral estimate.
%
h = (b-a)/n; I = 0;
x = a:h:b;
for i = 2:n,
    I = I + 2*f(x(i));
```

```
end
I=I+f(a)+f(b);
I=I*h/2;
```

Example 6.6: Composite Trapezoidal Rule

a. Evaluate the definite integral in Example 6.5 using the composite trapezoidal rule with $n = 8$:

$$\int_{-1}^{1} \frac{1}{x+2}\,dx$$

b. Confirm the result by executing the user-defined function `TrapComp`.

SOLUTION

a. The spacing size is $h = (b - a)/n = 0.25$ and the nine nodes are defined as $x_1 = -1, -0.75, -0.5, \ldots, 0.75, 1 = x_9$. Letting $f(x) = 1/(x+2)$, the integral estimate is found by Equation 6.23 as follows:

$$\int_{-1}^{1} \frac{1}{x+2}\,dx = \frac{0.25}{2}\left[f(-1) + 2f(-0.75) + 2f(-0.5) + \cdots + 2f(0.75) + f(1)\right]$$
$$= 1.1032$$

Recalling the actual value 1.0986, the relative error is calculated as 0.42%. As expected, and stated earlier, the accuracy of composite trapezoidal rule is compatible with the midpoint method and better than the composite rectangular rule using either endpoint.

b.

```
>> f = inline('1/(x+2)');
>> I = TrapComp(f,-1,1,8)

I =
    1.1032
```

6.3.5 Simpson's Rules

The trapezoidal rule estimates the value of a definite integral by approximating the integrand with a first-degree polynomial, the line connecting the points $(a, f(a))$ and $(b, f(b))$. Any method that uses a higher-degree polynomial to connect these points will provide a more accurate estimate. Simpson's 1/3 and 3/8 rules, respectively, use second- and third-degree polynomials to approximate the integrand.

6.3.5.1 Simpson's 1/3 Rule

In evaluating $\int_a^b f(x)dx$, the Simpson's 1/3 rule uses a second-degree polynomial to approximate the integrand $f(x)$. The three points that are needed to determine this polynomial are picked as $x_1 = a$, $x_2 = (a + b)/2$, and $x_3 = b$ as shown in Figure 6.6a. Consequently, the second-degree Lagrange interpolating polynomial (Section 5.5) is constructed as

$$p_2(x) = \frac{(x - x_2)(x - x_3)}{(x_1 - x_2)(x_1 - x_3)} f(x_1) + \frac{(x - x_1)(x - x_3)}{(x_2 - x_1)(x_2 - x_3)} f(x_2) + \frac{(x - x_1)(x - x_2)}{(x_3 - x_1)(x_3 - x_2)} f(x_3)$$

The definite integral will then be evaluated with this polynomial replacing the integrand:

$$\int_a^b f(x)\,dx \cong \int_a^b p_2(x)\,dx$$

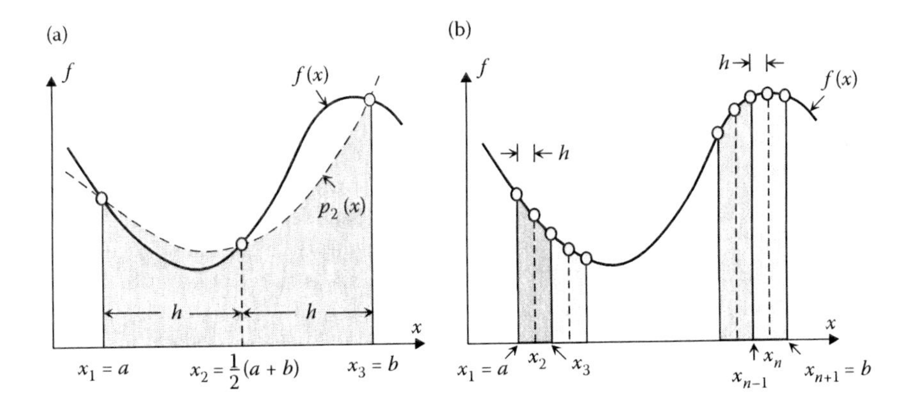

FIGURE 6.6
(a) Simpson's 1/3 rule and (b) composite Simpson's 1/3 rule.

Substituting for $p_2(x)$, with $x_1 = a$, $x_2 = (a+b)/2$, $x_3 = b$, integrating from a to b, and simplifying, yields

$$\int_a^b f(x)\,dx \cong \frac{h}{3}\left[f(x_1) + 4f(x_2) + f(x_3)\right], \quad h = \frac{b-a}{2} \tag{6.25}$$

The method is known as the 1/3 rule because h is multiplied by 1/3. The estimation error, which can be large depending on the nature of the integrand, can be improved significantly by repeated applications of the Simpson's 1/3 rule.

6.3.5.2 Composite Simpson's 1/3 Rule

In the composite Simpson's 1/3 rule, the interval $[a, b]$ is divided into n subintervals defined by $n+1$ points labeled $a = x_1, x_2, \ldots, x_n, x_{n+1} = b$. Although the subintervals can have different widths, the results that follow are based on the assumption that the points are equally spaced with spacing size $h = (b-a)/n$. Since three points are needed to construct a second-degree polynomial, the Simpson's 1/3 rule must be applied to two adjacent subintervals at a time. For example, the first application will be to the first two subintervals $[x_1, x_2]$ and $[x_2, x_3]$ so that the three points corresponding to x_1, x_2, and x_3 are used for polynomial construction. The next application will be to $[x_3, x_4]$ and $[x_4, x_5]$, so that x_3, x_4, and x_5 are used for construction. Continuing this pattern, the very last interval is composed of $[x_{n-1}, x_n]$ and $[x_n, x_{n+1}]$; see Figure 6.6b. Therefore, $[a, b]$ *must be divided into an even number of subintervals for the composite 1/3 rule to be implemented.* As a result

$$\int_a^b f(x)\,dx \cong \frac{h}{3}\left[f(x_1) + 4f(x_2) + f(x_3)\right] + \frac{h}{3}\left[f(x_3) + 4f(x_4) + f(x_5)\right] + \cdots$$
$$+ \frac{h}{3}\left[f(x_{n-1}) + 4f(x_n) + f(x_{n+1})\right]$$

The even-indexed points $(x_2, x_4, \ldots, x_n)$ are the middle terms in each application of the 1/3 rule, and therefore, by Equation 6.25 have a coefficient of 4. The odd-indexed terms $(x_3, x_5, \ldots, x_{n-1})$ are the common points to adjacent intervals and thus count twice and have a coefficient of 2. The two terms $f(x_1)$ and $f(x_{n+1})$ on the far left and far right each has a coefficient of 1. In summary

$$\int_a^b f(x)\,dx \cong \frac{h}{3}\left\{f(x_1) + 4\sum_{i=2,4,6}^{n} f(x_i) + 2\sum_{j=3,5,7}^{n-1} f(x_j) + f(x_{n+1})\right\} \tag{6.26}$$

6.3.5.3 Error Estimate for Composite Simpson's 1/3 Rule

The error for the composite Simpson's 1/3 rule can be shown to be

$$E = \left[-\frac{1}{180}(b-a)\overline{f^{(4)}} \right] h^4 = O(h^4) \tag{6.27}$$

where $\overline{f^{(4)}}$ is the estimated average value of $f^{(4)}$ over $[a, b]$. Therefore, the error $O(h^4)$ is superior to the composite trapezoidal rule, which has an error of $O(h^2)$.

The user-defined function `Simpson` uses the composite Simpson's 1/3 rule to estimate the value of a definite integral.

```
function I = Simpson(f,a,b,n)
%
% Simpson estimates the value of the integral of f(x)
% from a to b by using the composite Simpson's 1/3 rule
% applied to n equal-length subintervals.
%
%   I = Simpson(f,a,b,n) where
%
%      f is an inline function representing the integrand,
%      a, b are the limits of integration,
%      n is the (even) number of subintervals,
%
%      I is the integral estimate.

h = (b-a)/n;
x = a:h:b;
I = 0;
for i = 1:2:n,
    I = I + f(x(i)) + 4*f(x(i+1)) + f(x(i+2));
end
    I = (h/3)*I;
```

Example 6.7: Composite Simpson's 1/3 Rule

a. Evaluate the definite integral in Examples 6.5 and 6.6 using the composite Simpson's 1/3 rule with $n = 8$:

$$\int_{-1}^{1} \frac{1}{x+2}\, dx$$

b. Confirm the result by executing the user-defined function `Simpson`.

SOLUTION

a. The spacing size is $h = (b - a)/n = 0.25$ and the nine nodes are defined as $x_1 = -1, -0.75, -0.5, \ldots, 0.75, 1 = x_9$. Letting $f(x) = 1/(x+2)$, the integral estimate is found by Equation 6.26 as follows:

$$\int_{-1}^{1} \frac{1}{x+2}\,dx = \frac{0.25}{3}[f(-1) + 4f(-0.75) + 2f(-0.5) + 4f(-0.25) + 2f(0)$$

$$+ 4f(0.25) + 2f(0.5) + 4f(0.75) + f(1)] = 1.0987$$

Knowing the actual value is 1.0986, the relative error is calculated as 0.10%. As expected, the accuracy of the composite Simpson's 1/3 rule is superior to the composite trapezoidal rule. Recall that the relative error associated with the composite trapezoidal rule was calculated in Example 6.6 as 0.42%.

b.

```
>> f = inline('1/(x+2)');
>> I = Simpson(f, -1,1,8)

I =

    1.0987
```

6.3.5.4 Simpson's 3/8 Rule

The Simpson's 3/8 rule uses a third-degree polynomial to approximate the integrand $f(x)$. The four points that are needed to form this polynomial are picked as the four equally spaced points $x_1 = a$, $x_2 = (2a + b)/3$, $x_3 = (a + 2b)/3$, and $x_4 = b$ with spacing size $h = (b - a)/3$ as shown in Figure 6.7. The third-degree Lagrange interpolating polynomial (Section 5.5) is then constructed as

$$p_3(x) = \frac{(x - x_2)(x - x_3)(x - x_4)}{(x_1 - x_2)(x_1 - x_3)(x_1 - x_4)} f(x_1) + \frac{(x - x_1)(x - x_3)(x - x_4)}{(x_2 - x_1)(x_2 - x_3)(x_2 - x_4)} f(x_2)$$

$$+ \frac{(x - x_1)(x - x_2)(x - x_4)}{(x_3 - x_1)(x_3 - x_2)(x_3 - x_4)} f(x_3) + \frac{(x - x_1)(x - x_2)(x - x_3)}{(x_4 - x_1)(x_4 - x_2)(x_4 - x_3)} f(x_4)$$

The definite integral will be evaluated with this polynomial replacing the integrand:

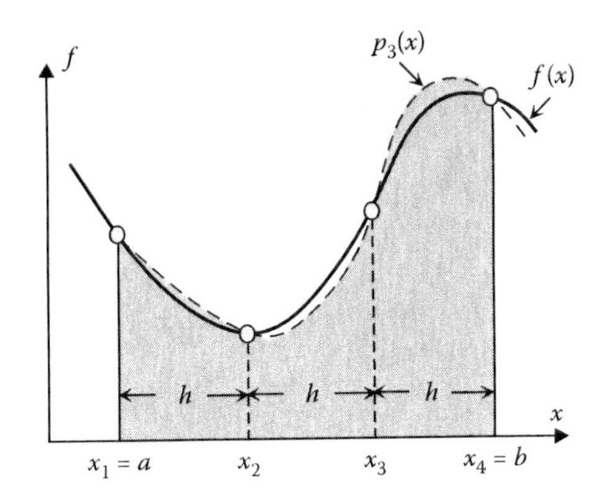

FIGURE 6.7
Simpson's 3/8 rule.

$$\int_a^b f(x)\,dx \cong \int_a^b p_3(x)\,dx$$

Substituting for $p_3(x)$, with x_1, x_2, x_3, x_4, integrating from a to b, and simplifying, yields

$$\int_a^b f(x)\,dx \cong \frac{3h}{8}\left[f(x_1) + 3f(x_2) + 3f(x_3) + f(x_4)\right], \quad h = \frac{b-a}{3} \qquad (6.28)$$

The method is known as the 3/8 rule because h is multiplied by 3/8. As before, the estimation error can be improved significantly by repeated applications of the Simpson's 3/8 rule.

6.3.5.5 *Composite Simpson's 3/8 Rule*

In the composite Simpson's 3/8 rule, the interval $[a, b]$ is divided into n subintervals defined by $n + 1$ points labeled $a = x_1, x_2, \ldots, x_n, x_{n+1} = b$. The subintervals can have different widths, but the results presented here are based on the assumption that they are equally spaced with spacing size $h = (b - a)/n$. Since four points are needed to construct a third-degree polynomial, the Simpson's 3/8 rule is applied to three adjacent subintervals at a time. For example, the first application will be to the first three subintervals $[x_1, x_2]$, $[x_2, x_3]$, and $[x_3, x_4]$ so that the four points corresponding to x_1, x_2, x_3, and x_4 are used for polynomial construction. The next application will be to $[x_4, x_5]$, $[x_5, x_6]$, and $[x_6, x_7]$ so that x_4, x_5, x_6, and x_7 are used for construction.

Continuing this pattern, the very last interval is composed of $[x_{n-2}, x_{n-1}]$, $[x_{n-1}, x_n]$, and $[x_n, x_{n+1}]$. Therefore, $[a, b]$ *must be divided into a number of subintervals that is a multiple of 3 for the composite 3/8 rule to be implemented.* As a result

$$\int_a^b f(x)\,dx \cong \frac{3h}{8}[f(x_1) + 3f(x_2) + 3f(x_3) + f(x_4)] + \frac{3h}{8}[f(x_4) + 3f(x_5) + 3f(x_6)$$

$$+ f(x_7)] + \cdots + \frac{3h}{8}[f(x_{n-2}) + 3f(x_{n-1}) + 3f(x_n) + f(x_{n+1})], \quad h = \frac{b-a}{n}$$

The middle terms in each application of 3/8 rule have a coefficient of 3 by Equation 6.28, while the common points to adjacent intervals are counted twice and have a coefficient of 2. The two terms $f(x_1)$ and $f(x_{n+1})$ on the far left and far right each has a coefficient of 1. In summary

$$\int_a^b f(x)\,dx \cong \frac{3h}{8}\left\{ f(x_1) + 3\sum_{i=2,5,8}^{n-1}\left[f(x_i) + f(x_{i+1})\right] + 2\sum_{j=4,7,10}^{n-2} f(x_j) + f(x_{n+1})\right\} \quad (6.29)$$

We summarize the implementation of the composite Simpson rules as follows: If the number of subintervals is even, then Simpson's 1/3 rule is applied. If the number of subintervals is odd, then Simpson's 3/8 rule is applied to the last three subintervals and the 1/3 rule is applied to all previous ones; see Problem Set 6.2.

6.3.5.6 Error Estimate for Composite Simpson's 3/8 Rule

The error for the composite Simpson's 3/8 rule can be shown to be

$$E = \left[-\frac{1}{80}(b-a)\overline{f^{(4)}}\right]h^4 = O(h^4) \quad (6.30)$$

where $\overline{f^{(4)}}$ is the estimated average value of $f^{(4)}$ over the interval $[a, b]$. Therefore, the error $O(h^4)$ is compatible with that of the composite 1/3 rule.

The rectangular rule, trapezoidal rule, and the Simpsons' 1/3 and 3/8 rules all belong to a class of integration techniques known as Newton–Cotes formulas. Although there are higher-order formulas, which need more than four points to form the interpolating polynomial and naturally offer better accuracy, Simpson's rules are adequate for most applications in engineering. To improve estimation accuracy, the composite Simpson's rules are preferred to higher-order formulas. In the event that the integrand is given analytically, other methods such as Romberg integration and Gaussian quadrature (Section 6.4) are practical alternatives.

6.3.6 MATLAB® Built-In Functions `quad` and `trapz`

MATLAB has two built-in functions to compute definite integrals: `quad` and `trapz`. The `quad` function handles cases where the integrand is given analytically, while `trapz` is used when the integrand is given as a discrete set of data.

```
QUAD  Numerically evaluate integral, adaptive Simpson
      quadrature.

   Q=quad(FUN,A,B) tries to approximate the integral of
   scalar-valued function FUN from A to B to within an error
   of 1.e-6 using recursive adaptive Simpson quadrature. FUN
   is a function handle. The function Y=FUN(X) should accept
   a vector argument X and return a vector result Y, the
   integrand evaluated at each element of X.
```

Note that `quad` uses adaptive Simpson quadrature. Adaptive integration methods adjust the number of subintervals needed to meet a desired accuracy by using more function evaluations in regions where the integrand shows rapid changes and less in areas where the integrand is well approximated by a quadratic function. In particular, adaptive Simpson quadrature uses an error estimate associated with the Simpson's rule, and if the error exceeds the desired tolerance, it divides the interval into two and applies Simpson's rule to each subinterval recursively.

The integral $\int_{-1}^{1}[1/(x+2)]\,dx$ considered throughout this section, can be evaluated as follows:

```
function y=myfunc(x)
y=1./(x+2);
end

>> Q=quad(@myfunc,-1,1)

Q =

    1.0986
```

For situations where the integrand is defined as a set of discrete data, the built-in function `trapz` is used.

```
TRAPZ Trapezoidal numerical integration.

   Z=trapz(X,Y) computes the integral of Y with respect to X
   using the trapezoidal method. X and Y must be vectors of
   the same length, or X must be a column vector and Y an
   array whose first non-singleton dimension is length(X).
   trapz operates along this dimension.
```

In Example 6.6, we used the composite trapezoidal rule with $n = 8$ to evaluate $\int_{-1}^{1}[1/(x+2)]dx$. To confirm the result of that example using `trapz`, we must first generate a discrete set of data (x, y) equally spaced on $[-1, 1]$ with spacing size of $h = 0.25$.

```
>> f=inline('1/(x+2)');
>> x=-1:0.25:1;
>> for i=1:9,
y(i)=f(x(i)); % Generate the y coordinates
end

>> I=trapz(x,y)
I =
    1.1032      % Result agrees with that in Example 6.6
```

6.4 Numerical Integration of Analytical Functions: Romberg Integration, Gaussian Quadrature

Throughout Section 6.3, we presented numerical methods to evaluate integrals of analytical functions, as well as tabulated data. When the function is given analytically, it can be discretized at as many points as desired and these points are subsequently used to estimate the value of the integral. When the integrand is in tabulated form, only the given points in the data can be used for integral estimation and the number of points cannot be increased.

In this section, we introduce two methods that are exclusively developed to estimate the value of $\int_a^b f(x)dx$, where $f(x)$ is an analytical function. The first method is based on Richardson's extrapolation, which combines two numerical estimates of an integral to find a third, more accurate estimate. Richardson's extrapolation can be efficiently implemented using Romberg integration. The second method is the Gaussian quadrature, which approximates the value of the integral by using a weighted sum of values of $f(x)$ at several points in $[a,b]$. These points and the weights are determined such that the error is minimized.

6.4.1 Richardson's Extrapolation, Romberg Integration

The errors associated with the composite trapezoidal and Simpson's rules were shown in Equations 6.24 and 6.27 to be

$$E_{\text{trapezoid}} = \left[-\frac{1}{12}(b-a)\overline{f''}\right]h^2 \overset{h=(b-a)/n}{=} -\frac{(b-a)^3}{12n^2}\overline{f''}$$

and

$$E_{\text{Simpson}} = \left[-\frac{1}{180}(b-a)\overline{f^{(4)}}\right]h^4 \overset{h=(b-a)/n}{=} -\frac{(b-a)^5}{180n^4}\overline{f^{(4)}}$$

This means in both cases the error is reduced as n increases. Therefore, to achieve high levels of precision, a large number n of subintervals of $[a,b]$ are needed, requiring greater computational effort as n gets larger. Consequently, as an alternative to composite trapezoidal and Simpson's rules with large n, Romberg integration can be used to attain more accurate estimates more efficiently.

6.4.1.1 *Richardson's Extrapolation*

Richardson's extrapolation combines two numerical estimates of an integral to find a third, more accurate estimate. For example, two estimates each with error $O(h^2)$ can be combined to obtain an estimate with error $O(h^4)$. Similarly, two estimates each with error $O(h^4)$ can be combined to obtain an estimate with error $O(h^6)$. In general, Richardson's extrapolation combines two integral estimates each with order $O(h^{\text{even}})$ to obtain a third, more accurate estimate with error $O(h^{\text{even}+2})$.

As an estimate with error $O(h^2)$, consider the composite trapezoidal rule applied to n subintervals with spacing $h = (b - a)/n$, and let the corresponding integral estimate be I_h. Noting the error as given in Equation 6.24, the true value of the integral is expressed as

$$I \cong I_h - \left(\frac{b-a}{12}\overline{f''}\right)h^2$$

But since $\overline{f''}$ is the estimated average value of f'' over $[a, b]$, it is independent of h and we can rewrite the above as

$$I \cong I_h + Ch^2, \quad C = \text{constant}$$

Suppose the composite trapezoidal rule is used with two different spacing sizes h_1 and h_2 to find two estimates I_{h_1} and I_{h_2} of the same integral. Then

$$I \cong I_{h_1} + Ch_1^2$$
$$I \cong I_{h_2} + Ch_2^2$$

Eliminating C between the two equations, we find

$$I \cong \frac{(h_1/h_2)^2 I_{h_2} - I_{h_1}}{(h_1/h_2)^2 - 1}$$

It can be shown* that this new estimate has an error of $O(h^4)$. In particular, two estimates given by the composite trapezoidal rule applied with $h_1 = h$ and $h_2 = \frac{1}{2}h$ can be combined to obtain

$$I \cong \frac{2^2 I_{h/2} - I_h}{2^2 - 1} \tag{6.31}$$

Simplifying the above, and realizing the error of the estimate is $O(h^4)$, we have

$$I = \frac{4}{3} I_{h/2} - \frac{1}{3} I_h + O(h^4) \tag{6.32}$$

Note that Equation 6.32 can be used in connection with any integration formula that has an error of $O(h^2)$. Also note that the coefficients in Equation 6.32 add up to 1, hence act as weights attached to each estimate. With increasing accuracy, they place greater weight on the better estimate. For instance, using spacing size $\frac{1}{2}h$ generates a better estimate than the one using h, and consequently $I_{h/2}$ has a larger weight attached to it than I_h does.

Similarly, it can readily be shown that two integral estimates with spacing sizes h_1 and h_2, each with error $O(h^4)$, can be combined to obtain a more accurate estimate

$$I \cong \frac{(h_1/h_2)^4 I_{h_2} - I_{h_1}}{(h_1/h_2)^4 - 1}$$

with an error of $O(h^6)$. In particular, two estimates corresponding to $h_1 = h$ and $h_2 = \frac{1}{2}h$ can be combined to obtain

$$I \cong \frac{2^4 I_{h/2} - I_h}{2^4 - 1} \tag{6.33}$$

* Refer to Ralston, A. and P. Rabinowitz, *A First Course in Numerical Analysis*, 2nd edition, McGraw-Hill, NY, 1978.

Simplifying, and realizing the error of the estimate is $O(h^6)$, we find

$$I = \frac{16}{15} I_{h/2} - \frac{1}{15} I_h + O(h^6) \tag{6.34}$$

Continuing in this fashion, two integral estimates corresponding to $h_1 = h$ and $h_2 = \frac{1}{2}h$, each with error $O(h^6)$, can be combined to obtain

$$I \cong \frac{2^6 I_{h/2} - I_h}{2^6 - 1} \tag{6.35}$$

so that

$$I = \frac{64}{63} I_{h/2} - \frac{1}{63} I_h + O(h^8) \tag{6.36}$$

Example 6.8: Richardson's Extrapolation

Consider

$$\int_0^1 (x^2 + 3x)^2 dx$$

Application of the trapezoidal rule with $n = 2$, $n = 4$, and $n = 8$ yields three estimates with error $O(h^2)$, as listed in the column labeled "Level 1" in Table 6.5, together with their respective relative errors. Combining the first two estimates in Level 1 via Equation 6.32, we find a new estimate with error $O(h^4)$:

TABLE 6.5

Integral Estimates at Three Levels of Accuracy; Example 6.8

n	Estimate $O(h^2)$ Level 1	Estimate $O(h^4)$ Level 2	Estimate $O(h^6)$ Level 3
2	5.531250000000000 (17.69%)		
		4.700520833333333 (0.011%)	
4	4.908203125000000 (4.43%)		4.700000000000000 (0%)
		4.700032552083333 (0.0007%)	
8	4.752075195312500 (1.11%)		

$$I \cong \frac{4}{3}(4.908203125000000) - \frac{1}{3}(5.531250000000000)$$
$$= 4.700520833333333$$

Combining the second and third estimates in Level 1 also yields a new estimate with $O(h^4)$:

$$I \cong \frac{4}{3}(4.752075195312500) - \frac{1}{3}(4.908203125000000)$$
$$= 4.700032552083333$$

These two better estimates are listed in Level 2 of Table 6.5. Combining these two via Equation 6.34 gives a new estimate with error $O(h^6)$, in Level 3:

$$I \cong \frac{16}{15}(4.700032552083333) - \frac{1}{15}(4.700520833333333)$$
$$= 4.700000000000000$$

This last estimate happens to match the exact value of the integral.

6.4.1.2 Romberg Integration

In the foregoing analysis, Richardson's extrapolation was employed to combine two trapezoidal rule integral estimates corresponding to spacing sizes h and $\frac{1}{2}h$, each with error $O(h^{\text{even}})$, to obtain a third, more accurate estimate with error $O(h^{\text{even}+2})$. The first three such results were shown in Equations 6.31, 6.33, and 6.35. These equations also follow a definite pattern that allows us to create a general formula as

$$I_{i,j} = \frac{4^{j-1} I_{i+1,j-1} - I_{i,j-1}}{4^{j-1} - 1} \tag{6.37}$$

The entries $I_{1,1}, I_{2,1}, \ldots, I_{m,1}$ are placed in the first column and represent the estimates by the composite trapezoidal rule with the number of subintervals $n, 2n, \ldots, 2^{m-1}n$. For example, $I_{4,1}$ is the trapezoidal estimate applied to $2^3 n = 8n$ subintervals. The second column has one element fewer than the first column, with entries $I_{1,2}, I_{2,2}, \ldots, I_{m-1,2}$, which are obtained by combining every two successive entries of the first column and represent more accurate estimates. This continues until the very last column, whose only entry is $I_{1,m}$. This scheme is depicted in Figure 6.8.

The user-defined function `Romberg` uses the scheme described in Figure 6.8 to find integral estimates at various levels of accuracy.

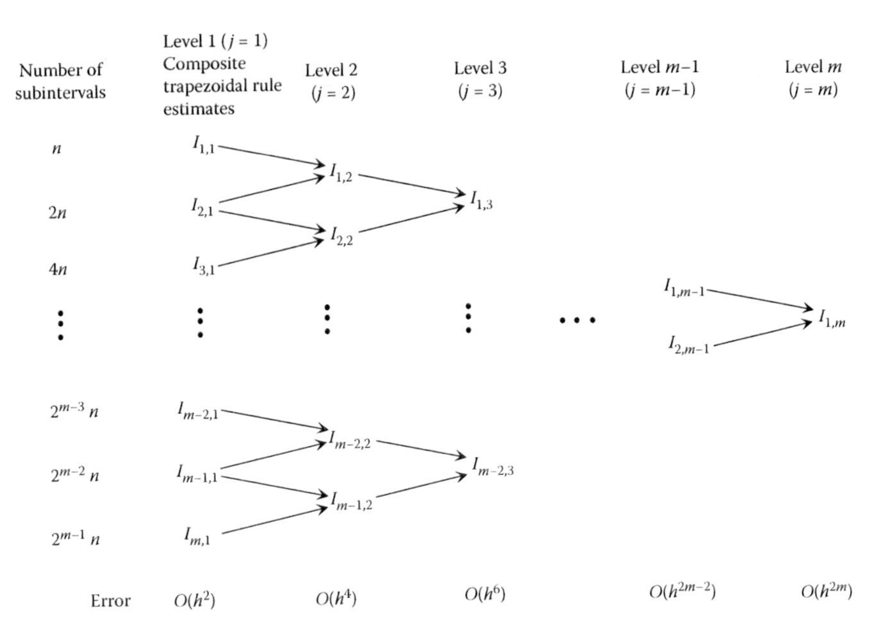

FIGURE 6.8
Romberg integration scheme.

```
function I=Romberg(f,a,b,n,n_levels)
%
% Romberg uses the Romberg integration scheme to find
% integral estimates at different levels of accuracy.
%
%   I=Romberg(f,a,b,n,n_levels) where
%
%     f is an inline function representing the integrand,
%     a and b are the limits of integration,
%     n is the initial number of equal-length
%     subintervals in [a,b],
%     n_levels is the number of accuracy levels,
%
%     I is the matrix of integral estimates.
%
I=zeros(n_levels,n_levels);    % Pre-allocate

% Calculate the first-column entries by using the
% composite trapezoidal rule, where the number of
% subintervals is doubled going from one element
% to the next.
```

```
for i=1:n_levels,
    n_intervals=2^(i-1)*n;
    I(i,1)=TrapComp(f,a,b,n_intervals);
end
% Starting with the second level, use Romberg scheme to
% generate the remaining entries of the table.
for j=2:n_levels,
    for i=1:n_levels-j+1,
        I(i,j)=(4^(j-1)*I(i+1,j-1)-I(i,j-1))/(4^(j-1)-1);
    end
end
```

The results of Example 6.8 can be verified by executing `Romberg`. Note that the initial number of subintervals for the application of composite trapezoidal rule was $n = 2$ and three levels of accuracy are desired.

```
>> format long
>> f=inline('(x^2+3*x)^2');
>> I=Romberg(f,0,1,2,3)

I =

   5.531250000000000   4.700520833333333   4.700000000000000
   4.908203125000000   4.700032552083333                   0
   4.752075195312500                   0                   0
```

As mentioned earlier, the Romberg integration scheme is more efficient than the trapezoidal and Simpson rules. Referring to the above example, if only Simpson's 1/3 rule were to be used, it would have to be applied with 192 subintervals to achieve 10-decimal accuracy.

6.4.2 Gaussian Quadrature

In estimating the value of $\int_a^b f(x)dx$ all the numerical integration methods discussed up to now have been based on approximating $f(x)$ with a polynomial and function evaluations at fixed, equally spaced points. But if these points were not fixed, we could pick them in such a way that the estimation error is minimized. Consider, for instance, the trapezoidal rule, Figure 6.9a, where the (fixed) points on the curve must correspond to a and b. Without this limitation, we could select two points on the curve so that the area of the resulting trapezoid is a much better estimate of the area under the curve; Figure 6.9b.

The Gaussian quadrature is based on this general idea, and estimates the integral value by using a weighted sum of values of $f(x)$ at several points in

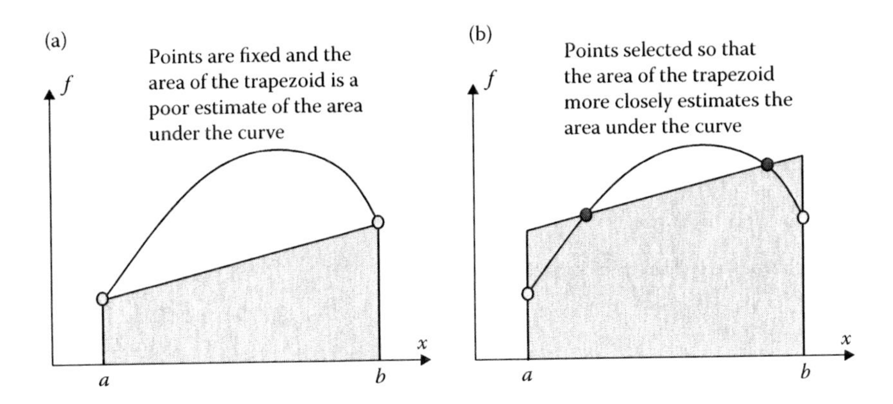

FIGURE 6.9
(a) Integral estimate by trapezoidal rule and (b) improved integral estimate.

$[a, b]$ that are not fixed or equally spaced. These points and the weights are determined such that the error is minimized.

The Gaussian quadrature is presented as applied to an integral in the explicit form

$$\int_{-1}^{1} f(x)\,dx$$

Note that any integral in the general form $\int_a^b f(\sigma)\,d\sigma$ can be converted to $\int_{-1}^{1} f(x)\,dx$ via a linear transformation

$$\sigma = \frac{b-a}{2}(x+1) + a \quad \text{so that} \quad d\sigma = \frac{b-a}{2}dx$$

and the new limits of integration are −1 and 1. Upon substitution, the original integral is transformed into

$$\int_{-1}^{1} f\left(\frac{b-a}{2}(x+1) + a\right)\frac{b-a}{2}dx \tag{6.38}$$

Gaussian quadrature estimates the integral as

$$\int_{-1}^{1} f(x)\,dx \cong \sum_{i=1}^{n} c_i f(x_i) \tag{6.39}$$

where the weights c_i and the Gauss nodes x_i ($i = 1, 2, \ldots, n$) are determined by assuming that Equation 6.39 fits exactly the above integral for functions

$f(x) = 1, x, x^2, \ldots$. How many of these functions need to be used depends on the value of n. For the simple case of $n = 2$, for example, we have

$$\int_{-1}^{1} f(x)dx \cong c_1 f(x_1) + c_2 f(x_2) \tag{6.40}$$

so that there are four unknowns: weights c_1 and c_2, and nodes x_1 and x_2. The required four equations will be provided by fitting the integral for functions $f(x) = 1, x, x^2, x^3$:

$$c_1 + c_2 = \int_{-1}^{1} 1 \cdot dx = 2$$

$$c_1 x_1 + c_2 x_2 = \int_{-1}^{1} x dx = 0$$

$$c_1 x_1^2 + c_2 x_2^2 = \int_{-1}^{1} x^2 dx = \frac{2}{3}$$

$$c_1 x_1^3 + c_2 x_2^3 = \int_{-1}^{1} x^3 dx = 0$$

Solving this system of four equations in four unknowns yields

$$c_1 = 1 = c_2$$

$$x_1 = -\frac{1}{\sqrt{3}} = -0.5773502692, \quad x_2 = \frac{1}{\sqrt{3}} = 0.5773502692$$

As a result, by Equation 6.40

$$\int_{-1}^{1} f(x)\,dx \cong f\left(-\frac{1}{\sqrt{3}}\right) + f\left(\frac{1}{\sqrt{3}}\right) \tag{6.41}$$

This provides the exact value of the integral as long as the integrand is any of the functions $f(x) = 1, x, x^2, x^3$ or their linear combination. Otherwise, it yields an approximate value of the integral.

The accuracy of approximation can be improved by increasing the value of n. For example, for the case of $n = 3$

$$\int_{-1}^{1} f(x)dx \cong c_1 f(x_1) + c_2 f(x_2) + c_3 f(x_3)$$

and there are now six unknowns: weights c_1, c_2, c_3, and nodes x_1, x_2, x_3. The six equations needed to solve for the unknowns are generated by fitting the integral for functions $f(x) = 1, x, x^2, x^3, x^4, x^5$. Proceeding as before, and solving the system of six equations, we arrive at

$$c_1 = 0.5555555556 = c_3, \quad c_2 = 0.8888888889$$

$$x_1 = -0.7745966692, \quad x_2 = 0, \quad x_3 = 0.7745966692$$

In general

$$\int_{-1}^{1} f(x)dx \cong \sum_{i=1}^{n} c_i f(x_i) = c_1 f(x_1) + c_2 f(x_2) + \cdots + c_n f(x_n)$$

which contains $2n$ unknowns: n weights and n nodes. The needed equations are generated by fitting the integral for functions $f(x) = 1, x, x^2, \ldots, x^{2n-1}$. The resulting values of c_i and x_i are tabulated in Table 6.6 for $n = 1, 2, \ldots, 6$. It turns out that the weights $c_1, c_2, \ldots, c_n$ can be calculated via

$$c_i = \int_{-1}^{1} \prod_{\substack{j=1 \\ j\neq i}}^{n} \frac{x - x_j}{x_i - x_j} dx \tag{6.42}$$

and the Gauss nodes $x_1, x_2, \ldots, x_n$ are the zeros of the nth-degree Legendre polynomial.* For example, for $n = 3$, the nodes are the zeros of $P_3(x) = \frac{1}{2}(5x^3 - 3x)$, that is, $0, \pm\sqrt{3/5}$, which agree with the values given above. The weights are computed via Equation 6.42 and will agree with those given earlier, as well as in Table 6.6.

* The first five Legendre polynomials are $P_0(x) = 1$, $P_1(x) = x$, $P_2(x) = \frac{1}{2}(3x^2 - 1)$, $P_3(x) = \frac{1}{2}(5x^3 - 3x)$, $P_4(x) = \frac{1}{8}(35x^4 - 30x^2 + 3)$.

TABLE 6.6

Weights and Nodes Used in the Gaussian Quadrature

n	Weights c_i	Gauss Nodes x_i
2	$c_1 = 1.000000000$	$x_1 = -0.577350269$
	$c_2 = 1.000000000$	$x_2 = 0.577350269$
3	$c_1 = 0.555555556$	$x_1 = -0.774596669$
	$c_2 = 0.888888889$	$x_2 = 0$
	$c_3 = 0.555555556$	$x_3 = 0.774596669$
4	$c_1 = 0.347854845$	$x_1 = -0.861136312$
	$c_2 = 0.652145155$	$x_2 = -0.339981044$
	$c_3 = 0.652145155$	$x_3 = 0.339981044$
	$c_4 = 0.347854845$	$x_4 = 0.861136312$
5	$c_1 = 0.236926885$	$x_1 = -0.906179846$
	$c_2 = 0.478628670$	$x_2 = -0.538469310$
	$c_3 = 0.568888889$	$x_3 = 0$
	$c_4 = 0.478628670$	$x_4 = 0.538469310$
	$c_5 = 0.236926885$	$x_5 = 0.906179846$
6	$c_1 = 0.171324492$	$x_1 = -0.932469514$
	$c_2 = 0.360761573$	$x_2 = -0.661209386$
	$c_3 = 0.467913935$	$x_3 = -0.238619186$
	$c_4 = 0.467913935$	$x_4 = 0.238619186$
	$c_5 = 0.360761573$	$x_5 = 0.661209386$
	$c_6 = 0.171324492$	$x_6 = 0.932469514$

Example 6.9: Gaussian Quadrature

Consider

$$\int_1^4 e^{-x^2/2}dx$$

a. Find an approximate value for the integral using the Gaussian quadrature with $n = 3$, $n = 4$, and $n = 5$.

b. How many subintervals must Simpson's 1/3 rule be applied to so that the accuracy is at the same level as that offered by the quadrature with $n = 5$? Use the user-defined function `Simpson` (Section 6.3) for this purpose. The actual value of the integral is 0.397610357.

SOLUTION

a. First rewrite the integral as $\int_1^4 e^{-\sigma^2/2}d\sigma$ and then convert it into the standard form using the linear transformation

$$\sigma = \frac{b-a}{2}(x+1) + a = \frac{3}{2}(x+1) + 1 = \frac{3x+5}{2} \quad \text{so that } d\sigma = \frac{3}{2}dx$$

Consequently, the integral in the desired form is

$$\int_{-1}^{1} \frac{3}{2} e^{-(3x+5)^2/8} dx \quad \text{so that } f(x) = \frac{3}{2} e^{-(3x+5)^2/8}$$

For the case of $n = 3$:

$$\begin{aligned}
\int_{-1}^{1} \frac{3}{2} e^{-(3x+5)^2/8} dx &\cong c_1 f(x_1) + c_2 f(x_2) + c_3 f(x_3) \\
&= 0.555555556 \cdot f(-0.774596669) + 0.888888889 \cdot f(0) \\
&\quad + 0.555555556 \cdot f(0.774596669) \\
&= 0.400020454
\end{aligned}$$

For the case of $n = 4$:

$$\begin{aligned}
\int_{-1}^{1} \frac{3}{2} e^{-(3x+5)^2/8} dx &\cong c_1 f(x_1) + c_2 f(x_2) + c_3 f(x_3) + c_4 f(x_4) \\
&= 0.347854845 \cdot f(-0.861136312) + 0.652145155 \\
&\quad \cdot f(-0.339981044) + 0.652145155 \cdot f(0.339981044) \\
&\quad + 0.347854845 \cdot f(0.861136312) = 0.397441982
\end{aligned}$$

Similarly, for $n = 5$:

$$\begin{aligned}
\int_{-1}^{1} \frac{3}{2} e^{-(3x+5)^2/8} dx &\cong c_1 f(x_1) + c_2 f(x_2) + c_3 f(x_3) + c_4 f(x_4) + c_5 f(x_5) \\
&= 0.397613225
\end{aligned}$$

Therefore, $n = 5$ yields an integral estimate that is exact to five decimal places.

b.

```
>> format long
>> f = inline('exp(-x^2/2)');
>> I = Simpson(f,1,4,50)

I =
   0.397610268688848
```

Simpson's 1/3 rule must be applied to 50 subintervals to produce an estimate that is five-decimal places accurate.

6.5 Improper Integrals

All numerical integration techniques introduced in Sections 6.3 and 6.4 were designed to estimate integrals in the form $\int_a^b f(x)\,dx$, where the limits a and b are finite. While it is quite common to see these types of integrals in engineering applications, there are situations where improper integrals are encountered and must be approximated numerically. Some of these integrals appear in the following forms:

$$\int_a^{\infty} f(x)dx \quad (a > 0), \quad \int_{-\infty}^{-b} f(x)dx \quad (b > 0), \quad \int_{-\infty}^{\infty} f(x)dx \tag{6.43}$$

Consider $\int_a^{\infty} f(x)dx, a > 0$. If the integrand reduces to zero at least as fast as x^{-2} does as $x \to \infty$, then the integral is handled by a simple change of variable:

$$x = \frac{1}{\sigma} \quad \text{so that } dx = -\frac{1}{\sigma^2}d\sigma$$

Then

$$\int_a^{\infty} f(x)dx = \int_{1/a}^{0} f\left(\frac{1}{\sigma}\right)\cdot\left(\frac{-1}{\sigma^2}\right)d\sigma = \int_0^{1/a} \frac{1}{\sigma^2} f\left(\frac{1}{\sigma}\right)d\sigma \tag{6.44}$$

The only concern is that the integrand is singular at the lower limit. Because of this, an open Newton–Cotes formula such as the composite midpoint rule (Section 6.3) can be utilized so that the integral is estimated without using the data at the endpoint(s).

The integral $\int_{-\infty}^{-b} f(x)dx$, $b > 0$ can be dealt with in a similar manner, including the condition on the rate of reduction of $f(x)$ to zero. The last form $\int_{-\infty}^{\infty} f(x)dx$ is treated as follows: we first decompose it as

$$\int_{-\infty}^{\infty} f(x)dx = \int_{-\infty}^{-b} f(x)dx + \int_{-b}^{a} f(x)dx + \int_a^{\infty} f(x)dx \tag{6.45}$$

In the first integral, we choose $-b$ so that $f(x)$ has started to asymptotically converge to zero at least as fast as x^{-2}. In the last integral, a is chosen such that the condition on the rate of reduction of $f(x)$ to zero is met as well. The integral in the middle can be approximated using a closed Newton–Cotes formula such as Simpson's 1/3 rule (Section 6.3).

Example 6.10: Improper Integral

Consider

$$\int_2^{\infty} \frac{\sin x}{x^2} dx$$

This is in the form of the first integral in Equation 6.43; hence, we use the change of variable $x = 1/\sigma$ leading to Equation 6.44:

$$\int_2^{\infty} \frac{\sin x}{x^2} dx = \int_0^{1/2} \frac{1}{\sigma^2} \frac{\sin(1/\sigma)}{(1/\sigma)^2} d\sigma = \int_0^{1/2} \sin\left(\frac{1}{\sigma}\right) d\sigma$$

Noting that the integrand is singular at the lower limit, we will use the composite midpoint method with $h = 0.0125$ to estimate this last integral.

```
>> f = inline('sin(1/x)'); h = 0.0125; x = 0:h:0.5;
>> n = length(x) - 1; m = zeros(n,1); I = 0;
>> for i = 1:n,
m(i) = (x(i+1) + x(i))/2;
I = I + f(m(i));
end
>> I = I*h;
>> I =

        0.0277
```

Accuracy can be improved by reducing the spacing size h.

PROBLEM SET

Finite-Difference Formulas for Numerical Differentiation (Section 6.2)

1. ✍ Consider $g(t) = te^{-t/3}$, $t = 0, 0.4, 0.8, 1.2$. Approximate $\dot{g}(0.4)$ using
   - Two-point backward difference formula
   - Two-point forward difference formula
   - Two-point central difference formula
   - Three-point forward difference formula

   Find the relative error in each case.

2. ✍ Consider $g(t) = e^{t/2} \cos t$, $t = 1.3, 1.6, 1.9, 2.2$. Approximate $\dot{g}(1.9)$ using
   - Two-point backward difference formula
   - Two-point forward difference formula
   - Two-point central difference formula
   - Three-point backward difference formula

   Find the relative error in each case.

3. ✍ Consider $f(x) = x^{-1}\, 3^x$, $x = 1.7, 2.0, 2.3, 2.6, 2.9$. Approximate $f'(2.3)$ using
   - Three-point backward difference formula
   - Three-point forward difference formula
   - Four-point central difference formula

   Find the relative error in each case.

4. ✍ Consider $f(x) = 2^x \log x$, $x = 2.0, 2.4, 2.8, 3.2, 3.6$. Find an estimate for $f'(2.8)$ using
   - Three-point backward difference formula
   - Three-point forward difference formula
   - Four-point central difference formula

   Find the relative error in each case.

5. ✍ The data in the following table shows the population of Canada recorded every 10 years between 1960 and 2010.
   a. Find the rate of population growth in 2010 using the three-point backward difference formula,
   b. Using the result of (a), and applying the two-point central difference formula, predict the population in 2020.

Year t	Population, p (millions)
1960	17.9
1970	21.3
1980	24.6
1990	27.8
2000	30.8
2010	34.1

6. ✍ The position of a moving object has been recorded as shown in the following table.
   a. Find the velocity of the object at $t = 3$ s using the three-point backward difference formula.
   b. Using the result of (a), and applying the two-point central difference formula, predict the position at $t = 3.5$ s.

Time, t (s)	Position, x (m)
1	0.65
1.5	1.275
2	2.15
2.5	3.275
3	4.65

7. ✍ The position of a moving object has been recorded as shown in the following table.
   a. Find the acceleration of the object at $t = 1.9$ s using the four-point backward difference formula.
   b. Using the result of (a), and the three-point central difference formula, predict the position at $t = 2.2$ s.

Time, t (s)	Position, x (m)
0.7	0.58
1	0.73
1.3	0.94
1.6	1.20
1.9	1.52

8. ✍ The deflection u of a beam along its longitudinal (x) axis has been recorded as shown in the following table. The bending moment at any point along this beam is modeled as $M(x) = 1.05u''(x)$. All parameters are in consistent physical units. Find an estimate for the bending moment at $x = 0.6$ using
   a. The three-point central difference formula
   b. The three-point backward difference formula

Position x	Deflection u
0.2	−0.15
0.4	−0.20
0.6	−0.20
0.8	−0.15

9. ✍ Consider $f(x) = (x + 1)e^{-2x/3}$, $x = 1, 1.5, 2, 2.5, 3$. Approximate $f''(2)$ using
   - Three-point central difference formula
   - Five-point central difference formula
10. ✍ Let $f(x) = 2^x - 3$.
   a. Approximate $f'(2.4)$ using the two-point central difference formula with $h = 0.2$.
   b. Approximate $f'(2.4)$ using the two-point central difference formula with $h = 0.1$.
   c. Apply an appropriate form of Richardson's extrapolation to the results of (a) and (b) to obtain a superior estimate.

11. ✍ Let $f(x) = x^2 - 2^x$.
    a. Approximate $f'(3.4)$ using the four-point central difference formula with $h = 0.2$.
    b. Approximate $f'(3.4)$ using the four-point central difference formula with $h = 0.1$.
    c. Apply an appropriate form of Richardson's extrapolation to the results of (a) and (b) to obtain a superior estimate.

    Display six decimal places for all numerical results.

12. ✍ Let $f(x) = e^{-x} + x$.
    a. Approximate $f''(1.5)$ using the three-point central difference formula with $h = 0.3$.
    b. Approximate $f''(1.5)$ using the three-point central difference formula with $h = 0.15$.
    c. Apply an appropriate form of Richardson's extrapolation to the results of (a) and (b) to obtain a superior estimate.
13. ✍ For the unevenly spaced data in the following table, estimate the first derivative at $x = 1.65$ by
    a. Fitting a second-degree Lagrange interpolating polynomial to a set of three suitable data points
    b. Fitting a third-degree Lagrange interpolating polynomial to the entire set

x_i	y_i
1.3	1.27
1.5	1.37
2	1.72
2.4	2.12

14. ✍ Given the unequally spaced data in the following table, estimate the first derivative at $x = 2.45$ by
    a. Fitting a second-degree Lagrange interpolating polynomial to a set of three suitable data points
    b. Fitting a third-degree Lagrange interpolating polynomial to the entire set

x_i	y_i
2	5.8432
2.4	7.5668
2.6	8.5643
3	10.8731

15. Given the unequally spaced data in the following table, estimate the first derivative at $x = 2.7$ by
    a. ✍ Fitting a second-degree Lagrange interpolating polynomial to a set of three suitable data points
    b. Applying the MATLAB built-in function `diff` to the entire set

x_i	y_i
2	1.0827
2.3	1.2198
2.7	1.3228
3	1.3443

16. Given the equally spaced data in the following table, estimate the second derivative at $x = 3.6$ by
    a. ✍ Fitting a second-degree Lagrange interpolating polynomial to a set of three suitable data points
    b. Applying the MATLAB built-in function `diff` to the entire set

x_i	y_i
3	0.4817
3.3	0.9070
3.6	1.4496
3.9	2.1287

Numerical Integration: Newton–Cotes Formulas (Section 6.3)

Composite Rectangular Rule

✍ In Problems 17 through 20, evaluate the definite integral using all three composite rectangular rule strategies with the indicated number of *equally spaced* data, and calculate the relative errors.

17. $\int_1^3 e^{-x^2}\,dx,\ n = 8$ (actual value 0.139383)

18. $\int_1^5 \sqrt{1+x}\,dx,\ n = 10$

19. $\int_2^3 \frac{\sin x}{x} dx$, $n = 10$ (actual value 0.24323955)

20. $\int_{0.4}^1 xe^x dx$, $n = 8$

✍ In Problems 21 through 24, evaluate the definite integral using all three composite rectangular rule strategies with the indicated *nonequally spaced* data, and calculate the relative errors.

21. $\int_1^3 x^{1/3} dx$, $x_1 = 1, 1.2, 1.5, 1.7, 1.8, 2.2, 2.5, 3 = x_8$

22. $\int_0^{2.5} (x+1)\sin x\, dx$, $x_1 = 0, 0.5, 0.8, 1.1, 1.6, 2, 2.3, 2.5 = x_8$

23. $\int_{0.3}^{1.3} e^x \cos x\, dx$, $x_1 = 0.3, 0.4, 0.6, 0.9, 1, 1.1, 1.2, 1.3 = x_8$

24. $\int_0^3 \frac{5}{2x^2+1} dx$, $x_1 = 0, 0.4, 0.9, 1.2, 1.8, 2.3, 2.6, 3 = x_8$

Composite Trapezoidal Rule

In Problems 25 through 28

a. ✍ Evaluate the integral using the composite trapezoidal rule with the indicated number of *equally spaced* data.

b. Confirm the results by executing the user-defined function `TrapComp`.

25. $\int_1^4 \frac{1}{x} dx$, $n = 5$

26. $\int_0^{2.1} xe^{-2x} dx$, $n = 7$

27. $\int_{0.2}^{1.4} 3^{2.5x} dx$, $n = 6$

28. $\int_{0.3}^3 \sqrt{1+x^2}\, dx$, $n = 9$

29. Write a user-defined function with function call `I = TrapComp_Gen(f,x)` that uses the composite trapezoidal rule to estimate the value of the integral of `f` from `a` to `b`, which are the first and last entries of the vector `x`, and the entries of `x` are not necessarily equally spaced. Then apply `TrapComp_Gen` to estimate the value of

$$\int_1^3 e^{-x/2}\sin^2 2x\,dx$$

where

   a. $x = 1, 1.2, 1.6, 1.8, 1.9, 2, 2.1, 2.3, 2.6, 2.8, 3$
   b. $x = 1{:}0.2{:}3$ (equally spaced with increments of 0.2)

30. Consider $\int_{-2}^{0} e^{-x}\cos x\,dx$.
   a. Evaluate by executing `TrapComp_Gen` (see Problem 29) with $x = -2, -1.8, -1.3, -0.9, -0.3, 0$.
   b. Evaluate by executing `TrapComp` (see Section 6.3) with $x = -2{:}0.4{:}0$ (equally spaced with increments of 0.4).
   c. Knowing the actual integral value is 1.3220, calculate the relative errors associated with the results of (a) and (b).

31. Consider $\int_0^2 \left[x/(x^3+1)\right]dx$.
   a. Evaluate by executing `TrapComp_Gen` (see Problem 29) with $x = 0, 0.2, 0.35, 0.6, 0.7, 0.9, 1.2, 1.5, 1.7, 1.8, 2$.
   b. Evaluate by executing `TrapComp` (see Section 6.3) with $x = 0{:}0.2{:}2$ (equally spaced with increments of 0.2).
   c. Knowing the actual integral value is 0.7238, calculate the relative errors associated with the results of (a) and (b).

32. Write a user-defined function with function call `I = TrapComp_Data(x,y)` that estimates the value of the integral of a tabular data `(x,y)`, which is not necessarily equally spaced, using the composite trapezoidal rule. Apply `TrapComp_Data` to the data in the following table.

x_i	0	0.20	0.25	0.35	0.50	0.60	0.65	0.8	0.90	1
y_i	0	0.19	0.23	0.30	0.37	0.39	0.4	0.38	0.34	0.31

33. Write a user-defined function with function call `I = TrapComp_ESData(x,y)` that estimates the value of the integral of a tabular, equally spaced data `(x,y)` using the composite trapezoidal rule. Apply `TrapComp_ESData` to the data in the following table.

x_i	1	1.1	1.2	1.3	1.4	1.5	1.6	1.7	1.8	1.9	2
y_i	0.31	0.27	0.22	0.17	0.14	0.10	0.08	0.05	0.04	0.02	0.01

34. A fragile instrument is placed inside a package to be protected during shipping and handling. The characteristics of the packing material are available experimentally, as shown in the following table. Assume that the force $F(x)$ exerted on the instrument is not to exceed 10 lbs. To determine the maximum safe drop height for the package, we first need to compute

$$\int_0^3 F(x)\,dx$$

   a. Evaluate this integral by executing the user-defined function `TrapComp_ESData` (see Problem 33).
   b. Determine the sixth-degree interpolating polynomial for the data in the following table using `polyfit`, and then integrate this polynomial from 0 to 3 to approximate the above integral.

x (inches)	0	0.5	1	1.5	2	2.5	3
F (lbs)	0	0.5	1	1.75	3	5	10

Composite Simpson's 1/3 Rule

In Problems 35 through 39

a. Evaluate the integral using the composite Simpson's 1/3 rule with the indicated number of *equally spaced* data.

b. Confirm the results by executing the user-defined function `Simpson`.

35. $\int_{0.2}^{1.4} x^2 e^{-x}\,dx, \quad n = 6$

36. $\int_{2.05}^{4.15} 2^{1-3x}\,dx, \quad n = 6$

37. $\int_{3.1}^{4.3} 4^{2-x}\,dx, \quad n = 8$

38. $\int_{3}^{7} \frac{x+1}{2x^2+3x}\,dx, \quad n = 8$

39. $\int_{-1}^{3} \left(x^2 + \frac{1}{2}\right)^2 dx, \quad n = 10$

40. Write a user-defined function with syntax `I = Simpson_Gen(f,x)` that uses the composite Simpson's 1/3 rule to estimate the value of the integral of `f` from `a` to `b`, which are the first and last entries of the vector `x`, and the entries of `x` are not necessarily equally spaced. Then apply `Simpson_Gen` to estimate the value of

$$\int_{0}^{1} \left(2 + \frac{1}{3}\cos 2x\right)^3 dx$$

where

a. $x = 0, 0.1, 0.25, 0.3, 0.4, 0.55, 0.6, 0.7, 0.85, 0.9, 1$

b. $x = 0:0.1:1$ (equally spaced with increments of 0.1)

41. Consider $\int_{-1}^{1} \left[(x-1)/(x^4+3)\right] dx$.

a. Evaluate by executing `Simpson_Gen` (see Problem 40) with $x = -1, -0.85, -0.6, -0.4, -0.25, -0.1, 0, 0.25, 0.6, 0.8, 1$.

b. Evaluate by executing `Simpson` (see Section 6.3) with $x = -1:0.2:1$ (equally spaced with increments of 0.2).

42. Consider $\int_{1}^{4} x \ln x \, dx$.

a. Evaluate by executing `Simpson_Gen` (see Problem 40) with $x = 1, 1.4, 1.7, 1.9, 2.3, 2.5, 2.6, 2.8, 3.3, 3.8, 4$.

b. Evaluate by executing `Simpson` (see Section 6.3) with $x = 1:0.3:4$ (equally spaced with increments of 0.3).

c. Calculate the relative errors associated with the results of (a) and (b).

43. Write a user-defined function with function call `I = Simpson_Data(x,y)` that estimates the value of the integral of a tabular data `(x,y)`, not necessarily equally spaced, using the composite Simpson's 1/3 rule. Apply `Simpson_Data` to the data in the following table.

x_i	0	0.25	0.35	0.50	0.60	0.65	0.8	0.90	1
y_i	0	0.23	0.30	0.37	0.39	0.4	0.38	0.34	0.31

44. Write a user-defined function with function call `I = Simpson_ESData(x,y)` that estimates the value of the integral of a tabular, equally spaced data `(x,y)` using the composite Simpson's 1/3 rule. Apply `Simpson_ESData` to the data in the following table.

x_i	1	1.1	1.2	1.3	1.4	1.5	1.6	1.7	1.8	1.9	2
y_i	0.31	0.27	0.22	0.17	0.14	0.10	0.08	0.05	0.04	0.02	0.01

Composite Simpson's 3/8 Rule

✍ In Problems 45 through 48, evaluate each integral using the composite Simpson's 3/8 rule.

45. $\int_0^{1.8} 3^{2.5x}\,dx, \quad n = 9$

46. $\int_{0.2}^{2} x^2 e^{-x/3}\,dx, \quad n = 6$

47. $\int_2^5 \dfrac{\ln x}{x+1}\,dx, \quad n = 6$

48. $\int_{1.2}^{4.8} x\sin(2x+1)\,dx, \quad n = 9$

49. Write a user-defined function with function call `I = Simpson_38(f,a,b,n)` that estimates the value of the integral of `f` from `a` to `b` using the composite Simpson's 3/8 rule applied to `n` subintervals of equal length. Apply `Simpson_38` to estimate the value of

$$\int_0^4 \cos^2(x-1)\,dx, \quad n = 9$$

50. Consider $\int_{-1}^{3} e^{-x}\sin x\,dx$. Evaluate by executing
    a. `Simpson_38` (see Problem 49) with $n = 6$
    b. `Simpson` with $n = 6$
    c. `quad`

51. Consider $\int_{0.2}^{0.5} x^{-2}e^{2x}\,dx$. Evaluate by executing
    a. `Simpson_38` (see Problem 49) with $n = 9$
    b. `quad`

52. Write a user-defined function with function call `I = Simpson_38_ESData(x,y)` that estimates the integral of a tabular, equally

spaced data (x,y) using the composite Simpson's 3/8 rule. Apply Simpson_38_ESData to

$$\int_1^2 \frac{x}{1+\ln x}\,dx$$

where y is the vector of discretized values of the integrand at the specified values of x = linspace(1,2,10). Also evaluate the integral using the quad function and compare the results.

53. Apply the user-defined function Simpson_38_ESData (see Problem 52) to

$$\int_{-3}^{1} 2^{-x}\cos(\tfrac{1}{2}x)\,dx$$

where y is the vector of discretized values of the integrand at the specified values of x = linspace(–3,1,10). Also evaluate the integral using the quad function and compare the results.

54. Write a user-defined function with function call I = Simpson_13_38(f,a,b,n) that estimates the integral of f from a to b as follows: If n = even, it applies composite Simpson's 1/3 rule throughout, and if n = odd, it applies the 3/8 rule to the last three subintervals and the 1/3 rule to all the previous ones. Execute this function to evaluate

$$\int_0^5 \frac{\sin 2x}{x+2}\,dx$$

using $n = 20$ and $n = 25$. Also evaluate using the quad function and compare the results. Use format long.

55. Execute the user-defined function Simpson_13_38 (see Problem 54) to evaluate

$$\int_1^4 e^{-x^2}\,dx$$

using $n = 20$ and $n = 25$. Also evaluate using the quad function and compare the results. Use format long.

56. In estimating $\int_a^b f(x)dx$, Boole's rule uses five equally spaced points to form a fourth-degree polynomial that approximates the integrand $f(x)$. The formula for Boole's rule is derived as

$$\int_a^b f(x)\,dx \cong \frac{2h}{45}\left[7f(x_1) + 32f(x_2) + 12f(x_3) + 32f(x_4) + 7f(x_5)\right],$$

$$h = \frac{b-a}{4}$$

The composite Boole's rule applies the above formula to four adjacent intervals at a time. Therefore, interval $[a,b]$ must be divided into a number of subintervals that is a multiple of 4. Write a user-defined function with function call `I = Boole_Comp(f,a,b,n)` that estimates the integral of `f` from `a` to `b` using the composite Boole's rule. Execute this function to evaluate

$$\int_0^1 \frac{2x-1}{1+\cos x}\,dx$$

using $n = 20$. Also evaluate using the `quad` function and compare the results. Use `format long`.

57. Evaluate the following integral by executing `Boole_Comp` (see Problem 56) with $n = 20$. Also evaluate using the `quad` function. Use `format long`.

$$\int_1^2 e^{1-x}x^{-1}\,dx$$

58. Consider

$$\int_0^1 \frac{1}{x^2+1}\,dx$$

Evaluate the integral using three different methods, using $n = 20$ in all cases: composite trapezoidal rule, composite Simpson's 1/3 rule, and composite Boole's rule (Problem 56). User-defined functions to be implemented are

`TrapComp` (Section 6.3)
`Simpson` (Section 6.3)
`Boole_Comp` (Problem 56)

Find the actual value of the integral using the MATLAB built-in function `int` (symbolic integration). Calculate the relative errors corresponding to all three strategies used above. Discuss the results. Use `format long` throughout.

Numerical Integration of Analytical Expressions
Romberg Integration, Gaussian Quadrature (Section 6.4)

Richardson's Extrapolation, Romberg Integration

In Problems 59 through 62

a. ✍ Apply the trapezoidal rule with the indicated values of n to yield estimates with error $O(h^2)$. Then calculate all subsequent higher-order estimates using Richardson's extrapolation, and tabulate the results as in Table 6.5. Also compute the relative error associated with each integral estimate.

b. Confirm the results by executing the user-defined function `Romberg`. Use `format long`.

59. $\displaystyle\int_0^{\pi/3} e^{-x/3}\sin 3x\,dx, \quad n = 2,4,8$

60. $\displaystyle\int_0^1 e^{x^2+1}, \quad n = 2,4,8,16, \text{ Actual value} = 3.975899662263389$

61. $\displaystyle\int_0^4 \frac{\cos x}{x^2+2}\,dx, \quad n = 2,4,8, \text{ Actual value} = 0.247005259638691$

62. $\displaystyle\int_{-1}^1 \sqrt{1-x^2}\,dx, \quad n = 2,4,8$

Gaussian Quadrature

✍ In Problems 63 through 68, evaluate each integral using the Gaussian quadrature with the indicated number(s) of nodes.

63. $\displaystyle\int_{-4}^4 \frac{\sin x}{x}\,dx, \quad n = 4$

64. $\displaystyle\int_0^1 2^x x^2\,dx, \quad n = 3, n = 4$

65. $\displaystyle\int_0^5 \left(\sin 3x + \tfrac{1}{2}\cos 4x\right)^2 dx, \quad n = 3, n = 4$

66. $\int_{-1}^{3} e^{\sin x} \cos x \, dx, \quad n=3, n=5$

67. $\int_{0}^{4} \frac{x}{2x^3+1} dx, \quad n=2, n=3, n=4$

68. $\int_{0}^{3/2} \left(x^2+3x+1\right)^2 dx, \quad n=4, n=5$

69. Write a user-defined function with syntax `I = Gauss_Quad_4(f,a,b)` that evaluates the integral of `f` from `a` to `b` using the Gaussian quadrature with $n=4$. Execute `Gauss_Quad_4` to evaluate the integral in Problem 67:

$$\int_{0}^{4} \frac{x}{2x^3+1} dx$$

70. Write a user-defined function with function call `I = Gauss_Quad_5(f,a,b)` that evaluates the integral of `f` from `a` to `b` using the Gaussian quadrature with $n=5$. Execute `Gauss_Quad_5` to evaluate the integral in Problem 68:

$$\int_{0}^{3/2} (x^2+3x+1)^2 \, dx$$

71. Consider

$$\int_{1}^{2} x \ln x \, dx$$

a. Find the actual value of the integral analytically.

b. Evaluate by executing the user-defined function `Simpson` (Section 6.3) using $n=4$.

c. Evaluate by executing the user-defined function `Gauss_Quad_4` (see Problem 69).

d. Calculate the relative error in (b) and (c) and comment.

72. Evaluate

$$\int_{1}^{2} \frac{e^{-x} \sin x}{x^3+1} dx$$

a. By executing the user-defined function `Gauss_Quad_4` (see Problem 69).
b. By executing the user-defined function `Romberg` (Section 6.4) with `n` = 2 and `n_levels` = 3. Use `format long`.
c. Knowing the actual value is 0.062204682443299, calculate the relative errors in (a) and (b) and comment.

Improper Integrals (Section 6.5)

73. Estimate

$$\int_0^\infty e^{-x}x^{1/2}dx$$

as follows: Decompose it as $\int_0^\infty \cdots = \int_0^1 \cdots + \int_1^\infty \cdots$. Evaluate the first integral using composite Simpson's 1/3 rule with $n = 6$, and evaluate the second integral by first changing the variable and subsequently using the composite midpoint method with $h = 0.05$.

74. Estimate

$$\int_{-\infty}^\infty \frac{1}{2x^2+1}dx$$

as follows: Decompose it as $\int_{-\infty}^\infty \cdots = \int_{-\infty}^{-1} \cdots + \int_{-1}^1 \cdots + \int_1^\infty \cdots$. Evaluate the middle integral using composite Simpson's 1/3 rule with $n = 6$. Evaluate the first and third integrals by first changing the variable and subsequently using the composite midpoint method with $h = 0.05$.

75. The cumulative distribution function (CDF) for the standard normal variable z (zero mean and standard deviation of 1) is defined as

$$\Phi(z) = \frac{1}{\sqrt{2\pi}}\int_{-\infty}^{z} e^{-z^2/2}dz$$

and gives the probability that an event is less than z. Approximate $\Phi(0.5)$.

76. Evaluate

$$\frac{1}{\sqrt{2\pi}}\int_{-\infty}^{\infty} e^{-z^2/2}dz$$

7

Numerical Solution of Initial-Value Problems

An nth-order differential equation is accompanied by n auxiliary conditions, which are needed to determine the n constants of integration that arise in the solution process. When these conditions are provided at the same value of the independent variable, we speak of an initial-value problem (IVP). In other situations, the supplementary conditions are specified at different values of the independent variable. And since these values are usually stated at the extremities of the system, these types of problems are referred to as boundary-value problems (BVPs).

In this chapter, we will discuss various methods to numerically solve IVPs. Stability and stiffness of differential equations will also be covered. Treatment of BVPs is presented in Chapter 8. Numerical methods for a single, first-order IVP will be studied first; Figure 7.1. Some of these methods will then be extended and used to solve higher-order and systems of differential equations.

A single, first-order IVP is represented as

$$y' = f(x,y), \quad y(x_0) = y_0, \quad a = x_0 \le x \le x_n = b \tag{7.1}$$

where y_0 is the specified initial condition, the independent variable x assumes values in $[a, b]$, and it is assumed that a unique solution $y(x)$ exists in the interval $[a, b]$. The interval is divided into n segments of equal width h so that

$$x_1 = x_0 + h, \quad x_2 = x_0 + 2h, \ldots, \quad x_n = x_0 + nh$$

The solution at the point x_0 is available from the initial condition. The objective is to find estimates of the solution at the subsequent points $x_1, x_2, \ldots, x_n$.

7.1 One-Step Methods

One-step methods find the solution estimate y_{i+1} at the location x_{i+1} by extrapolating from the solution estimate y_i at the previous location x_i. Exactly, how this new estimate is extrapolated from the previous estimate depends on the numerical method used. Figure 7.2 describes a very simple one-step method, where the slope φ is used to extrapolate from y_i to the new estimate y_{i+1}

$$y_{i+1} = y_i + h\varphi \quad (i = 0, 1, 2, \ldots, n-1) \tag{7.2}$$

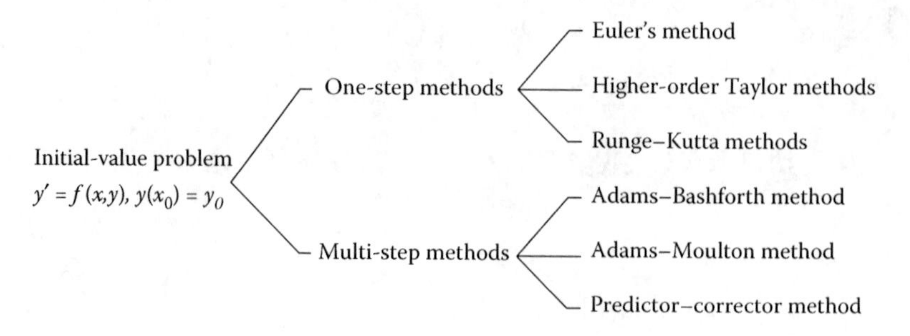

FIGURE 7.1
Classification of methods to solve an initial-value problem.

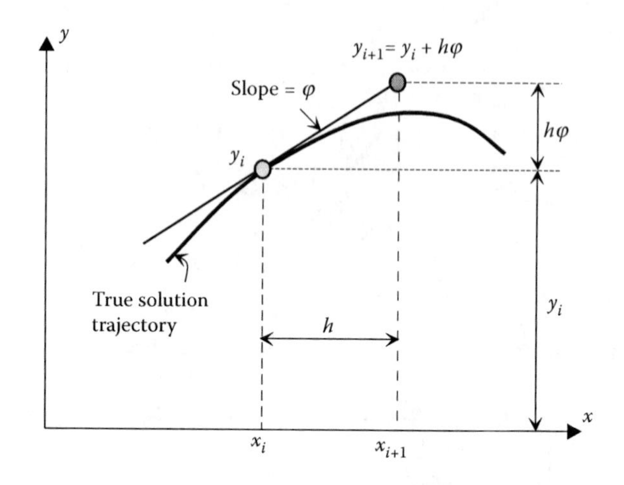

FIGURE 7.2
A simple one-step method.

Starting with the prescribed initial condition y_0, Equation 7.2 is applied in every subinterval $[x_i, x_{i+1}]$ to find solution estimates at $x_1, x_2, \ldots, x_n$. The general form in Equation 7.2 describes all one-step methods, with each method using a specific approach to estimate the slope φ. The simplest of all one-step methods is Euler's method, explained below.

7.2 Euler's Method

The expansion of $y(x_1)$ in a Taylor series about x_0 yields

$$y(x_1) = y(x_0 + h) = y(x_0) + hy'(x_0) + \frac{1}{2!}h^2 y''(x_0) + \cdots$$

Retaining the linear terms only, the above is rewritten as

$$y(x_1) = y(x_0) + hy'(x_0) + \frac{1}{2!}h^2y''(\xi_0)$$

for some ξ_0 between x_0 and x_1. In general, expanding $y(x_{i+1})$ about x_i yields

$$y(x_{i+1}) = y(x_i) + hy'(x_i) + \frac{1}{2!}h^2y''(\xi_i)$$

for some ξ_i between x_i and x_{i+1}. Note that $y'(x_i) = f(x_i, y_i)$ by Equation 7.1. Introducing notations $y_i = y(x_i)$ and $y_{i+1} = y(x_{i+1})$, the estimated solution y_{i+1} can be found via

$$y_{i+1} = y_i + hf(x_i, y_i), \quad i = 0, 1, 2, \ldots, n-1 \tag{7.3}$$

known as Euler's method. Comparing with the description of the general one-step method, Equation 7.2, we see that the slope φ at x_i is simply estimated by $f(x_i, y_i)$, which is the first derivative at x_i, namely, $y'(x_i)$. Equation 7.3 is called the difference equation for Euler's method.

The user-defined function EulerODE uses Euler's method to estimate the solution of an IVP.

```
function y = EulerODE(f,x,y0)
%
% EulerODE uses Euler's method to solve a first-order
% ODE given in the form y' = f(x,y) subject to initial
% condition y0.
%
%   y = EulerODE(f,x,y0) where
%
%     f is an inline function representing f(x,y),
%     x is a vector representing the mesh points,
%     y0 is a scalar representing the initial value of y,
%
%     y is the vector of solution estimates at the mesh
%     points.

y = 0*x;          % Pre-allocate
y(1) = y0; h = x(2)-x(1);
for n = 1:length(x)-1,
   y(n+1) = y(n)+h*f(x(n),y(n));
end
```

Example 7.1: Euler's Method

Consider the IVP

$$y' + y = 2x, \quad y(0) = 1, \quad 0 \le x \le 1$$

The exact solution is derived as $y_{exact}(x) = 2x + 3e^{-x} - 2$. We will solve the IVP numerically using Euler's method with step size $h = 0.1$. Comparing with Equation 7.1, we find $f(x,y) = -y + 2x$. Starting with $y_0 = 1$, we use Equation 7.3 to find the estimate at the next point, $x = 0.1$, as

$$y_1 = y_0 + hf(x_0, y_0) = 1 + 0.1f(0,1) = 1 + 0.1(-1) = 0.9$$

The exact solution at $x = 0.1$ is calculated as

$$y_{exact}(0.1) = 2(0.1) + 3e^{-0.1} - 2 = 0.914512$$

Therefore, the relative error is 1.59%. Similar computations may be performed at the subsequent points 0.2, 0.3, . . . , 1. The following MATLAB® script file uses the user-defined function `EulerODE` to find the numerical solution of the IVP and returns the results, including the exact values, in tabulated form. Figure 7.3 shows that Euler estimates capture the trend of the actual solution.

```
disp('   x           yEuler          yExact')
h = 0.1; x = 0:h:1; y0 = 1;
f = inline('-y+2*x','x','y');
yEuler = EulerODE(f,x,y0);
```

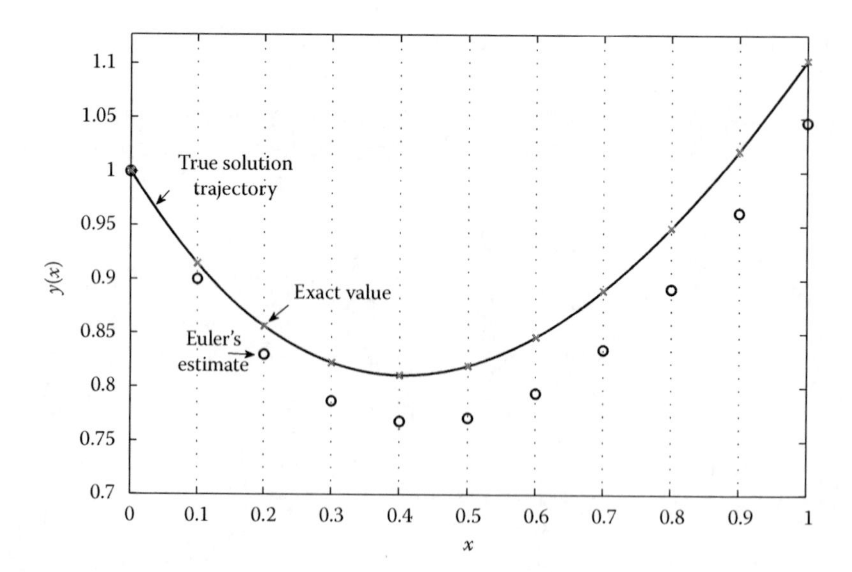

FIGURE 7.3
Comparison of Euler's and exact solutions in Example 7.1.

```
yExact = inline('2*x+3*exp(-x)-2');

for k = 1:length(x),
    x_coord = x(k);
    yE = yEuler(k);
    yEx = yExact(x(k));

  fprintf('%6.2f    %11.6f   %11.6f\n',x_coord,yE,yEx)

end
```

```
  x          yEuler       yExact
 0.00       1.000000     1.000000
 0.10       0.900000     0.914512
 0.20       0.830000     0.856192
 0.30       0.787000     0.822455
 0.40       0.768300     0.810960
 0.50       0.771470     0.819592
 0.60       0.794323     0.846435
 0.70       0.834891     0.889756
 0.80       0.891402     0.947987
 0.90       0.962261     1.019709
 1.00       1.046035     1.103638
```

The largest relative error is roughly 6.17% and occurs at $x = 0.7$. Using a smaller step size h will reduce the errors. Executing the above script file with $h = 0.05$, for example, shows a maximum relative error of 3.01% at $x = 0.65$.

7.2.1 Error Analysis for Euler's Method

Two sources of error are involved in the numerical solution of ordinary differential equations: round-off and truncation. Round-off errors are caused by the number of significant digits retained and used for calculations by the computer. Truncation errors are caused by the way a numerical method approximates the solution, and comprise two parts. The first part is a local truncation error resulting from the application of the numerical method in each step. The second part is a propagated truncation error caused by the approximations made in the previous steps. Adding the local and propagated truncation errors yields the global truncation error. It can be shown that the local truncation error is $O(h^2)$, while the global truncation error is $O(h)$.

7.2.2 Calculation of Local and Global Truncation Errors

The global truncation error at the point x_{i+1} is simply the difference between the actual solution y_{i+1}^a and the computed solution y_{i+1}^c. This contains the local truncation error, as well as the effects of all the errors accumulated in the steps prior to the current location x_{i+1}:

$$\text{Global truncation error at } x_{i+1} = \underbrace{\boxed{y_{i+1}^a}}_{\substack{\text{Actual}\\ \text{solution}\\ \text{at } x_{i+1}}} - \underbrace{\boxed{y_i^c + hf(x_i, y_i^c)}}_{\substack{\text{Euler's estimate at } x_{i+1}\\ \text{using computed solution at } x_i}} \tag{7.4}$$

The local truncation error at x_{i+1} is the difference between the actual solution y_{i+1}^a and the solution that would have been computed had the actual solution been used by Euler's method going from x_i to x_{i+1},

$$\text{Local truncation error at } x_{i+1} = \underbrace{\boxed{y_{i+1}^a}}_{\substack{\text{Actual}\\ \text{solution}\\ \text{at } x_{i+1}}} - \underbrace{\boxed{y_i^a + hf(x_i, y_i^a)}}_{\substack{\text{Euler's estimate at } x_{i+1}\\ \text{using actual solution at } x_i}} \tag{7.5}$$

Example 7.2: Local and Global Truncation Errors

In Example 7.1, calculate the local and global truncation errors at each point and tabulate the results.

SOLUTION

Starting with the initial condition $y_0 = 1$, the Euler's computed value at $x_1 = 0.1$ is $y_1^c = 0.9$ while the actual value is $y_1^a = 0.914512$. At this stage, the global and local truncation errors are the same because Euler's method used the initial condition, which is exact, to find the estimate. At $x_2 = 0.2$, the computed value is $y_2^c = 0.83$, which was calculated by Euler's method using the estimated value $y_1^c = 0.9$ from the previous step. If instead of y_1^c we use the actual value $y_1^a = 0.914512$, the computed value at x_2 is

$$\tilde{y}_2 = y_1^a + hf(x_1, y_1^a) = 0.914512 + 0.1f(0.1, 0.914512) = 0.843061$$

Therefore

$$\text{Local truncation error at } x_2 = y_2^a - \tilde{y}_2$$

$$= 0.856192 - 0.843061 = 0.013131$$

The global truncation error at x_2 is simply calculated as

$$\text{Global truncation error at } x_2 = y_2^a - y_2^c$$

$$= 0.856192 - 0.830000 = 0.026192$$

It is common to express these errors in the form of percent relative errors; hence, at each point we evaluate

$$\frac{\text{(Local or global) Truncation error}}{\text{Actual value}} \times 100$$

With this, the (local) percent relative error at x_2 is

$$\frac{y_2^a - \tilde{y}_2}{y_2^a} \times 100 = \frac{0.013131}{0.856192} \times 100 = 1.53$$

The (global) percent relative error at x_2 is

$$\frac{y_2^a - y_2^c}{y_2^a} \times 100 = \frac{0.026192}{0.856192} \times 100 = 3.06$$

The following MATLAB script file uses this approach to find the percent relative errors at all x_i, and completes the table presented earlier in Example 7.1.

```
disp('   x        yEuler        yExact     e_local    e_global')
h=0.1; x=0:h:1; y0=1; f=inline('-y+2*x','x','y');
yEuler=EulerODE(f,x,y0); yExact=inline('2*x+3*exp(-x)-2');

ytilda=0*x; ytilda(1)=y0;
for n=1:length(x)-1,
    ytilda(n+1)=yExact(x(n))+h*f(x(n),yExact(x(n)));
end

for k=1:length(x),
    x_coord=x(k);
    yE=yEuler(k);
    yEx=yExact(x(k));
    e_local=(yEx-ytilda(k))/yEx*100;
    e_global=(yEx-yE)/yEx*100;

   fprintf('%6.2f %11.6f %11.6f %6.2f
   %6.2f\n',x_coord,yE,yEx,e_local,e_global)
end
```

```
  x       yEuler       yExact     e_local   e_global

0.00    1.000000     1.000000     0.00       0.00
0.10    0.900000     0.914512     1.59       1.59
0.20    0.830000     0.856192     1.53       3.06    Calculated by
                                                     hand earlier
0.30    0.787000     0.822455     1.44       4.31
0.40    0.768300     0.810960     1.33       5.26
0.50    0.771470     0.819592     1.19       5.87
0.60    0.794323     0.846435     1.04       6.16
0.70    0.834891     0.889756     0.90       6.17
0.80    0.891402     0.947987     0.76       5.97
0.90    0.962261     1.019709     0.64       5.63
1.00    1.046035     1.103638     0.53       5.22
```

7.2.3 Higher-Order Taylor Methods

Euler's method was developed by retaining only the linear terms in a Taylor series. Retaining more terms in the series is the premise of higher-order Taylor methods. Expanding $y(x_{i+1})$ in a Taylor series about x_i yields

$$y(x_{i+1}) = y(x_i) + hy'(x_i) + \frac{1}{2!}h^2y''(x_i) + \cdots + \frac{1}{k!}h^k y^{(k)}(x_i) + \frac{1}{(k+1)!}h^{k+1}y^{(k+1)}(\xi_i)$$

where ξ_i is between x_i and x_{i+1}. The kth-order Taylor method is defined as

$$y_{i+1} = y_i + hp_k(x_i, y_i), \quad i = 0, 1, 2, \ldots, n-1 \tag{7.6}$$

where

$$p_k(x_i, y_i) = f(x_i, y_i) + \frac{1}{2!}hf'(x_i, y_i) + \cdots + \frac{1}{k!}h^{k-1}f^{(k-1)}(x_i, y_i)$$

It is then clear that Euler's method is a first-order Taylor method. Recall that Euler's method has a local truncation error $O(h^2)$ and a global truncation error $O(h)$. The kth-order Taylor method has a local truncation error $O(h^{k+1})$ and a global truncation error $O(h^k)$. Therefore, the higher the order of the Taylor method, the more accurately it estimates the solution of the IVP. However, this reduction in error demands the calculation of the derivatives of $f(x,y)$, which is an obvious drawback.

Example 7.3: Second-Order Taylor Method

Solve the IVP in Examples 7.1 and 7.2 using the second-order Taylor method with the same step size $h = 0.1$ as before, and compare the numerical results with those produced by Euler's method.

SOLUTION

The problem is

$$y' + y = 2x, \quad y(0) = 1, \quad 0 \le x \le 1$$

so that $f(x,y) = -y + 2x$. Implicit differentiation with respect to x yields

$$f'(x,y) = -y' + 2 \overset{y'=f(x,y)}{=} -(-y + 2x) + 2 = y - 2x + 2$$

By Equation 7.6, the second-order Taylor method is defined as

$$y_{i+1} = y_i + hp_2(x_i, y_i), \quad i = 0, 1, 2, \ldots, n-1$$

where

$$p_2(x_i, y_i) = f(x_i, y_i) + \frac{1}{2!}hf'(x_i, y_i)$$

Therefore

$$y_{i+1} = y_i + h\left[f(x_i, y_i) + \frac{1}{2!}hf'(x_i, y_i)\right]$$

Starting with $y_0 = 1$, the solution estimate at the next location $x_1 = 0.1$ is calculated as

$$y_1 = y_0 + h\left[f(x_0, y_0) + \frac{1}{2}hf'(x_0, y_0)\right] = 0.9150$$

Noting the actual value at $x_1 = 0.1$ is 0.914512, this estimate has a (global) percent relative error of −0.05%, which is a significant improvement over the 1.59% offered by Euler's method at the same location. This upgrading of accuracy was expected because the second-order Taylor method has a (global) truncation error $O(h^2)$ compared with $O(h)$ for Euler's method. As mentioned before, this came at the expense of the evaluation of the first derivative $f'(x, y)$. The following MATLAB script tabulates the solution estimates generated by the second-order Taylor method:

```
disp('   x        yEuler     yTaylor2      e_Euler    e_Taylor2')
h=0.1; x=0:h:1; y0=1;
f=inline('-y+2*x','x','y'); fp=inline('y-2*x+2','x','y');
yEuler=EulerODE(f,x,y0); yExact=inline('2*x+3*exp(-x)-2');

yTaylor2=0*x; yTaylor2(1)=y0;
for n=1:length(x)-1,
   yTaylor2(n+1)=yTaylor2(n)+h*(f(x(n),yTaylor2(n))+(1/2)*h
   *fp(x(n),yTaylor2(n)));
end

for k=1:length(x),
    x_coord=x(k);
    yE=yEuler(k);
    yEx=yExact(x(k));
    yT=yTaylor2(k);
    e_Euler=(yEx-yE)/yEx*100;
    e_Taylor2=(yEx-yT)/yEx*100;
  fprintf('%6.2f  %11.6f  %11.6f  %6.2f
  %6.2f\n',x_coord,yE,yT,e_Euler,e_Taylor2)
end
```

```
 x       yEuler      yTaylor2   e_Euler  e_Taylor2

0.00    1.000000    1.000000     0.00      0.00
0.10    0.900000    0.915000     1.59     -0.05    Calculated by
                                                   hand earlier
0.20    0.830000    0.857075     3.06     -0.10
0.30    0.787000    0.823653     4.31     -0.15
0.40    0.768300    0.812406     5.26     -0.18
0.50    0.771470    0.821227     5.87     -0.20
0.60    0.794323    0.848211     6.16     -0.21
0.70    0.834891    0.891631     6.17     -0.21
0.80    0.891402    0.949926     5.97     -0.20
0.90    0.962261    1.021683     5.63     -0.19
1.00    1.046035    1.105623     5.22     -0.18
```

7.3 Runge–Kutta Methods

In the last section, we learned that a kth-order Taylor method has a global truncation error $O(h^k)$ but requires the calculation of derivatives of $f(x,y)$. Runge–Kutta methods generate solution estimates with the accuracy of Taylor methods without having to calculate these derivatives. Recall from Equation 7.2 that all one-step methods to solve the IVP

$$y' = f(x,y), \quad y(x_0) = y_0, \quad a = x_0 \le x \le x_n = b$$

are expressed as

$$y_{i+1} = y_i + h\varphi(x_i, y_i)$$

where $\varphi(x_i, y_i)$ is an increment function and is essentially a suitable slope over the interval $[x_i, x_{i+1}]$ that is used for extrapolating y_{i+1} from y_i. The order of the Runge–Kutta method is the number of points that are used in $[x_i, x_{i+1}]$ to determine this suitable slope. For example, second-order Runge–Kutta methods use two points in each subinterval to find the representative slope, and so on.

7.3.1 Second-Order Runge–Kutta Methods

For the second-order Runge–Kutta (RK2) methods, the increment function is expressed as $\varphi(x_i, y_i) = a_1k_1 + a_2k_2$ so that

$$y_{i+1} = y_i + h(a_1k_1 + a_2k_2) \tag{7.7}$$

with

$$\begin{aligned} k_1 &= f(x_i, y_i) \\ k_2 &= f(x_i + b_1h, y_i + c_{11}k_1h) \end{aligned} \tag{7.8}$$

where a_1, a_2, b_1, and c_{11} are constants, each set determined separately for each specific RK2 method. These constants are evaluated by setting Equation 7.7 equal to the first three terms in a Taylor series, neglecting terms with h^3 and higher:

$$y_{i+1} = y_i + hy'\big|_{x_i} + \frac{1}{2}h^2y''\big|_{x_i} + O(h^3) \tag{7.9}$$

The term $y'\big|_{x_i}$ is simply $f(x_i, y_i)$, while

$$y''\big|_{x_i} = f'(x_i, y_i) \overset{\text{Chain rule}}{=} \left.\frac{\partial f}{\partial x}\right|_{(x_i,y_i)} + \left.\frac{\partial f}{\partial y}\right|_{(x_i,y_i)} \left.\frac{dy}{dx}\right|_{x_i}$$

$$\overset{y'=f(x,y)}{=} \left.\frac{\partial f}{\partial x}\right|_{(x_i,y_i)} + \left.\frac{\partial f}{\partial y}\right|_{(x_i,y_i)} f(x_i, y_i)$$

Substituting for $y'|_{x_i}$ and $y''|_{x_i}$ in Equation 7.9, we have

$$y_{i+1} = y_i + hf(x_i, y_i) + \frac{1}{2}h^2 \left.\frac{\partial f}{\partial x}\right|_{(x_i,y_i)} + \frac{1}{2}h^2 \left.\frac{\partial f}{\partial y}\right|_{(x_i,y_i)} f(x_i, y_i) + O(h^3) \tag{7.10}$$

Next, we will calculate y_{i+1} using a different approach as follows. In Equation 7.7, the term $k_2 = f(x_i + b_1 h, y_i + c_{11} k_1 h)$ is a function of two variables, and can be expanded about (x_i, y_i) as

$$k_2 = f(x_i + b_1 h, y_i + c_{11} k_1 h) = f(x_i, y_i) + b_1 h \left.\frac{\partial f}{\partial x}\right|_{(x_i,y_i)}$$

$$+ c_{11} k_1 h \left.\frac{\partial f}{\partial y}\right|_{(x_i,y_i)} + O(h^2) \tag{7.11}$$

Substituting Equation 7.11 and $k_1 = f(x_i, y_i)$ in Equation 7.7, we find

$$y_{i+1} = y_i + h\left\{a_1 f(x_i, y_i) + a_2\left[f(x_i, y_i) + b_1 h \left.\frac{\partial f}{\partial x}\right|_{(x_i,y_i)} + c_{11} k_1 h \left.\frac{\partial f}{\partial y}\right|_{(x_i,y_i)} + O(h^2)\right]\right\}$$

$$\overset{k_1 = f(x_i,y_i)}{=} y_i + (a_1 + a_2) hf(x_i, y_i) + a_2 b_1 h^2 \left.\frac{\partial f}{\partial x}\right|_{(x_i,y_i)} + a_2 c_{11} h^2 \left.\frac{\partial f}{\partial y}\right|_{(x_i,y_i)} f(x_i, y_i) + O(h^3) \tag{7.12}$$

The right-hand sides of Equations 7.10 and 7.12 represent the same quantity, y_{i+1}; hence, they must be equal. That yields

$$a_1 + a_2 = 1, \quad a_2 b_1 = \frac{1}{2}, \quad a_2 c_{11} = \frac{1}{2} \tag{7.13}$$

Since there are four unknowns and only three equations, a unique set of solutions does not exist. But if a value is assigned to one of the constants, the other three can be calculated. This is why there are several versions of RK2 methods, three of which are presented below. RK2 methods have local truncation error $O(h^3)$ and global truncation error $O(h^2)$, as did the second-order Taylor methods.

7.3.1.1 Improved Euler's Method

Assuming $a_2 = 1$, the other three constants in Equation 7.13 are determined as $a_1 = 0$, $b_1 = \frac{1}{2}$, and $c_{11} = \frac{1}{2}$. Inserting into Equations 7.7 and 7.8, the improved Euler's method is described by

$$y_{i+1} = y_i + hk_2 \tag{7.14}$$

where

$$\begin{aligned} k_1 &= f(x_i, y_i) \\ k_2 &= f\left(x_i + \frac{1}{2}h, y_i + \frac{1}{2}k_1 h\right) \end{aligned} \tag{7.15}$$

7.3.1.2 Heun's Method

Assuming $a_2 = \frac{1}{2}$, the other three constants in Equation 7.13 are determined as $a_1 = \frac{1}{2}$, $b_1 = 1$, and $c_{11} = 1$. Inserting into Equations 7.7 and 7.8, Heun's method is described by

$$y_{i+1} = y_i + \frac{1}{2}h(k_1 + k_2) \tag{7.16}$$

where

$$\begin{aligned} k_1 &= f(x_i, y_i) \\ k_2 &= f(x_i + h, y_i + k_1 h) \end{aligned} \tag{7.17}$$

7.3.1.3 Ralston's Method

Assuming $a_2 = \frac{2}{3}$, the other three constants in Equation 7.13 are determined as $a_1 = \frac{1}{3}$, $b_1 = \frac{3}{4}$, and $c_{11} = \frac{3}{4}$. Inserting into Equations 7.7 and 7.8, Ralston's method is described by

$$y_{i+1} = y_i + \frac{1}{3}h(k_1 + 2k_2) \tag{7.18}$$

where

$$\begin{aligned} k_1 &= f(x_i, y_i) \\ k_2 &= f\left(x_i + \frac{3}{4}h,\ y_i + \frac{3}{4}k_1h\right) \end{aligned} \tag{7.19}$$

Note that each of these RK2 methods produces estimates with the accuracy of a second-order Taylor method without calculating the derivative of $f(x,y)$. Instead, each method requires two function evaluations per step.

7.3.1.4 Graphical Representation of Heun's Method

Equations 7.16 and 7.17 can be combined as

$$y_{i+1} = y_i + h\frac{f(x_i, y_i) + f(x_i + h,\ y_i + k_1h)}{2} \tag{7.20}$$

Since $k_1 = f(x_i, y_i)$, we have $y_i + k_1h = y_i + hf(x_i, y_i)$. But this is the estimate y_{i+1} given by Euler's method at x_{i+1}, which we denote by y_{i+1}^{Euler} to avoid confusion with y_{i+1} in Equation 7.20. With this, and the fact that $x_i + h = x_{i+1}$, Equation 7.20 is rewritten as

$$y_{i+1} = y_i + h\frac{f(x_i, y_i) + f(x_{i+1}, y_{i+1}^{\text{Euler}})}{2} \tag{7.21}$$

The fraction multiplying h is the average of two quantities: the first one is the slope at the left end x_i of the interval; the second one is the estimated slope at the right end x_{i+1} of the interval. This is illustrated in Figure 7.4.

In Figure 7.4a, the slope at the left end of the interval is shown as $f(x_i, y_i)$. Figure 7.4b shows the estimated slope at the right end of the interval to be $f(x_{i+1}, y_{i+1}^{\text{Euler}})$. The line whose slope is the average of these two slopes, Figure 7.4c, yields an estimate that is superior to y_{i+1}^{Euler}. In Heun's method, y_{i+1} is extrapolated from y_i using this line.

The user-defined function `HeunODE` uses Heun's method to estimate the solution of an IVP.

```
function y = HeunODE(f,x,y0)
%
% HeunODE uses Heun's method to solve a first-order ODE
% given in the form y' = f(x,y) subject to initial
% condition y0.
%
%   y = HeunODE(f,x,y0) where
%
%     f is an inline function representing f(x,y),
%     x is a vector representing the mesh points,
%     y0 is a scalar representing the initial value of y,
%
%     y is the vector of solution estimates at the mesh
%     points.
y = 0*x;    % Pre-allocate
y(1) = y0; h = x(2)-x(1);
for n = 1:length(x)-1,
    k1 = f(x(n),y(n));
    k2 = f(x(n)+h,y(n)+h*k1);
    y(n+1) = y(n)+h*(k1+k2)/2;
end
```

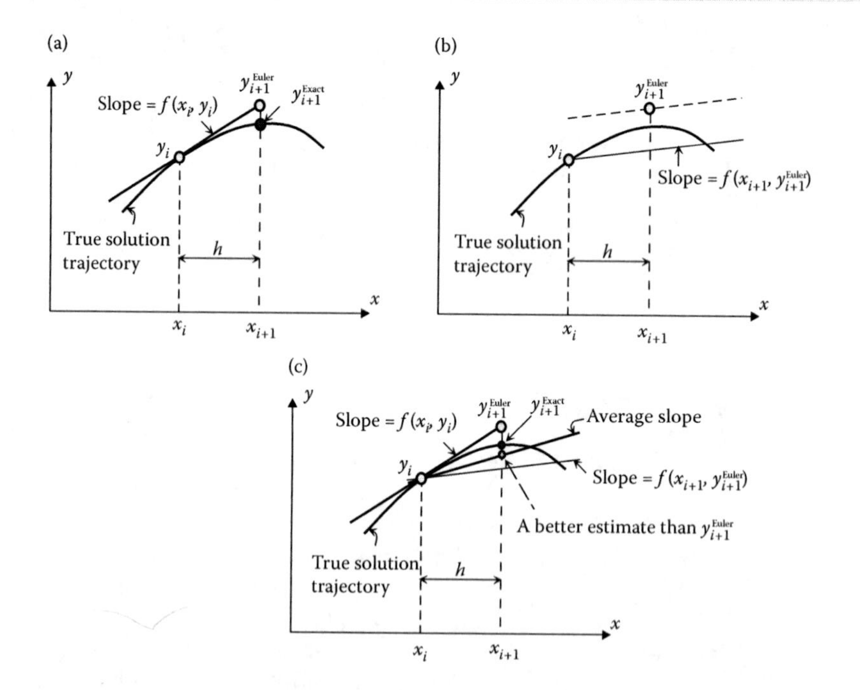

FIGURE 7.4
Graphical representation of Heun's method.

Example 7.4: RK2 Methods

Consider the IVP

$$y' - x^2 y = 2x^2, \quad y(0) = 1, \quad 0 \le x \le 1, \quad h = 0.1$$

Compute the estimated value of $y_1 = y(0.1)$ using each of the three RK2 methods discussed earlier.

SOLUTION

Noting that $f(x, y) = x^2(2 + y)$, the value of $y_1 = y(0.1)$ estimated by each RK2 method is calculated as follows:

Improved Euler's Method

$$k_1 = f(x_0, y_0) = f(0,1) = 0$$
$$k_2 = f\left(x_0 + \frac{1}{2}h,\, y_0 + \frac{1}{2}k_1 h\right) = f(0.05,1) = 0.0075 \quad \Rightarrow \quad y_1 = y_0 + hk_2$$
$$= 1 + 0.1(0.0075) = 1.0008$$

Heun's Method

$$k_1 = f(x_0, y_0) = f(0,1) = 0$$
$$k_2 = f(x_0 + h,\, y_0 + k_1 h) = f(0.1,1) = 0.0300 \quad \Rightarrow \quad y_1 = y_0 + \frac{1}{2}h(k_1 + k_2)$$
$$= 1 + 0.05(0.0300) = 1.0015$$

Ralston's Method

$$k_1 = f(x_0, y_0) = f(0,1) = 0$$
$$k_2 = f\left(x_0 + \frac{3}{4}h,\, y_0 + \frac{3}{4}k_1 h\right) = f(0.075,1) = 0.0169 \quad \Rightarrow \quad y_1 = y_0 + \frac{1}{3}h(k_1 + 2k_2)$$
$$= 1 + \frac{1}{3}(0.1)(0.0338) = 1.0011$$

Continuing this process, the estimates given by the three methods at the remaining points will be obtained and tabulated as in Table 7.1. The exact solution is $y = 3e^{x^3/3} - 2$. The global percent relative errors for all three methods are also listed, where it is readily observed that all RK2 methods perform better than Euler, and that Ralston's method is producing the most accurate estimates.

7.3.2 Third-Order Runge–Kutta Methods

For the third-order Runge–Kutta (RK3) methods, the increment function is expressed as $\varphi(x_i, y_i) = a_1k_1 + a_2k_2 + a_3k_3$, so that

$$y_{i+1} = y_i + h(a_1k_1 + a_2k_2 + a_3k_3) \tag{7.22}$$

TABLE 7.1

Summary of Calculations in Example 7.4

x	y_{Euler}	y_{Heun}	$y_{\text{Imp\_Euler}}$	y_{Ralston}	e_{Euler}	e_{Heun}	$e_{\text{Imp\_Euler}}$	e_{Ralston}
0.0	1.0000	1.0000	1.0000	1.0000	0.00	0.00	0.00	0.00
0.1	1.0000	1.0015	1.0008	1.0011	0.10	−0.05	0.02	−0.01
0.2	1.0030	1.0090	1.0075	1.0083	0.50	−0.10	0.05	−0.02
0.3	1.0150	1.0286	1.0263	1.0275	1.18	−0.15	0.08	−0.03
0.4	1.0421	1.0667	1.0636	1.0651	2.12	−0.19	0.10	−0.04
0.5	1.0908	1.1302	1.1261	1.1281	3.27	−0.23	0.14	−0.04
0.6	1.1681	1.2271	1.2219	1.2245	4.56	−0.25	0.17	−0.04
0.7	1.2821	1.3671	1.3604	1.3637	5.96	−0.27	0.22	−0.03
0.8	1.4430	1.5626	1.5541	1.5583	7.40	−0.28	0.27	−0.00
0.9	1.6633	1.8301	1.8191	1.8246	8.87	−0.27	0.34	0.04
1.0	1.9600	2.1922	2.1777	2.1849	10.37	−0.25	0.42	0.09

with

$$
\begin{aligned}
k_1 &= f(x_i, y_i)\\
k_2 &= f(x_i + b_1 h,\ y_i + c_{11} k_1 h)\\
k_3 &= f(x_i + b_2 h,\ y_i + c_{21} k_1 h + c_{22} k_2 h)
\end{aligned}
$$

where a_1, a_2, a_3, b_1, b_2, c_{11}, c_{21}, and c_{22} are constants, each set determined separately for each specific RK3 method. These constants are found by setting Equation 7.22 equal to the first four terms in a Taylor series, neglecting terms with h^4 and higher. Proceeding as with RK2 methods, we will end up with six equations and eight unknowns. By assigning values to two of the constants, the other six can be determined. Because of this, there are several RK3 methods, two of which are presented here. RK3 methods have local truncation error $O(h^4)$ and global truncation error $O(h^3)$, as did the third-order Taylor methods.

7.3.2.1 Classical RK3 Method

The classical RK3 method is described by

$$
y_{i+1} = y_i + \frac{1}{6} h (k_1 + 4k_2 + k_3) \tag{7.23}
$$

where

$$
\begin{aligned}
k_1 &= f(x_i, y_i)\\
k_2 &= f\left(x_i + \frac{1}{2} h,\ y_i + \frac{1}{2} k_1 h\right)\\
k_3 &= f(x_i + h, y_i - k_1 h + 2k_2 h)
\end{aligned}
$$

7.3.2.2 Heun's RK3 Method

Heun's RK3 method is described by

$$y_{i+1} = y_i + \frac{1}{4}h(k_1 + 3k_3) \tag{7.24}$$

where

$$\begin{aligned} k_1 &= f(x_i, y_i) \\ k_2 &= f\left(x_i + \frac{1}{3}h,\, y_i + \frac{1}{3}k_1 h\right) \\ k_3 &= f\left(x_i + \frac{2}{3}h,\, y_i + \frac{2}{3}k_2 h\right) \end{aligned}$$

Each of these RK3 methods produces estimates with the accuracy of a third-order Taylor method without calculating the derivatives of $f(x,y)$. Instead, each method requires three function evaluations per step.

Example 7.5: RK3 Methods

Consider the IVP in Example 7.4:

$$y' - x^2 y = 2x^2, \quad y(0) = 1, \quad 0 \le x \le 1, \quad h = 0.1$$

Compute the estimated value of $y_1 = y(0.1)$ using the two RK3 methods presented above.

SOLUTION

Noting that $f(x, y) = x^2\,(2 + y)$, the calculations are carried out as follows.

Classical RK3 Method

$$\begin{aligned} k_1 &= f(x_0, y_0) = f(0,1) = 0 \\ k_2 &= f(x_0 + \frac{1}{2}h,\, y_0 + \frac{1}{2}k_1 h) = f(0.05,1) = 0.0075 \\ k_3 &= f(x_0 + h,\, y_0 - k_1 h + 2k_2 h) = f(0.1,1.0015) = 0.0300 \end{aligned}$$

$$\begin{aligned} y_1 &= y_0 + \frac{1}{6}h(k_1 + 4k_2 + k_3) \\ &= 1 + \frac{1}{6}(0.1)(4 \times 0.0075 + 0.0300) = 1.0010 \end{aligned}$$

TABLE 7.2

Summary of Calculations in Example 7.5

x	y_{Euler}	y_{RK3}	y_{Heun_RK3}	e_{Euler}	e_{RK3}	e_{Heun_RK3}
0.0	1.0000	1.0000	1.0000	0.00	0.0000	0.0000
0.1	1.0000	1.0010	1.0010	0.10	−0.0000	0.0000
0.2	1.0030	1.0080	1.0080	0.50	−0.0001	0.0001
0.3	1.0150	1.0271	1.0271	1.18	−0.0004	0.0003
0.4	1.0421	1.0647	1.0647	2.12	−0.0010	0.0007
0.5	1.0908	1.1277	1.1276	3.27	−0.0018	0.0014
0.6	1.1681	1.2240	1.2239	4.56	−0.0030	0.0024
0.7	1.2821	1.3634	1.3633	5.96	−0.0044	0.0038
0.8	1.4430	1.5584	1.5582	7.40	−0.0059	0.0059
0.9	1.6633	1.8253	1.8250	8.87	−0.0074	0.0086
1.0	1.9600	2.1870	2.1866	10.37	−0.0087	0.0124

Heun's RK3 Method

$$k_1 = f(x_0, y_0) = f(0,1) = 0$$

$$k_2 = f(x_0 + \frac{1}{3}h, y_0 + \frac{1}{3}k_1h) = f(0.0333,1) = 0.0033$$

$$k_3 = f(x_0 + \frac{2}{3}h, y_0 + \frac{2}{3}k_2h) = f(0.0667,1.0002) = 0.0133$$

$$y_1 = y_0 + \frac{1}{4}h(k_1 + 3k_3) = 1 + 0.05(0.0300) = 1.0010$$

A summary of all calculations is given in Table 7.2 where it is easily seen that the global percent relative errors for the two RK3 methods are considerably lower than all previous methods covered up to this point.

7.3.3 Fourth-Order Runge–Kutta Methods

For the fourth-order Runge–Kutta (RK4) methods, the increment function is expressed as

$$\varphi(x_i, y_i) = a_1k_1 + a_2k_2 + a_3k_3 + a_4k_4$$

so that

$$y_{i+1} = y_i + h(a_1k_1 + a_2k_2 + a_3k_3 + a_4k_4) \tag{7.25}$$

with

$$k_1 = f(x_i, y_i)$$
$$k_2 = f(x_i + b_1h, y_i + c_{11}k_1h)$$
$$k_3 = f(x_i + b_2h, y_i + c_{21}k_1h + c_{22}k_2h)$$
$$k_4 = f(x_i + b_3h, y_i + c_{31}k_1h + c_{32}k_2h + c_{33}k_3h)$$

where a_j, b_j, and c_{ij} are constants, each set determined separately for each specific RK4 method. These constants are found by setting Equation 7.25 equal to the first five terms in a Taylor series, neglecting terms with h^5 and higher. Proceeding as before leads to 10 equations and 13 unknowns. By assigning values to three of the constants, the other 10 can be determined. This is why there are many RK4 methods, but only the classical RK4 is presented here. RK4s have local truncation error $O(h^5)$ and global truncation error $O(h^4)$.

7.3.3.1 Classical RK4 Method

The classical RK4 method is described by

$$y_{i+1} = y_i + \frac{1}{6}h(k_1 + 2k_2 + 2k_3 + k_4) \tag{7.26}$$

where

$$\begin{aligned} k_1 &= f(x_i, y_i) \\ k_2 &= f\left(x_i + \frac{1}{2}h,\, y_i + \frac{1}{2}k_1 h\right) \\ k_3 &= f\left(x_i + \frac{1}{2}h,\, y_i + \frac{1}{2}k_2 h\right) \\ k_4 &= f(x_i + h,\, y_i + k_3 h) \end{aligned} \tag{7.27}$$

RK4 methods produce estimates with the accuracy of a fourth-order Taylor method without calculating the derivatives of $f(x,y)$. Instead, four function evaluations per step are performed. The classical RK4 method is the most commonly used technique for numerical solution of first-order IVPs, as it offers the most acceptable balance of accuracy and computational effort.

The user-defined function RK4 uses the classical RK4 method to estimate the solution of an IVP.

```
function y=RK4(f,x,y0)
%
% RK4 uses the classical RK4 method to solve a first-
% order ODE given in the form y'=f(x,y) subject to
% initial condition y0.
%
%   y=RK4(f,x,y0) where
%
%   f is an inline function representing f(x,y),
```

```
%      x is a vector representing the mesh points,
%      y0 is a scalar representing the initial value of y,
%
%      y is the vector of solution estimates at the mesh
%      points.

y=0*x;        % Pre-allocate
y(1)=y0; h=x(2)-x(1);
for n=1:length(x)-1,
    k1=f(x(n),y(n));
    k2=f(x(n)+h/2,y(n)+h*k1/2);
    k3=f(x(n)+h/2,y(n)+h*k2/2);
    k4=f(x(n)+h,y(n)+h*k3);
  y(n+1)=y(n)+h*(k1+2*k2+2*k3+k4)/6;
end
```

Example 7.6: RK4 methods

Consider the IVP in Examples 7.4 and 7.5:

$$y' - x^2 y = 2x^2, \quad y(0) = 1, \quad 0 \le x \le 1, \quad h = 0.1$$

Compute the estimated value of $y_1 = y(0.1)$ using the classical RK4 method.

SOLUTION

Noting that $f(x, y) = x^2 (2 + y)$, the calculations are carried out as follows.

Classical RK4 Method

$$\begin{aligned}
k_1 &= f(x_0, y_0) = f(0,1) = 0 \\
k_2 &= f(x_0 + \frac{1}{2}h, y_0 + \frac{1}{2}k_1 h) = f(0.05,1) = 0.0075 \\
k_3 &= f(x_0 + \frac{1}{2}h, y_0 + \frac{1}{2}k_2 h) = f(0.05,1.0004) = 0.0075 \\
k_4 &= f(x_0 + h, y_0 + k_3 h) = f(0.1,1.0008) = 0.0300 \\
y_1 &= y_0 + \frac{1}{6}h(k_1 + 2k_2 + 2k_3 + k_4) = 1.0010
\end{aligned}$$

A summary of all calculations is provided in Table 7.3 where it is easily seen that the global percent relative error for the classical RK4 method is significantly lower than all previous methods covered up to this point. As expected, starting with Euler's method, which is indeed a first-order Runge–Kutta method, the accuracy improves with the order of the RK method.

TABLE 7.3

Summary of Calculations in Example 7.6

		RK4	RK3	RK2	RK1
x	y_{RK4}	e_{RK4}	e_{RK3}	e_{Heun}	e_{Euler}
0.0	1.000000	0.000000	0.0000	0.00	0.00
0.1	1.001000	0.000001	−0.0000	−0.05	0.10
0.2	1.008011	0.000002	−0.0001	−0.10	0.50
0.3	1.027122	0.000003	−0.0004	−0.15	1.18
0.4	1.064688	0.000004	−0.0010	−0.19	2.12
0.5	1.127641	0.000005	−0.0018	−0.23	3.27
0.6	1.223966	0.000006	−0.0030	−0.25	4.56
0.7	1.363377	0.000007	−0.0044	−0.27	5.96
0.8	1.558286	0.000010	−0.0059	−0.28	7.40
0.9	1.825206	0.000016	−0.0074	−0.27	8.87
1.0	2.186837	0.000028	−0.0087	−0.25	10.37

7.3.3.2 Higher-Order Runge–Kutta Methods

The classical RK4 is the most commonly used numerical method for solving first-order IVPs. If higher levels of accuracy are desired, the recommended technique is Butcher's fifth-order Runge–Kutta method (RK5), which is defined as

$$y_{i+1} = y_i + \frac{1}{90}h(7k_1 + 32k_3 + 12k_4 + 32k_5 + 7k_6) \tag{7.28}$$

where

$$\begin{aligned}
k_1 &= f(x_i, y_i)\\
k_2 &= f\left(x_i + \frac{1}{4}h, y_i + \frac{1}{4}k_1h\right)\\
k_3 &= f\left(x_i + \frac{1}{4}h, y_i + \frac{1}{8}k_1h + \frac{1}{8}k_2h\right)\\
k_4 &= f\left(x_i + \frac{1}{2}h, y_i - \frac{1}{2}k_2h + k_3h\right)\\
k_5 &= f\left(x_i + \frac{3}{4}h, y_i + \frac{3}{16}k_1h + \frac{9}{16}k_4h\right)\\
k_6 &= f\left(x_i + h, y_i - \frac{3}{7}k_1h + \frac{2}{7}k_2h + \frac{12}{7}k_3h - \frac{12}{7}k_4h + \frac{8}{7}k_5h\right)
\end{aligned}$$

Therefore, Butcher's RK5 method requires six function evaluations per step.

7.3.4 Runge–Kutta–Fehlberg Method

One way to estimate the local truncation error for Runge–Kutta methods is to use two RK methods of different order and subtract the results. For cases involving variable step size, the error estimate can be used to decide when the step size needs to be adjusted. Naturally, a drawback of this approach is the number of function evaluations required per step. For example, we consider a common approach that uses a fourth-order and a fifth-order RK. This requires a total of 10 (four for RK4 and six for RK5) function evaluations per step. To get around the computational burden, the Runge–Kutta–Fehlberg (RKF) method utilizes an RK5 method that uses the function evaluations provided by its accompanying RK4 method.* This will reduce the number of function evaluations per step from 10 to 6.

$$y_{i+1} = y_i + h\left(\frac{25}{216}k_1 + \frac{1408}{2565}k_3 + \frac{2197}{4104}k_4 - \frac{1}{5}k_5\right) \tag{7.29}$$

together with a fifth-order method

$$y_{i+1} = y_i + h\left(\frac{16}{135}k_1 + \frac{6656}{12825}k_3 + \frac{28561}{56430}k_4 - \frac{9}{50}k_5 + \frac{2}{55}k_6\right) \tag{7.30}$$

where

$$\begin{aligned}
k_1 &= 2f(x_i, y_i)\\
k_2 &= f\left(x_i + \frac{1}{4}h, y_i + \frac{1}{4}k_1 h\right)\\
k_3 &= f\left(x_i + \frac{3}{8}h, y_i + \frac{3}{32}k_1 h + \frac{9}{32}k_2 h\right)\\
k_4 &= f\left(x_i + \frac{12}{13}h, y_i + \frac{1932}{2197}k_1 h - \frac{7200}{2197}k_2 h + \frac{7296}{2197}k_3 h\right)\\
k_5 &= f\left(x_i + h, y_i + \frac{439}{216}k_1 h - 8k_2 h + \frac{3680}{513}k_3 h - \frac{845}{4104}k_4 h\right)\\
k_6 &= f\left(x_i + \frac{1}{2}h, y_i - \frac{8}{27}k_1 h + 2k_2 h - \frac{3544}{2565}k_3 h + \frac{1859}{4104}k_4 h - \frac{11}{40}k_5 h\right)
\end{aligned}$$

* Refer to K.E. Atkinson, *An Introduction to Numerical Analysis*. 2nd edition, John Wiley, New York, 1989.

Subtracting Equation 7.29 from Equation 7.30 yields the estimate of the local truncation error:

$$\text{Error} = h\left(\frac{1}{360}k_1 - \frac{128}{4275}k_3 - \frac{2197}{75240}k_4 + \frac{1}{50}k_5 + \frac{2}{55}k_6\right) \tag{7.31}$$

In each step, Equation 7.29 gives the fourth-order accurate estimate, Equation 7.30 gives the fifth-order accurate estimate, and Equation 7.31 returns the estimated local truncation error.

7.4 Multistep Methods

In single-step methods, the solution estimate y_{i+1} at the point x_{i+1} is obtained by using information at a single previous point x_i. Multistep methods are based on the idea that a more accurate estimate for y_{i+1} at x_{i+1} can be attained by utilizing information on two or more previous points rather than x_i only. Consider $y' = f(x, y)$ subject to initial condition $y(x_0) = y_0$. To use a multistep method to find an estimate for y_1, information on at least two previous points are needed. However, the only available information is y_0. This means that such methods cannot self-start and the estimates at the first few points—depending on the order of the method—must be found using either a single-step method such as the classical RK4 or another multistep method that uses fewer previous points.

Multistep methods can be explicit or implicit. Explicit methods employ an explicit formula to calculate the estimate. For example, if an explicit method uses two previous points, the estimate y_{i+1} at x_{i+1} is in the form

$$y_{i+1} = F(x_{i+1}, x_i, y_i, x_{i-1}, y_{i-1})$$

This way, only known values appear on the right-hand side. In implicit methods, the unknown estimate y_{i+1} is involved on both sides of the equation

$$y_{i+1} = \tilde{F}(x_{i+1}, y_{i+1}, x_i, y_i, x_{i-1}, y_{i-1})$$

and must be determined iteratively using the methods described in Chapter 3.

7.4.1 Adams–Bashforth Method

Adams–Bashforth method is an explicit multistep method to estimate the solution y_{i+1} of an IVP at x_{i+1} by using the solution estimates at two or more previous points. Several formulas can be derived depending on the number of previous points used. The order of each formula is the number of previous points it uses. For example, a second-order formula finds y_{i+1} by utilizing the estimates y_i and y_{i-1} at the two prior points x_i and x_{i-1}.

To derive the Adams–Bashforth formulas, we integrate $y' = f(x, y)$ over an arbitrary interval $[x_i, x_{i+1}]$

$$\int_{x_i}^{x_{i+1}} y'\,dx = \int_{x_i}^{x_{i+1}} f(x,y)\,dx$$

Because $\int_{x_i}^{x_{i+1}} y'\,dx = y(x_{i+1}) - y(x_i)$, the above can be rewritten as

$$y(x_{i+1}) = y(x_i) + \int_{x_i}^{x_{i+1}} f(x,y)\,dx$$

or

$$y_{i+1} = y_i + \int_{x_i}^{x_{i+1}} f(x,y)\,dx \tag{7.32}$$

But since $y(x)$ is unknown, $f(x,y)$ cannot be integrated. To remedy this, $f(x,y)$ is approximated by a polynomial that interpolates the data at (x_i, y_i) and a few previous points. The number of the previous points that end up being used depends on the order of the formula to be derived. For example, for a second-order Adams–Bashforth formula, we use the polynomial that interpolates the data at (x_i, y_i) and one previous point, (x_{i-1}, y_{i-1}), and so on.

7.4.1.1 Second-Order Adams–Bashforth Formula

The polynomial that interpolates the data at (x_i, y_i) and (x_{i-1}, y_{i-1}) is linear and in the form

$$p_1(x) = f(x_i, y_i) + \frac{f(x_i, y_i) - f(x_{i-1}, y_{i-1})}{x_i - x_{i-1}}(x - x_i)$$

Letting $f_i = f(x_i, y_i)$ and $f_{i-1} = f(x_{i-1}, y_{i-1})$ for brevity, using $p_1(x)$ in Equation 7.32

$$y_{i+1} = y_i + \int_{x_i}^{x_{i+1}} p_1(x)\,dx$$

and assuming equally spaced data with spacing h, we arrive at

$$y_{i+1} = y_i + \frac{1}{2}h(3f_i - f_{i-1}) \tag{7.33}$$

As mentioned earlier, this formula cannot self-start because finding y_1 requires y_0 and y_{-1}, the latter not known. First, a single-step method such as RK4 is used to find y_1 from the initial condition y_0. The first application of Equation 7.33 is when $i = 1$ so that y_2 can be obtained using the information on y_0 and y_1.

7.4.1.2 Third-Order Adams–Bashforth Formula

Approximating the integrand $f(x,y)$ in Equation 7.32 by the second-degree polynomial $p_2(x)$ (Section 5.5) that interpolates the data at (x_i, y_i), (x_{i-1}, y_{i-1}), and (x_{i-2}, y_{i-2}), and carrying out the integration yields

$$y_{i+1} = y_i + \frac{1}{12}h(23f_i - 16f_{i-1} + 5f_{i-2}) \tag{7.34}$$

Since only y_0 is known, we first apply a method such as RK4 to find y_1 and y_2. The first application of Equation 7.34 is when $i = 2$ to obtain y_3 by using the information on y_0, y_1, and y_2.

7.4.1.3 Fourth-Order Adams–Bashforth Formula

Approximating the integrand $f(x,y)$ in Equation 7.32 by the third-degree polynomial $p_3(x)$ (Section 5.5) that interpolates the data at (x_i, y_i), (x_{i-1}, y_{i-1}), (x_{i-2}, y_{i-2}), and (x_{i-3}, y_{i-3}), and carrying out the integration yields

$$y_{i+1} = y_i + \frac{1}{24}h(55f_i - 59f_{i-1} + 37f_{i-2} - 9f_{i-3}) \tag{7.35}$$

Since only y_0 is known, we first apply a method such as RK4 to find y_1, y_2, and y_3. The first application of Equation 7.35 is when $i = 3$ to obtain y_4 by using the information on y_0, y_1, y_2, and y_3.

Adams–Bashforth formulas are primarily used in conjunction with the Adams–Moulton formulas, which are also multistep but implicit, to be presented next. A weakness of higher-order Adams–Bashforth formulas is that stability requirements place limitations on the step size that is necessary for desired accuracy. The user-defined function `AdamsBashforth4` uses the fourth-order Adams–Bashforth formula to estimate the solution of an IVP.

```
function y=AdamsBashforth4(f,x,y0)
%
% AdamsBashforth4 uses the fourth-order Adams-Bashforth
% formula to solve a first-order ODE in the form y'=f(x,y)
% subject to initial condition y0.
%
%   y=AdamsBashforth4(f,x,y0) where
%
%     f is an inline function representing f(x,y),
%     x is a vector representing the mesh points,
%     y0 is a scalar representing the initial value of y,
%
%     y is the vector of solution estimates at the mesh
%     points.

y(1:4)=RK4(f,x(1:4),y0);
for n=4:length(x)-1,
    h=x(n+1)-x(n);
    y(n+1)=y(n)+h*(55*f(x(n),y(n))-59*f(x(n-1),y(n-1))
    +37*f(x(n-2),y(n-2))-9*f(x(n-3),y(n-3)))/24;
end
```

7.4.2 Adams–Moulton Method

Adams–Moulton method is an implicit multistep method to estimate the solution y_{i+1} of an IVP at x_{i+1} by using the solution estimates at two or more previous points, as well as (x_{i+1}, y_{i+1}), where the solution is to be determined. Several formulas can be derived depending on the number of points used. The order of each formula is the total number of points it uses. For example, a second-order formula finds y_{i+1} by utilizing the estimates y_i and y_{i+1} at the points x_i and x_{i+1}. This makes the formula implicit because the unknown y_{i+1} will appear on both sides of the ensuing equation.

Derivation of Adams–Moulton formulas is similar to Adams–Bashforth where the integrand in Equation 7.32 is approximated by a polynomial that interpolates the data at prior points, as well as the point where the solution is being determined.

7.4.2.1 Second-Order Adams–Moulton Formula

The polynomial that interpolates the data at (x_i, y_i) and (x_{i+1}, y_{i+1}) is linear and in the form

$$p_1(x) = f_i + \frac{f_{i+1} - f_i}{x_{i+1} - x_i}(x - x_i)$$

where $f_i = f(x_i, y_i)$ and $f_{i+1} = f(x_{i+1}, y_{i+1})$. Replacing $f(x,y)$ in Equation 7.32 with $p_1(x)$ and carrying out the integration yields

$$y_{i+1} = y_i + \frac{1}{2}h(f_i + f_{i+1}) \tag{7.36}$$

This formula is implicit because $f_{i+1} = f(x_{i+1}, y_{i+1})$ contains y_{i+1}, which is the solution being sought. In this type of a situation, y_{i+1} must be found iteratively using the techniques listed in Chapter 3. This formula has a global truncation error $O(h^2)$.

7.4.2.2 *Third-Order Adams–Moulton Formula*

Approximating the integrand $f(x,y)$ in Equation 7.32 by the second-degree polynomial that interpolates the data at (x_{i+1}, y_{i+1}), (x_i, y_i), and (x_{i-1}, y_{i-1}), and carrying out the integration yields

$$y_{i+1} = y_i + \frac{1}{12}h(5f_{i+1} + 8f_i - f_{i-1}) \tag{7.37}$$

Since only y_0 is initially known, a method such as RK4 is first applied to find y_1. The first application of Equation 7.37 is when $i = 1$ to obtain y_2 implicitly. This formula has a global truncation error $O(h^3)$.

7.4.2.3 *Fourth-Order Adams–Moulton Formula*

Approximating the integrand $f(x,y)$ in Equation 7.32 by the third-degree polynomial that interpolates the data at (x_{i+1}, y_{i+1}), (x_i, y_i), (x_{i-1}, y_{i-1}), and (x_{i-2}, y_{i-2}), and carrying out the integration we find

$$y_{i+1} = y_i + \frac{1}{24}h(9f_{i+1} + 19f_i - 5f_{i-1} + f_{i-2}) \tag{7.38}$$

Since only y_0 is initially known, a method such as RK4 is first applied to find y_1 and y_2. The first application of Equation 7.38 is when $i = 2$ to obtain y_3 implicitly. This formula has a global truncation error $O(h^4)$.

7.4.3 Predictor–Corrector Methods

Predictor–corrector methods are a class of techniques that employ a combination of an explicit formula and an implicit formula to solve an IVP. First, the explicit formula is used to predict the value of y_{i+1}. This predicted value is denoted by $\tilde{y}_{i+1}$. The predicted $\tilde{y}_{i+1}$ is then used on the right-hand side of an implicit formula to obtain a new, more accurate value for y_{i+1} on the left-hand side.

The simplest predictor–corrector method is Heun's method, presented in Section 7.3. Heun's method first uses Euler's method—an explicit formula—as the predictor to obtain y_{i+1}^{Euler}. This predicted value is then used in Equation 7.21, which is the corrector

$$y_{i+1} = y_i + h\frac{f(x_i, y_i) + f(x_{i+1}, y_{i+1}^{\text{Euler}})}{2}$$

to find a more accurate value for y_{i+1}. A modified version of this approach is derived next so that a desired accuracy may be achieved through repeated applications of the corrector formula.

7.4.3.1 Heun's Predictor–Corrector Method

The objective is to find an estimate for y_{i+1}. The method is implemented as follows:

1. Find a first estimate for y_{i+1}, denoted by $y_{i+1}^{(1)}$, using Euler's method, which is an explicit formula

   Predictor $$y_{i+1}^{(1)} = y_i + hf(x_i, y_i) \tag{7.39}$$

2. Improve the predicted estimate by solving Equation 7.21 iteratively

   Corrector $$y_{i+1}^{(k+1)} = y_i + h\frac{f(x_i, y_i) + f(x_{i+1}, y_{i+1}^{(k)})}{2}, \quad k = 1, 2, 3, \ldots \tag{7.40}$$

   Therefore, $y_{i+1}^{(1)}$ is used in Equation 7.40 to obtain $y_{i+1}^{(2)}$, and so on.

3. The iterations in Step 2 are terminated when the following criterion is satisfied:

   Tolerance $$\left|\frac{y_{i+1}^{(k+1)} - y_{i+1}^{(k)}}{y_{i+1}^{(k+1)}}\right| < \varepsilon \tag{7.41}$$

   where ε is a prescribed tolerance.

4. If the tolerance criterion is met, increment i by 1 and set y_i equal to this last $y_{i+1}^{(k+1)}$ and go to Step 1.

7.4.3.2 Adams–Bashforth–Moulton Predictor–Corrector Method

Several predictor–corrector formulas can be created by combining one of the (explicit) Adams–Bashforth formulas of a particular order as the predictor with the (implicit) Adams–Moulton formula of the same order as the corrector. The fourth-order formulas of these two methods, for example, can be combined to create the fourth-order Adams–Bashforth–Moulton (ABM4) predictor–corrector:

Predictor $$y_{i+1}^{(1)} = y_i + \frac{1}{24}h(55f_i - 59f_{i-1} + 37f_{i-2} - 9f_{i-3}), \quad i = 3, 4, \ldots, n \tag{7.42}$$

Corrector $$y_{i+1}^{(k+1)} = y_i + \frac{1}{24}h(9f_{i+1}^{(k)} + 19f_i - 5f_{i-1} + f_{i-2}), \quad k = 1, 2, 3, \ldots \tag{7.43}$$

where $f_{i+1}^{(k)} = f(x_{i+1}, y_{i+1}^{(k)})$. This method cannot self-start and is implemented as follows: starting with the initial condition y_0, apply a method such as RK4 to find estimates for y_1, y_2, and y_3 and calculate their respective $f(x, y)$ values. At this stage, the predictor (Equation 7.42) is applied to find $y_4^{(1)}$, which is then used to calculate $f_4^{(1)}$. The corrector (Equation 7.43) is next applied to obtain $y_4^{(2)}$. The estimate can be substituted back into Equation 7.43 for iterative correction. The process is repeated for the remaining values of the index i.

The user-defined function `ABM4PredCorr` uses the fourth-order ABM predictor–corrector method to estimate the solution of an IVP. Note that the function does not perform the iterative correction mentioned above.

```
function y = ABM4PredCorr(f,x,y0)
%
% ABM4PredCorr uses the fourth-order Adams-Bashforth-
% Moulton predictor-corrector formula to solve y' = f(x,y)
% subject to initial condition y0.
%
%   y = ABM4PredCorr(f,x,y0) where
%
%     f is an inline function representing f(x,y),
%     x is a vector representing the mesh points,
%     y0 is a scalar representing the initial value of y,
%
%     y is the vector of solution estimates at the mesh
%     points.
%
py = zeros(4,1);    % Pre-allocate
```

```
y(1:4)=RK4(f,x(1:4),y0); % Find the first 4 elements by RK4
h=x(2)-x(1);

% Start ABM4
for n=4:length(x)-1,
    py(n+1)=y(n)+(h/24)*(55*f(x(n),y(n))-59*f(x(n-1),
    y(n-1))+37*f(x(n-2),y(n-2))-9*f(x(n-3),y(n-3)));
    y(n+1)=y(n)+(h/24)*(9*f(x(n+1),py(n+1))+19*f(x(n),
    y(n))-5*f(x(n-1),y(n-1))+f(x(n-2),y(n-2)));
end
```

Example 7.7: ABM4 Predictor–Corrector Method

Consider the IVP in Examples 7.4 through 7.6:

$$y' - x^2 y = 2x^2, \quad y(0) = 1, \quad 0 \le x \le 1, \quad h = 0.1$$

Compute the estimated value of $y_4 = y(0.4)$ using the ABM4 predictor–corrector method.

SOLUTION

$f(x, y) = x^2(2 + y)$. The first element y_0 is given by the initial condition. The next three are obtained by RK4 as

$$y_1 = 1.001000, \quad y_2 = 1.008011, \quad y_3 = 1.027122$$

The respective $f(x, y)$ values are calculated next:

$$f_0 = f(x_0, y_0) = f(0,1) = 0$$

$$f_1 = f(x_1, y_1) = f(0.1, 1.001000) = 0.030010$$

$$f_2 = f(x_2, y_2) = f(0.2, 1.008011) = 0.120320$$

$$f_3 = f(x_3, y_3) = f(0.3, 1.027122) = 0.272441$$

Prediction

Equation 7.42 yields

$$y_4^{(1)} = y_3 + \frac{1}{24} h(55f_3 - 59f_2 + 37f_1 - 9f_0) = 1.064604$$

Calculate $f_4^{(1)} = f(x_4, y_4^{(1)}) = f(0.4, 1.064604) = 0.490337$.

Correction

Equation 7.43 yields

$$y_4^{(2)} = y_3 + \frac{1}{24}h(9f_4^{(1)} + 19f_3 - 5f_2 + f_1)$$

$$= 1.064696 \quad \text{Rel error} = 0.0008\%$$

This corrected value may be improved by substituting $y_4^{(2)}$ and the corresponding $f_4^{(2)} = f(x_4, y_4^{(2)})$ into Equation 7.43

$$y_4^{(3)} = y_3 + \frac{1}{24}h(9f_4^{(2)} + 19f_3 - 5f_2 + f_1)$$

and inspecting the accuracy. In the present analysis, we perform only one correction so that $y_4^{(2)}$ is regarded as the value that will be used for y_4. This estimate is then used in Equation 7.42 with the index i incremented from 3 to 4. Continuing this process, we generate the numerical results in Table 7.4.

Another well-known predictor–corrector method is the fourth-order Milne's method:

$$\text{Predictor} \quad y_{i+1}^{(1)} = y_{i-3} + \frac{4}{3}h(2f_i - f_{i-1} + 2f_{i-2}), \quad i = 3, 4, \ldots, n$$

$$\text{Corrector} \quad y_{i+1}^{(k+1)} = y_{i-1} + \frac{1}{3}h(f_{i+1}^{(k)} + 4f_i + f_{i-1}), \quad k = 1, 2, 3, \ldots$$

where $f_{i+1}^{(k)} = f(x_{i+1}, y_{i+1}^{(k)})$. As with the fourth-order ABM, this method cannot self-start and needs a method such as RK4 for estimating y_1, y_2, and y_3 first.

TABLE 7.4

Summary of Calculations in Example 7.7

x	y_{RK4}	$\tilde{y}_i$ Predicted	y_i Corrected
0.0	1.000000		
0.1	1.001000		
0.2	1.008011		
0.3	1.027122	Start ABM4	
0.4		1.064604	1.064696
0.5		1.127517	1.127662
0.6		1.223795	1.224004
0.7		1.363143	1.363439
0.8		1.557958	1.558381
0.9		1.824733	1.825350
1.0		2.186134	2.187052

7.5 Systems of Ordinary Differential Equations

Mathematical models of most systems in various engineering disciplines comprise one or more first- or higher-order differential equations subject to an appropriate number of initial conditions. In this section, we will achieve two tasks: (1) transforming the model into a *system* of first-order differential equations, and (2) numerically solving the system of first-order differential equations thus obtained.

7.5.1 Transformation into a System of First-Order ODEs

The first task is to show how a single higher-order IVP or a system of various-order IVPs may be transformed into a system of first-order IVPs. The most important tools that facilitate this process are the state variables.

7.5.1.1 State Variables

State variables form the smallest set of linearly independent variables that completely describe the state of a system. Given the mathematical model of a system, the state variables are determined as follows:

- How many state variables are there?

  The number of state variables is the same as the number of initial conditions required to completely solve the model.
- What are selected as state variables?

  The state variables are selected to be exactly those variables for which initial conditions are required.

7.5.1.2 Notation

State variables are represented by u_i; for example, if there are three state variables, they will be denoted u_1, u_2, and u_3.

7.5.1.3 State-Variable Equations

If a system has m state variables $u_1, u_2, \ldots, u_m$, then there are exactly m state-variable equations. Each of these equations is a first-order differential equation in the form

$$\dot{u}_i = f_i(t, u_1, u_2, \ldots, u_m), \quad i = 1, 2, \ldots, m \tag{7.44}$$

Therefore, the left side is the first derivative (with respect to time in most applications) of the state variable u_i, and the right side is an algebraic function of the state variables, and possibly time t explicitly. Note that only state variables and functions of time that are part of the system model are allowed to appear on the right side of Equation 7.44. The system of first-order differential equations in Equation 7.44 can be conveniently expressed in vector form as

$$\dot{\mathbf{u}} = \mathbf{f}(t,\mathbf{u}), \quad \mathbf{u} = \begin{Bmatrix} u_1 \\ u_2 \\ \dots \\ u_m \end{Bmatrix}, \quad \mathbf{f} = \begin{Bmatrix} f_1(t,\mathbf{u}) \\ f_2(t,\mathbf{u}) \\ \dots \\ f_m(t,\mathbf{u}) \end{Bmatrix} \tag{7.45}$$

Example 7.8: A Single Higher-Order IVP

The mathematical model of a mechanical system is derived as

$$2\ddot{x} + 3\dot{x} + 10x = \sin t, \quad x(0) = 0, \quad \dot{x}(0) = 1$$

Since two initial conditions are required, there are two state variables: u_1 and u_2. The state variables are exactly those variables for which initial conditions are required. Therefore

$$\begin{aligned} u_1 &= x \\ u_2 &= \dot{x} \end{aligned}$$

This means there are two state-variable equations in the form

$$\begin{aligned} \dot{u}_1 &= \dots \\ \dot{u}_2 &= \dots \end{aligned}$$

Note that only state variables and functions of time such as $\sin t$ that are already in the model are allowed on the right sides. Since $u_1 = x$, we have $\dot{u}_1 = \dot{x}$ but $\dot{x}$ is not allowed on the right side. However, we know $\dot{x} = u_2$, which is a state variable. Therefore, the first of the state-variable equations is

$$\dot{u}_1 = u_2$$

The second equation has $\dot{u}_2$ on the left side. Since $u_2 = \dot{x}$, we have $\dot{u}_2 = \ddot{x}$, and $\ddot{x}$ is not eligible for the right side. But $\ddot{x}$ can be solved for in the system model

$$\ddot{x} = \frac{1}{2}(-3\dot{x} - 10x + \sin t) \underset{x=u_1}{\overset{\dot{x}=u_2}{=}} \frac{1}{2}(-3u_2 - 10u_1 + \sin t)$$

Therefore, the state-variable equations are formed as

$$\begin{aligned} \dot{u}_1 &= u_2 \\ \dot{u}_2 &= \frac{1}{2}(-3u_2 - 10u_1 + \sin t) \end{aligned} \quad \text{subject to initial conditions} \quad \begin{aligned} u_1(0) &= 0 \\ u_2(0) &= 1 \end{aligned}$$

Expressing these in vector form, we find

$$\dot{\mathbf{u}} = \mathbf{f}(t,\mathbf{u}), \quad \mathbf{u} = \begin{Bmatrix} u_1 \\ u_2 \end{Bmatrix}, \quad \mathbf{f} = \begin{Bmatrix} u_2 \\ \frac{1}{2}(-3u_2 - 10u_1 + \sin t) \end{Bmatrix}, \quad \mathbf{u}_0 = \begin{Bmatrix} 0 \\ 1 \end{Bmatrix}$$

Example 7.9: A Single Higher-Order IVP

Consider the third-order IVP described by

$$y''' - y'' - 5y' - 3y = e^{-x}, \quad y(0) = 0, \quad y'(0) = 1, \quad y''(0) = -1$$

Since three initial conditions are required, there are three state variables: u_1, u_2, and u_3. The state variables are those variables for which initial conditions are required. Therefore:

$$u_1 = y, \quad u_2 = y', \quad u_3 = y''$$

There are three state-variable equations:

$$\begin{aligned} u_1' &= y' = u_2 \\ u_2' &= y'' = u_3 \\ u_3' &= y''' = y'' + 5y' + 3y + e^{-x} = u_3 + 5u_2 + 3u_1 + e^{-x} \end{aligned} \quad \text{subject to} \quad \begin{aligned} u_1(0) &= 0 \\ u_2(0) &= 1 \\ u_3(0) &= -1 \end{aligned}$$

In vector form:

$$\mathbf{u}' = \mathbf{f}(x,\mathbf{u}), \quad \mathbf{u} = \begin{Bmatrix} u_1 \\ u_2 \\ u_3 \end{Bmatrix}, \quad \mathbf{f} = \begin{Bmatrix} u_2 \\ u_3 \\ u_3 + 5u_2 + 3u_1 + e^{-x} \end{Bmatrix}, \quad \mathbf{u}_0 = \begin{Bmatrix} 0 \\ 1 \\ -1 \end{Bmatrix}$$

Example 7.10: A System of Different-Order IVPs

Consider

$$\begin{aligned} \ddot{x}_1 + 2\dot{x}_1 + 2(x_1 - x_2) &= e^{-t}\sin t \\ \dot{x}_2 - 2(x_1 - x_2) &= 0 \end{aligned} \quad \begin{aligned} &\text{subject to initial conditions} \\ &x_1(0) = 0, x_2(0) = 0, \dot{x}_1(0) = -2 \end{aligned}$$

Three initial conditions are required; hence, there are three state variables: u_1, u_2, and u_3. The state variables are those for which initial conditions are required. Therefore

$$u_1 = x_1, \quad u_2 = x_2, \quad u_3 = \dot{x}_1$$

This is the natural order for selecting state variables, as the derivatives of variables are chosen after all nonderivatives have been used up. There are three state-variable equations

$$\begin{aligned} \dot{u}_1 &= \dot{x}_1 = u_3 & & u_1(0) = 0 \\ \dot{u}_2 &= \dot{x}_2 = 2(x_1 - x_2) = 2(u_1 - u_2) \quad & \text{subject to} \quad & u_2(0) = 0 \\ \dot{u}_3 &= \ddot{x}_1 = -2\dot{x}_1 - 2(x_1 - x_2) + e^{-t}\sin t & & u_3(0) = -2 \\ &= -2u_3 - 2(u_1 - u_2) + e^{-t}\sin t & & \end{aligned}$$

In vector form

$$\dot{\mathbf{u}} = \mathbf{f}(t, \mathbf{u}), \quad \mathbf{u} = \begin{Bmatrix} u_1 \\ u_2 \\ u_3 \end{Bmatrix}, \quad \mathbf{f} = \begin{Bmatrix} u_3 \\ 2(u_1 - u_2) \\ -2u_3 - 2(u_1 - u_2) + e^{-t}\sin t \end{Bmatrix}, \quad \mathbf{u}_0 = \begin{Bmatrix} 0 \\ 0 \\ -2 \end{Bmatrix}$$

7.5.2 Numerical Solution of a System of First-Order ODEs

In Examples 7.8 through 7.10, we learned how to transform a single higher-order IVP or a system of different-order IVPs into one system of first-order differential equations subject to a set of initial conditions in the general form

$$\mathbf{u}' = \mathbf{f}(x, \mathbf{u}), \quad \mathbf{u}(x_0) = \mathbf{u}_0, \quad a = x_0 \le x \le x_n = b \tag{7.46}$$

In many applications, the independent variable x is replaced with time t. Equation 7.46 is exactly in the form of Equation 7.1, where except for the independent variable x, all other quantities are vectors. And as such, the numerical methods presented so far in this chapter for solving Equation 7.1, a single first-order IVP, can be extended and applied to a system of first-order IVPs in Equation 7.46. We will present three of these methods here: Euler's method, Heun's method, and the classical RK4 method.

7.5.2.1 Euler's Method for Systems

As before, the interval $[a, b]$ is divided into subintervals of equal length h such that

$$x_1 = x_0 + h, \quad x_2 = x_0 + 2h, \ldots, \quad x_n = x_0 + nh$$

Euler's method for a system in the form of Equation 7.46 is defined as

$$\mathbf{u}_{i+1} = \mathbf{u}_i + h\mathbf{f}(x_i, \mathbf{u}_i), \quad i = 0, 1, 2, \ldots, n-1 \tag{7.47}$$

The user-defined function EulerODESystem uses Euler's method as outlined in Equation 7.47 to estimate the solution vector of a system of IVPs in the form of Equation 7.46.

```
function u=EulerODESystem(f,x,u0)
%
% EulerODESystem uses Euler's method to solve a system of
% first-order ODEs given in the form u'=f(x,u) subject to
% initial condition vector u0.
%
%   u=EulerODESystem(f,x,u0) where
%
%     f is an inline (m-dim. vector) function
%     representing f(x,u),
%     x is an (n+1)-dim. vector representing the mesh
%     points,
%     u0 is an m-dim. vector representing the initial
%     condition of u,
%
%     u is an m-by-(n+1) matrix, each column being the
%     vector of solution estimates at a mesh point.

u(:,1)=u0;
% The first column is set to be the initial vector u0
h=x(2)-x(1);

for i=1:length(x)-1,
  u(:,i+1)=u(:,i)+h*f(x(i),u(:,i));
end
```

Example 7.11: Euler's Method for Systems

Consider the third-order IVP in Example 7.9:

$$y''' - y'' - 5y' - 3y = e^{-x}, \quad y(0) = 0, \quad y'(0) = 1, \quad y''(0) = -1, \quad 0 \le x \le 1$$

Using Euler's method for systems, with step size $h = 0.1$, find an estimate for $y_2 = y(0.2)$. Confirm by executing the user-defined function EulerODESystem.

SOLUTION

In Example 7.9, the IVP was transformed into the standard form of a system of first-order IVPs as

$$\mathbf{u}' = \mathbf{f}(x,\mathbf{u}), \quad \mathbf{u} = \begin{Bmatrix} u_1 \\ u_2 \\ u_3 \end{Bmatrix} = \begin{Bmatrix} y \\ y' \\ y'' \end{Bmatrix}, \quad \mathbf{f}(x,\mathbf{u}) = \begin{Bmatrix} u_2 \\ u_3 \\ u_3 + 5u_2 + 3u_1 + e^{-x} \end{Bmatrix},$$

$$\mathbf{u}_0 = \begin{Bmatrix} u_1(0) \\ u_2(0) \\ u_3(0) \end{Bmatrix} = \begin{Bmatrix} 0 \\ 1 \\ -1 \end{Bmatrix}$$

To find $y(0.2)$, we need to find the solution vector $\mathbf{u}_2 = \mathbf{u}(0.2)$ and then extract its first component, which is $y(0.2)$. By Equation 7.47

$$\mathbf{u}_1 = \mathbf{u}_0 + h\mathbf{f}(x_0,\mathbf{u}_0)$$

But

$$\mathbf{f}(x_0,\mathbf{u}_0) = \begin{Bmatrix} u_2(0) \\ u_3(0) \\ u_3(0) + 5u_2(0) + 3u_1(0) + e^{-x_0} \end{Bmatrix} \overset{x_0=0}{=} \begin{Bmatrix} 1 \\ -1 \\ -1 + 5(1) + 3(0) + 1 \end{Bmatrix} = \begin{Bmatrix} 1 \\ -1 \\ 5 \end{Bmatrix}$$

Therefore

$$\mathbf{u}_1 = \mathbf{u}_0 + h\mathbf{f}(x_0,\mathbf{u}_0) = \begin{Bmatrix} 0 \\ 1 \\ -1 \end{Bmatrix} + 0.1\begin{Bmatrix} 1 \\ -1 \\ 5 \end{Bmatrix} = \begin{Bmatrix} 0.1 \\ 0.9 \\ -0.5 \end{Bmatrix}$$

In the next step, $\mathbf{u}_2 = \mathbf{u}_1 + h\mathbf{f}(x_1, \mathbf{u}_1)$ where

$$\mathbf{f}(x_1,\mathbf{u}_1) = \begin{Bmatrix} u_2(0.1) \\ u_3(0.1) \\ u_3(0.1) + 5u_2(0.1) + 3u_1(0.1) + e^{-x_1} \end{Bmatrix}$$

$$\overset{x_1=0.1}{=} \begin{Bmatrix} 0.9 \\ -0.5 \\ -0.5 + 5(0.9) + 3(0.1) + e^{-0.1} \end{Bmatrix} = \begin{Bmatrix} 0.9 \\ -0.5 \\ 5.2048 \end{Bmatrix}$$

Therefore

$$\mathbf{u}_2 = \mathbf{u}_1 + h\mathbf{f}(x_1, \mathbf{u}_1) = \begin{Bmatrix} 0.1 \\ 0.9 \\ -0.5 \end{Bmatrix} + 0.1 \begin{Bmatrix} 0.9 \\ -0.5 \\ 5.2048 \end{Bmatrix} = \begin{Bmatrix} 0.19 \\ 0.85 \\ 0.0205 \end{Bmatrix}$$

The first component represents $y(0.2)$; thus $y(0.2) = 0.19$. The exact solution happens to be 0.1869; hence, our estimate has a relative error of 1.67%. The result may be confirmed in MATLAB as follows:

```
>> f = inline('[u(2,1);u(3,1);u(3,1)+5*u(2,1)+3*u(1,1)+
exp(-x)]','x','u');
>> x = 0:0.1:1;    % 11 mesh points
>> u0 = [0;1;-1];
>> u = EulerODESystem(f,x,u0);    % Returns a 3-by-11 matrix
>> y = u(1,:)
% Retain the first row of u: estimates of y at 11 mesh points

y =

  Columns 1 through 7

        0   0.1000   0.1900   0.2750   0.3602   0.4513   0.5546

  Columns 8 through 11

   0.6778   0.8298   1.0221   1.2687
```

The (shaded) value of $y(0.2)$ agrees with our earlier finding.

7.5.2.2 Heun's Method for Systems

Heun's method for a system in the form of Equation 7.46 is defined as

$$\mathbf{u}_{i+1} = \mathbf{u}_i + \frac{1}{2}h(\mathbf{k}_1 + \mathbf{k}_2), \quad i = 0, 1, 2, \ldots, n-1 \tag{7.48}$$

where

$$\begin{aligned} \mathbf{k}_1 &= \mathbf{f}(x_i, \mathbf{u}_i) \\ \mathbf{k}_2 &= \mathbf{f}(x_i + h, \mathbf{u}_i + h\mathbf{k}_1) \end{aligned}$$

The user-defined function `HeunODESystem` uses Heun's method as outlined in Equation 7.48 to estimate the solution vector of a system of IVPs in the form of Equation 7.46.

```
function u=HeunODESystem(f,x,u0)
%
% HeunODESystem uses Heun's method to solve a system of
% first-order ODEs given in the form u'=f(x,u) subject
% to initial condition vector u0.
%
%   u=HeunODESystem(f,x,u0) where
%
%       f is an inline (m-dim. vector) function
%       representing f(x,u),
%       x is an (n+1)-dim. vector representing the mesh
%       points,
%       u0 is an m-dim. vector representing the initial
%       condition of u,
%
%       u is an m-by-(n+1) matrix, each column being the
%       vector of solution estimates at a mesh point.

u(:,1)=u0;
% The first column is set to be the initial vector u0
h=x(2)-x(1);

for i=1:length(x)-1,
    k1=f(x(i),u(:,i));
    k2=f(x(i)+h,u(:,i)+h*k1);
    u(:,i+1)=u(:,i)+h*(k1+k2)/2;
end
```

Example 7.12: Heun's Method for Systems

In Example 7.11, find an estimate for $y(0.2)$ by executing the user-defined function HeunODESystem.

SOLUTION

```
>> x=0:0.1:1;
>> u0=[0;1;-1];
>> f=inline('[u(2,1);u(3,1);u(3,1)+5*u(2,1)+3*u(1,1)+
exp(-x)]','x','u');
>> u=HeunODESystem(f,x,u0);
>> y=u(1,:)

y =

  Columns 1 through 7

  0    0.0950    0.1851    0.2756    0.3731    0.4848    0.6201
```

```
Columns 8 through 11

 0.7907    1.0117    1.3031    1.6911
```

The (shaded) computed value is $y(0.2) = 0.1851$. Since the exact value is 0.1869, given in Example 7.11, our estimate has a relative error of 0.97% compared with 1.67% by Euler's estimate. As expected, Heun's method returns a more accurate approximation.

7.5.2.3 Classical RK4 Method for Systems

The classical RK4 method for a system in the form of Equation 7.46 is defined as

$$\mathbf{u}_{i+1} = \mathbf{u}_i + \frac{1}{6}h(\mathbf{k}_1 + 2\mathbf{k}_2 + 2\mathbf{k}_3 + \mathbf{k}_4), \quad i = 0, 1, 2, \ldots, n-1 \tag{7.49}$$

where

$$\begin{aligned}
\mathbf{k}_1 &= \mathbf{f}(x_i, \mathbf{u}_i) \\
\mathbf{k}_2 &= \mathbf{f}\left(x_i + \frac{1}{2}h, \mathbf{u}_i + \frac{1}{2}h\mathbf{k}_1\right) \\
\mathbf{k}_3 &= \mathbf{f}\left(x_i + \frac{1}{2}h, \mathbf{u}_i + \frac{1}{2}h\mathbf{k}_2\right) \\
\mathbf{k}_4 &= \mathbf{f}(x_i + h, \mathbf{u}_i + h\mathbf{k}_3)
\end{aligned}$$

The user-defined function `RK4System` uses the RK4 method as outlined in Equation 7.49 to estimate the solution vector of a system of IVPs in the form of Equation 7.46.

```
function u = RK4System(f,x,u0)
%
% RK4System uses RK4 method to solve a system of first-
% order ODEs given in the form u' = f(x,u) subject to
% initial condition vector u0.
%
%    u = RK4System(f,x,u0) where
%
%       f is an inline (m-dim. vector) function
%       representing f(x,u),
%       x is an (n+1)-dim. vector representing the mesh
%       points,
%       u0 is an m-dim. vector representing the initial
%       condition of u,
%
```

```
%       u is an m-by-(n+1) matrix, each column being the
%       vector of solution estimates at a mesh point.

u(:,1) = u0;
% The first column is set to be the initial vector u0
h = x(2) - x(1);

for i = 1:length(x)-1,
    k1 = f(x(i),u(:,i));
    k2 = f(x(i)+h/2,u(:,i)+h*k1/2);
    k3 = f(x(i)+h/2,u(:,i)+h*k2/2);
    k4 = f(x(i)+h,u(:,i)+h*k3);
    u(:,i+1) = u(:,i)+h*(k1+2*k2+2*k3+k4)/6;
end
```

Example 7.13: RK4 Method for Systems

Reconsider Example 7.11. Using RK4 method for systems, with $h = 0.1$, find an estimate for $y_1 = y(0.1)$. Confirm by executing the user-defined function `RK4System`.

SOLUTION

Recall

$$\mathbf{u}' = \mathbf{f}(x,\mathbf{u}), \quad \mathbf{u} = \begin{Bmatrix} u_1 \\ u_2 \\ u_3 \end{Bmatrix} = \begin{Bmatrix} y \\ y' \\ y'' \end{Bmatrix}, \quad \mathbf{f}(x,\mathbf{u}) = \begin{Bmatrix} u_2 \\ u_3 \\ u_3 + 5u_2 + 3u_1 + e^{-x} \end{Bmatrix},$$

$$\mathbf{u}_0 = \begin{Bmatrix} u_1(0) \\ u_2(0) \\ u_3(0) \end{Bmatrix} = \begin{Bmatrix} 0 \\ 1 \\ -1 \end{Bmatrix}$$

We first need to find $\mathbf{u}_1 = \mathbf{u}_0 + \frac{1}{6}h(\mathbf{k}_1 + 2\mathbf{k}_2 + 2\mathbf{k}_3 + \mathbf{k}_4)$.

$$\mathbf{k}_1 = \mathbf{f}(x_0,\mathbf{u}_0) = \begin{Bmatrix} u_2(0) \\ u_3(0) \\ u_3(0) + 5u_2(0) + 3u_1(0) + e^{-x_0} \end{Bmatrix}$$

$$\overset{x_0=0}{=} \begin{Bmatrix} 1 \\ -1 \\ -1 + 5(1) + 3(0) + 1 \end{Bmatrix} = \begin{Bmatrix} 1 \\ -1 \\ 5 \end{Bmatrix}$$

To calculate $\mathbf{k}_2 = \mathbf{f}(x_0 + \frac{1}{2}h, \mathbf{u}_0 + \frac{1}{2}h\mathbf{k}_1) = \mathbf{f}(0.05, \mathbf{u}_0 + \frac{1}{2}h\mathbf{k}_1)$, we first find

$$\mathbf{u}_0 + \frac{1}{2}h\mathbf{k}_1 = \begin{Bmatrix} 0 \\ 1 \\ -1 \end{Bmatrix} + 0.05\begin{Bmatrix} 1 \\ -1 \\ 5 \end{Bmatrix} = \begin{Bmatrix} 0.05 \\ 0.95 \\ -0.75 \end{Bmatrix}$$

Then

$$\mathbf{k}_2 = \mathbf{f}\left(0.05, \mathbf{u}_0 + \frac{1}{2}h\mathbf{k}_1\right) = \begin{Bmatrix} 0.95 \\ -0.75 \\ -0.75 + 5(0.95) + 3(0.05) + e^{-0.05} \end{Bmatrix} = \begin{Bmatrix} 0.95 \\ -0.75 \\ 5.1012 \end{Bmatrix}$$

To calculate $\mathbf{k}_3 = \mathbf{f}(x_0 + \frac{1}{2}h, \mathbf{u}_0 + \frac{1}{2}h\mathbf{k}_2) = \mathbf{f}(0.05, \mathbf{u}_0 + \frac{1}{2}h\mathbf{k}_2)$, we first find

$$\mathbf{u}_0 + \frac{1}{2}h\mathbf{k}_2 = \begin{Bmatrix} 0 \\ 1 \\ -1 \end{Bmatrix} + 0.05\begin{Bmatrix} 0.95 \\ -0.75 \\ 5.1012 \end{Bmatrix} = \begin{Bmatrix} 0.0475 \\ 0.9625 \\ -0.7449 \end{Bmatrix}$$

Then

$$\mathbf{k}_3 = \mathbf{f}\left(0.05, \mathbf{u}_0 + \frac{1}{2}h\mathbf{k}_2\right) = \begin{Bmatrix} 0.9625 \\ -0.7449 \\ -0.7449 + 5(0.9625) + 3(0.0475) + e^{-0.05} \end{Bmatrix}$$

$$= \begin{Bmatrix} 0.9625 \\ -0.7449 \\ 5.1613 \end{Bmatrix}$$

To find $\mathbf{k}_4 = \mathbf{f}(x_0 + h, \mathbf{u}_0 + h\mathbf{k}_3) = \mathbf{f}(0.1, \mathbf{u}_0 + h\mathbf{k}_3)$, we first calculate

$$\mathbf{u}_0 + h\mathbf{k}_3 = \begin{Bmatrix} 0 \\ 1 \\ -1 \end{Bmatrix} + 0.1\begin{Bmatrix} 0.9625 \\ -0.7449 \\ 5.1613 \end{Bmatrix} = \begin{Bmatrix} 0.0963 \\ 0.9255 \\ -0.4839 \end{Bmatrix}$$

Then

$$\mathbf{k}_4 = \mathbf{f}(0.1, \mathbf{u}_0 + h\mathbf{k}_3) = \begin{Bmatrix} 0.9255 \\ -0.4839 \\ -0.4839 + 5(0.9255) + 3(0.0963) + e^{-0.1} \end{Bmatrix}$$

$$= \begin{Bmatrix} 0.9255 \\ -0.4839 \\ 5.3373 \end{Bmatrix}$$

Finally

$$\mathbf{u}_1 = \mathbf{u}_0 + \frac{1}{6}h(\mathbf{k}_1 + 2\mathbf{k}_2 + 2\mathbf{k}_3 + \mathbf{k}_4) = \begin{Bmatrix} 0.0958 \\ 0.9254 \\ -0.4856 \end{Bmatrix}$$

Therefore, our estimate is $y(0.1) = 0.0958$. The result may be confirmed in MATLAB as follows:

```
>> x=0:0.1:1;
>> u0=[0;1;-1];
>> f=inline('[u(2,1);u(3,1);u(3,1)+5*u(2,1)+3*u(1,1)+
exp(-x)]','x','u');
>> u=RK4System(f,x,u0);
>> y=u(1,:)

y =

  Columns 1 through 7

       0    0.0958    0.1869    0.2787    0.3779    0.4923    0.6314

  Columns 8 through 11

    0.8077    1.0372    1.3410    1.7471
```

The shaded value agrees with the hand calculations.

7.6 Stability

Numerical stability is a desirable property of numerical methods. A numerical method is said to be stable if errors incurred in one step do not magnify in later steps. Stability analysis of a numerical method often boils down to the error analysis of the method when applied to a basic IVP, which serves as a model, in the form

$$y' = -\lambda y, \quad y(0) = y_0, \quad \lambda = \text{constant} > 0 \tag{7.50}$$

Since the exact solution $y(x) = y_0 e^{-\lambda x}$ exponentially decays toward zero, it is desired that the error also approaches zero as x gets sufficiently large. If a method is unstable when applied to this model, it is likely to have difficulty when applied to other differential equations.

7.6.1 Euler's Method

Suppose Euler's method (Section 7.2) with step size h is applied to this model. This means each subinterval $[x_i, x_{i+1}]$ has length $h = x_{i+1} - x_i$. Noting that $f(x,y) = -\lambda y$, the solution estimate y_{i+1} at x_{i+1} by Euler's method is

$$y_{i+1} = y_i + hf(x_i, y_i) = y_i - h\lambda y_i = (1 - \lambda h)y_i \tag{7.51}$$

At x_{i+1}, the exact solution is

$$y_{i+1}^{\text{Exact}} = y_0 e^{-\lambda x_{i+1}} \overset{x_{i+1}=x_i+h}{=} \left[y_0 e^{-\lambda x_i}\right] e^{-\lambda h} = y^{\text{Exact}}(x_i) e^{-\lambda h} \tag{7.52}$$

Comparison of Equations 7.51 and 7.52 reveals that $1 - \lambda h$ in the computed solution is an approximation for $e^{-\lambda h}$ in the exact solution.* From Equation 7.51, it is also observed that error will not be magnified if $|1 - \lambda h| < 1$. This implies that the numerical method (in this case, Euler's method) is stable if

$$|1 - \lambda h| < 1 \;\Rightarrow\; -1 < 1 - \lambda h < 1 \;\Rightarrow\; 0 < \lambda h < 2 \tag{7.53}$$

which acts as a stability criterion for Euler's method when applied to the IVP in Equation 7.50. Equation 7.53 describes a region of absolute stability for Euler's method. The wider the region of stability, the less limitation imposed on the step size h.

7.6.2 Euler's Implicit Method

Euler's implicit method is described by

$$y_{i+1} = y_i + hf(x_{i+1}, y_{i+1}), \quad i = 0, 1, 2, \ldots, n-1 \tag{7.54}$$

so that y_{i+1} appears on both sides of the equation. As with other implicit methods discussed so far, y_{i+1} is normally found numerically via the methods of Chapter 3, and can only be solved analytically if the function $f(x,y)$ has a simple structure. This is certainly the case when applied to the model in Equation 7.50 where $f(x,y) = -\lambda y$.

$$y_{i+1} = y_i + hf(x_{i+1}, y_{i+1}) = y_i + h(-\lambda y_{i+1}) \;\Rightarrow\; y_{i+1} = \frac{1}{1 + \lambda h} y_i$$

Therefore, Euler's implicit method is stable if

$$\left|\frac{1}{1 + \lambda h}\right| < 1 \;\Rightarrow\; |1 + \lambda h| > 1 \;\overset{h,\lambda>0}{\Rightarrow}\; h > 0$$

which implies it is stable regardless of the step size.

* Taylor series of $e^{-\lambda h}$ is $e^{-\lambda h} = 1 - \lambda h + \frac{1}{2!}(\lambda h)^2 - \cdots$ so that for small λh, we have $e^{-\lambda h} \cong 1 - \lambda h$.

Example 7.14: Stability of Euler's Methods

Consider the IVP

$$y' = -4y, \quad y(0) = 2, \quad 0 \le x \le 5$$

a. Solve using Euler's method with step size $h = 0.3$ and again with $h = 0.55$, plot the estimated solutions together with the exact solution $y(x) = 2e^{-4x}$, and discuss stability.
b. Repeat using the Euler's implicit method, Equation 7.54.

SOLUTION

a. Comparing the IVP at hand with the model in Equation 7.50, we have $\lambda = 4$. The stability criterion for Euler's method is

$$4h < 2 \quad \Rightarrow \quad h < \frac{1}{2}$$

Therefore, Euler's method is stable if $h < 0.5$, and is unstable otherwise. As observed from the first plot in Figure 7.5, Euler's method with $h = 0.3 < 0.5$ produces estimates that closely follow the exact solution, while those generated by $h = 0.55 > 0.5$ grow larger in each step, indicating instability of the method.

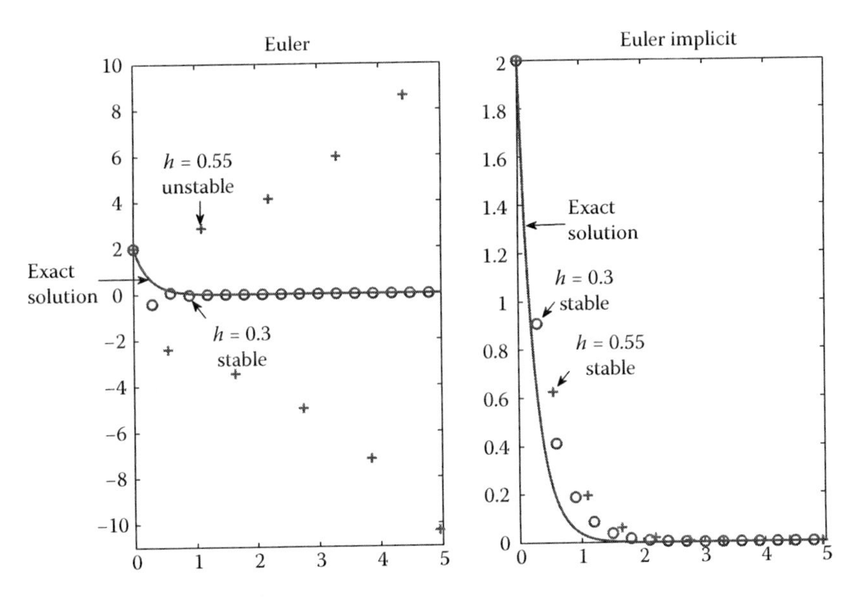

FIGURE 7.5
Stability analysis of Euler's and Euler's implicit methods in Example 7.14.

b. As mentioned earlier, when applied to a simple IVP such as the one here, Euler's implicit method is stable for all step sizes. The second plot in Figure 7.5 clearly shows that the errors associated with both step sizes decay to zero as x increases, indicating stability.

```
y0=2; f=inline('-4*y','x','y');
yExact=inline('2*exp(-4*x)');
h1=0.3; x1=0:h1:5; h2=0.55; x2=0:h2:5;
y1=EulerODE(f,x1,y0); y2=EulerODE(f,x2,y0);

y1I(1)=y0;
for i=2:length(x1),
    y1I(i)=y1I(i-1)/(1+4*h1);          % Implicit (h=0.3)
end
y2I(1)=y0;
for i=2:length(x2),
    y2I(i)=y2I(i-1)/(1+4*h2);          % Implicit (h=0.55)
end
x3=linspace(0,5);
ye=zeros(100,1);
for i=1:100,
    ye(i)=yExact(x3(i)); % Exact
end
subplot (1,2,1), plot(x1,y1,'o',x2,y2,'+',x3,ye,'-')
title('Euler')
subplot (1,2,2), plot(x1,y1I,'o',x2,y2I,'+',x3,ye,'-')
title('Euler implicit')
```

7.7 Stiff Differential Equations

In many engineering applications, such as chemical kinetics, mass–spring–damper systems, and control system analysis, we encounter systems of differential equations whose solutions contain terms with magnitudes that vary at rates that are considerably different. For example, if a solution includes the terms e^{-at} and e^{-bt}, with $a,b > 0$, where the magnitude of a is much larger than b, then e^{-at} decays to zero at a much faster rate than e^{-bt} does. In the presence of a rapidly decaying transient solution, certain numerical methods become unstable unless the step size is unreasonably small. Explicit methods generally are subjected to this stability constraint, which requires them to use an extremely small step size for accuracy. But using a very small step size not only substantially increases the number of operations to find a solution, it also causes round-off error to grow, thus causing limited accuracy. Implicit methods, on the other hand, are free of stability restrictions and are therefore preferred for solving stiff differential equations.

Example 7.15: Stiff System of ODEs

Consider

$$\begin{matrix} \dot{v} = 790v - 1590w \\ \dot{w} = 793v - 1593w \end{matrix} \quad \text{subject to} \quad \begin{matrix} v_0 = v(0) = 1 \\ w_0 = w(0) = -1 \end{matrix}$$

The exact solution of this system can be found in closed form as

$$v(t) = \frac{3180}{797}e^{-3t} - \frac{2383}{797}e^{-800t}$$

$$w(t) = \frac{1586}{797}e^{-3t} - \frac{2383}{797}e^{-800t}$$

Exact solution at $t = 0.1$:

$$v_1 = v(0.1) = \frac{3180}{797}e^{-3(0.1)} - \frac{2383}{797}e^{-800(0.1)} = 2.955836815$$

$$w_1 = w(0.1) = \frac{1586}{797}e^{-3(0.1)} - \frac{2383}{797}e^{-800(0.1)} = 1.474200374$$

Euler's method with $h = 0.1$: We first express the system in vector form

$$\dot{\mathbf{u}} = \mathbf{f}(t,\mathbf{u}), \quad \mathbf{u} = \begin{Bmatrix} v \\ w \end{Bmatrix}, \quad \mathbf{f} = \begin{Bmatrix} 790v - 1590w \\ 793v - 1593w \end{Bmatrix}, \quad \mathbf{u}_0 = \begin{Bmatrix} 1 \\ -1 \end{Bmatrix}$$

Then, by Euler's method

$$\mathbf{u}_1 = \mathbf{u}_0 + h\mathbf{f}(x_0,\mathbf{u}_0) = \begin{Bmatrix} 1 \\ -1 \end{Bmatrix} + 0.1\begin{Bmatrix} 790v_0 - 1590w_0 \\ 793v_0 - 1593w_0 \end{Bmatrix}$$

$$= \begin{Bmatrix} 239 \\ 237.6 \end{Bmatrix} \quad \Rightarrow \quad \begin{matrix} v_1 = 239 \\ w_1 = 237.6 \end{matrix}$$

which are totally incorrect. Euler's method is an explicit method, and as such, it is unstable unless a very small step size is selected. The main contribution to numerical instability is indeed done by the rapidly decaying e^{-800t}. Reduction of the step size to $h = 0.001$ results in

$$\mathbf{u}_1 = \mathbf{u}_0 + h\mathbf{f}(x_0,\mathbf{u}_0) = \begin{Bmatrix} 1 \\ -1 \end{Bmatrix} + 0.001\begin{Bmatrix} 790v_0 - 1590w_0 \\ 793v_0 - 1593w_0 \end{Bmatrix}$$

$$= \begin{Bmatrix} 3.380 \\ 1.386 \end{Bmatrix} \quad \Rightarrow \quad \begin{matrix} v_1 = 3.380 \\ w_1 = 1.386 \end{matrix}$$

which are much more accurate than before. Further reduction to $h = 0.0001$ allows round-off error to grow, causing the estimates to be much less accurate.

Euler's implicit method with $h = 0.1$: Using the vector form of Equation 7.54

$$\mathbf{u}_1 = \mathbf{u}_0 + h\mathbf{f}(x_1, \mathbf{u}_1) = \begin{Bmatrix} 1 \\ -1 \end{Bmatrix} + 0.1 \begin{Bmatrix} 790v_1 - 1590w_1 \\ 793v_1 - 1593w_1 \end{Bmatrix}$$

$$\Rightarrow \begin{matrix} v_1 = 1 + 79v_1 - 159w_1 \\ w_1 = -1 + 79.3v_1 - 159.3w_1 \end{matrix} \quad \Rightarrow \quad \begin{matrix} v_1 = 3.032288699 \\ w_1 = 1.493827160 \end{matrix}$$

These are closer to the exact values than those generated by Euler's method with a much smaller h.

7.8 MATLAB® Built-In Functions for Initial-Value Problems

There are several MATLAB built-in functions designed for solving a single first-order IVP, as well as a system of first-order IVPs. These are known as ODE solvers and include `ode23`, `ode45`, and `ode113` for nonstiff equations, and `ode15s` for stiff equations. Most of these built-in functions use highly developed techniques that allow them to use an optimal step size, or in some cases, adjust the step size for error minimization in each step.

7.8.1 Nonstiff Equations

We will first show how the built-in functions can be used to solve a single first-order IVP, then extend their applications to systems of first-order IVPs.

7.8.1.1 Single First-Order IVP

The IVP is

$$\dot{y} = f(t, y), \quad y(t_0) = y_0 \tag{7.55}$$

Note that, as mentioned previously, we are using t in place of x since in most applied problems time is the independent variable. For nonstiff equations, MATLAB built-in ODE solvers are `ode23`, `ode45`, and `ode113`.

`ode23` is a single-step method based on second- and third-order Runge–Kutta methods. As always, MATLAB help file should be regularly consulted for detailed information.

```
ode23 Solve non-stiff differential equations, low order
method.

    [TOUT,YOUT]=ode23(ODEFUN,TSPAN,Y0) with TSPAN=[T0 TFINAL]
    integrates the system of differential equations y'=f(t,y)
    from time T0 to TFINAL with initial conditions Y0. ODEFUN is
    a function handle. For a scalar T and a vector Y, ODEFUN(T,Y)
    must return a column vector corresponding to f(t,y). Each row
    in the solution array YOUT corresponds to a time returned in
    the column vector TOUT. To obtain solutions at specific times
    T0,T1,...,TFINAL (all increasing or all decreasing), use
    TSPAN = [T0 T1... TFINAL].
```

Note that if `TSPAN` has only two elements (the left and right endpoints for t), then vector `YOUT` will contain solutions calculated by `ode23` at all steps.

`ode45` is a single-step method based on fourth- and fifth-order Runge–Kutta methods. Refer to MATLAB help file for more detailed information.

```
ode45 Solve non-stiff differential equations, medium order
method.
```

The function call is

```
[TOUT,YOUT] = ode45(ODEFUN,TSPAN,Y0)
```

with all input and output argument descriptions as in `ode23` above. `ode45` can handle most IVPs and should be the first solver attempted for a given problem.

`ode113` is a multistep method based on ABM methods. Refer to MATLAB help file for more detailed information.

```
ode113 Solve non-stiff differential equations, variable order
method.
```

The function call is

```
[TOUT,YOUT] = ode113(ODEFUN,TSPAN,Y0)
```

with all input and output argument descriptions as in `ode23` and `ode45` above.

Example 7.16: `ode23, ode45`

Write a MATLAB script file that solves the following IVP using `ode23` and `ode45` solvers and returns a table that includes the solution estimates at `t = 0:0.1:1`, as well as the exact solution and the percent relative error for both methods at each point. The exact solution is $y = (t+1)e^{-t}/2$.

$$2\dot{y} + y = 2e^{-t/2}, \quad y(0) = 1, \quad 0 \le t \le 1$$

SOLUTION

```
disp('    t      yode23      yode45      yExact      e_23      e_45')

t = 0:0.1:1; y0 = 1;
[t,y23] = ode23(@myfunc,t,y0);    % 'myfunc' defined below
[t,y45] = ode45(@myfunc,t,y0);

yExact = inline('(t+1)*exp(-t/2)');

for k=1:length(t),
    t_coord = t(k);
    yode23 = y23(k);
    yode45 = y45(k);
    yEx = yExact(t(k));
    e_23 = (yEx - yode23)/yEx*100;
    e_45 = (yEx - yode45)/yEx*100;

   fprintf('%6.2f %11.6f%11.6f %11.6f %11.6f
   %11.8f\n',t_coord,yode23,yode45,yEx,e_23,e_45)
end

function dydt = myfunc(t,y)
dydt = (2*exp(-t/2)-y)/2;
```

t	yode23	yode45	yExact	e_23	e_45
0.00	1.000000	1.000000	1.000000	0.000000	0.00000000
0.10	1.046353	1.046352	1.046352	−0.000077	0.00000000
0.20	1.085806	1.085805	1.085805	−0.000140	0.00000000
0.30	1.118922	1.118920	1.118920	−0.000190	0.00000000
0.40	1.146226	1.146223	1.146223	−0.000232	0.00000000
0.50	1.168204	1.168201	1.168201	−0.000266	0.00000000
0.60	1.185313	1.185309	1.185309	−0.000294	0.00000000
0.70	1.197974	1.197970	1.197970	−0.000317	0.00000001
0.80	1.206580	1.206576	1.206576	−0.000336	0.00000001
0.90	1.211498	1.211493	1.211493	−0.000352	0.00000001
1.00	1.213066	1.213061	1.213061	−0.000365	0.00000001

As expected, `ode45` produces much more accurate estimates but is often slower than `ode23`. Note that even though we could have used an inline function, we opted for a function handle because it is more efficient.

7.8.1.2 Setting ODE Solver Options

The `ode23`, `ode45`, and `ode113` functions accept an optional input argument called an "options structure" that allows many properties of the solution method to be specified. Two examples of such properties are the relative tolerance and the minimum step size. The `odeset` function creates these options structures.

```
odeset Create/alter ODE OPTIONS structure.
```

```
OPTIONS = odeset('NAME1',VALUE1,'NAME2',VALUE2,...)
creates an integrator options structure OPTIONS in
which the named properties have the specified values.
Any unspecified properties have default values. It is
sufficient to type only the leading characters that
uniquely identify the property. Case is ignored for
property names.
```

A complete list of the various properties and their descriptions is available in the help function. A default options structure is created using

```
>> options = odeset;
```

The default relative tolerance `RelTol` is 10^{-3}. `RelTol` can either be specified upon creation of the options structure

```
>> options = odeset('RelTol', 1e-7);
```

or can be changed by including the current options as the first input argument.

```
>> options = odeset(options, 'RelTol', 1e-6);
```

The option structure is then specified as the fourth input argument to the ODE solvers.

```
>> [t y] = ode45(@myfunc,t,y0,options);
```

Therefore, we can trade speed for accuracy by specifying the relative tolerance. Other parameters may also be specified.

7.8.1.3 System of First-Order IVPs

The ODE solvers discussed here can also handle systems of first-order IVPs.

Example 7.17: `ode45` for a System

Using `ode45` solve the following system:

$$\begin{aligned}\dot{u}_1 &= u_2 \\ \dot{u}_2 &= \frac{1}{2}(-3u_2 - 10u_1 + \sin t)\end{aligned} \quad \text{subject to initial conditions} \quad \begin{aligned}u_1(0) &= 0 \\ u_2(0) &= 1\end{aligned}$$

SOLUTION

We first express the system in vector form

$$\dot{\mathbf{u}} = \mathbf{f}(t,\mathbf{u}), \quad \mathbf{u} = \begin{Bmatrix} u_1 \\ u_2 \end{Bmatrix}, \quad \mathbf{f} = \begin{Bmatrix} u_2 \\ \frac{1}{2}(-3u_2 - 10u_1 + \sin t) \end{Bmatrix}, \quad \mathbf{u}_0 = \begin{Bmatrix} 0 \\ 1 \end{Bmatrix}$$

```
>> u0 = [0;1]; t = 0:0.1:1;
>> [t,u45] = ode45(@myfunc,t,u0)

t =
        0
   0.1000
   0.2000
   0.3000
   0.4000
   0.5000
   0.6000
   0.7000
   0.8000
   0.9000
   1.0000

u45 =
% u45 has two columns: 1st column is u1, 2nd column is u2
        0    1.0000
   0.0922    0.8405
   0.1677    0.6690
   0.2259    0.4950
   0.2669    0.3267
   0.2917    0.1710
   0.3017    0.0332
   0.2990   -0.0832
   0.2858   -0.1760
   0.2646   -0.2445
   0.2377   -0.2893

function dudt = myfunc(t,u)
dudt = [u(2);(-3*u(2)-10*u(1)+sin(t))/2];
```

7.8.2 Stiff Equations

The ODE solver `ode15s` can be used to solve stiff equations. `ode15s` is a multistep, variable-order method. Refer to MATLAB help file for more detailed information.

```
ode15s Solve stiff differential equations and DAEs, variable
order method.
```

The function call is

```
[TOUT,YOUT] = ode15s(ODEFUN,TSPAN,Y0)
```

with all input and output argument descriptions as in `ode23` and others.

Example 7.18: `ode15s`

Consider the stiff system in Example 7.15:

$$\begin{aligned} \dot{v} &= 790v - 1590w \\ \dot{w} &= 793v - 1593w \end{aligned} \quad \text{subject to} \quad \begin{aligned} v_0 &= v(0) = 1 \\ w_0 &= w(0) = -1 \end{aligned}$$

Write a MATLAB script file that solves the system using ode15s and ode45 and returns a table that includes the solution estimates for $v(t)$ at t = 0:0.1:1, as well as the exact solution and the percent relative error for both methods at each point. The exact solution was provided in Example 7.15.

SOLUTION

```
disp('  t       v15s       v45       vExact       e_15s       e_45')

t = 0:0.1:1; u0 = [1;-1];
[t,u15s] = ode15s(@myfunc,t,u0);
% Values of v are in the 1st column of u15s
[t,u45] = ode45(@myfunc,t,u0);
% Values of v are in the 1st column of u45

uExact = inline('[3180/797*exp(-3*t)-2383/797*exp
(-800*t);1586/797*exp(-3*t)-2383/797*exp(-800*t)]');
for i = 1:length(t),
    uex(:,i) = uExact(t(i));
% Evaluate exact solution vector at each t
end

for k = 1:length(t),
   t_coord = t(k);
   v15s = u15s(k,1); % Retain the 1st column of u15s: values of v
   v45 = u45(k,1);   % Retain the 1st column of u45: values of v
   vExact = uex(1,k); % Retain the exact values of v

   e_15s = (vExact - v15s)/vExact*100;
   e_45 = (vExact - v45)/vExact*100;

   fprintf('%6.2f %11.6f%11.6f %11.6f %11.6f
   %11.8f\n',t_coord,v15s,v45,vExact,e_15s,e_45)

end

function dudt = myfunc(t,u)
dudt = [790*u(1)-1590*u(2);793*u(1)-1593*u(2)];
```

t	v15s	v45	vExact	e_15s	e_45
0.00	1.000000	1.000000	1.000000	0.000000	0.00000000
0.10	2.955055	2.955511	2.955837	0.026464	0.01102274
0.20	2.187674	2.188567	2.189738	0.094262	0.05345813
0.30	1.620781	1.622657	1.622198	0.087342	-0.02830662
0.40	1.201627	1.201785	1.201754	0.010535	-0.00261497
0.50	0.890826	0.890278	0.890281	-0.061233	0.00033632
0.60	0.660191	0.659508	0.659536	-0.099254	0.00427387
0.70	0.489188	0.488525	0.488597	-0.121031	0.01461519
0.80	0.362521	0.361836	0.361961	-0.154754	0.03468108
0.90	0.268708	0.268171	0.268147	-0.209208	-0.00866986
1.00	0.199060	0.198645	0.198649	-0.207140	0.00178337

It is easy to see that even for this stiff system of ODEs, the solver `ode45` still outperforms `ode15s`, which is specially designed to handle such systems.

PROBLEM SET

Euler's Method (Section 7.2)

In Problems 1 through 6, given each IVP

a. ✍ Using Euler's method with the indicated step size h, calculate the solution estimates at x_1, x_2, and x_3, as well as the local and global truncation errors at those locations,

b. Write a MATLAB script file that uses `EulerODE` to find the approximate values produced by Euler's method and returns a table that includes these values, as well as the exact values and the global percent relative error, at all mesh points in the given interval.

1. $y' + 2xy = 0,\quad y(0) = 1,\quad 0 \le x \le 1,\quad h = 0.1$

   Exact solution is $y = e^{-x^2}$.

2. $xy' = y - x,\quad y(1) = 0,\quad 1 \le x \le 2,\quad h = 0.1$

   Exact solution is $y = -x\ln x$.

3. $e^x y' = y^2,\quad y(0) = 1,\quad 0 \le x \le 0.5,\quad h = 0.05$

   Exact solution is $y = e^x$.

4. $y' = y^2 \cos x,\quad y(0) = \dfrac{1}{3},\quad 0 \le x \le 1,\quad h = 0.1$

   Exact solution is $y = 1/(3 - \sin x)$.

5. $xy' = y + y^2,\quad y(2) = 1,\quad 2 \le x \le 3,\quad h = 0.1$

   Exact solution is $y = x/(4 - x)$.

6. $e^x y' = x^2 y^2,\quad y(0) = \dfrac{1}{3},\quad 0 \le x \le 1,\quad h = 0.2$

   Exact solution is $y = [e^{-x}(x^2 + 2x + 2) + 1]^{-1}$.

7. Write a MATLAB script to solve the following IVP using `EulerODE` with $h = 0.4$, and again with $h = 0.2$. The file must return a table showing the estimated solutions at 0, 0.4, 0.8, 1.2, 1.6, 2 produced by both scenarios, as well as the exact values at these points. The exact solution is $y = e^{-x/10}[4 - 3e^{-7x/30}]$.

$$y' = -0.1y + 0.7e^{-x/3},\quad y(0) = 1,\quad 0 \le x \le 2$$

8. Write a MATLAB script to solve the following IVP using `EulerODE` with $h = 0.2$, and again with $h = 0.1$. The file must return a table showing the estimated solutions at 0, 0.2, 0.4, 0.6, 0.8, 1 produced by both scenarios, as well as the exact values at these points. The exact solution is $y = 3e^{x^2/2} - x^2 - 2$.

$$y' - xy = x^3,\quad y(0) = 1,\quad 0 \le x \le 1$$

9. The free fall of a light particle of mass m released from rest and moving with a velocity v is governed by

$$m\dot{v} = mg - bv, \quad v(0) = 0$$

where $g = 9.81$ m/s² is the gravitational acceleration and $b = 0.2m$ is the coefficient of viscous damping. Write a MATLAB script that solves the IVP using `EulerODE` with $h = 0.05$ and returns a table of the estimated solutions at 0, 0.2, 0.4, 0.6, 0.8, 1, as well as the exact values at these points. The exact solution is $v(t) = 49.05(1 - e^{-t/5})$.

10. The free fall of a heavy particle of mass m released from rest and moving with a velocity v is governed by

$$m\dot{v} = mg - bv^2, \quad v(0) = 0$$

where $g = 9.81$ m/s² is the gravitational acceleration and $b = 0.2m$ is the coefficient of viscous damping. Write a MATLAB script that solves the IVP using `EulerODE` with $h = 0.05$ and $h = 0.1$, and returns a table of estimated solutions at 0, 0.2, 0.4, 0.6, 0.8, 1 for both step sizes, as well as the exact values at these points. The exact solution is

$$v(t) = \sqrt{5g}\,\tanh\left(\sqrt{0.2g}\; t\right)$$

Higher-Order Taylor Methods

11. Solve the following IVP using the second-order Taylor method with step size $h = 0.3$ to calculate the solution estimates at x_1, x_2, and x_3.

$$y' = -y + x^2 y, \quad y(0) = 1, \quad 0 \le x \le 1.2$$

12. Write a user-defined function with function call `y = Taylor_2_ODE(f,fp,x,y0)` that solves a first-order IVP in the form $y' = f(x,y)$, $y(x_0) = y_0$ using the second-order Taylor method. The input argument `fp` represents the first derivative of the function `f`, while the others are as in `EulerODE`. Execute the function to solve

$$y' = y/x + x^2 e^x, \quad y(1) = 0, \quad 1 \le x \le 2, \quad h = 0.1$$

In Problems 13 through 16, for each IVP

a. Using the second-order Taylor method with the indicated step size h, calculate the solution estimates at x_1, x_2, and x_3.
b. Write a MATLAB script file that uses `EulerODE` and `Taylor_2_ODE` (see Problem 12) to find the approximate values produced by

Euler's and second-order Taylor methods and returns a table that includes these values, as well as the exact values and the global percent relative error for both methods, at all mesh points in the given interval.

13. $xy' = x - y,\quad y(1) = 0,\quad 1 \le x \le 2,\quad h = 0.1$

    Exact solution is $y = x/2 - 1/(2x)$.

14. $2y' = -y + 2e^{-x/2},\quad y(0) = 1,\quad 0 \le x \le 2,\quad h = 0.2$

    Exact solution is $y = (x+1)e^{-x/2}$.

15. $\dot{x} = 10(1-t)(2-t),\quad x(0) = 0,\quad 0 \le t \le 1,\quad h = 0.1$

    Exact solution is $x(t) = \frac{5}{3}t(2t^2 - 9t + 12)$.

16. $t\dot{x} = x + x^3,\quad x(2) = 1,\quad 2 \le t \le 2.5,\quad h = 0.05$

    Exact solution is $x(t) = t\sqrt{1/(8 - t^2)}$.

17. Write a MATLAB script file that solves the IVP below using `EulerODE` with $h = 0.04$ and `Taylor_2_ODE` (see Problem 12) with $h = 0.08$. The file must return a table that includes the estimated solutions at $x = 0{:}0.08{:}0.8$, as well as the exact values and the global percent relative error for both methods, at all mesh points in the given interval. The exact solution is $y = e^{-x}/(x-1) + 2$. Discuss the results.

$$(x-1)y' + xy = 2x,\quad y(0) = 1,\quad 0 \le x \le 0.8$$

18. Write a MATLAB script file that solves the IVP below using `EulerODE` with $h = 0.05$ and `Taylor_2_ODE` (see Problem 12) with $h = 0.1$. The file must return a table that includes the estimated solutions at $x = 1{:}0.1{:}2$, as well as the exact values and the global percent relative error for both methods, at all mesh points in the given interval. The exact solution is $y = \frac{1}{2}x\ln x - \frac{1}{4}x + x^{-1}$. Discuss the results.

$$xy' + y = x\ln x,\quad y(1) = \frac{3}{4},\quad 1 \le x \le 2$$

Runge–Kutta Methods (Section 7.3)

RK2 Methods

In Problems 19 through 22, for each IVP and the indicated step size h, use the following methods to compute the solution estimates at x_1 and x_2.

a. Improved Euler
b. Heun
c. Ralston

19. $2y' + xy = 0, \quad y(0) = 1, \quad 0 \le x \le 1, \; h = 0.1$

20. $xy' = y - 3x, \quad y(1) = 1, \quad 1 \le x \le 2, \; h = 0.1$

21. $y' = e^{-2x}y^2, \quad y(0) = \dfrac{1}{2}, \; 0 \le x \le 1, \; h = 0.05$

22. $y' = y^2 \sin\left(\dfrac{1}{2}x\right), \quad y(0) = 1, \; 0 \le x \le 2, \; h = 0.2$

23. Write a user-defined function with function call `y = Imp_EulerODE(f,x,y0)` that solves a first-order IVP in the form $y' = f(x,y)$, $y(x_0) = y_0$ using the improved Euler's method. The input arguments are as in `HeunODE`. Execute the function to solve

$$xy' = y + x^2 e^x, \quad y(1) = 1, \quad 1 \le x \le 2, \quad h = 0.1$$

24. Write a user-defined function with function call `y = Ralston_RK2(f,x,y0)` that solves a first-order IVP in the form $y' = f(x,y)$, $y(x_0) = y_0$ using Ralston's method. The input arguments are as in `HeunODE`. Execute the function to solve

$$(x-1)y' + xy = x, \quad y(2) = 2, \quad 2 \le x \le 5, \quad h = 0.3$$

25. Write a MATLAB script file that uses `EulerODE` (Section 7.2), `Taylor_2_ODE` (see Problem 12), and `HeunODE` (Section 7.3) to find the approximate values produced by Euler's, second-order Taylor, and Heun's methods and returns a table that includes these values, as well as the exact values and the global percent relative error for all three methods, at all mesh points in the given interval. Discuss the results.

$$xy' = -y + xe^{-x/2}, \quad y(1) = 0, \quad 1 \le x \le 1.5, \quad h = 0.05$$

The exact solution is given as $y = [6e^{-1/2} - 2(x+2)e^{-x/2}]/x$.

26. Write a MATLAB script file to generate Table 7.1 of Example 7.4. The file must call functions `HeunODE` (Section 7.3), `Imp_EulerODE` (see Problem 23), and `Ralston_RK2` (see Problem 24).

RK3 Methods

In Problems 27 and 28, for each IVP and the indicated step size h, use the following methods to compute the solution estimates at x_1 and x_2.

a. Classical RK3
b. Heun's RK3

27. $y' = y \sin 2x, \quad y(0) = 1, \quad 0 \le x \le 1, \quad h = 0.1$
28. $y' = x^2 + y, \quad y(0) = 1, \quad 0 \le x \le 1, \quad h = 0.1$
29. Write a user-defined function with function call `y = Classical_RK3(f,x,y0)` that solves a first-order IVP in the form $y' = f(x,y)$, $y(x_0) = y_0$ using the classical RK3 method. The input arguments are as in `HeunODE`.
30. Write a user-defined function with function call `y = Heun_RK3(f,x,y0)` that solves a first-order IVP $y' = f(x,y)$, $y(x_0) = y_0$ using Heun's RK3 method. The input arguments are as in `HeunODE`.
31. Write a MATLAB script file to generate Table 7.2 of Example 7.5. The file must call functions `EulerODE` (Section 7.2), `Classical_RK3` (see Problem 29), and `Heun_RK3` (see Problem 30).

RK4 Methods

In Problems 32 through 35, for each IVP and the indicated step size h, use the classical RK4 method to compute the solution estimates at x_1 and x_2.

32. $xy' + 3y = x, \quad y(1) = 1, \quad 1 \le x \le 2, \quad h = 0.2$
33. $y' + 0.65y = 1.3e^{-x/4}, \quad y(0) = 1, \quad 0 \le x \le 1.5, \quad h = 0.15$
34. $y' = 12(2 - x)(3 - x), \quad y(0) = 0, \quad 0 \le x \le 0.8, \quad h = 0.08$
35. $xy' + 2y = x \ln x, \quad y(1) = 0, \quad 1 \le x \le 2, \quad h = 0.1$
36. Write a MATLAB script file to generate Table 7.3 of Example 7.6. The file must call functions `EulerODE` (Section 7.2), `Classical_RK3` (see Problem 29), `HeunODE` (Section 7.3), and `RK4` (Section 7.3).
37. Write a MATLAB script file that uses `EulerODE` (Section 7.2), `HeunODE` (Section 7.3), and `RK4` (Section 7.3) to find the approximate values produced by Euler's method, Heun's method, and the classical RK4 method and returns a table that includes these values at all mesh points in the given interval.

$$y' = y^2 + e^x, \quad y(0) = 0, \quad 0 \le x \le 0.2, \quad h = 0.02$$

38. Write a MATLAB script file that uses `EulerODE` (Section 7.2), `HeunODE` (Section 7.3), and `RK4` (Section 7.3) to find the approximate values produced by Euler's method, Heun's method, and the classical RK4 method and returns a table that includes the estimated solutions, as well as the global percent relative error for all three methods at all mesh points in the given interval. Discuss the results.

$$xy' = y + e^{y/x}, \quad y(1) = 1, \quad 1 \le x \le 1.5, \quad h = 0.05$$

The exact solution is $y = -x \ln(e^{-1} + 1/x - 1)$.

39. Write a MATLAB script file that solves the IVP in Problem 38 using `EulerODE` with $h = 0.025$, `HeunODE` with $h = 0.05$, and `RK4` with $h = 0.1$. The file must return a table that includes the estimated solutions, as well as the global percent relative error, at $x = 1, 1.1, 1.2, 1.3, 1.4, 1.5$ for all three methods. Discuss the results.

$$xy' = y + e^{y/x}, \quad y(1) = 1, \quad 1 \le x \le 1.5$$

40. Write a MATLAB script file that solves the following IVP using `EulerODE` with $h = 0.0125$, `HeunODE` with $h = 0.025$, and `RK4` with $h = 0.1$. The file must return a table that includes the estimated solutions, as well as the global percent relative error, at $x = 1{:}0.1{:}2$ for all three methods. Discuss the results.

$$(x + 1)y' + y = x \ln x, \quad y(1) = 0, \quad 1 \le x \le 2$$

The exact solution is $y = \left(\frac{1}{2}x^2\left(\ln x - \frac{1}{2}\right) + \frac{1}{4}\right)/(x + 1)$.

RKF Method

41. Write a user-defined function with function call `[y5 y4 err] = RK_Fehlberg(f,x,y0)` that solves an IVP $y' = f(x,y)$, $y(x_0) = y_0$ using the RKF method. The output arguments are, in order, the vector of fifth-order accurate estimates, vector of fourth-order accurate estimates, and the vector of estimated truncation errors.

42. Write a MATLAB script file that employs the user-defined function `RK_Fehlberg` (see Problem 41) to solve the following IVP. The file must return a table that includes the fourth-order and fifth-order accurate estimates, as well as the estimated local trunction error.

$$3y' + y = e^{-x/2} \sin x, \quad y(0) = 0, \quad 0 \le x \le 1, \quad h = 0.1$$

Multistep Methods (Section 7.4)

Adams–Bashforth Method

43. Consider the IVP

$$x^2 y' = y, \quad y(1) = 1, \quad 1 \le x \le 2, \quad h = 0.1$$

a. Using RK4, find an estimate for y_1.
b. Using y_0 (initial condition) and y_1 from (a), apply the second-order Adams–Bashforth formula to find y_2.

c. Apply the third-order Adams–Bashforth formula to find an estimate for y_3.

d. Apply the fourth-order Adams–Bashforth formula to find an estimate for y_4.

44. ✍ Repeat Problem 43 for the IVP

$$y' + y^2 = 0, \quad y(0) = 1, \quad 0 \le x \le 1, \quad h = 0.1$$

45. ✍ Consider the IVP

$$yy' = 2x - xy^2, \quad y(0) = 2, \quad 0 \le x \le 1, \quad h = 0.1$$

a. Using RK4, find an estimate for y_1 and y_2.

b. Using y_0 (initial condition) and y_1 from (a), apply the second-order Adams–Bashforth formula to find y_2.

c. Compare the relative errors associated with the y_2 estimates found in (a) and (b), and comment. The exact solution is $y = [2(e^{-x^2} + 1)]^{1/2}$.

46. ✍ Consider the IVP

$$y' = y \sin x, \quad y(1) = 1, \quad 1 \le x \le 2, \quad h = 0.1$$

a. Using RK4, find an estimate for y_1, y_2, and y_3.

b. Using y_0 (initial condition), and y_1 and y_2 from (a), apply the third-order Adams–Bashforth formula to find y_3.

c. Compare the relative errors associated with the y_3 estimates found in (a) and (b), and comment. The exact solution is $y = e^{\cos 1 - \cos x}$.

47. ✍ Consider the IVP

$$y' + y = \cos x, \quad y(0) = 1, \quad 0 \le x \le 1, \quad h = 0.1$$

a. Using RK4, find an estimate for y_1, y_2, and y_3.

b. Apply the fourth-order Adams–Bashforth formula to find y_4.

48. Write a user-defined function `y = AB_3(f,x,y0)` that solves an IVP $y' = f(x,y)$, $y(x_0) = y_0$ using the third-order Adams–Bashforth method. Solve the IVP in Problem 46 by executing `AB_3`. Compare the value of y_3 with that obtained in Problem 46.

49. Write a user-defined function with function call `y = ABM_3 (f,x,y0)` that solves an IVP $y' = f(x,y)$, $y(x_0) = y_0$ using the third-order

ABM predictor–corrector method. Solve the following IVP by executing `ABM_3`.

$$y' = y^2 \sin x, \quad y(0) = \frac{1}{2}, \quad 0 \le x \le 0.5, \quad h = 0.05$$

50. Write a MATLAB script file that utilizes the user-defined function `ABM4PredCorr` (Section 7.4) and generates Table 7.4 in Example 7.7.

Systems of Ordinary Differential Equations (Section 7.5)

In Problems 51 through 54, given each second-order IVP

a. Transform into a system of first-order IVPs using state variables.

b. Apply Euler, Heun, and classical RK4 methods for systems, with the indicated step size h, to compute the solution estimate y_1, and compare with the exact value.

51. $y'' + 3y' + 2y = x + 1, \quad y(0) = 0, \quad y'(0) = 0, \quad 0 \le x \le 1, \quad h = 0.1$

   Exact solution is $y(x) = \frac{1}{2}x + \frac{1}{4}(e^{-2x} - 1)$.

52. $y'' - y = 2x, \quad y(0) = 0, \quad y'(0) = 1, \quad 0 \le x \le 1, \quad h = 0.1$

   Exact solution is $y(x) = 3 \sinh x - 2x$.

53. $y'' + 4y = 0, \quad y(0) = 1, \quad y'(0) = -1, \quad 0 \le x \le 1, \quad h = 0.1$

   Exact solution is $y(x) = \cos 2x - \frac{1}{2}\sin 2x$.

54. $y'' + 2y' + 2y = 0, \quad y(0) = 1, \quad y'(0) = 0, \quad 0 \le x \le 1, \quad h = 0.1$

   Exact solution is $y(x) = e^{-x}(\cos x + \sin x)$.

55. Consider the nonlinear, second-order IVP

$$y'' + 2yy' = 0, \quad y(0) = 0, \quad y'(0) = 1$$

   Using Heun's method for systems, with step size $h = 0.1$, find the estimated values of $y(0.1)$ and $y(0.2)$.

56. Consider the nonlinear, second-order IVP

$$yy'' + (y')^2 = 0, \quad y(0) = 1, \quad y'(0) = -1$$

   Using RK4 method for systems, with step size $h = 0.1$, find the estimated value of $y(0.1)$.

57. Consider the linear, second-order IVP

$$x^2 y'' = 2y, \quad y(1) = 1, \quad y'(1) = 0, \quad 1 \le x \le 2, \quad h = 0.1$$

   a. Transform into a system of first-order IVPs.

   b. Write a MATLAB script file that employs the user-defined functions `EulerODESystem`, `HeunODESystem`, and `RK4System` to solve the system in (a). The file must return a table of values for y generated by the three methods, as well as the exact values, at all the mesh points `x = 1:0.1:2`. The exact solution is $y = \frac{2}{3}x^{-1} + \frac{1}{3}x^2$.

58. Consider the linear, second-order IVP

$$x^2 y'' + 5xy' + 3y = x^2 - x, \quad y(1) = 0, \quad y'(1) = 0, \quad 1 \le x \le 1.5, \quad h = 0.05$$

   a. ✍ Transform into a system of first-order IVPs.
   b. Write a MATLAB script file that employs the user-defined functions `EulerODESystem`, `HeunODESystem`, and `RK4System` to solve the system in (a). The file must return a table of values for y generated by the three methods, as well as the exact values, at all the mesh points `x = 1:0.05:1.5`. The exact solution is

$$y = \frac{8x^3 - 15x^2 + 10}{120x} - \frac{1}{40x^3}$$

59. Consider the mechanical system in translational motion shown in the following figure, where m is mass, k_1 and k_2 are stiffness coefficients, c_1 and c_2 are the coefficients of viscous damping, x_1 and x_2 are displacements, and $F(t)$ is the applied force. Assume, in consistent physical units, the following parameter values

$$m = 1, \quad c_1 = 1, \quad c_2 = 1, \quad k_1 = 2, \quad k_2 = \frac{1}{2}, \quad F(t) = 10e^{-t}\sin t$$

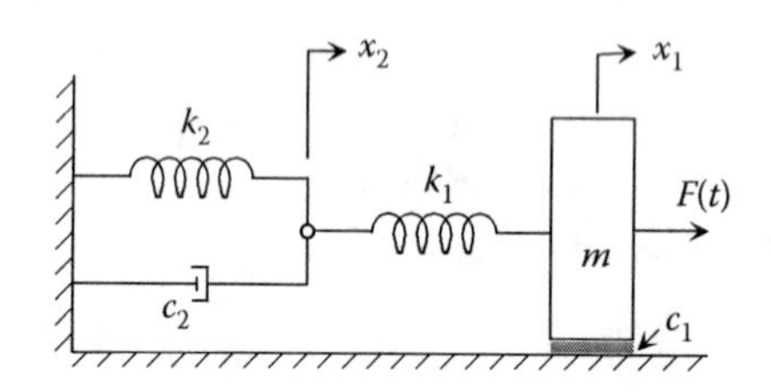

The system's equations of motion are then expressed as

$$\ddot{x}_1 + \dot{x}_1 + 2(x_1 - x_2) = 10e^{-t}\sin t$$

$$\dot{x}_2 + \frac{1}{2}x_2 - 2(x_1 - x_2) = 0$$

subject to initial conditions

$$x_1(0) = 0, \; x_2(0) = 0, \; \dot{x}_1(0) = 1$$

a. ✍ Transform into a system of first-order IVPs.

b. Write a MATLAB script file that employs the user-defined function `RK4System` to solve the system in (a). The file must return the plot of x_1 and x_2 versus $0 \le t \le 5$ in the same graph. It is recommended to use at least 100 points for smoothness of curves.

60. In the mechanical system shown in the following figure, m is mass, c is the coefficient of viscous damping, $f_s = x^3$ is the nonlinear spring force, x is displacement, and $F(t)$ is the applied force. Assume, in consistent physical units, the following parameter values

$$m = 1, \quad c = 0.6, \quad F(t) = 100e^{-t/3}$$

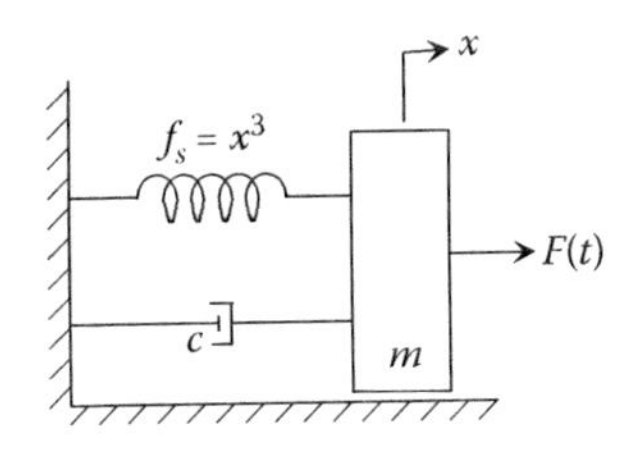

The system's equation of motion is expressed as

$$\ddot{x} + 0.6\dot{x} + x^3 = 100e^{-t/3} \quad \text{subject to} \quad x(0) = 0, \ \dot{x}(0) = 1$$

a. ✍ Transform into a system of first-order IVPs.

b. Write a MATLAB script file that employs the user-defined function `RK4System` to solve the system in (a). The file must return the plot of x versus $0 \le t \le 5$. At least 100 points are recommended for smoothness of curves.

61. Consider the mechanical system shown in the following figure, where m_1 and m_2 represent mass, k_1 and k_2 are stiffness coefficients, c is the coefficient of viscous damping, x_1 and x_2 are displacements, and $F_1(t)$ and $F_2(t)$ are the applied forces. Assume, in consistent physical units, the following parameter values

$$m_1 = 1, \quad m_2 = 1, \quad c = 2, \quad k_1 = 1, \quad k_2 = 1, \quad F_1(t) = \sin 2t, \quad F_2(t) = e^{-t}$$

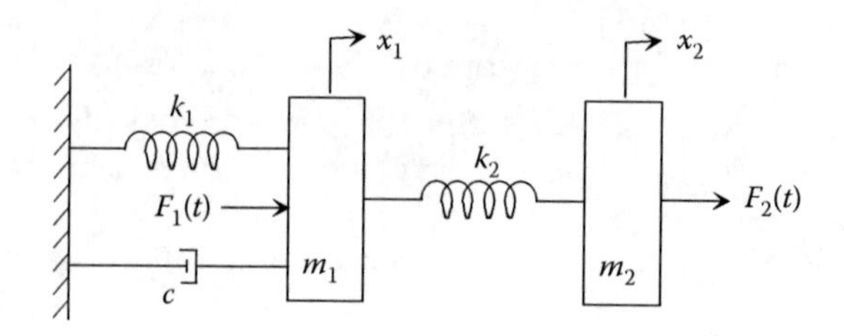

The system's equations of motion are then expressed as

$$\ddot{x}_1 + 2\dot{x}_1 + 2x_1 - x_2 = \sin 2t \qquad \text{subject to initial conditions}$$
$$\ddot{x}_2 - x_1 + x_2 = e^{-t} \qquad x_1(0) = 0,\, x_2(0) = 0,\, \dot{x}_1(0) = 0,\, \dot{x}_2(0) = 0$$

a. ✍ Transform into a system of first-order IVPs.

b. Write a MATLAB script file that employs the user-defined function `RK4System` to solve the system in (a). The file must return the plot of x_1 and x_2 versus $0 \le t \le 6$ in the same graph. It is recommended to use at least 100 points for smoothness of curves.

62. The pendulum system in the following figure consists of a uniform thin rod of length l and a concentrated mass m at its tip. The friction at the pivot causes the system to be damped. When the angular displacement θ is not very small, the system is described by a nonlinear model in the form

$$\frac{2}{3}ml^2\ddot{\theta} + 0.09\dot{\theta} + \frac{1}{2}mgl\sin\theta = 0$$

Assume, in consistent physical units, that $ml^2 = 1.28$, $g/l = 7.45$.

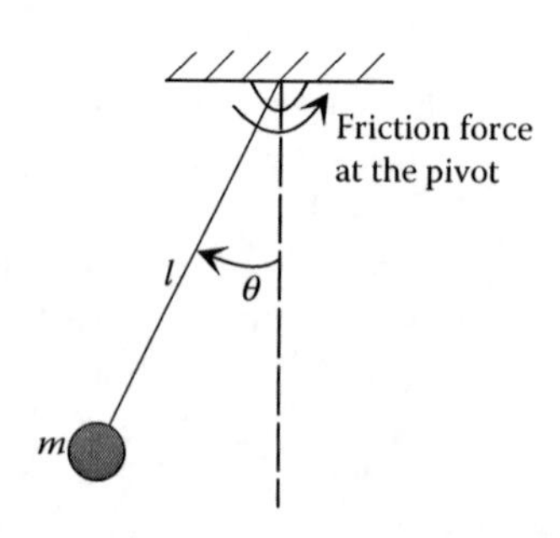

a. Transform into a system of first-order IVPs.

b. Write a MATLAB script file that utilizes the user-defined function `RK4System` to solve the system in (a). Two sets of initial conditions are to be considered: (1) $\theta(0) = 15°$, $\dot{\theta}(0) = 0$, and (2) $\theta(0) = 30°$, $\dot{\theta}(0) = 0$. The file must return the plots of the two angular displacements θ_1 and θ_2 corresponding to the two sets of initial conditions versus $0 \le t \le 5$ in the same graph. Angle measures must be converted to radians. Use at least 100 points for plotting.

63. The governing equations for an armature-controlled DC motor with a rigid shaft are derived as

$$J\frac{d\omega}{dt} + B\omega - K_t i = T_L, \qquad \omega(0) = 0$$

$$L\frac{di}{dt} + Ri + K_e\omega = v, \qquad i(0) = 0$$

where

ω = angular velocity of the rotor
i = armature current
T_L = torque load
v = armature voltage

and J, B, L, R, K_t, and K_e represent constant parameters. Suppose, in consistent physical units

$$J = 1 = L, \; B = 0.5, \; K_t = 0.25, \; K_e = 1, \; R = 1.5, \; T_L = te^{-t}, v = \sin t$$

Write a MATLAB script file that employs the user-defined function `RK4System` to solve the system of governing equations. The file must return two separate plots (use `subplot`): angular velocity ω versus $0 \le t \le 10$, and the current i versus $0 \le t \le 10$.

64. The governing equations for a field-controlled DC motor are derived as

$$J\frac{d\omega}{dt} + B\omega - K_t i = 0, \qquad \omega(0) = 0$$

$$L\frac{di}{dt} + Ri = v, \qquad i(0) = 0$$

where

ω = angular velocity of the rotor
i = armature current
v = armature voltage

and J, B, L, R, and K_t represent constant parameters. Suppose, in consistent physical units

$$J = 1 = L, \quad B = 1, \quad K_t = 0.75, \quad R = 0.4, \quad v = \cos t$$

Write a MATLAB script file that employs the user-defined function `RK4System` to solve the system of governing equations. The file must return two separate plots (use `subplot`): angular velocity ω versus $0 \le t \le 10$, and the current i versus $0 \le t \le 10$.

65. Write a MATLAB script file that solves the following second-order IVP using `EulerODESystem` with $h = 0.025$, `HeunODESystem` with $h = 0.05$, and `RK4System` with $h = 0.1$. The file must return a table that includes the estimated solutions, as well as the global percent relative error, at $x = 1{:}0.1{:}2$ for all three methods. Discuss the results. The exact solution is $y = x - 2x^{-1/2}$.

$$2x^2 y'' + xy' - y = 0, \quad y(1) = -1, \quad y'(1) = 2, \quad 1 \le x \le 2$$

66. The two-loop electrical network shown in the following figure is governed by

$$L\ddot{q}_1 + R_1\dot{q}_1 + \frac{1}{C}(q_1 - q_2) = v_1(t)$$
$$R_2\dot{q}_2 - \frac{1}{C}(q_1 - q_2) = v_2(t)$$

subject to $q_1(0) = 0,\ q_2(0) = 0,\ \dot{q}_1(0) = 0$

where q_1 and q_2 are electric charges, L is inductance, R_1 and R_2 are resistances, C is capacitance, and $v_1(t)$ and $v_2(t)$ are the applied voltages. The electric charge and current are related through $i = dq/dt$. Assume, in consistent physical units, that the physical parameter values are

$$L = 0.1, \quad R_1 = 1, \quad R_2 = 1, \quad C_1 = 0.4, \quad v_1(t) = \sin t, \quad v_2(t) = \sin 2t$$

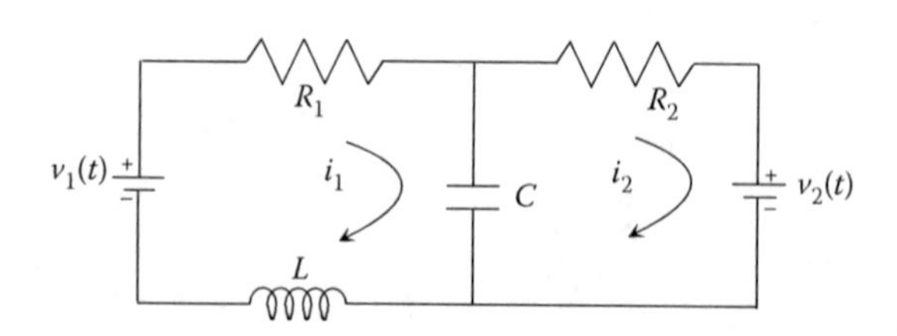

a. ✍ Transform into a system of first-order IVPs.

b. Write a MATLAB script file that utilizes the user-defined function `RK4System` to solve the system in (a). The file must return the plot of q_1 and q_2 versus $0 \le t \le 5$ in the same graph. At least 100 points are recommended for plotting.

67. Write a MATLAB script that employs the user-defined function `RK4System` with $h = 0.1$ to solve the following system of first-order IVPs. The file must return the plots of x_2 and x_3 versus $0 \le t \le 5$ in the same graph.

$$\begin{aligned} \dot{x}_1 &= x_2 - (t+7) \\ \dot{x}_2 &= -x_1 + 1 - 3t \\ \dot{x}_3 &= x_1 - x_3 + 3t + \sin t \end{aligned} \qquad \text{subject to} \qquad \begin{aligned} x_1(0) &= 0 \\ x_2(0) &= 5 \\ x_3(0) &= -1 \end{aligned} \qquad \text{and} \quad 0 \le t \le 5$$

68. An inverted pendulum of length L and mass m, mounted on a motor-driven cart of mass M is shown in the following figure, where x is the linear displacement of the cart, φ is the angular displacement of the pendulum from the vertical, and μ is the force applied to the cart by the motor. The equations of motion are derived as

$$\begin{aligned} \ddot{x} &= \mu - \dot{x} \\ \ddot{\varphi} &= \frac{g}{\bar{L}}\sin\varphi - \frac{1}{\bar{L}}\ddot{x}\cos\varphi \end{aligned}, \qquad M = 1, \quad \bar{L} = \frac{J + mL^2}{mL} = 0.85, \quad g = 9.81$$

where J is the moment of inertia of the rod.

a. ✍ By choosing the state variables $u_1 = x$, $u_2 = \dot{x}$, $u_3 = x + \bar{L}\varphi$, $u_4 = \dot{x} + \bar{L}\dot{\varphi}$, obtain the state-variable equations. Then insert the following into the state-variable equations:

$$\mu = -90u_1 - 10u_2 + 120u_3 + 30u_4$$

b. Write a MATLAB script file that utilizes the user-defined function `RK4System` to solve the system in (a). Three sets of initial conditions are to be considered: $\varphi(0) = 10°, 20°, 30°$, while $x(0) = 0 = \dot{x}(0) = \dot{\varphi}(0)$ in all three cases. The file must return the plots of the three angular displacements corresponding to the three sets of initial conditions versus $0 \le t \le 1.5$ in the same graph. Angle measures must be converted to radians. Use at least 100 points for plotting.

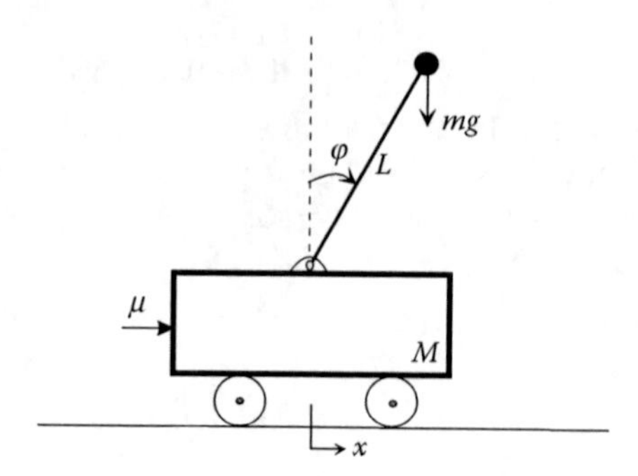

69. Write a user-defined function with function call `u=ABM_3System(f,x,u0)` that solves a system of first-order IVPs

$$\mathbf{u}' = \mathbf{f}(x,\mathbf{u}), \quad \mathbf{u}(x_0) = \mathbf{u}_0, \quad a = x_0 \le x \le x_n = b$$

using the third-order ABM predictor–corrector method.

70. Write a MATLAB script file that employs the user-defined function `ABM_3System` (see Problem 69) to solve the following IVP. The file must return a table of values for y generated by the method, as well as the exact values at all the mesh points `x=1:0.05:1.5`.

$$x^2 y'' + 5xy' + 3y = x^2 - x, \quad y(1) = 0, \quad y'(1) = 0, \quad 1 \le x \le 1.5, \quad h = 0.05$$

The exact solution is

$$y = \frac{8x^3 - 15x^2 + 10}{120x} - \frac{1}{40x^3}$$

Stability (Section 7.6)

71. The second-order Adams–Bashforth (AB2) method can be shown to have a stability region described by $0 < \lambda h < 1$ when applied to $y' = -\lambda y$ (λ = constant > 0) subject to an initial condition. Solve the following IVP using AB2 with step size $h = 0.2$ and again with $h = 0.55$, plot the estimated solutions together with the exact solution $y(x) = e^{-2x}$, and discuss stability. Use 100 points for plotting.

$$y' + 2y = 0, \quad y(0) = 1, \quad 0 \le x \le 5$$

72. Consider the non-self-starting method

$$y_{i+1} = y_{i-1} + 2hf(x_i, y_i), \quad i = 1, 2, \ldots, n-1$$

for solving $y' = f(x,y)$ subject to initial condition y_0. A method such as RK4 can be used to start the iterations. Investigate the stability of this method as follows. Apply the method to $y' = -3y$, $y(0) = 1$ with $h = 0.1$ and plot over the range $0 \le x \le 2$. Repeat with a substantially reduced step size $h = 0.02$ and plot over the range $0 \le x \le 4$. Fully discuss the findings.

Stiff Differential Equations (Section 7.7)

73. Consider the IVP

$$\dot{y} + 100y = (99t + 1)e^{-t}, \quad y(0) = 1$$

The exact solution is $y = e^{-100t} + te^{-t}$ so that the first term quickly becomes negligible relative to the second term, but continues to govern stability. Apply Euler's method with $h = 0.1$ and plot the solution estimates versus $0 \le t \le 1$. Repeat with $h = 0.01$ and plot versus $0 \le t \le 5$. Fully discuss the results as related to the stiffness of the differential equation.

74. Consider the IVP

$$\dot{y} + 250y = 250 \sin t + \cos t, \quad y(0) = 1$$

The exact solution is $y = e^{-250t} + \sin t$ so that the first term quickly becomes negligible relative to the second term, but will continue to govern stability. Apply Euler's method with $h = 0.1$ and plot the solution estimates versus $0 \le t \le 1$. Repeat with $h = 0.01$ and again with $h = 0.001$ and plot these two sets versus $0 \le t \le 5$. Fully discuss the results as related to the stiffness of the differential equation.

MATLAB® Built-In Functions for Initial-Value Problems (Section 7.8)

75. Write a MATLAB script file that solves the following IVP using `ode45` and returns a table that includes the solution estimates at `x = 0:0.1:1`.

$$(1 - x^2)y'' - xy' + 4y = 0, \quad y(0) = 1, \quad y'(0) = 1$$

76. A device that plays an important role in the study of nonlinear vibrations is a van der Pol oscillator,* a system with a damping mechanism. When the amplitude of the motion is small, it acts to

* A van der Pol oscillator is an electrical circuit, which consists of two dc power sources, a capacitor, resistors, inductors, and a triode composed of a cathode, an anode, and a grid controlling the electron flow.

increase the energy. And, for large motion amplitude, it decreases the energy. The governing equation for the oscillator is

$$\ddot{y} - \mu(1 - y^2)\dot{y} + y = 0, \quad \mu = \text{constant} > 0$$

Write a MATLAB script file that solves the van der Pol equation with $\mu = 0.5$ and initial conditions $y(0) = 0.1$ and $\dot{y}(0) = 0$ using `ode23` and `ode45` and returns a table that includes the solution estimates for $y(t)$ at `t = 0:0.1:1` at each point.

77. The motion of an object is described by

$$\begin{aligned} \ddot{x} &= -0.006v\dot{x} \\ \ddot{y} &= -0.006v\dot{y} - 9.81 \end{aligned} \quad \text{subject to} \quad \underbrace{\begin{aligned} x(0) &= 0 \\ y(0) &= 0 \end{aligned}}_{\text{initial positions}} \quad \underbrace{\begin{aligned} \dot{x}(0) &= 30 \\ \dot{y}(0) &= 25 \end{aligned}}_{\text{initial velocities}}$$

where 9.81 represents the gravitational acceleration, and $v = \sqrt{\dot{x}^2 + \dot{y}^2}$ is the speed of the object. Initial positions of zero indicate that the object is placed at the origin of the xy-plane at the initial time. Determine the positions x and y at `t = 0:0.1:1` using `ode45`.

78. Consider

$$\begin{aligned} \dot{x} &= 10(y - x) \\ \dot{y} &= 15x - xz - y \\ \dot{z} &= xy - 3z \end{aligned} \quad \text{subject to} \quad \begin{aligned} x(0) &= 1 \\ y(0) &= -1 \\ z(0) &= 1 \end{aligned}$$

Solve the system using any MATLAB solver and plot each variable versus $0 \le t \le 10$. What are the steady-state values of the three dependent variables?

79. Consider the double-pendulum system shown in the following figure consisting of two identical rods and bobs attached to them, coupled with a linear spring. The system's free motion is described by

$$\begin{aligned} mL^2\ddot{\theta}_1 + (mgL + kl^2)\theta_1 - kl^2\theta_2 &= 0 \\ mL^2\ddot{\theta}_2 - kl^2\theta_1 + (mgL + kl^2)\theta_2 &= 0 \end{aligned} \quad \text{subject to} \quad \underbrace{\begin{aligned} \theta_1(0) &= 0 \\ \theta_2(0) &= 0 \end{aligned}}_{\text{initial angular positions}} \quad \underbrace{\begin{aligned} \dot{\theta}_1(0) &= -1 \\ \dot{\theta}_2(0) &= 0 \end{aligned}}_{\text{initial angular velocities}}$$

Assume

$$\frac{g}{L} = 8, \quad \frac{k}{m}\left(\frac{l}{L}\right)^2 = 2$$

Solve using `ode45`, and plot the angular displacements θ_1 and θ_2 versus $0 \le t \le 2$, and angular velocities $\dot{\theta}_1$ and $\dot{\theta}_2$ versus $0 \le t \le 2$, each pair in one figure.

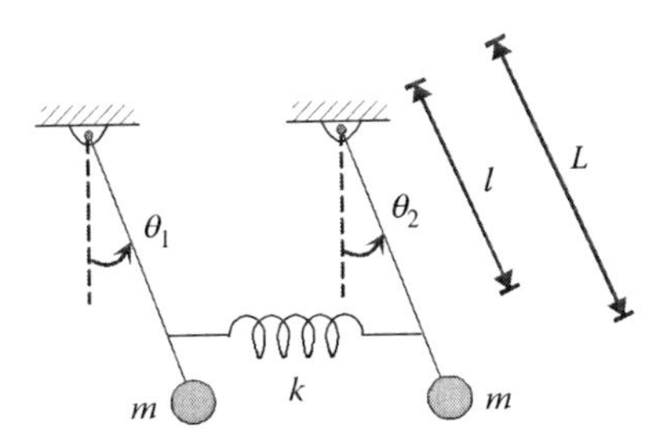

80. The mechanical system in the following figure undergoes translational and rotational motions, described by

$$mL^2\ddot{\theta} + mL\ddot{x} + mgL\theta = 0$$
$$mL\ddot{\theta} + (m + M)\ddot{x} + c\dot{x} + kx = F(t)$$

subject to

$\theta(0) = 0$	$\dot{\theta}(0) = 0$
$x(0) = 0$	$\dot{x}(0) = -1$
initial positions	initial velocities

where the applied force is $F(t) = e^{-t/2}$. Assume, in consistent physical units, the following parameter values:

$$m = 0.4, \quad M = 1, \quad L = 0.5, \quad g = 9.81, \quad k = 1, \quad c = 1$$

Solve using `ode45`, and plot the angular displacement θ and linear displacement x versus $0 \le t \le 10$, in two separate figures. Determine the steady-state values of θ and x.

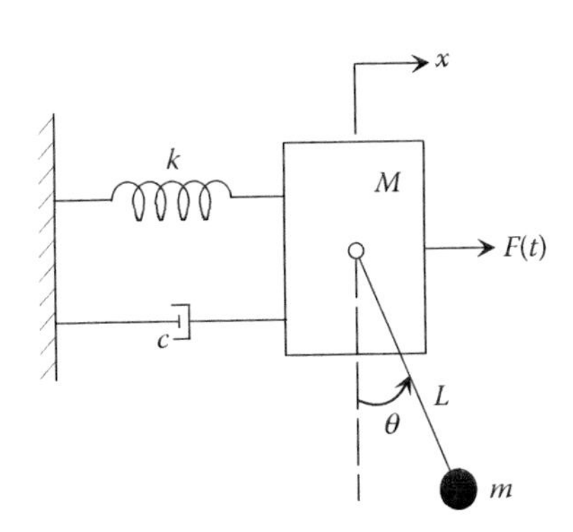

8

Numerical Solution of Boundary-Value Problems

In Chapter 7, we learned that an initial-value problem refers to the situation when an nth-order differential equation is accompanied by n initial conditions, specified at the same value of the independent variable. We also mentioned that in other applications, these auxiliary conditions may be specified at different values of the independent variable, usually at the extremities of the system, and the problem is known as a boundary-value problem (BVP).

BVPs can be solved numerically by either using the shooting method or the finite-difference method. The shooting method is based on making an initial-value problem (IVP) out of the BVP by guessing at the initial condition(s) that are obviously absent in the description of a BVP, solving the IVP just created, and testing to see if the ensuing solution satisfies the original boundary conditions. Therefore, the shooting method relies on techniques such as the fourth-order Runge–Kutta method (RK4, Chapter 7) for solving IVPs. The finite-difference method is based on dividing the system interval into several subintervals and replacing the derivatives by the finite-difference approximations that were discussed in Chapter 6. As a result, a system of algebraic equations will be generated and solved using the techniques of Chapter 4.

Second-order BVP: Consider a second-order differential equation in its most general form

$$y'' = f(x, y, y'), \quad a \le x \le b$$

subject to two boundary conditions, normally specified at the endpoints a and b. Because a boundary condition can be a given value of y or a value of y', different forms of boundary conditions may be encountered.

Boundary conditions: The most common boundary conditions are as follows:

Dirichlet boundary conditions (values of y at the endpoints are given):

$$y(a) = y_a, \quad y(b) = y_b$$

Neumann boundary conditions (values of y' at the endpoints are given):

$$y'(a) = y'_a, \quad y'(b) = y'_b$$

Mixed boundary conditions:

$$c_1 y'(a) + c_2 y(a) = B_a, \quad c_3 y'(b) + c_4 y(b) = B_b$$

Higher-order BVP: BVPs can be based on differential equations with orders higher than two, which require additional boundary conditions. As an example, in the analysis of free transverse vibrations of a uniform beam of length L, simply supported (pinned) at both ends, we encounter a fourth-order ODE

$$\frac{d^4 X}{dx^4} - \gamma^4 X = 0 \quad (\gamma = \text{constant} > 0)$$

subject to

$$X(0) = 0, \quad X(L) = 0$$
$$X''(0) = 0, \quad X''(L) = 0$$

8.1 Shooting Method

A BVP involving an nth-order ODE comes with n boundary conditions. Using state variables, the nth-order ODE is readily transformed into a system of first-order ODEs. Solving this new system requires exactly n initial conditions. The boundary condition given at the left end of the interval also serves as an initial condition at that point, while the rest of the initial conditions must be guessed. The system is then solved numerically via RK4 and the value of the resulting solution at the other endpoint is compared with the given boundary condition there. If the accuracy is not acceptable, the initial values are guessed again and the ensuing system is solved one more time. The procedure is repeated until the solution at that end agrees with the prescribed boundary condition there.

Example 8.1: Linear BVP

Solve the BVP

$$\ddot{u} = 0.02u + 1, \quad u(0) = 30, \quad u(10) = 120 \qquad \dot{u} = du/dt$$

SOLUTION

Since the ODE is of the second order, there are two state variables, selected as $x_1 = u$ and $x_2 = \dot{u}$. Subsequently, the state-variable equations are

$$\begin{cases} \dot{x}_1 = x_2 \\ \dot{x}_2 = 0.02x_1 + 1 \end{cases}$$

This system could then be solved numerically via RK4, but that would require initial conditions $x_1(0)$ and $x_2(0)$. Of these two, only $x_1(0) = u(0) = 30$ is available. For the other one, we make a first estimate (or guess)

$$x_2(0) = \dot{u}(0) = 10$$

So, the problem to solve is

$$\begin{cases} \dot{x}_1 = x_2 \\ \dot{x}_2 = 0.02x_1 + 1 \end{cases}, \quad \begin{matrix} x_1(0) = 30 \\ x_2(0) = 10 \end{matrix} \tag{8.1}$$

```
>> f = inline('[x(2,1);0.02*x(1,1)+1]','t','x');
>> x0 = [30;10];
>> t = linspace(0,10,20);
>> xfirst = RK4system(f, t, x0);
>> xfirst(1,end)

ans =

  261.0845             % u(10) = 261.0845
```

The result does not agree with the given boundary value $u(10) = 120$. Therefore, we make a second guess, say

$$x_2(0) = \dot{u}(0) = -1$$

and solve the ensuing system (similar to Equation 8.1 except for $x_2(0) = -1$):

```
>> x0 = [30;-1];
>> xsecond = RK4system(f, t, x0);
>> xsecond(1,end)

ans =

  110.5716          % u(10) = 110.5716
```

In summary:

$$\begin{matrix} x_2(0) = 10 & u(10) = 261.0845 \\ x_2(0) = -1 & u(10) = 110.5716 \end{matrix} \tag{8.2}$$

Note that one of the resulting values for $u(10)$ must be above the targeted 120, and the other must be below 120. Because the original ODE is linear, these values are linearly related as well. As a result, a linear

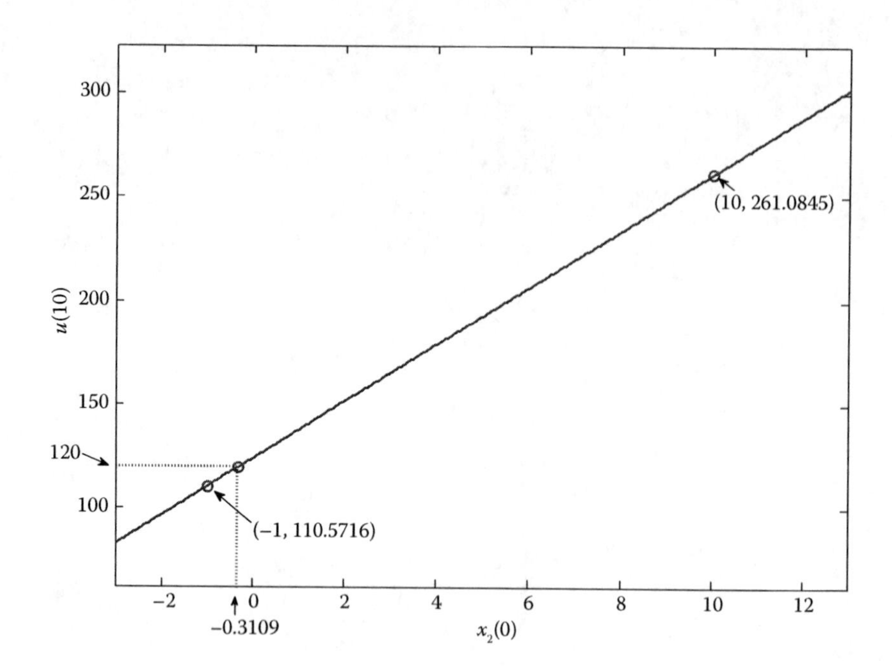

FIGURE 8.1
Linear interpolation of the data in Equation 8.2.

interpolation (Figure 8.1) of this data will provide a value for $x_2(0)$ that will result in the correct solution.

$$u(10) - 110.5716 = \frac{261.0845 - 110.5716}{10-(-1)}[x_2(0)+1]$$

$$\underset{\text{when } u(10)=120}{\overset{\text{Solve}}{\Rightarrow}} \quad x_2(0) = -0.3109$$

Therefore, the solution to the original BVP can now be obtained by solving the following initial-value problem:

$$\begin{cases} \dot{x}_1 = x_2 \\ \dot{x}_2 = 0.02x_1 + 1 \end{cases}, \quad \begin{matrix} x_1(0) = 30 \\ x_2(0) = -0.3109 \end{matrix}$$

This fact is readily confirmed in MATLAB® as follows:

```
>> x0 = [30; -0.3109];
>> xthird = RK4system(f, t, x0);
>> xthird(1,end)

ans =

  120.0006
```

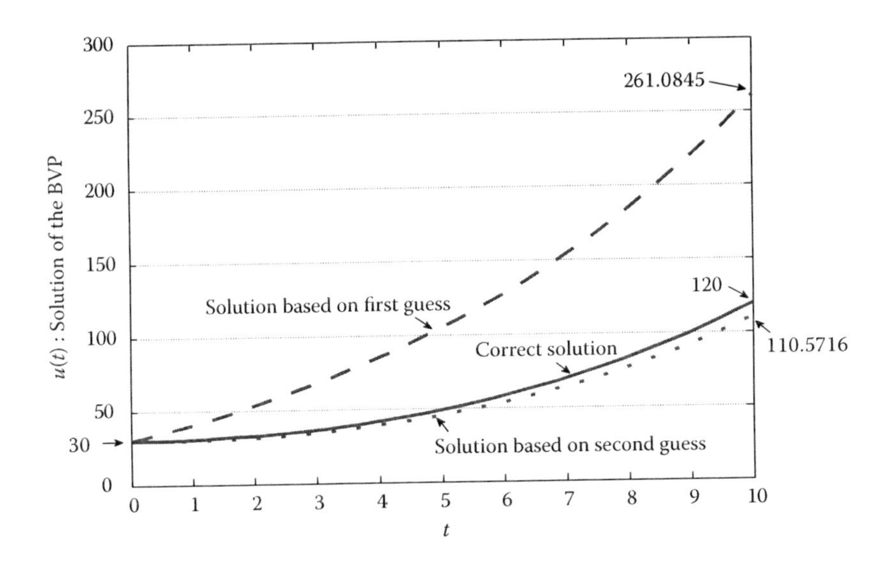

FIGURE 8.2
Solution of the BVP in Example 8.1.

As expected, the correct result is achieved and no further action is needed. The solutions obtained based on the first two initial-condition estimates, as well as the correct solution of the BVP are generated using the code below and are depicted in Figure 8.2.

```
>> x1first=xfirst(1,:);
>> x1second=xsecond(1,:);
>> x1third=xthird(1,:);
>> plot(t,x1first,t,x1second,t,x1third)     % Figure 8.2
```

The correct solution obtained here closely matches the actual solution. We will do this below while suppressing the plot.

```
>> u_ex=dsolve('D2u-0.02*u=1','u(0)=30','u(10)=120');
>> ezplot(u_ex,[0 10])     % Plot not shown here!
>> hold on
>> plot(t,x1third)              % Plot not shown here!
```

A linear BVP such as in Example 8.1 is relatively easy to solve using the shooting method because the data generated by two initial-value estimates can be interpolated linearly leading to the correct estimate. This, however, will not necessarily be enough in the case of a nonlinear BVP. One remedy would be to apply the shooting method three times and then use a quadratic interpolating polynomial to estimate the initial value. But it is not very likely that this approach would generate the correct solution, and further iterations would probably be required. Another option is to use the *bisection method* (Chapter 3) to obtain the new estimate for the unknown initial value and to repeat until a desired accuracy is achieved. The following example illustrates this approach.

Example 8.2: Nonlinear BVP

The temperature distribution along a fin can be modeled as a BVP

$$T_{xx} - \beta T^4 - \alpha T + \gamma = 0, \quad T(0) = 500, \quad T(0.15) = 300, \quad T_x = dT/dx$$

where $\alpha = 20$, $\beta = 10^{-8}$, and $\gamma = 5 \times 10^3$ with all parameters in consistent physical units. Find $T(x)$, $0 \le x \le 0.15$.

SOLUTION

Choosing the state variables $\eta_1 = T$ and $\eta_2 = T_x$, we have

$$\begin{cases} \eta_1' = \eta_2 \\ \eta_2' = \alpha\eta_1 + \beta\eta_1^4 - \gamma \end{cases}$$

Of the two required initial conditions, only $\eta_1(0) = 500$ is available. For the other, we use the estimate

$$\eta_2(0) = T_x(0) = -3000$$

so that the system to solve is

$$\begin{cases} \eta_1' = \eta_2 \\ \eta_2' = \alpha\eta_1 + \beta\eta_1^4 - \gamma \end{cases}, \quad \begin{matrix} \eta_1(0) = 500 \\ \eta_2(0) = -3000 \end{matrix}$$

```
>> f = inline('[eta(2,1);20*eta(1,1)+1e-8*eta(1,1)^4-5e3]',
'x','eta');
>> x = linspace(0,0.15,20);
>> eta0 = [500;-3000];
>> eta_first = RK4system(f,x,eta0);
>> eta1first = eta_first(1,end)

eta1first =

   76.6313          % Lower than the target 300
```

As a second estimate, we pick $\eta_2(0) = T_x(0) = -500$ and solve the system:

```
>> eta0 = [500;-500];
>> eta_second = RK4system(f,x,eta0);
>> eta1second = eta_second(1,end)
```

```
eta1second =

  484.1188      % Higher than the target 300
```

Note that the resulting values are such that one is higher than the targeted 300, while the other is lower. From this point onward, we use the bisection to generate the subsequent estimates until a prescribed tolerance is met.

```
eta20L=-3000; eta20R=-500;
% Left and right values for initial-value estimate
kmax=20;          % Maximum number of iterations
tol=1e-3;         % Set tolerance
T_right=300;
% Actual boundary condition (target) on the right end

for k=1:kmax,
eta2=(eta20L+eta20R)/2;       % Bisection
eta0=[500;eta2];              % Set initial state vector
eta=RK4system(f,x,eta0);      % Solve the system
T(k)=eta(1,end);              % Extract T(0.15)
err=T(k) - T_right;           % Compare with target

% Adjust the left and/or right value of initial value based
% on whether error is positive or negative

if abs(err)<tol
break
end
if err>0
eta20R=eta2;
else
eta20L=eta2;
end
end

>> k

k =

   17         % Number of iterations needed to meet tolerance

>> T(17)

ans =

   300.0006  % Agrees with target T(0.15)=300

>> eta2

eta2 =

-1.6260e+003 % Final estimate for the missing initial value
```

In summary, the set of initial conditions that led to the correct solution are $T(0) = 300$, $T_x(0) = -1626$. It took 17 applications of bisection to arrive

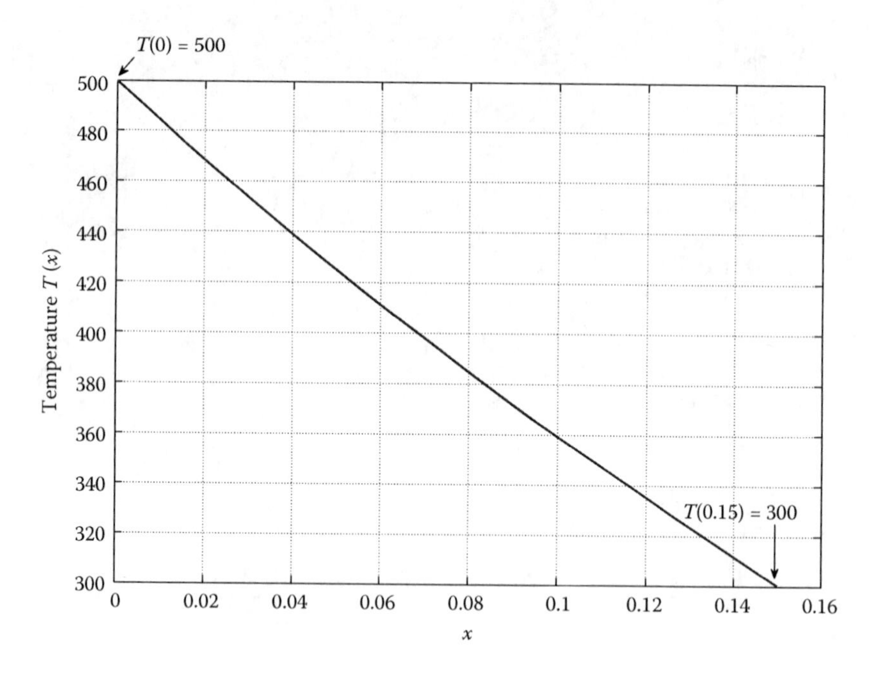

FIGURE 8.3
Temperature distribution in Example 8.2.

at the (missing) initial condition. The solution T can then be found and plotted versus x as follows:

```
>> eta0 = [500;eta2];
>> eta = RK4system(f,x,eta0);
>> eta1 = eta(1,:);
>> plot(x,eta1)          % Figure 8.3
```

The shooting method loses its efficiency when applied to higher-order BVPs, which will require at least two guesses (estimates) for the initial values. For those cases, other techniques, such as the finite-difference method, need to be employed.

8.2 Finite-Difference Method

The finite-difference method is the most commonly used alternative to the shooting method. The interval $[a,b]$ over which the differential equation is to be solved is first divided into N subintervals of length $h = (b - a)/N$. As a result, a total of $N + 1$ grid points are generated, including $x_1 = a$ and $x_{N+1} = b$, which are the left and right endpoints. The other $N - 1$ points $x_2, \ldots, x_N$ are the interior grid points. At each interior grid point, the derivatives involved in the differential equation are replaced with finite divided differences. This way, the differential equation is transformed into a system of $N - 1$ algebraic equations that can then be solved using the methods previously discussed in Chapter 4.

Because of their accuracy, central-difference formulas are often used in finite-difference methods. Specifically, for the first and second derivatives of y with respect to x

$$\left.\frac{dy}{dx}\right|_{x_i} \cong \frac{y_{i+1} - y_{i-1}}{2h}, \quad \left.\frac{d^2y}{dx^2}\right|_{x_i} \cong \frac{y_{i-1} - 2y_i + y_{i+1}}{h^2} \tag{8.3}$$

Several difference formulas are listed in Table 6.3 in Section 6.2.

Example 8.3: Finite-Difference Method: Linear BVP

Consider the BVP in Example 8.1:

$$\ddot{u} = 0.02u + 1, \quad u(0) = 30, \quad u(10) = 120$$

Solve by the finite-difference method using central-difference formulas and $h = \Delta t = 2$.

SOLUTION

Replace the second derivative in the differential equation with a central-difference formula to obtain

$$\frac{u_{i-1} - 2u_i + u_{i+1}}{\Delta t^2} = 0.02u_i + 1$$

This equation is applied at the interior grid points. Since the interval length is 10 and $\Delta t = 2$, we have $N = 10/2 = 5$, which means there are $N - 1 = 4$ interior points. Simplifying the above equation, we find

$$u_{i-1} - \left(2 + 0.02\Delta t^2\right)u_i + u_{i+1} = \Delta t^2, \quad i = 2,3,4,5 \tag{8.4}$$

Note that $u_1 = 30$ and $u_6 = 120$ are available from the boundary conditions. Applying Equation 8.4 yields

$$\begin{aligned} u_1 - 2.08u_2 + u_3 &= 4 \\ u_2 - 2.08u_3 + u_4 &= 4 \\ u_3 - 2.08u_4 + u_5 &= 4 \\ u_4 - 2.08u_5 + u_6 &= 4 \end{aligned} \quad \overset{u_1=30}{\underset{u_6=120}{\Rightarrow}} \quad \begin{bmatrix} -2.08 & 1 & & \\ 1 & -2.08 & 1 & \\ & 1 & -2.08 & 1 \\ & & 1 & -2.08 \end{bmatrix} \begin{Bmatrix} u_2 \\ u_3 \\ u_4 \\ u_5 \end{Bmatrix} = \begin{Bmatrix} -26 \\ 4 \\ 4 \\ -116 \end{Bmatrix}$$

TABLE 8.1

Comparison of Results: Shooting Method, Finite Difference, and Actual (Example 8.3)

t	Shooting Method	Finite Difference	Actual
0	30	30	30
2	32.5912	32.6757	32.5912
4	41.8333	41.9654	41.8338
6	58.4703	58.6123	58.4722
8	83.8422	83.9482	83.8464
10	120	120	120

As projected by the structure of Equation 8.4, the resulting coefficient matrix is tridiagonal, and thus will be solved using Thomas method (Section 4.3). That yields

$$u_2 = 32.6757, \quad u_3 = 41.9654, \quad u_4 = 58.6123, \quad u_5 = 83.9482$$

Comparison with the Shooting Method

Using the results of Example 8.1, adjusting the step size to 2, the numerical values at the interior grids are calculated and shown in Table 8.1. Also included in Table 8.1 are the actual values provided by the actual solution

$$u(t) = 41.0994e^{-0.1414t} + 38.9006e^{0.1414t} - 50$$

The accuracy of both techniques can be improved by reducing the step size Δt. While both methods yield acceptable results for the current example, which happens to be linear and second order, the finite-difference method is generally preferred because it enables us to treat higher-order and nonlinear systems.

Example 8.4: Finite-Difference Method: Nonlinear BVP

Consider the BVP in Example 8.2:

$$T_{xx} - \beta T^4 - \alpha T + \gamma = 0, \quad T(0) = 500, \quad T(0.15) = 300$$

where $\alpha = 20$, $\beta = 10^{-8}$, and $\gamma = 5 \times 10^3$ with all parameters in consistent physical units. Solve by the finite-difference method using central-difference formulas and $h = \Delta x = 0.025$.

SOLUTION

The interval is 0.15 in length and $\Delta x = 0.025$; hence, $N = 0.15/0.025 = 6$, and there are 5 interior grid points. Replace the second derivative in the differential equation with a central-difference formula to obtain

$$\frac{T_{i-1} - 2T_i + T_{i+1}}{\Delta x^2} - \beta T_i^4 - \alpha T_i + \gamma = 0$$

This equation is applied at the five interior grid points. Simplifying the above

$$T_{i-1} - (2 + \alpha\Delta x^2)T_i + T_{i+1} - \beta\Delta x^2 T_i^4 + \gamma\Delta x^2 = 0, \quad i = 2,3,4,5,6 \tag{8.5}$$

Applying Equation 8.5, keeping in mind that $T_1 = 500$ and $T_7 = 300$, we find

$$\begin{aligned}
f_1(T_2,T_3,T_4,T_5,T_6) &= 503.1250 - 2.0125T_2 + T_3 - 6.25\times10^{-12}T_2^4 = 0\\
f_2(T_2,T_3,T_4,T_5,T_6) &= 3.1250 + T_2 - 2.0125T_3 + T_4 - 6.25\times10^{-12}T_3^4 = 0\\
f_3(T_2,T_3,T_4,T_5,T_6) &= 3.1250 + T_3 - 2.0125T_4 + T_5 - 6.25\times10^{-12}T_4^4 = 0\\
f_4(T_2,T_3,T_4,T_5,T_6) &= 3.1250 + T_4 - 2.0125T_5 + T_6 - 6.25\times10^{-12}T_5^4 = 0\\
f_5(T_2,T_3,T_4,T_5,T_6) &= 303.1250 + T_5 - 2.0125T_6 - 6.25\times10^{-12}T_6^4 = 0
\end{aligned}$$

Newton's method for systems (Chapter 4) will be used to solve this nonlinear system of algebraic equations. This means we must eventually solve the system

$$\begin{bmatrix} \dfrac{\partial f_1}{\partial T_2} & \cdot & \cdot & \cdot & \dfrac{\partial f_1}{\partial T_6} \\ \cdot & & & & \cdot \\ \cdot & & & & \cdot \\ \cdot & & & & \cdot \\ \dfrac{\partial f_5}{\partial T_2} & \cdot & \cdot & \cdot & \dfrac{\partial f_5}{\partial T_6} \end{bmatrix} \begin{Bmatrix} \Delta T_2 \\ \cdot \\ \cdot \\ \cdot \\ \Delta T_6 \end{Bmatrix} = \begin{Bmatrix} -f_1 \\ \cdot \\ \cdot \\ \cdot \\ -f_5 \end{Bmatrix} \tag{8.6}$$

The coefficient matrix in the above equation is tridiagonal and is in the form

$$\begin{bmatrix} A_{11} & 1 & & & \\ 1 & A_{22} & 1 & & \\ & 1 & A_{33} & 1 & \\ & & 1 & A_{44} & 1 \\ & & & 1 & A_{55} \end{bmatrix} \quad \text{where } A_{ii} = -2.0125 - 2.5\times10^{-11}T_{i+1}^3$$

In using Newton's method, initial values must be assigned to the unknown variables. Suppose the initial value for each of the five unknowns is 400. The MATLAB code below will generate the solution iteratively.

```
kmax=10; b=zeros(5,1); A=zeros(5,5);

for i=1:4,
    A(i,i+1)=1; A(i+1,i)=1;
end

f1=inline('503.1250-2.0125*T2+T3-6.25e-12*T2^4',
'T2','T3','T4','T5','T6');
f2=inline('3.1250+T2-2.0125*T3+T4-6.25e-12*T3^4',
'T2','T3','T4','T5','T6');
f3=inline('3.1250+T3-2.0125*T4+T5-6.25e-12*T4^4',
'T2','T3','T4','T5','T6');
f4=inline('3.1250+T4-2.0125*T5+T6-6.25e-12*T5^4',
'T2','T3','T4','T5','T6');
f5=inline('303.1250+T5-2.0125*T6-6.25e-12*T6^4',
'T2','T3','T4','T5','T6');

% Function to generate the diagonal entries of A
g=inline('-2.0125-2.5e-11*y^3','y');

T(1,:)=400*ones(5,1)';    % Assign initial values
tol=1e-3;          % Set tolerance

for k=1:kmax,
  T2=T(k,1); T3=T(k,2); T4=T(k,3); T5=T(k,4); T6=T(k,5);

for i=1:5,
A(i,i)=g(T(k,i));
% Diagonal entries of the coefficient matrix A
end

b(1)=-f1(T2,T3,T4,T5,T6);
b(2)=-f2(T2,T3,T4,T5,T6);
b(3)=-f3(T2,T3,T4,T5,T6);
b(4)=-f4(T2,T3,T4,T5,T6);
b(5)=-f5(T2,T3,T4,T5,T6);

delT=ThomasMethod(A,b); delT=delT';
% Solve by Thomas method
T(k+1,:)=T(k,:)+delT;
if norm(T(k+1,:)' - T(k,:)')<tol,    % See if tolerance is met
   break
end
end
```

Execution of this code results in

```
>> T

T =

  400.0000   400.0000   400.0000   400.0000   400.0000
  461.0446   424.9850   391.3126   359.5528   329.2576
  461.0149   424.9499   391.2755   359.5135   329.2249
  461.0149   424.9498   391.2755   359.5135   329.2249
```

TABLE 8.2

Comparison of Results: Shooting Method and Finite Difference (Example 8.4)

x	Shooting Method	Finite Difference
0	500	500
0.025	461.0054	461.0149
0.05	424.9362	424.9498
0.075	391.2618	391.2755
0.1	359.5028	359.5135
0.1250	329.2194	329.2249
0.15	300	300

Therefore, the tolerance is met after four iterations. Note that these are approximate solutions at the interior grid points. The process will be completed once the boundary points are also attached.

```
>> T_complete = [500 T(k+1,:) 300]
% Include the boundary points

500.0000  461.0149  424.9498  391.2755  359.5135  329.2249  300.0000
```

Comparison with the Shooting Method

This problem was solved in Example 8.2 via the shooting method. Using the final result in Example 8.2, adjusting the step size to 0.025, the numerical values at the interior grids are calculated and shown in Table 8.2.

8.3 BVPs with Mixed Boundary Conditions

In the case of mixed boundary conditions, information involving the derivative of the dependent function is prescribed at one or both of the endpoints of the domain of solution. In these situations, the finite-difference method can be used as seen earlier, but the resulting system of equations cannot be solved because the values of the dependent function at both endpoints are not available. This means there are more unknowns than there are equations. The additional equations are derived by using finite differences to discretize the one or two boundary conditions that involve the derivative. The combination of the equations already obtained at the interior grid points and those just generated at the endpoint(s) form a system of algebraic equations that can be solved as before.

Example 8.5: Finite-Difference Method: Linear BVP with Mixed Boundary Conditions

Consider the BVP

$$\ddot{w} + \frac{1}{t}\dot{w} + 2 = 0, \quad w(1) = 1, \quad \dot{w}(4) = -1 \qquad \dot{w} = dw/dt$$

Solve by the finite-difference method and $\Delta t = 0.5$. Use central-difference approximations for all derivatives. Compare the results with the exact solution values at the grid points. Also, confirm that the boundary condition at the right end is satisfied by the computed solution.

SOLUTION

The interval is 3 in length and $\Delta t = 0.5$; hence, $N = 3/0.5 = 6$, and there are $N - 1 = 5$ interior grid points. Replacing the first and second derivatives with central-difference formulas yields

$$\frac{w_{i-1} - 2w_i + w_{i+1}}{\Delta t^2} + \frac{1}{t_i}\frac{w_{i+1} - w_{i-1}}{2\Delta t} + 2 = 0$$

Simplify the above equation and apply at interior grid points so that

$$(2t_i - \Delta t)w_{i-1} - 4t_i w_i + (2t_i + \Delta t)w_{i+1} = -4t_i\Delta t^2, \quad i = 2,3,4,5,6$$

Consequently

$$\begin{array}{ll} i = 2 & (2t_2 - \Delta t)w_1 - 4t_2w_2 + (2t_2 + \Delta t)w_3 = -4t_2\Delta t^2 \\ i = 3 & (2t_3 - \Delta t)w_2 - 4t_3w_3 + (2t_3 + \Delta t)w_4 = -4t_3\Delta t^2 \\ i = 4 & (2t_4 - \Delta t)w_3 - 4t_4w_4 + (2t_4 + \Delta t)w_5 = -4t_4\Delta t^2 \\ i = 5 & (2t_5 - \Delta t)w_4 - 4t_5w_5 + (2t_5 + \Delta t)w_6 = -4t_5\Delta t^2 \\ i = 6 & (2t_6 - \Delta t)w_5 - 4t_6w_6 + (2t_6 + \Delta t)w_7 = -4t_6\Delta t^2 \end{array} \tag{8.7}$$

$w_1 = w(1) = 1$ is provided by the boundary condition at the left endpoint. At the right endpoint, however, $w_7 = w(4)$ is not directly available, but $\dot{w}(4)$ is. To approximate $\dot{w}$ at the right end, we will use a one-sided, three-point backward difference formula so that the values at the previous points are utilized. Note that this has second-order accuracy (see Table 6.3 in Section 6.2), which is in line with the central-difference formulas used in the earlier stages.

$$\dot{w}\Big|_{t_i} \cong \frac{w_{i-2} - 4w_{i-1} + 3w_i}{2\Delta t} \tag{8.8}$$

Applying Equation 8.8 at the right end ($i = 7$)

$$\frac{w_5 - 4w_6 + 3w_7}{2\Delta t} = -1 \overset{\text{Solve for } w_7}{\Rightarrow} w_7 = \tfrac{1}{3}(-2\Delta t + 4w_6 - w_5) \tag{8.9}$$

Substitution into Equation 8.7 for w_7, and expressing the system in matrix form, yields

$$\begin{bmatrix} -4t_2 & 2t_2 + \Delta t & 0 & 0 & 0 \\ 2t_3 - \Delta t & -4t_3 & 2t_3 + \Delta t & 0 & 0 \\ 0 & 2t_4 - \Delta t & -4t_4 & 2t_4 + \Delta t & 0 \\ 0 & 0 & 2t_5 - \Delta t & -4t_5 & 2t_5 + \Delta t \\ 0 & 0 & 0 & \frac{4}{3}(t_6 - \Delta t) & -\frac{4}{3}(t_6 - \Delta t) \end{bmatrix} \begin{Bmatrix} w_2 \\ w_3 \\ w_4 \\ w_5 \\ w_6 \end{Bmatrix}$$

$$= \begin{Bmatrix} -4t_2\Delta t^2 - (2t_2 - \Delta t)w_1 \\ -4t_3\Delta t^2 \\ -4t_4\Delta t^2 \\ -4t_5\Delta t^2 \\ -4t_6\Delta t^2 + \frac{2}{3}(2t_6 + \Delta t)\Delta t \end{Bmatrix} \tag{8.10}$$

The coefficient matrix is once again tridiagonal. This system is solved for the solution vector to obtain the approximate values at the interior grid points. The process is completed by attaching w_1, which is available, and w_7, which is found by Equation 8.9, to the vector of the interior grid values. The MATLAB code that follows performs these tasks.

```
a=1; b=4;    % Interval endpoints
dt=0.5;
N=(b-a)/dt;  % No. of interior grid points=N-1
t=a:dt:b;
w1=1;        % Given BC at the left end

A=zeros(N-1,N-1); b=zeros(N-1,1);
b(1)=-4*t(2)*dt^2-(2*t(2)-dt)*w1;
for i=1:N-2,
    A(i,i+1)=2*t(i+1)+dt;
end
for i=1:N-3,
    A(i+1,i)=2*t(i+2)-dt;
    b(i+1)=-4*t(i+2)*dt^2;
end
A(N-1,N-2)=(4/3)*(t(N)-dt);
for i=1:N-2,
    A(i,i)=-4*t(i+1);
end
A(N-1,N-1)=-(4/3)*(t(N)-dt);
b(N-1)=-4*t(N)*dt^2+(2/3)*(2*t(N)+dt)*dt;

w_interior=ThomasMethod(A,b);
w=[w1 w_interior'];
w(N+1)=(1/3)*(-2*dt+4*w(N)-w(N-1));

>> w
```

```
w =
     1.0000    5.2500    7.8571    9.4405    10.2814    10.5314    10.2814

% Exact solution

>> w_ex = dsolve('D2w+(1/t)*Dw+2=0','w(1)=1','Dw(4)=-1');
>> w_exact = vectorize(inline(char(w_ex)));
>> w_e = w_exact(t)

w_e =
     1.0000    5.2406    7.8178    9.3705    10.1833    10.4082    10.1355
```

Table 8.3 below summarizes the computed and exact values at the grid points.

To confirm that the solution obtained here satisfies the boundary condition at the right end, we run the same code with $\Delta t = 0.05$ to generate a smooth solution curve. The result is shown in Figure 8.4.

TABLE 8.3

Comparison of Exact and Computed Values at Grid Points (Example 8.5)

Solution	$t=1$	$t=1.5$	$t=2$	$t=2.5$	$t=3$	$t=3.5$	$t=4$
Exact	1.0000	5.2406	7.8178	9.3705	10.1833	10.4082	10.1355
Computed	1.0000	5.2500	7.8571	9.4405	10.2814	10.5314	10.2814

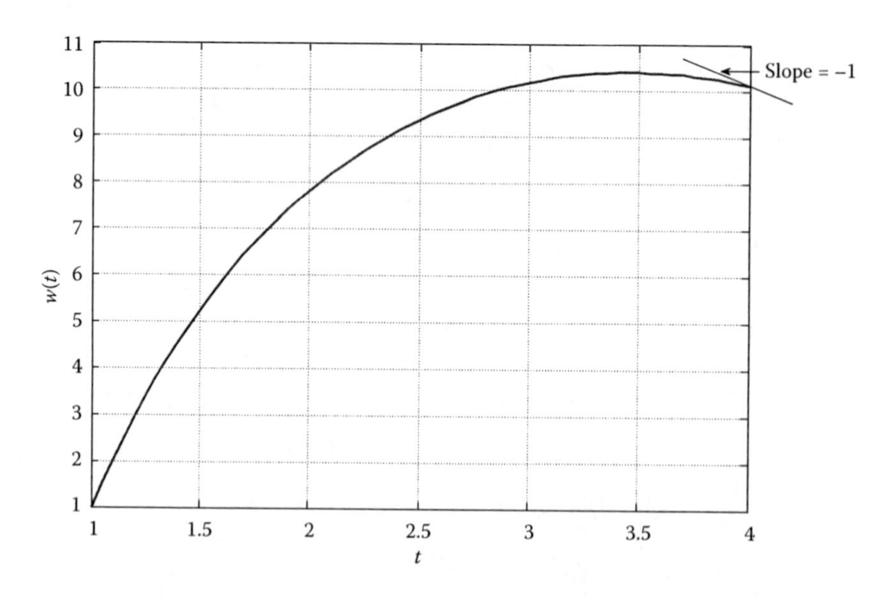

FIGURE 8.4
Approximate solution in Example 8.5.

8.4 MATLAB® Built-In Function bvp4c for BVPs

MATLAB has a built-in function, `bvp4c`, which can numerically solve two-point BVPs. We will present the simplest form of this function here. The more intricate form includes "options," and solves the same problem with default parameters replaced by user-specified values, a structure created with the `bvpset` function, similar to the `odeset` function that we used in connection with the ode solvers in Section 7.8.

To use `bvp4c`, we need to have the BVP in the proper form. This process is best illustrated when applied to a basic, second-order problem.

8.4.1 Second-Order BVP

Consider, once again, a second-order differential equation in its most general form

$$y'' = f\left(x, y, y'\right), \quad a \le x \le b$$

subject to two boundary conditions, specified at the endpoints a and b. Using state variables $y_1 = y$ and $y_2 = y'$ we find the state-variable equations in vector form as

$$\mathbf{y}' = \mathbf{f}(x, \mathbf{y}), \quad \mathbf{y} = \begin{Bmatrix} y_1 \\ y_2 \end{Bmatrix}, \quad \mathbf{f} = \begin{Bmatrix} y_2 \\ f(x, y_1, y_2) \end{Bmatrix} \tag{8.11}$$

Once the system of first-order ODEs is formed, as it is here, the function `bvp4c` may be applied.

```
bvp4c  Solve boundary value problems for ODEs by collocation.

SOL = bvp4c(ODEFUN,BCFUN,SOLINIT) integrates a system of
ordinary differential equations of the form y' = f(x,y) on the
interval [a,b], subject to general two-point boundary
conditions of the form bc(y(a),y(b)) = 0. ODEFUN and BCFUN are
function handles. For a scalar X and a column vector Y,
ODEFUN(X,Y) must return a column vector representing f(x,y).
For column vectors YA and YB, BCFUN(YA,YB) must return a
column vector representing bc(y(a),y(b)). SOLINIT is a
structure with fields
        x -- ordered nodes of the initial mesh with
             SOLINIT.x(1) = a, SOLINIT.x(end) = b
        y -- initial guess for the solution with SOLINIT.y(:,i)
             a guess for y(x(i)), the solution at the node
             SOLINIT.x(i)
```

`odefun` This is a user-defined function with function call `dydx = odefun(x,y)` where `x` is a scalar and `y` is the column vector composed of the state variables, vector **y** in Equation 8.11. The function returns `dydx`, which is the column vector **f** in Equation 8.11.

`bcfun` This is a user-defined function with function call `res = bcfun(ya,yb)`, where `ya` and `yb` are the column vectors of numerical solution estimates at **y**(a) and **y**(b). Note that

$$\texttt{ya} = \mathbf{y}(a) = \begin{Bmatrix} y_1(a) \\ y_2(a) \end{Bmatrix} = \begin{Bmatrix} y(a) \\ y'(a) \end{Bmatrix}, \quad \texttt{yb} = \mathbf{y}(b) = \begin{Bmatrix} y_1(b) \\ y_2(b) \end{Bmatrix} = \begin{Bmatrix} y(b) \\ y'(b) \end{Bmatrix}$$

Therefore, `ya(1)` and `yb(1)` represent the values of solution y at $x = a$ and $x = b$, while `ya(2)` and `yb(2)` represent the values of y' at $x = a$ and $x = b$. The function returns the so-called residual vector `res`, which is composed of the residuals, that is, the differences between the numerical solution estimates and the prescribed boundary conditions. The function `bcfun` can be used in connection with any of the boundary conditions listed at the outset of this section:

Dirichlet boundary conditions: $y(a) = y_a$, $y(b) = y_b$
In this case, the `res` vector is in the form

$$\texttt{res} = \begin{Bmatrix} \texttt{ya}(1) - y_a \\ \texttt{yb}(1) - y_b \end{Bmatrix}$$

Neumann boundary conditions: $y'(a) = y'_a$, $y'(b) = y'_b$
In this case, the `res` vector is in the form

$$\texttt{res} = \begin{Bmatrix} \texttt{ya}(2) - y'_a \\ \texttt{yb}(2) - y'_b \end{Bmatrix}$$

Mixed boundary conditions: $c_1\, y'(a) + c_2\, y(a) = B_a$, $c_3\, y'(b) + c_4\, y(b) = B_b$
In this case, the `res` vector is in the form

$$\texttt{res} = \begin{Bmatrix} \texttt{ya}(2) + \dfrac{c_2}{c_1}\texttt{ya}(1) - \dfrac{B_a}{c_1} \\ \texttt{yb}(2) + \dfrac{c_4}{c_3}\texttt{yb}(1) - \dfrac{B_b}{c_3} \end{Bmatrix}, \quad c_1, c_3 \neq 0$$

For more simple mixed boundary conditions, such as $y(a) = y_a$, $y'(b) = y'_b$, we have

$$\texttt{res} = \begin{Bmatrix} \texttt{ya}(1) - y_a \\ \texttt{yb}(2) - y'_b \end{Bmatrix}$$

`solinit` This contains the initial guess of the solution vector and is created by the MATLAB built-in function `bvpinit` with the syntax `solinit = bvpinit(x,yinit)`. The first input `x` is the vector of the initial points in the interval `[a,b]`. Normally, 10 points will suffice; hence, `x = linspace(a,b,10)`. The second input `yinit` is the vector of assigned initial guesses for the variables. In the case of the two-dimensional system in Equation 8.11, for example, `yinit` has two components: the first component is the initial guess for y, the second component is the initial guess for y'. The initial guesses can also be assigned using a user-defined function. In that case, the function call modifies to `solinit = bvpinit(x,@yinit)`, and as before, `x = linspace(a,b,10)`.

`sol` This is the output of `sol = bvp4c(@odefun,@bcfun,solinit)` and consists of two fields:

`sol.x` is the vector of the x coordinates of the interior points used for calculations by MATLAB, generally different from the user-specified values in `bvpinit`.

`Sol.y` is the matrix whose columns are the variables we are solving for. For example, in the case of the two-dimensional system in Equation 8.11, there are two variables to solve: y and y'. Then, the first column in `sol.y` represents the values for y at the interior points, and the second column consists of the values of y' at those points. The number of rows in `sol.y` is the same as the number of components in `sol.x`, namely, the points at which `bvp4c` solved the BVP.

Example 8.6: bvp4c with Dirichlet Boundary Conditions

Consider the BVP (with Dirichlet boundary conditions) in Examples 8.1 and 8.3:

$$\ddot{u} = 0.02u + 1, \quad u(0) = 30, \quad u(10) = 120$$

Replacing t by x, the ODE is written as $u'' = 0.02u + 1$. Selecting state variables $y_1 = u$ and $y_2 = u'$, we have

$$\mathbf{y}' = \mathbf{f}(x,\mathbf{y}), \quad \mathbf{y} = \begin{Bmatrix} y_1 \\ y_2 \end{Bmatrix}, \quad \mathbf{f} = \begin{Bmatrix} y_2 \\ 0.02y_1 + 1 \end{Bmatrix}, \quad 0 \le x \le 10$$

The residual vector in this problem is

$$\text{res} = \begin{Bmatrix} \text{ya}(1) - 30 \\ \text{yb}(1) - 120 \end{Bmatrix}$$

We now write a script file that solves the BVP using bvp4c and follows the instructions given above.

```
x = linspace(0,10,10);
solinit = bvpinit(x,[0, 0.1]);
% Assign initial values to y1 and y2
sol = bvp4c(@odefun8_6,@bcfun8_6,solinit);
% Solve using bvp4c
t = sol.x; % Change the name of sol.x to t
y = sol.y; % Change the name of sol.y to y
% Note that y is a 2-by-10 vector
u = y(1,:);
% First row of y contains the values of y1, which is u
udot = y(2,:);
% Second row of y gives the values of y2, which is udot
plot(t,u)      % Figure 8.5
```

```
function dydx = odefun8_6(x,y)
dydx = [y(2);0.02*y(1)+1];
```

```
function res = bcfun8_6(ya,yb)
res = [ya(1)-30;yb(1)-120];
```

Executing the script file generates the plot of the solution $u(t)$ in Figure 8.5, which clearly agrees with what we obtained in Figure 8.2, Example 8.1. It is also worth mentioning that the first element of the vector udot is

```
>> udot(1)

ans =

  -0.3109
```

This of course is precisely the value we obtained in Example 8.1, while applying the shooting method. This was the value for the missing initial condition $\dot{u}(0)$ that was needed to create an IVP whose solution agrees with the original BVP.

Example 8.7: bvp4c with Mixed Boundary Conditions

Consider the BVP (with mixed boundary conditions) in Example 8.5:

$$\ddot{w} + \frac{1}{t}\dot{w} + 2 = 0, \quad w(1) = 1, \quad \dot{w}(4) = -1$$

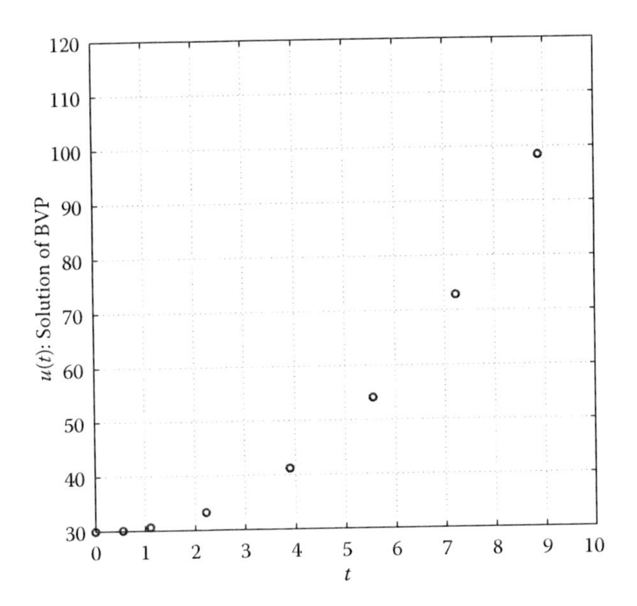

FIGURE 8.5
Solution $u(t)$ of the BVP in Example 8.6 using bvp4c.

Replacing t by x, the ODE is written as $w'' + (1/x)w' + 2 = 0$. Selecting the state variables $y_1 = w$ and $y_2 = w'$, we have

$$\mathbf{y}' = \mathbf{f}(x,\mathbf{y}), \quad \mathbf{y} = \begin{Bmatrix} y_1 \\ y_2 \end{Bmatrix}, \quad \mathbf{f} = \begin{Bmatrix} y_2 \\ -y_2/x - 2 \end{Bmatrix}, \quad 1 \le x \le 4$$

The residual vector in this problem is

$$\texttt{res} = \begin{Bmatrix} \mathrm{ya}(1) - 1 \\ \mathrm{yb}(2) + 1 \end{Bmatrix}$$

We now write a script file that solves the BVP using bvp4c and follows the instructions given above.

```
x = linspace(1,4,10);
solinit = bvpinit(x, [0.1, 0]);
% Assign initial values to y1 and y2
sol = bvp4c(@odefun8_7,@bcfun8_7,solinit); % Solve using bvp4c
t = sol.x;      % Change the name of sol.x to t
y = sol.y;      % Change the name of sol.y to y
% Note that y is a 2-by-10 vector
w = y(1,:);
% First row of y contains the values of y1, which is w
wdot = y(2,:);
```

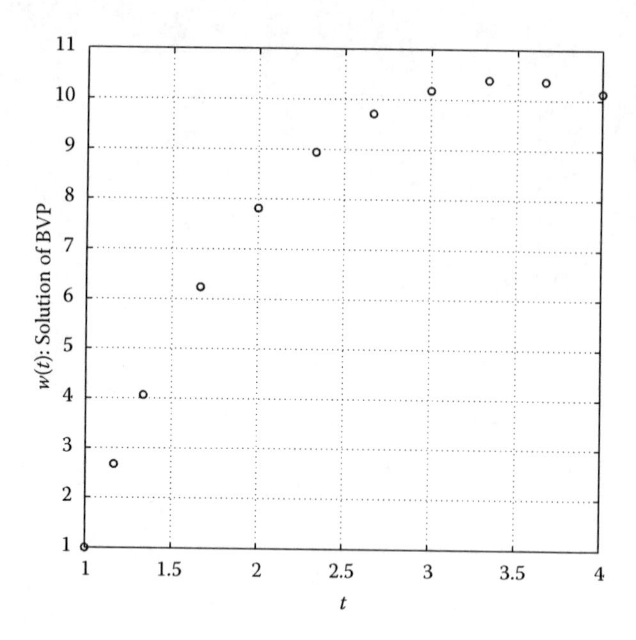

FIGURE 8.6
Solution $w(t)$ of the BVP in Example 8.7 using bvp4c.

```
% Second row of y gives the values of y2, which is wdot
plot(t,w,'o')      % Figure 8.6
```

```
function dydx=odefun8_7(x,y)
dydx=[y(2);-y(2)/x-2];
```

```
function res=bcfun8_7(ya,yb)
res=[ya(1)-1;yb(2)+1];
```

It is readily seen that the results are in agreement with those obtained in Example 8.5 using the finite-difference method.

PROBLEM SET

Shooting Method (Section 8.1)

In Problems 1 through 4, solve the linear BVP using the shooting method.

1. $\ddot{u} + 3\dot{u} = t, \quad u(0) = 1, \quad u(5) = 0.2$

2. $2\ddot{u} + 3\dot{u} + u = \sin(t/3), \quad u(0) = 0, \quad u(1) = -1$

3. $w'' + \dfrac{1}{2x}w' = 0, \quad w(1) = 3, \quad w(3) = 1$

4. $(2t - 1)\ddot{w} - 4t\dot{w} + 4w = 0, \quad w(1) = -1, \quad w(2) = 1$

In Problems 5 through 10, solve the nonlinear BVP using the shooting method combined with the bisection method. Use the specified tolerance and maximum of 20 iterations. Plot the solution trajectory.

5. $u\ddot{u} + \dot{u}^2 = 0, \quad u(0) = 2, \quad u(2) = 3$

   Tolerance: $\varepsilon = 10^{-6}$

6. $\ddot{u} + u^2\dot{u} = t, \quad u(0) = 1, \quad u(3) = 2$

   Tolerance: $\varepsilon = 10^{-6}$

7. $2\ddot{w} - e^{w} = 0, \quad w(0) = 1, \quad w(2) = 0$

   Tolerance: $\varepsilon = 10^{-4}$

8. $\ddot{y} + \dot{y} - 2y^2 = 0, \quad y(0) = 1, \quad y(3) = 2$

   Tolerance: $\varepsilon = 10^{-4}$

9. $\ddot{x} + t\dot{x}^2 + 3t^2x = 0, \quad x(0) = -1, \quad x(1) = 1$

   Tolerance: $\varepsilon = 10^{-4}$

10. $\ddot{w} - w^3 = 0, \quad w(1) = \frac{1}{2}, \quad w(3) = \frac{5}{2}$

    Tolerance: $\varepsilon = 10^{-3}$

Finite-Difference Method (Section 8.2)

Linear Boundary-Value Problems

11. Consider the linear BVP in Problem 1:

$$\ddot{u} + 3\dot{u} = t, \quad u(0) = 1, \quad u(5) = 0.2$$

   a. Solve using the finite-difference method with $\Delta t = 1$.
   b. Solve using the shooting method (see Problem 1). In doing so, apply `RK4system` with step size of 0.25.
   c. Solve using the finite-difference method with $\Delta t = 0.5$. Tabulate and compare all calculated solution values at $t = 1$, 2, 3, and 4. Also include the exact solution values at those points.

12. Consider the linear BVP in Problem 2:

$$2\ddot{u} + 3\dot{u} + u = \sin(t/3), \quad u(0) = 0, \quad u(1) = -1$$

   a. Solve using the finite-difference method with $\Delta t = 0.25$.
   b. Solve using the shooting method (see Problem 2). In doing so, apply `RK4system` with step size of 0.05.
   c. Solve using the finite-difference method with $\Delta t = 0.125$. Tabulate and compare all calculated solution values at $t = 0.25$, 0.5, and 0.75. Also include the exact solution values at those points.

13. Consider the linear BVP in Problem 3:

$$w'' + \frac{1}{2x}w' = 0, \quad w(1) = 3, \quad w(3) = 1$$

   a. Solve using the finite-difference method with $\Delta x = 0.5$.
   b. Solve using the finite-difference method with $\Delta x = 0.25$. Tabulate and compare all calculated solution values at $x = 1.5$, 2.0, and 2.5. Also include the exact solution values at those points.

14. Consider the linear BVP in Problem 4:

$$(2t-1)\ddot{w} - 4t\dot{w} + 4w = 0, \quad w(1) = -1, \quad w(2) = 1$$

   a. Solve using the finite-difference method with $\Delta t = 0.25$.
   b. Solve using the finite-difference method with $\Delta t = 0.125$. Tabulate and compare all calculated solution values at $t = 1.25$, 1.5, and 1.75. Also include the exact solution values at those points.

Nonlinear Boundary-Value Problems: Use Central-Difference Formulas

15. Consider the nonlinear BVP in Problem 5:

$$u\ddot{u} + \dot{u}^2 = 0, \quad u(0) = 2, \quad u(2) = 3$$

   Solve using the finite-difference method with $\Delta t = 0.5$, $k$max = 10, and tol = 10^{-3}. When using Newton's method, set the initial values of the unknown quantities to 2.

16. Consider the nonlinear BVP in Problem 6:

$$\ddot{u} + u^2\dot{u} = t, \quad u(0) = 1, \quad u(3) = 2$$

   Solve using the finite-difference method with $\Delta t = 0.5$, $k$max = 10, and tol = 10^{-3}. When using Newton's method, set the initial values of the unknown quantities to 1.

17. Consider the nonlinear BVP in Problem 7:

$$2\ddot{w} - e^{w} = 0, \quad w(0) = 1, \quad w(2) = 0$$

   Solve using the finite-difference method with $\Delta t = 0.5$, $k$max = 10, and tol = 10^{-3}. When using Newton's method, set the initial values of the unknown quantities to 1.

18. Consider the nonlinear BVP in Problem 8:

$$\ddot{y} + \dot{y} - 2y^2 = 0, \quad y(0) = 1, \quad y(3) = 2$$

Solve using the finite-difference method with $\Delta t = 0.5$, $k\text{max} = 10$, and tol $= 10^{-3}$. When using Newton's method, set the initial values of the unknown quantities to 1.

19. Consider the nonlinear BVP in Problem 9:

$$\ddot{x} + t\dot{x}^2 + 3t^2 x = 0, \quad x(0) = -1, \quad x(1) = 1$$

Solve using the finite-difference method with $\Delta t = 0.25$, $k\text{max} = 10$, and tol $= 10^{-4}$. When using Newton's method, set the initial values of the unknown quantities to 0.2.

20. Consider the nonlinear BVP in Problem 10:

$$\ddot{w} - w^3 = 0, \quad w(1) = \frac{1}{2}, \quad w(3) = \frac{5}{2}$$

a. Solve using the finite-difference method with $\Delta t = 0.5$, $k\text{max} = 10$, and tol $= 10^{-4}$. When using Newton's method, set the initial values of the unknown quantities to 2.

b. Repeat (a) for $\Delta t = 0.25$. Tabulate the results of the two parts comparing approximate values at $t = 1.5, 2, 2.5$.

Mixed Boundary Conditions (Section 8.3)

21. Consider

$$\ddot{u} - \frac{1}{3}u = e^{-t/3}, \quad u(0) = -1, \quad \dot{u}(0.5) = 1$$

Solve by the finite-difference method and $\Delta t = 0.1$. Use central-difference formula for the second-order derivative, and a three-point, backward-difference formula for the first-order derivative in the right boundary condition, which also has second-order accuracy. Compare the computed results with the exact solution values at the grid points.

22. Consider

$$\ddot{y} + \frac{2}{t}\dot{y} = e^{-t}, \quad y(2) = -0.8, \quad \dot{y}(4) = 1.3$$

a. Solve by the finite-difference method with $\Delta t = 0.5$. Use central-difference formula for the second-order derivative, and approximate the first-order derivative in the right boundary condition by a three-point, backward-difference formula, which also has second-order accuracy. Compare the computed results with the exact solution values at the grid points.

b. Repeat (a) for $\Delta t = 0.25$. Tabulate the values obtained in (a) and (b), as well as the exact values, at the interior grid points $t = 2.5, 3, 3.5$.

23. Consider

$$\ddot{y} + 2y = 0, \quad \dot{y}(0) = 0, \quad y(1) = 1$$

a. Solve by the finite-difference method with $\Delta t = 0.125$. Use central-difference formula for the second-order derivative, and approximate the first-order derivative in the left boundary condition by a three-point, forward-difference formula, which also has second-order accuracy. Compare the results with the exact solution values at the grid points.

b. Repeat (a) for $\Delta t = 0.0625$. Tabulate the values obtained in (a) and (b), as well as the exact values, at the interior grid points $t = 0.25, 0.5, 0.75$.

24. Consider

$$\ddot{w} + \frac{1}{t}w + t = 0, \quad w(0) = 0, \quad \dot{w}(1) = w(1)$$

a. Solve by the finite-difference method with $\Delta t = 0.25$. Use central-difference formula for the second-order derivative, and approximate the first-order derivative in the right boundary condition by a three-point, backward-difference formula. Compare the results with the exact solution values at the grid points.

b. Repeat (a) for $\Delta t = 0.125$. Tabulate the values obtained in (a) and (b), as well as the exact values, at the interior grid points $t = 0.25, 0.5, 0.75$.

25. Consider the nonlinear BVP with mixed boundary conditions

$$\ddot{w} - w^3 = 0, \quad w(1) = 0, \quad \dot{w}(2) + w(2) = 1$$

a. Solve by the finite-difference method with $\Delta t = 0.25$. Use central-difference formula for the second-order derivative, and approximate the first-order derivative in the right boundary condition by a three-point, backward-difference formula. Solve the ensuing nonlinear system of equations via Newton's method with all initial values set to 1, $k$max = 10, and tol = 10^{-3}.

b. Repeat (a) for $\Delta t = 0.125$. Tabulate the values obtained in (a) and (b) at grid points $t = 1.25, 1.5, 1.75$.

26. Consider the nonlinear BVP with mixed boundary conditions

$$2\ddot{w} - e^{w} = 0, \quad \dot{w}(0) = 1, \quad \dot{w}(2) + w(2) = 0$$

Solve by the finite-difference method with $\Delta t = 0.5$. Use central-difference formula for the second-order derivative, and approximate the first-order derivatives in the left and right boundary conditions by a three-point, forward- and a three-point backward-difference formula, respectively. Solve the ensuing nonlinear system of equations via Newton's method with all initial values set to 1, $k$max = 10, and tol = 10^{-3}.

MATLAB® Built-In Function bvp4c for Boundary-Value Problems (Section 8.4)

In Problems 27 through 30, solve the BVP using `bvp4c`, and plot the dependent variable versus the independent variable.

27. The BVP in Example 8.2:

$$T_{xx} - \beta T^4 - \alpha T + \gamma = 0, \quad T(0) = 500, \quad T(0.15) = 300 \qquad T_x = dT/dx$$

where $\alpha = 20$, $\beta = 10^{-8}$, and $\gamma = 5 \times 10^3$ with all parameters in consistent physical units.

28. $y''' = y^4 - y', \quad y(0) = 0, \quad y'(0) = 1, \quad y(1) = 20$
29. $y'' - 0.8y = 0, \quad y(0) + y'(0) = 0, \quad y'(1) = 0$
30. $xy'' + y' - x^{-1} y = 0, \quad y(1) = 0, \quad y'(e^2) = 0$

31. The deflection y and rotation ψ of a uniform beam of length $L = 8$, pinned at both ends, are governed by

$$\begin{aligned} y' &= \psi \\ \psi' &= \frac{5x - x^2}{EI} \end{aligned} \quad \text{subject to} \quad \begin{aligned} y(0) &= 0 \\ y(8) &= 0 \end{aligned}$$

where EI is the flexural rigidity and is assumed to be $EI = 4000$. All parameter values are in consistent physical units. Using `bvp4c`, find y and ψ, and plot them versus $0 \le x \le 8$ in two separate figures.

32. In the analysis of free transverse vibrations of a uniform beam of length $L = 10$, simply supported (pinned) at both ends, we encounter a fourth-order ODE

$$\frac{d^4X}{dx^4} - 16X = 0$$

subject to

$$X(0) = 0, \quad X(L) = 0$$
$$X''(0) = 0, \quad X''(L) = 0$$

Using bvp4c, find and plot $X(x)$.

9

Matrix Eigenvalue Problem

Matrix eigenvalue problem plays a significant role in engineering applications and is frequently encountered in numerical methods. In vibration analysis, for example, eigenvalues are directly related to the system's natural frequencies, while the eigenvectors represent the modes of vibration. Eigenvalues play an equally important role in numerical methods. Consider, for example, the iterative solution of linear systems via Jacobi and Gauss–Seidel methods; Chapter 4. The eigenvalues of the Jacobi iteration matrix, or the Gauss–Seidel iteration matrix, not only determine whether or not the respective iteration will converge to a solution, they also establish their rate of convergence. In this chapter, we present numerical techniques to approximate the eigenvalues and eigenvectors of a matrix.

Matrix Eigenvalue Problem:
Recall from Chapter 1 that the eigenvalue problem associated with a square matrix $\mathbf{A}_{n\times n}$ is defined as

$$\mathbf{Av} = \lambda\mathbf{v}, \quad \mathbf{v} \neq \mathbf{0}_{n\times 1} \tag{9.1}$$

A number λ for which Equation 9.1 has a nontrivial solution ($\mathbf{v} \neq \mathbf{0}_{n\times 1}$) is called an eigenvalue or characteristic value of matrix $\mathbf{A}$. The corresponding solution $\mathbf{v} \neq \mathbf{0}$ of Equation 9.1 is the eigenvector or characteristic vector of $\mathbf{A}$ corresponding to λ. The set of all eigenvalues of $\mathbf{A}$ is called the spectrum of $\mathbf{A}$.

9.1 Power Method: Estimation of the Dominant Eigenvalue

The power method is an iterative technique to estimate the dominant eigenvalue of a matrix $\mathbf{A}_{n\times n}$. The basic assumptions are

- $\mathbf{A}_{n\times n}$ is a real matrix.
- $\mathbf{A}$ has n eigenvalues $\lambda_1, \lambda_2, \ldots, \lambda_n$, where λ_1 is the dominant eigenvalue,

  $$|\lambda_1| > |\lambda_2| \geq |\lambda_3| \geq \cdots \geq |\lambda_n|$$

  and the corresponding eigenvectors $\mathbf{v}_1, \mathbf{v}_2, \ldots, \mathbf{v}_n$ are linearly independent.
- The dominant eigenvalue is real.

We present the power method as follows. Since $\mathbf{v}_1, \mathbf{v}_2, \ldots, \mathbf{v}_n$ is a linearly independent set, by assumption, there exist constants $c_1, c_2, \ldots, c_n$ such that an arbitrary $n \times 1$ vector $\mathbf{x}_0$ can be uniquely expressed as

$$\mathbf{x}_0 = c_1\mathbf{v}_1 + c_2\mathbf{v}_2 + \cdots + c_n\mathbf{v}_n \tag{9.2}$$

The eigenvalue problem is

$$\mathbf{A}\mathbf{v}_i = \lambda_i\mathbf{v}_i \ (i = 1, 2, \ldots, n) \quad \overset{\text{Pre-multiply by } \mathbf{A}}{\Rightarrow} \quad \mathbf{A}(\mathbf{A}\mathbf{v}_i) = \mathbf{A}(\lambda_i\mathbf{v}_i) \Rightarrow \mathbf{A}^2\mathbf{v}_i = \lambda_i^2\mathbf{v}_i$$

In general,

$$\mathbf{A}^k\mathbf{v}_i = \lambda_i^k\mathbf{v}_i \tag{9.3}$$

Define the sequence of vectors

$$\begin{aligned} \mathbf{x}_1 &= \mathbf{A}\mathbf{x}_0 \\ \mathbf{x}_2 &= \mathbf{A}\mathbf{x}_1 = \mathbf{A}^2\mathbf{x}_0 \\ &\cdots \\ \mathbf{x}_k &= \mathbf{A}\mathbf{x}_{k-1} = \mathbf{A}^k\mathbf{x}_0 \end{aligned}$$

Therefore, by Equations 9.2 and 9.3,

$$\begin{aligned} \mathbf{x}_k &= \mathbf{A}^k\mathbf{x}_0 = \mathbf{A}^k\left[c_1\mathbf{v}_1 + c_2\mathbf{v}_2 + \cdots + c_n\mathbf{v}_n\right] = c_1\lambda_1^k\mathbf{v}_1 + c_2\lambda_2^k\mathbf{v}_2 + \cdots + c_n\lambda_n^k\mathbf{v}_n \\ &= \lambda_1^k\left[c_1\mathbf{v}_1 + c_2\left(\frac{\lambda_2}{\lambda_1}\right)^k\mathbf{v}_2 + \cdots + c_n\left(\frac{\lambda_n}{\lambda_1}\right)^k\mathbf{v}_n\right] \end{aligned} \tag{9.4}$$

Since λ_1 is the dominant eigenvalue, the ratios $\lambda_2/\lambda_1, \lambda_3/\lambda_1, \ldots, \lambda_n/\lambda_1$ are less than 1, hence for a sufficiently large k, we have $\mathbf{x}_k \cong c_1\lambda_1^k\mathbf{v}_1$. Equation 9.4 can also be used to obtain $\mathbf{x}_{k+1} \cong c_1\lambda_1^{k+1}\mathbf{v}_1$. Thus,

$$\mathbf{x}_{k+1} \cong \lambda_1\mathbf{x}_k \tag{9.5}$$

Estimation of λ_1 will be based on Equation 9.5. Pre-multiply Equation 9.5 by $\mathbf{x}_k^T$ to create scalars on both sides, and

$$\lambda_1 \cong \frac{\mathbf{x}_k^T\mathbf{x}_{k+1}}{\mathbf{x}_k^T\mathbf{x}_k} \tag{9.6}$$

The power method states that the sequence of scalars

$$\alpha_{k+1} = \frac{\mathbf{x}_k^T \mathbf{x}_{k+1}}{\mathbf{x}_k^T \mathbf{x}_k}, \quad \mathbf{x}_{k+1} = \mathbf{A}\mathbf{x}_k \tag{9.7}$$

converges to the dominant eigenvalue λ_1 for sufficiently large k. In fact, it can be shown that the sequence $\{\alpha_k\}$ converges to λ_1 at roughly the same rate as $(\lambda_2/\lambda_1)^k \to 0$. Therefore, the larger the magnitude of λ_1 is compared to the next largest eigenvalue λ_2, the faster the convergence. Also, because the components of vector $\mathbf{x}_k$ grow rapidly, it is common practice to normalize $\mathbf{x}_k$, that is, divide each vector $\mathbf{x}_k$ by its 2-norm, $\|\mathbf{x}_k\|_2$. Consequently, the denominator in Equation 9.7 simply becomes $\mathbf{x}_k^T\mathbf{x}_k = 1$ in each step. It should be mentioned that the power method can handle situations where a dominant eigenvalue λ_1 is repeated (see Problem Set), but not when there are two dominant eigenvalues with the same magnitude and opposite signs.

Algorithm for Power Method

Starting with a matrix $\mathbf{A}_{n\times n}$, an initial $n \times 1$ vector $\mathbf{x}_0$, and an initial $\alpha_1 = 0$,

1. Normalize $\mathbf{x}_0$ to construct a unit vector $\mathbf{x}_1 = \mathbf{x}_0/\|\mathbf{x}_0\|_2$.
   For $k = 1$ to $k$max (maximum number of iterations),
2. Find $\mathbf{x}_{k+1} = \mathbf{A}\mathbf{x}_k$.
3. Calculate $\alpha_{k+1} = \mathbf{x}_k^T\mathbf{x}_{k+1}$.
4. Normalize $\mathbf{x}_{k+1}$ and do not rename, $\mathbf{x}_{k+1} = \mathbf{x}_{k+1}/\|\mathbf{x}_{k+1}\|_2$.
5. Terminating condition: if $|\alpha_{k+1} - \alpha_k| < \varepsilon$ (prescribed tolerance), STOP.
   Otherwise, increment k to $k + 1$ and go to Step 2.

When the iterations stop, α_{k+1} is the approximate dominant eigenvalue, and the unit vector $\mathbf{x}_{k+1}$ is the corresponding eigenvector. Note that other terminating conditions may also be used in Step 5. For example, we can force the iterations to stop if two consecutive vectors satisfy $\|\mathbf{x}_{k+1} - \mathbf{x}_k\|_\infty < \varepsilon$.

The user-defined function PowerMethod uses the power method to estimate the dominant eigenvalue and the associated eigenvector for a square matrix.

```
function [e_val, e_vec, k]=PowerMethod(A, x0, tol, kMax)
%
% PowerMethod approximates the dominant eigenvalue and
% corresponding eigenvector of a square matrix.
%
%  [e_val, e_vec, k]=PowerMethod(A, x0, tol, kMax) where
%
```

```
%    A is an n-by-n matrix,
%    x0 is the n-by-1 initial guess (default ones),
%    tol is the tolerance (default 1e-4),
%    kMax is the maximum number of iterations (default 50),
%
%    e_val is the approximated dominant eigenvalue,
%    e_vec is the corresponding eigenvector,
%    k is the number of iterations required for convergence.
%
%    Note: The terminating condition applies to successive
%    alpha values!

n = size(A, 1);
if nargin<2 || isempty(x0), x0 = ones(n,1); end
if nargin<3 || isempty(tol), tol = 1e-4; end
if nargin<4 || isempty(kMax), kMax = 50; end
x(:,1) = x0./norm(x0, 2); x(:,2) = A*x(:,1);
alpha(2) = x(:,1).'*x(:,2);
x(:,2) = x(:,2)./norm(x(:,2), 2);

for k = 2:kMax,
    x(:,k+1) = A*x(:,k);    % Generate next vector
    alpha(k+1) = x(:,k).'*x(:,k+1);
    x(:,k+1) = x(:,k+1)./norm(x(:,k+1), 2);
    if abs(alpha(k+1) - alpha(k))< tol,
    % Check for convergence
        break
    end
end
e_val = alpha(end); e_vec = x(:, end);
```

If output k (number of iterations) is not needed in an application, the function can be executed as

```
[e_val,e_vec] = PowerMethod(A,x0,tol,kMax)
```

Also, if e_vec (eigenvector) is not needed, we can still execute

```
[e_val] = PowerMethod(A,x0,tol,kMax)
```

No other combination of outputs is possible. For instance, [e_val,k] still returns e_val and e_vec because when there are two outputs, the second output is automatically regarded as e_vec.

Example 9.1: Power Method

Consider

$$\mathbf{A} = \begin{bmatrix} -3 & -4 & -4 \\ 6 & 9 & 6 \\ -9 & -14 & -8 \end{bmatrix}$$

Starting with $\alpha_1 = 0, \mathbf{x}_0 = [0 \quad 1 \quad 1]^T$, and tolerance $\varepsilon = 10^{-4}$, we follow the algorithm outlined above:

$$\mathbf{x}_1 = \frac{\mathbf{x}_0}{\|\mathbf{x}_0\|_2} = \frac{1}{\sqrt{2}}\begin{Bmatrix} 0 \\ 1 \\ 1 \end{Bmatrix} = \begin{Bmatrix} 0 \\ 0.7071 \\ 0.7071 \end{Bmatrix},$$

$$\mathbf{x}_2 = \mathbf{A}\mathbf{x}_1 = \begin{bmatrix} -3 & -4 & -4 \\ 6 & 9 & 6 \\ -9 & -14 & -8 \end{bmatrix}\begin{Bmatrix} 0 \\ 0.7071 \\ 0.7071 \end{Bmatrix} = \begin{Bmatrix} -5.6569 \\ 10.6066 \\ -15.5563 \end{Bmatrix}$$

$$\alpha_2 = \mathbf{x}_1^T\mathbf{x}_2 = -3.5000$$

Normalize

$$\mathbf{x}_2 = \frac{\mathbf{x}_2}{\|\mathbf{x}_2\|_2} = \begin{Bmatrix} -0.2877 \\ 0.5395 \\ -0.7913 \end{Bmatrix}$$

Check the terminating condition: Since $|\alpha_2 - \alpha_1| = 3.5$, convergence is not observed and the iterations continue.

$$\mathbf{x}_3 = \mathbf{A}\mathbf{x}_2 = \begin{bmatrix} -3 & -4 & -4 \\ 6 & 9 & 6 \\ -9 & -14 & -8 \end{bmatrix}\begin{Bmatrix} -0.2877 \\ 0.5395 \\ -0.7913 \end{Bmatrix} = \begin{Bmatrix} 1.8703 \\ -1.6185 \\ 1.3668 \end{Bmatrix},$$

$$\alpha_3 = \mathbf{x}_2^T\mathbf{x}_3 = -2.4929$$

Since $|\alpha_3 - \alpha_2| = 1.0071$, the tolerance $\varepsilon = 10^{-4}$ is not met and iterations must continue. This procedure is repeated until $|\alpha_{k+1} - \alpha_k| < \varepsilon$ is achieved.

Execution of the user-defined function `PowerMethod` yields

```
>> A = [-3 -4 -4;6 9 6;-9 -14 -8];
>> x0 = [0;0;1];
```

```
>> [e_val, e_vec, k] = PowerMethod(A, x0)
% Using default tol=1e-4, kmax=50

e_val =

   -3.0000

e_vec =

   -0.5771
    0.5774
   -0.5776

k =

     7
```

It takes seven steps for the power method to generate an estimate for the dominant eigenvalue $\lambda_1 = -3$ and its eigenvector. Note that the returned (unit vector) eigenvector estimate is equivalent to $[-1 \quad 1 \quad -1]^T$. Direct solution of the eigenvalue problem reveals that $\lambda(\mathbf{A}) = -3,0,1$ and that the eigenvector associated with $\lambda = -3$ agrees with the one obtained here.

9.1.1 Inverse Power Method: Estimation of the Smallest Eigenvalue

The smallest (in magnitude) eigenvalue of $\mathbf{A}_{n\times n}$ can be approximated by applying the power method to $\mathbf{A}^{-1}$. This is simply because if the eigenvalues of $\mathbf{A}$ are $\lambda_1, \lambda_2, \ldots, \lambda_n$, then the eigenvalues of $\mathbf{A}^{-1}$ are $1/\lambda_1, 1/\lambda_2, \ldots, 1/\lambda_n$:

$$\mathbf{A}\mathbf{v}_i = \lambda_i \mathbf{v}_i (i = 1, 2, \ldots, n) \overset{\text{Pre-multiply by } \mathbf{A}^{-1}}{\Rightarrow} (\mathbf{A}^{-1}\mathbf{A})\mathbf{v}_i = \mathbf{A}^{-1}\lambda_i \mathbf{v}_i \underset{\text{Divide by } \lambda_i}{\overset{\mathbf{A}^{-1}\mathbf{A}=\mathbf{I}}{\Rightarrow}} \mathbf{A}^{-1}\mathbf{v}_i = \frac{1}{\lambda_i}\mathbf{v}_i$$

This last equation is the eigenvalue problem associated with $\mathbf{A}^{-1}$ so that its eigenvalues are $1/\lambda_i$ and the eigenvectors are $\mathbf{v}_i$, the same as those of $\mathbf{A}$.

Example 9.2: Inverse Power Method

The matrix $\mathbf{A}$ in Example 9.1 happens to be singular so that $\mathbf{A}^{-1}$ does not technically exist. But $\mathbf{A}$ being singular implies that at least one of its eigenvalues must be zero, which serves as the smallest eigenvalue of $\mathbf{A}$. Therefore, there is no need to apply the inverse power method, and it is decided that the smallest eigenvalue of $\mathbf{A}$ is indeed zero.

9.1.2 Shifted Inverse Power Method: Estimation of the Eigenvalue Nearest a Specified Value

Once the largest or smallest eigenvalue of a matrix is known, the remaining eigenvalues can be approximated using the shifted inverse power method. In order to establish this method, we first recognize the fact that if the eigenvalues of $\mathbf{A}$ are $\lambda_1, \lambda_2, \ldots, \lambda_n$, then the eigenvalues of the matrix $\mathbf{A} - \alpha\mathbf{I}$ are $\lambda_1 - \alpha, \lambda_2 - \alpha, \ldots, \lambda_n - \alpha$:

$$\mathbf{A}\mathbf{v}_i = \lambda_i \mathbf{v}_i \ (i = 1, 2, \ldots, n) \overset{\text{Subtract } \alpha\mathbf{v}_i}{\Rightarrow} \mathbf{A}\mathbf{v}_i - \alpha\mathbf{v}_i = \lambda_i \mathbf{v}_i - \alpha\mathbf{v}_i \Rightarrow (\mathbf{A} - \alpha\mathbf{I})\mathbf{v}_i = (\lambda_i - \alpha)\mathbf{v}_i$$

This last equation is the eigenvalue problem associated with the matrix $\mathbf{A} - \alpha\mathbf{I}$; therefore, its eigenvalues are $\lambda_i - \alpha$ and the eigenvectors are $\mathbf{v}_i$, the same as those of $\mathbf{A}$. Combining this with the fact regarding the eigenvalues of $\mathbf{A}^{-1}$, it is easily seen that the eigenvalues of $(\mathbf{A} - \alpha\mathbf{I})^{-1}$ are

$$\mu_1 = \frac{1}{\lambda_1 - \alpha}, \quad \mu_2 = \frac{1}{\lambda_2 - \alpha}, \ldots, \quad \mu_n = \frac{1}{\lambda_n - \alpha} \tag{9.8}$$

while the eigenvectors $\mathbf{v}_1, \mathbf{v}_2, \ldots, \mathbf{v}_n$ are the same as those for $\mathbf{A}$ corresponding to $\lambda_1, \lambda_2, \ldots, \lambda_n$. If the power method is applied to $(\mathbf{A} - \alpha\mathbf{I})^{-1}$, its dominant eigenvalue (say, μ_m) is estimated. Since $|\mu_m|$ is largest among all those listed in Equation 9.8, $|\lambda_m - \alpha|$ must be the smallest among its counterparts, that is,

$$|\lambda_m - \alpha| \le |\lambda_i - \alpha|, \quad i = 1, 2, \ldots, n$$

In conclusion, application of the power method to $(\mathbf{A} - \alpha\mathbf{I})^{-1}$ gives an estimate of λ_m which is closest to α than all the other eigenvalues of $\mathbf{A}$.

Inspired by this, we present the shifted inverse power method as follows. Let α be an arbitrary scalar, and $\mathbf{x}_0$ any initial vector. Generate the sequence of vectors

$$\mathbf{x}_{k+1} = (\mathbf{A} - \alpha\mathbf{I})^{-1}\,\mathbf{x}_k \tag{9.9}$$

and scalars

$$\beta_k = \frac{\mathbf{x}_k^T \mathbf{x}_{k+1}}{\mathbf{x}_k^T \mathbf{x}_k} \tag{9.10}$$

Then, $\beta_k \to \mu_m = 1/(\lambda_m - \alpha)$ so that $\lambda_m = (1/\mu_m) + \alpha$ is the eigenvalue of $\mathbf{A}$ that is closest to α. Also, the sequence of vectors $\mathbf{x}_k$ converges to the eigenvector corresponding to λ_m.

Notes on Shifted Inverse Power Method

- The initial vector $\mathbf{x}_0$ and the subsequent vectors $\mathbf{x}_k$ will be normalized to have a length of 1,
- Equation 9.9 is solved as $(\mathbf{A} - \alpha\mathbf{I})\mathbf{x}_{k+1} = \mathbf{x}_k$ using Doolittle's method (Section 4.4), which employs LU factorization of the coefficient matrix $\mathbf{A} - \alpha\mathbf{I}$. This proves useful, especially if α happens to be very close to an eigenvalue of $\mathbf{A}$, causing $\mathbf{A} - \alpha\mathbf{I}$ to be near singular,

- Setting $\alpha = 0$ leads to the estimation of the smallest (in magnitude) eigenvalue of **A**.

The user-defined function `InvPowerMethod` uses the shifted inverse power method to estimate the eigenvalue of a square matrix closest to a specified value. It also returns the eigenvector associated with the desired eigenvalue.

```
function [eigenval, eigenvec, k] = InvPowerMethod(A, alpha,
x0, tol, kMax)
%
% InvPowerMethod uses the shifted inverse power method to
% find the eigenvalue of a matrix closest to a specified
% value. It also returns the eigenvector associated with
% this eigenvalue.
%
% [eigenval, eigenvec, k]=InvPowerMethod(A, alpha, x0,
% tol, kMax) where
%
%   A is an n-by-n matrix,
%   alpha is a specified value,
%   x0 is the n-by-1 initial guess (default ones),
%   tol is the tolerance (default 1e-4),
%   kMax is the maximum number of iterations (default 50),
%
%   eigenval is the estimated eigenvalue,
%   eigenvec is the corresponding eigenvector,
%   k is the number of iterations required for convergence.
%
n = size(A, 1);
if nargin<3 || isempty(x0), x0 = ones(n,1); end
if nargin<4 || isempty(tol), tol = 1e-4; end
if nargin<5 || isempty(kMax), kMax = 50; end
x(:,1) = x0./norm(x0, 2);
betas(1) = 0;
for k = 1:kMax,
    x(:,k+1) = DoolittleMethod(A-alpha*eye(n,n),x(:,k));
    % Doolittle's method
    betas(k+1) = x(:,k).'*x(:,k+1);
    x(:,k+1) = x(:,k+1)./norm(x(:,k+1), 2);
    if abs(betas(k+1) - betas(k))< tol,
    % Check for convergence
       break
    end
end
betas = betas(end); eigenval = 1/betas+alpha;
eigenvec = x(:, end);
```

9.1.2.1 Shifted Power Method

The shifted power method is based on the fact that if the eigenvalues of **A** are $\lambda_1, \lambda_2, \ldots, \lambda_n$, then the eigenvalues of the matrix $\mathbf{A} - \alpha\mathbf{I}$ are $\lambda_1 - \alpha, \lambda_2 - \alpha, \ldots, \lambda_n - \alpha$. In particular, the eigenvalues of $\mathbf{A} - \lambda_1\mathbf{I}$ are $0, \lambda_2 - \lambda_1, \ldots, \lambda_n - \lambda_1$. If the power method is applied to $\mathbf{A} - \lambda_1\mathbf{I}$, the dominant eigenvalue of this matrix (say, μ_2) is found. But μ_2 is a member of the list $0, \lambda_2 - \lambda_1, \ldots, \lambda_n - \lambda_1$. Without loss of generality, suppose $\mu_2 = \lambda_2 - \lambda_1$ so that $\lambda_2 = \lambda_1 + \mu_2$. This way, another eigenvalue of **A** is estimated.

9.1.2.1.1 *Strategy to Estimate All Eigenvalues of a Matrix*

- Apply the power method to **A** to find the dominant λ_d.
- Apply the shifted inverse power method with $\alpha = 0$ to **A** to find the smallest λ_s.
- Apply the shifted power method to $\mathbf{A} - \lambda_d\mathbf{I}$ or $\mathbf{A} - \lambda_s\mathbf{I}$ to find at least one more eigenvalue.
- Apply the shifted inverse power method as many times as necessary with α adjusted so that an λ between any two available λ's may be found.

Example 9.3: Power Methods

Find all eigenvalues and eigenvectors of the following matrix using the power and the shifted inverse power methods.

$$\mathbf{A} = \begin{bmatrix} 4 & -3 & 3 & -9 \\ -3 & 6 & -3 & 11 \\ 0 & 8 & -5 & 8 \\ 3 & -3 & 3 & -8 \end{bmatrix}$$

SOLUTION

Apply the user-defined function `PowerMethod` to find the dominant eigenvalue of **A**:

```
>> A = [4 -3 3 -9;-3 6 -3 11;0 8 -5 8;3 -3 3 -8];
>> x0 = [0;1;0;1];     % Initial vector
>> [e_val, e_vec] = PowerMethod(A, x0)
% Default values for tol and kmax

e_val =

   -5.0000     % Dominant eigenvalue

e_vec =

    0.5000
   -0.5000
   -0.5000
    0.5000
```

The smallest eigenvalue of **A** is estimated by applying the user-defined function InvPowerMethod with $\alpha = 0$:

```
>> [eigenval, eigenvec] = InvPowerMethod(A, 0, x0) % Set alpha=0

eigenval =

    1.0001    % Smallest eigenvalue

eigenvec =

   -0.8165
    0.4084
    0.0001
   -0.4082
```

The largest and smallest eigenvalues of **A** are therefore –5 and 1. To see if the remaining two eigenvalues are between these two values, we set α to be the average $\alpha = (-5 + 1)/2 = -2$ and apply the shifted inverse power method. However, $\alpha = -2$ causes $\mathbf{A} - \alpha\mathbf{I}$ to be singular, meaning α is an eigenvalue of **A**. Since we also need the eigenvector associate with this eigenvalue, we set α to a value close to –2 and apply the shifted inverse power:

```
>> [eigenval, eigenvec] = InvPowerMethod(A, -1.5, x0)
% alpha=-1.5

  eigenval =

    -2.0000    % The third eigenvalue

eigenvec =

     0.5774
    -0.5774
     0.0000
     0.5774
```

We find the fourth eigenvalue using the shifted power method. Noting $\lambda = -2$ is an eigenvalue of **A**, we apply the power method to $\mathbf{A} + 2\mathbf{I}$ to find

```
>> A1 = A+2*eye(4,4);
>> [e_val, e_vec] = PowerMethod(A1, x0)

e_val =

    5.0000

e_vec =

   -0.0000
    0.7071
    0.7071
    0.0000
```

Following the reasoning behind the shifted power method, the fourth eigenvalue is $\lambda = -2 + 5 = 3$. In summary, all four eigenvalues and their eigenvectors are

$$\lambda_1 = -5, \quad \mathbf{v}_1 = \begin{Bmatrix} 1 \\ -1 \\ -1 \\ 1 \end{Bmatrix}, \quad \lambda_2 = 1, \quad \mathbf{v}_2 = \begin{Bmatrix} -2 \\ 1 \\ 0 \\ -1 \end{Bmatrix},$$

$$\lambda_3 = -2, \quad \mathbf{v}_3 = \begin{Bmatrix} 1 \\ -1 \\ 0 \\ 1 \end{Bmatrix}, \quad \lambda_4 = 3, \quad \mathbf{v}_4 = \begin{Bmatrix} 0 \\ 1 \\ 1 \\ 0 \end{Bmatrix}$$

9.1.3 MATLAB® Built-In Function `eig`

The built-in function `eig` in MATLAB® finds the eigenvalues and eigenvectors of a matrix. The function `eig` can be used in two different forms.

```
eig   Eigenvalues and eigenvectors.
    E=eig(X) is a vector containing the eigenvalues of a
    square matrix X.

    [V,D]=eig(X) produces a diagonal matrix D of eigenvalues
    and a full matrix V whose columns are the corresponding
    eigenvectors so that X*V=V*D.
```

The first form `E = eig(A)` is used when only the eigenvalues of a matrix **A** are needed. If the eigenvalues, as well as the eigenvectors of **A** are desired, `[V,D] = eig(A)` is used. This returns a matrix `V` whose columns are the eigenvectors of `A`, each a unit vector, and a diagonal matrix `D` whose entries are the eigenvalues of `A` and whose order agrees with the columns of `V`. Applying the latter to the matrix in Example 9.3 yields

```
>> A=[4 -3 3 -9;-3 6 -3 11;0 8 -5 8;3 -3 3 -8];
>> [V,D]=eig(A)

V =

   -0.8165     -0.5774      0.0000     -0.5000
    0.4082      0.5774      0.7071      0.5000
   -0.0000     -0.0000      0.7071      0.5000
   -0.4082     -0.5774      0.0000     -0.5000

D =

    1.0000           0           0           0
         0     -2.0000           0           0
         0           0      3.0000           0
         0           0           0     -5.0000
```

The results clearly agree with those obtained earlier in Example 9.3.

9.2 Deflation Methods

In the previous section, we learned how to estimate eigenvalues of a matrix by using different variations of the power method. Another tactic to find eigenvalues of a matrix involves the idea of deflation. Suppose $\mathbf{A}_{n\times n}$ has eigenvalues $\lambda_1, \lambda_2, \ldots, \lambda_n$, and one of them is available; for example, the dominant λ_1 obtained by the power method. The basic idea behind deflation is to generate a matrix $\mathbf{B}$, which is $(n-1)\times(n-1)$, one size smaller than $\mathbf{A}$, whose eigenvalues are $\lambda_2, \ldots, \lambda_n$, meaning all the eigenvalues of $\mathbf{A}$ excluding the dominant λ_1. We next focus on $\mathbf{B}$ and suppose its dominant eigenvalue is λ_2. The power method can then be used to estimate λ_2. With that, $\mathbf{B}$ is deflated to a yet smaller matrix, and so on. Although there are many proposed deflation methods, we will introduce the most common one, known as Wielandt's deflation method.

9.2.1 Wielandt's Deflation Method

To deflate an $n\times n$ matrix $\mathbf{A}$ to an $(n-1)\times(n-1)$ matrix $\mathbf{B}$, we must first construct an $n\times n$ matrix $\mathbf{A}_1$ whose eigenvalues are $0, \lambda_2, \ldots, \lambda_n$. This is explained in the following theorem.

Theorem 9.1

Suppose $\mathbf{A}_{n\times n}$ has eigenvalues $\lambda_1, \lambda_2, \ldots, \lambda_n$ and eigenvectors $\mathbf{v}_1, \mathbf{v}_2, \ldots, \mathbf{v}_n$. Assume that λ_1 and $\mathbf{v}_1$ are known and the first component of $\mathbf{v}_1$ is nonzero,* which can be made into 1. If $\mathbf{a}_1$ is the first row of $\mathbf{A}$, then

$$\mathbf{A}_1 = \mathbf{A} - \mathbf{v}_1\mathbf{a}_1 \tag{9.11}$$

has eigenvalues $0, \lambda_2, \ldots, \lambda_n$.

Proof

Since the first entry of $\mathbf{v}_1$ is nonzero, $\mathbf{v}_1$ can always be normalized so that its first component is made into 1. As a result, the first row of the matrix $\mathbf{v}_1\mathbf{a}_1$ is simply the first row of $\mathbf{A}$, and $\mathbf{A}_1$ has the form in Figure 9.1.

Because the entire first row of $\mathbf{A}_1$ is zero, $\mathbf{A}_1$ is singular and thus has at least one eigenvalue of 0. We next show that the remaining eigenvalues of $\mathbf{A}_1$ are $\lambda_2, \ldots, \lambda_n$. To do this, we realize that the eigenvectors of $\mathbf{A}$ are either in the form

$$\underset{\text{Case (1)}}{\mathbf{v}_i} = \begin{Bmatrix} 1 \\ v_{i2} \\ \cdots \\ v_{in} \end{Bmatrix} \quad \text{or} \quad \underset{\text{Case (2)}}{\mathbf{v}_i} = \begin{Bmatrix} 0 \\ v_{i2} \\ \cdots \\ v_{in} \end{Bmatrix}, \quad i = 2, 3, \ldots, n$$

* The case when the first entry of $\mathbf{v}_1$ is zero can be treated similarly, as in Example 9.5.

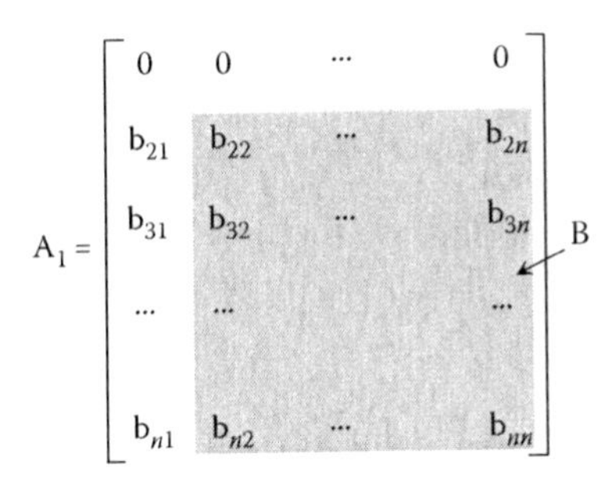

FIGURE 9.1
Matrix generated by Wielandt's deflation method.

Case (1) Consider

$$\mathbf{A}_1(\mathbf{v}_1 - \mathbf{v}_i) = (\mathbf{A} - \mathbf{v}_1\mathbf{a}_1)(\mathbf{v}_1 - \mathbf{v}_i)$$
$$= \mathbf{A}(\mathbf{v}_1 - \mathbf{v}_i) - \mathbf{v}_1\mathbf{a}_1(\mathbf{v}_1 - \mathbf{v}_i) \quad (9.12)$$

In the second term on the right hand side of Equation 9.12, note that $\mathbf{a}_1\mathbf{v}_k$ simply gives the first component of $\mathbf{Av}_k$. Noting $\mathbf{Av}_k = \lambda_k\mathbf{v}_k$ and the nature of the eigenvectors in Case (1), we have $\mathbf{a}_1\mathbf{v}_1 = \lambda_1$ and $\mathbf{a}_1\mathbf{v}_i = \lambda_i$. Using these in Equation 9.12,

$$\mathbf{A}_1(\mathbf{v}_1 - \mathbf{v}_i) = \lambda_1\mathbf{v}_1 - \lambda_i\mathbf{v}_i - \mathbf{v}_1(\lambda_1 - \lambda_i) = \lambda_i(\mathbf{v}_1 - \mathbf{v}_i)$$

Letting $\mathbf{u}_i = \mathbf{v}_1 - \mathbf{v}_i$, the above equation reads $\mathbf{A}_1\mathbf{u}_i = \lambda_i\mathbf{u}_i$. Since this is the eigenvalue problem associated with $\mathbf{A}_1$, the proof of Case (1) is complete. The eigenvectors of $\mathbf{A}_1$ corresponding to $\lambda_2, \ldots, \lambda_n$ are in the form $\mathbf{u}_i = \mathbf{v}_1 - \mathbf{v}_i$. Therefore, all of these eigenvectors have a first component of zero.

Case (2) Consider

$$\mathbf{A}_1\mathbf{v}_i = (\mathbf{A} - \mathbf{v}_1\mathbf{a}_1)\mathbf{v}_i$$
$$= \lambda_i\mathbf{v}_i - \mathbf{v}_1\mathbf{a}_1\mathbf{v}_i \quad (9.13)$$

Following an earlier reasoning, the term $\mathbf{a}_1\mathbf{v}_i$ is the first component of $\mathbf{Av}_i$. Noting $\mathbf{Av}_i = \lambda_i\mathbf{v}_i$ and the nature of the eigenvectors in Case (2), we conclude that $\mathbf{a}_1\mathbf{v}_i = 0$. Then Equation 9.13 becomes $\mathbf{A}_1\mathbf{v}_i = \lambda_i\mathbf{v}_i$, indicating $\lambda_2, \ldots, \lambda_n$ are eigenvalues of $\mathbf{A}_1$ with corresponding eigenvectors $\mathbf{v}_2, \ldots, \mathbf{v}_n$. With that, the proof is complete. Once again, note that the eigenvectors all have a first component of zero.

9.2.1.1 Deflation Process

While proving Theorem 9.1, we learned that in both cases the eigenvectors of $\mathbf{A}_1$ corresponding to $\lambda_2, \ldots, \lambda_n$ all have zeros in their first components.

Thus, the first column of $\mathbf{A}_1$ can be dropped all together. As a result, the $(n-1)\times(n-1)$ block of $\mathbf{A}_1$, called $\mathbf{B}$ in Figure 9.1, must have eigenvalues $\lambda_2, \ldots, \lambda_n$. Therefore, the problem reduces to finding the eigenvalues of $\mathbf{B}$, a matrix one size smaller than the original $\mathbf{A}$. The power method can be applied to estimate the dominant eigenvalue of $\mathbf{B}$ and its eigenvector, which in turn may be deflated further, and so on.

Example 9.4: Wielandt's Deflation Method

Consider the 4×4 matrix in Example 9.3:

$$\mathbf{A} = \begin{bmatrix} 4 & -3 & 3 & -9 \\ -3 & 6 & -3 & 11 \\ 0 & 8 & -5 & 8 \\ 3 & -3 & 3 & -8 \end{bmatrix}$$

Using the power method the dominant eigenvalue and its eigenvector are obtained as

$$\lambda_1 = -5, \quad \mathbf{v}_1 = \begin{Bmatrix} 1 \\ -1 \\ -1 \\ 1 \end{Bmatrix}$$

The first component of $\mathbf{v}_1$ is 1, hence $\mathbf{v}_1$ is already normalized. Proceeding with Wielandt's deflation method,

$$\mathbf{A}_1 = \mathbf{A} - \mathbf{v}_1\mathbf{a}_1 = \begin{bmatrix} 4 & -3 & 3 & -9 \\ -3 & 6 & -3 & 11 \\ 0 & 8 & -5 & 8 \\ 3 & -3 & 3 & -8 \end{bmatrix} - \begin{Bmatrix} 1 \\ -1 \\ -1 \\ 1 \end{Bmatrix} \begin{bmatrix} 4 & -3 & 3 & -9 \end{bmatrix}$$

$$= \begin{bmatrix} 4 & -3 & 3 & -9 \\ -3 & 6 & -3 & 11 \\ 0 & 8 & -5 & 8 \\ 3 & -3 & 3 & -8 \end{bmatrix} - \begin{bmatrix} 4 & -3 & 3 & -9 \\ -4 & 3 & -3 & 9 \\ -4 & 3 & -3 & 9 \\ 4 & -3 & 3 & -9 \end{bmatrix}$$

$$= \begin{bmatrix} 0 & 0 & 0 & 0 \\ 1 & 3 & 0 & 2 \\ 4 & 5 & -2 & -1 \\ -1 & 0 & 0 & 1 \end{bmatrix}$$

Eliminating the first column and the first row of $\mathbf{A}_1$ yields the new, smaller matrix **B**:

$$\mathbf{B} = \begin{bmatrix} 3 & 0 & 2 \\ 5 & -2 & -1 \\ 0 & 0 & 1 \end{bmatrix}$$

Application of the power method to **B** produces the dominant eigenvalue $\lambda = 3$ and its eigenvector $\mathbf{v} = [1 \quad 1 \quad 0]^T$. Note that this is not an eigenvector of **A** corresponding to $\lambda = 3$, which would have been a 4×1 vector. Repeating the deflation process, this time applied to **B**, we find

$$\mathbf{B}_1 = \mathbf{B} - \mathbf{v}\mathbf{b}_1 = \begin{bmatrix} 3 & 0 & 2 \\ 5 & -2 & -1 \\ 0 & 0 & 1 \end{bmatrix} - \begin{Bmatrix} 1 \\ 1 \\ 0 \end{Bmatrix} [3 \quad 0 \quad 2]$$

$$= \begin{bmatrix} 0 & 0 & 0 \\ 2 & -2 & -3 \\ 0 & 0 & 1 \end{bmatrix} \underset{\text{first row, first column}}{\overset{\text{Eliminate}}{\Rightarrow}} \mathbf{C} = \begin{bmatrix} -2 & -3 \\ 0 & 1 \end{bmatrix}$$

Since **C** is upper triangular, its eigenvalues are the diagonal entries −2 and 1. In summary, the four eigenvalues of the original matrix **A** are −5, 3, −2, 1, as asserted.

Example 9.5: The First Component of $\mathbf{v}_1$ is Zero

Consider

$$\mathbf{A} = \begin{bmatrix} 2 & 1 & -1 \\ 1 & 3 & 2 \\ -1 & 2 & 3 \end{bmatrix}$$

The power method gives the dominant eigenvalue $\lambda_1 = 5$ and eigenvector

$$\mathbf{v}_1 = \begin{Bmatrix} 0 \\ 1 \\ 1 \end{Bmatrix}$$

The first component is zero, and the second component is made into 1. The matrix $\mathbf{A}_1$ is now formed differently than Equation 9.11, as follows. Because the second component of $\mathbf{v}_1$ is 1, we consider $\mathbf{a}_2$, the second row of **A**, and perform

$$\mathbf{A}_1 = \mathbf{A} - \mathbf{v}_1\mathbf{a}_2 = \begin{bmatrix} 2 & 1 & -1 \\ 1 & 3 & 2 \\ -1 & 2 & 3 \end{bmatrix} - \begin{Bmatrix} 0 \\ 1 \\ 1 \end{Bmatrix} \begin{bmatrix} 1 & 3 & 2 \end{bmatrix} = \begin{bmatrix} 2 & 1 & -1 \\ 0 & 0 & 0 \\ -2 & -1 & 1 \end{bmatrix}$$

Eliminating the second row and the second column of $\mathbf{A}_1$ yields

$$\mathbf{B} = \begin{bmatrix} 2 & -1 \\ -2 & 1 \end{bmatrix}$$

The two eigenvalues of **B** are subsequently found to be 0,3. In conclusion, the three eigenvalues of the original matrix **A** are 5,3,0, which may be directly verified.

9.3 Householder Tridiagonalization and QR Factorization Methods

The methods presented so far to estimate eigenvalues and eigenvectors of a matrix can be tedious and are also prone to round-off errors, the latter particularly evident in the case of repeated application of power and deflation methods. The deflation process relies greatly on the available eigenvalue and its corresponding eigenvector, which are often provided by the power method. But because these are only estimates, the entries of the ensuing deflated matrix are also not exact. This approximated matrix is then subjected to the power method, which approximates its dominant eigenvalue, causing an accumulated round-off error. Therefore, repeated application of this process can pose serious round-off problems.

Other, more common, techniques to estimate the eigenvalues of a matrix are mostly two-step methods. In the first step, the original matrix is transformed into a simpler form, such as tridiagonal, which has the same eigenvalues as the original matrix. In the second step, the eigenvalues of this simpler matrix are found iteratively. A majority of these methods are designed to specifically handle symmetric matrices. Jacobi's method, for example, transforms a symmetric matrix into a diagonal one. This method, however, is not very efficient because as it zeros out an off-diagonal entry, it often creates a new, nonzero entry at the location where a zero was previously generated. A more polished technique is Givens' method, which transforms a symmetric matrix into a tridiagonal matrix. It should be mentioned that Givens' method can also be applied to a general, nonsymmetric matrix. In this event, the original matrix is transformed into a special matrix that is no longer tridiagonal but one known as the Hessenberg matrix, discussed later in this section.

The most efficient and commonly used method, however, is Householder's method, which also transforms a symmetric matrix into a tridiagonal matrix. Like Givens' method, Householder's method also applies to nonsymmetric matrices, transforming them into the Hessenberg form. The outcome of Householder's method is then subjected to repeated application of QR factorization—covered later in this section—to reduce it to a matrix whose off-diagonal elements are considerably smaller than its diagonal entries. Ultimately, these diagonal entries serve as estimates of the eigenvalues of the original matrix. Householder's method is a so-called similarity transformation method. Recall from Chapter 1 that matrices $\mathbf{A}_{n\times n}$ and $\mathbf{B}_{n\times n}$ are similar if there exists a nonsingular matrix $\mathbf{S}_{n\times n}$ such that

$$\mathbf{B} = \mathbf{S}^{-1}\mathbf{A}\mathbf{S}$$

We say $\mathbf{B}$ is obtained through a similarity transformation of $\mathbf{A}$. Similarity transformations preserve eigenvalues, that is, $\mathbf{A}$ and $\mathbf{B}$ have exactly the same characteristic polynomial, and hence the same eigenvalues.

9.3.1 Householder's Tridiagonalization Method (Symmetric Matrices)

Let $\mathbf{A} = [a_{ij}]_{n\times n}$ be a real, symmetric matrix whose eigenvalues are $\lambda_1, \lambda_2, \ldots, \lambda_n$. Householder's method uses $n - 2$ successive similarity transformations to reduce $\mathbf{A}$ into $\mathbf{T}$. Let $\mathbf{P}_1, \mathbf{P}_2, \ldots, \mathbf{P}_{n-2}$ denote the matrices used in this process where each $\mathbf{P}_k$ is symmetric and orthogonal, that is,

$$\mathbf{P}_k = \mathbf{P}_k^T = \mathbf{P}_k^{-1} \quad (k = 1, 2, \ldots, n-2)$$

Generate a sequence of matrices $\mathbf{A}_k$ $(k = 1, 2, \ldots, n - 2)$, as

$$\begin{aligned} \mathbf{A}_0 &= \mathbf{A} \\ \mathbf{A}_1 &= \mathbf{P}_1^{-1}\mathbf{A}_0\mathbf{P}_1 = \mathbf{P}_1\mathbf{A}_0\mathbf{P}_1 \\ \mathbf{A}_2 &= \mathbf{P}_2\mathbf{A}_1\mathbf{P}_2 \\ &\ldots \\ \mathbf{A}_{n-3} &= \mathbf{P}_{n-3}\mathbf{A}_{n-4}\mathbf{P}_{n-3} \\ \mathbf{A}_{n-2} &= \mathbf{P}_{n-2}\mathbf{A}_{n-3}\mathbf{P}_{n-2} \end{aligned} \tag{9.14}$$

In the first iteration, we create zeros in the appropriate slots in the first row and the first column of $\mathbf{A}$ to obtain a new matrix $\mathbf{A}_1 = \left[a_{ij}^{(1)}\right]$, as shown below. Then, in the second iteration, zeros are generated in the appropriate locations in the second row and the second column of $\mathbf{A}_1$ to obtain $\mathbf{A}_2 = \left[a_{ij}^{(2)}\right]$, shown below.

$$\mathbf{A}_1 = \begin{bmatrix} a_{11}^{(1)} & a_{12}^{(1)} & 0 & \dots & 0 \\ a_{21}^{(1)} & a_{22}^{(1)} & \dots & \dots & a_{2n}^{(1)} \\ 0 & a_{32}^{(1)} & \dots & \dots & a_{3n}^{(1)} \\ \dots & \dots & \dots & \dots & \dots \\ 0 & a_{n2}^{(1)} & \dots & \dots & a_{nn}^{(1)} \end{bmatrix}, \quad \mathbf{A}_2 = \begin{bmatrix} a_{11}^{(2)} & a_{12}^{(2)} & 0 & \dots & 0 \\ a_{21}^{(2)} & a_{22}^{(2)} & a_{23}^{(2)} & \dots & 0 \\ 0 & a_{32}^{(2)} & \dots & \dots & a_{3n}^{(2)} \\ \dots & 0 & \dots & \dots & \dots \\ 0 & 0 & a_{n3}^{(2)} & \dots & a_{nn}^{(2)} \end{bmatrix}$$

Conducting this process $n-2$ times, yields a tridiagonal matrix $\mathbf{A}_{n-2} = \left[a_{ij}^{(n-2)}\right]$, where

$$\mathbf{T} = \mathbf{A}_{n-2} = \begin{bmatrix} a_{11}^{(n-2)} & a_{12}^{(n-2)} & 0 & \dots & 0 \\ a_{21}^{(n-2)} & a_{22}^{(n-2)} & a_{23}^{(n-2)} & 0 & \dots \\ 0 & a_{32}^{(n-2)} & \dots & \dots & 0 \\ \dots & 0 & \dots & \dots & a_{n-1,n}^{(n-2)} \\ 0 & \dots & 0 & a_{n,n-1}^{(n-2)} & a_{nn}^{(n-2)} \end{bmatrix} \tag{9.15}$$

9.3.1.1 Determination of Symmetric Orthogonal $\mathbf{P}_k$ ($k = 1, 2, \dots, n-2$)

Each matrix $\mathbf{P}_k$ is defined by

$$\mathbf{P}_k = \mathbf{I} - 2\mathbf{v}_k\mathbf{v}_k^T, \quad k = 1, 2, \dots, n-2 \tag{9.16}$$

where $\mathbf{v}_k$ is a unit vector ($\mathbf{v}_k^T\mathbf{v}_k = 1$), the first k components of which are zero. Moreover, it can be verified that each $\mathbf{P}_k$ is symmetric and orthogonal. To further understand the structure of the unit vectors $\mathbf{v}_k$, let us consider $\mathbf{v}_1$ first:

$$\mathbf{v}_1 = \begin{Bmatrix} 0 \\ v_{21} \\ \dots \\ v_{n1} \end{Bmatrix} \quad \text{where}$$

$$v_{21} = \sqrt{\frac{1}{2}\left(1 + \frac{|a_{21}|}{\Sigma_1}\right)} \quad \text{where } \Sigma_1 = \sqrt{a_{21}^2 + a_{31}^2 + \dots + a_{n1}^2} \tag{9.17}$$

$$v_{i1} = \begin{cases} \dfrac{a_{i1}}{2v_{21}\Sigma_1} & \text{if } a_{21} \ge 0 \\ \dfrac{-a_{i1}}{2v_{21}\Sigma_1} & \text{if } a_{21} < 0 \end{cases}, \quad i = 3, 4, \dots, n$$

Note that since $\mathbf{v}_1$ is used to form $\mathbf{P}_1$, which in turn is involved in the first iteration of Equation 9.14, the entries of $\mathbf{A} = \mathbf{A}_0$ are used in Equation 9.17 in the construction of $\mathbf{v}_1$. Similarly, we construct the unit vector $\mathbf{v}_2$ this time by using the entries of $\mathbf{A}_1 = \left[a_{ij}^{(1)}\right]$ from the second iteration of Equation 9.14,

$$\mathbf{v}_2 = \begin{Bmatrix} 0 \\ 0 \\ v_{32} \\ \cdots \\ v_{n2} \end{Bmatrix} \quad \text{where}$$

$$v_{32} = \sqrt{\frac{1}{2}\left(1 + \frac{\left|a_{32}^{(1)}\right|}{\Sigma_2}\right)} \quad \text{where } \Sigma_2 = \sqrt{\left[a_{32}^{(1)}\right]^2 + \left[a_{42}^{(1)}\right]^2 + \cdots + \left[a_{n2}^{(1)}\right]^2} \tag{9.18}$$

$$v_{i2} = \begin{cases} \dfrac{a_{i2}^{(1)}}{2v_{32}\Sigma_2} & \text{if } a_{32}^{(1)} \geq 0 \\ \dfrac{-a_{i2}^{(1)}}{2v_{32}\Sigma_2} & \text{if } a_{32}^{(1)} < 0 \end{cases}, \quad i = 4, 5, \ldots, n$$

Continuing this way $\mathbf{v}_{n-2}$ is constructed, and subsequently, the tridiagonal matrix $\mathbf{T} = \mathbf{A}_{n-2}$ is obtained. Now, there are two possible scenarios: If the entries of $\mathbf{T}$ along the sub- and super-diagonals are notably smaller in magnitude than those along the main diagonal, then $\mathbf{T}$ is regarded as almost diagonal, and the diagonal entries roughly approximate its eigenvalues, hence those of the original matrix $\mathbf{A}$. If not, we proceed to further transform $\mathbf{T}$ into a tridiagonal matrix whose diagonal elements dominate all other entries. This will be accomplished using the QR factorization.

The user-defined function Householder uses Householder's method to transform a real, symmetric matrix into a tridiagonal matrix.

```
function T = Householder(A)
%
% Householder uses Householder's method to transform a
% symmetric matrix into a tridiagonal matrix.
%
%  T = Householder(A) where
%
%       A is an n-by-n real, symmetric matrix,
%
%       T is the n-by-n tridiagonal matrix.

N = size(A, 1);
```

```
for n = 1:N-2,
    G = sqrt(A(n + 1:end, n)'*A(n + 1:end, n)); % Compute gamma
    v(1:N, 1) = 0;    % Set initial set of entries to 0
    v(n+1) = sqrt((1+abs(A(n + 1, n))/G)/2);
    % First nonzero entry
    sn = sign(A(n+1, n));
    % Determine sign of relevant entry
    v(n+2:N) = sn*A(n+2:end, n)/2/v(n + 1)/G;
    % Compute remaining entries
    P = eye(N) - 2*(v*v');
    % Compute the symmetric, orthogonal matrices
    A = P\A*P;    % Compute sequence of matrices
end
T = A;
```

Example 9.6: Householder's Method

Consider

$$\mathbf{A} = \begin{bmatrix} 4 & 4 & 1 & 1 \\ 4 & 4 & 1 & 1 \\ 1 & 1 & 3 & 2 \\ 1 & 1 & 2 & 3 \end{bmatrix}$$

Since $n = 4$, matrices $\mathbf{A}_1$ and $\mathbf{A}_2$ are generated by Equation 9.14 using $\mathbf{P}_1$ and $\mathbf{P}_2$, as follows. Find the unit vectors $\mathbf{v}_1$ and $\mathbf{v}_2$ by Equations 9.17 and 9.18, respectively. First,

$$\mathbf{v}_1 = \begin{Bmatrix} 0 \\ v_{21} \\ v_{31} \\ v_{41} \end{Bmatrix}, \quad \Sigma_1 = \sqrt{18}, \quad \begin{matrix} v_{21} = 0.9856 \\ v_{31} = 0.1196 = v_{41} \end{matrix} \quad \Rightarrow \mathbf{v}_1 = \begin{Bmatrix} 0 \\ 0.9856 \\ 0.1196 \\ 0.1196 \end{Bmatrix}$$

By Equation 9.16, we find $\mathbf{P}_1 = \mathbf{I} - 2\mathbf{v}_1\mathbf{v}_1^T$, and subsequently,

$$\mathbf{A}_1 = \mathbf{P}_1\mathbf{A}\mathbf{P}_1 = \begin{bmatrix} 4 & -4.2426 & 0 & 0 \\ -4.2426 & 5 & -1 & -1 \\ 0 & -1 & 2.5 & 1.5 \\ 0 & -1 & 1.5 & 2.5 \end{bmatrix} \quad \text{so that} \quad \begin{cases} a_{32}^{(1)} = -1 \\ a_{42}^{(1)} = -1 \end{cases}$$

Using this in Equation 9.18,

$$\mathbf{v}_2 = \begin{Bmatrix} 0 \\ 0 \\ v_{32} \\ v_{42} \end{Bmatrix}, \quad \Sigma_2 = \sqrt{2}, \quad \begin{matrix} v_{32} = 0.9239 \\ v_{42} = 0.3827 \end{matrix} \quad \Rightarrow \quad \mathbf{v}_2 = \begin{Bmatrix} 0 \\ 0 \\ 0.9239 \\ 0.3827 \end{Bmatrix}$$

Form $\mathbf{P}_2 = \mathbf{I} - 2\mathbf{v}_2\mathbf{v}_2^T$, and

$$\mathbf{A}_2 = \mathbf{P}_2\mathbf{A}_1\mathbf{P}_2 = \begin{bmatrix} 4 & -4.2426 & 0 & 0 \\ -4.2426 & 5 & 1.4142 & 0 \\ 0 & 1.4142 & 4 & 0 \\ 0 & 0 & 0 & 1 \end{bmatrix}$$

which is symmetric and tridiagonal, as projected. Executing the user-defined function `Householder` will confirm this.

```
>> A= [4 4 1 1;4 4 1 1;1 1 3 2;1 1 2 3];
>> T=Householder(A)

T =
    4.0000   -4.2426    0.0000    0.0000
   -4.2426    5.0000    1.4142         0
   -0.0000    1.4142    4.0000    0.0000
    0.0000    0.0000    0.0000    1.0000
```

9.3.2 QR Factorization Method

Once a special matrix such as tridiagonal matrix $\mathbf{T}$ is obtained via Householder's method, the goal is to transform it into a new tridiagonal matrix whose off-diagonal entries are considerably smaller (in magnitude) than the diagonal ones. For this purpose, we employ the QR factorization (or decomposition) method. This is based on the fact that any matrix $\mathbf{M}$ can be decomposed into a product, $\mathbf{M} = \mathbf{QR}$ where $\mathbf{Q}$ is orthogonal and $\mathbf{R}$ is upper triangular.

Start the process by setting $\mathbf{T}_0 = \mathbf{T}$, and factorize it as $\mathbf{T}_0 = \mathbf{Q}_0\mathbf{R}_0$. Once $\mathbf{Q}_0$ and $\mathbf{R}_0$ have been identified, we multiply them in reverse order to form a new matrix $\mathbf{R}_0\mathbf{Q}_0 = \mathbf{T}_1$. Then, apply the QR factorization to $\mathbf{T}_1$ to achieve $\mathbf{Q}_1\mathbf{R}_1$, multiply $\mathbf{Q}_1$ and $\mathbf{R}_1$ in reverse order to create $\mathbf{T}_2 = \mathbf{R}_1\mathbf{Q}_1$, and so on. In this fashion, a sequence $\mathbf{T}_k$ of tridiagonal matrices is generated, as

$$\begin{array}{ll} \mathbf{T}_0 = \mathbf{T} = \mathbf{Q}_0\mathbf{R}_0 & \mathbf{T}_1 = \mathbf{R}_0\mathbf{Q}_0 \\ \mathbf{T}_1 = \mathbf{Q}_1\mathbf{R}_1 & \mathbf{T}_2 = \mathbf{R}_1\mathbf{Q}_1 \\ \cdots & \cdots \\ \mathbf{T}_k = \mathbf{Q}_k\mathbf{R}_k & \mathbf{T}_{k+1} = \mathbf{R}_k\mathbf{Q}_k \end{array} \tag{9.19}$$

Using the last relation in Equation 9.19, we find

$$\mathbf{R}_k = \mathbf{Q}_k^{-1}\mathbf{T}_k \quad \Rightarrow \quad \mathbf{T}_{k+1} = \mathbf{Q}_k^{-1}\mathbf{T}_k\mathbf{Q}_k$$

so that $\mathbf{T}_{k+1}$ and $\mathbf{T}_k$ are similar matrices, and have the same eigenvalues. In fact, it is easy to see that $\mathbf{T}_{k+1}$ is similar to the original tridiagonal matrix $\mathbf{T}$. If the eigenvalues of $\mathbf{T}$ have the property that $|\lambda_1| > |\lambda_2| > \cdots > |\lambda_n|$, it can then be shown that

$$\mathbf{T}_k \to \Lambda = \begin{bmatrix} \lambda_1 & & & \\ & \lambda_2 & & \\ & & \cdots & \\ & & & \lambda_n \end{bmatrix} \quad \text{as } k \to \infty$$

Therefore, theoretically speaking, the sequence generated by Equation 9.19 converges to a diagonal matrix consisting of the eigenvalues of $\mathbf{T}$. The primary challenge is the construction of matrices $\mathbf{Q}_k$ and $\mathbf{R}_k$ in each iteration step of Equation 9.19.

9.3.2.1 Determination of $\mathbf{Q}_k$ and $\mathbf{R}_k$ Matrices

We begin with the first relation in Equation 9.19, and determine $\mathbf{Q}_0$ and $\mathbf{R}_0$ for $\mathbf{T}_0 = [t_{ij}]$. Pre-multiply $\mathbf{T}_0$ by an $n \times n$ matrix $\mathbf{L}_2$, the result denoted by $\mathbf{L}_2\mathbf{T}_0 = \left[t_{ij}^{(2)}\right]$, such that $t_{21}^{(2)} = 0$. Then, pre-multiply this matrix by $\mathbf{L}_3$, denoting the product by $\mathbf{L}_3(\mathbf{L}_2\mathbf{T}_0) = \left[t_{ij}^{(3)}\right]$, so that $t_{32}^{(3)} = 0$. Performing $n - 1$ of these operations yields an upper-triangular matrix $\mathbf{R}_0$, that is,

$$\mathbf{L}_n\mathbf{L}_{n-1} \ldots \mathbf{L}_3\mathbf{L}_2\mathbf{T}_0 = \mathbf{R}_0 \tag{9.20}$$

We will see later that $\mathbf{L}_k$ ($k = 2, 3, \ldots, n$) are orthogonal. Manipulation of Equation 9.20 results in QR factorization of $\mathbf{T}_0$,

$$\mathbf{T}_0 = (\mathbf{L}_n\mathbf{L}_{n-1}\ldots\mathbf{L}_3\mathbf{L}_2)^{-1}\mathbf{R}_0 \quad \Rightarrow \quad \mathbf{T}_0 = \mathbf{Q}_0\mathbf{R}_0 \tag{9.21}$$

where

$$\mathbf{Q}_0 = \left(\mathbf{L}_n\mathbf{L}_{n-1}\ldots\mathbf{L}_3\mathbf{L}_2\right)^{-1} = \mathbf{L}_2^{-1}\mathbf{L}_3^{-1}\ldots\mathbf{L}_n^{-1} = \mathbf{L}_2^T\mathbf{L}_3^T\ldots\mathbf{L}_n^T$$

Note that the orthogonality of $\mathbf{L}_k$ has been utilized.

9.3.2.2 Structure of $\mathbf{L}_k$ ($k = 2, 3, \ldots, n$)

The $\mathbf{L}_k$ matrices are generally simple in nature in the sense that each $\mathbf{L}_k$ consists of a 2×2 submatrix that occupies rows k and $k-1$, and columns k and $k-1$, and ones along the remaining portion of the main diagonal, and zeros everywhere else. The 2×2 submatrix has the form of a clockwise rotation matrix,

$$\begin{bmatrix} \cos\theta_k & \sin\theta_k \\ -\sin\theta_k & \cos\theta_k \end{bmatrix} \quad \text{or simply} \quad \begin{bmatrix} c_k & s_k \\ -s_k & c_k \end{bmatrix} \quad \text{where} \quad \begin{aligned} c_k &= \cos\theta_k \\ s_k &= \sin\theta_k \end{aligned}$$

and θ_k is to be chosen appropriately. For instance, if the size of the matrices involved is $n = 5$, then

$$\mathbf{L}_2 = \begin{bmatrix} c_2 & s_2 & & & \\ -s_2 & c_2 & & & \\ & & 1 & & \\ & & & 1 & \\ & & & & 1 \end{bmatrix}, \quad \mathbf{L}_4 = \begin{bmatrix} 1 & & & & \\ & 1 & & & \\ & & c_4 & s_4 & \\ & & -s_4 & c_4 & \\ & & & & 1 \end{bmatrix}$$

We now address the selection of c_k and s_k. Recall that we must have $\mathbf{L}_2\mathbf{T}_0 = \left[t_{ij}^{(2)}\right]$ such that $t_{21}^{(2)} = 0$, hence in this new matrix the only element that needs to be analyzed is the (2,1) entry. But we determine the (2,1) entry by using the second row of $\mathbf{L}_2$ and first column of $\mathbf{T}_0$. Regardless of the size n, the second row of matrix $\mathbf{L}_2$ is always as shown above. Therefore, the (2,1) entry of $\mathbf{L}_2\mathbf{T}_0$ is given by

$$t_{21}^{(2)} = -s_2 t_{11} + c_2 t_{21}$$

Forcing it to be zero, we find

$$-s_2 t_{11} + c_2 t_{21} = 0 \quad \Rightarrow \quad \frac{s_2}{c_2} = \tan\theta_2 = \frac{t_{21}}{t_{11}} \tag{9.22}$$

Application of trigonometric identities $\cos\alpha = 1/\sqrt{1+\tan^2\alpha}$ and $\sin\alpha = \tan\alpha/\sqrt{1+\tan^2\alpha}$ to Equation 9.22 yields

$$c_2 = \cos\theta_2 = \frac{1}{\sqrt{1+(t_{21}/t_{11})^2}}, \quad s_2 = \sin\theta_2 = \frac{t_{21}/t_{11}}{\sqrt{1+(t_{21}/t_{11})^2}} \tag{9.23}$$

With this, matrix $\mathbf{L}_2$ is completely defined. Next, consider the matrix $\mathbf{L}_3(\mathbf{L}_2\mathbf{T}_0) = \left[t_{ij}^{(3)}\right]$, of which the (3,2) entry must be made into zero. Proceeding as above, we can obtain c_3 and s_3, and so on. This continues until $\mathbf{L}_n$ is determined, and ultimately, $\mathbf{L}_n\mathbf{L}_{n-1} \ldots \mathbf{L}_3\mathbf{L}_2\mathbf{T}_0 = \mathbf{R}_0$. Once all $\mathbf{L}_k$ matrices have been found, we form $\mathbf{Q}_0 = \mathbf{L}_2^T\mathbf{L}_3^T \ldots \mathbf{L}_n^T$, by Equation 9.21, and the first QR factorization is complete. Next, form the new symmetric tridiagonal matrix $\mathbf{T}_1 = \mathbf{R}_0\mathbf{Q}_0$. If the off-diagonal elements are much smaller than the diagonal ones, the process is terminated and the diagonal entries of $\mathbf{T}_1$ approximate the eigenvalues of $\mathbf{A}$. Otherwise, the QR factorization is repeated for $\mathbf{T}_1$ via the steps listed above until a desired tridiagonal matrix is achieved.

9.3.2.3 MATLAB® Built-In Function `qr`

MATLAB has a built-in function that performs the QR factorization of a matrix:

```
qr   Orthogonal-triangular decomposition.

     [Q,R] = qr(A), where A is m-by-n, produces an m-by-n upper
     triangular matrix R and an m-by-m unitary matrix Q so
     that A = Q*R.
```

The user-defined function `HouseholderQR` uses Householder's method to transform a real, symmetric matrix into a tridiagonal matrix, to which the QR factorization is repeatedly applied in order to obtain a tridiagonal matrix whose diagonal entries are much larger than those along the super- and sub-diagonals.

```
function [T Tfinal eigenvals] = HouseholderQR(A, tol, kmax)
%
% HouseholderQR uses Householder's method and repeated
% application of QR factorization to estimate the
% eigenvalues of a real, symmetric matrix.
%
%  [T Tfinal eigenvals] = HouseholderQR(A, tol, kmax) where
%
%    A is an n-by-n real, symmetric matrix,
%    tol is the tolerance used in the QR process
%    (default 1e-6),
%    kmax is the maximum number of QR iterations
%    (default 50),
%
%    T is the tridiagonal matrix created by Householder's
%    method,
%    Tfinal is the final tridiagonal matrix,
%    eigenvals is a list of estimated eigenvalues of
%    matrix A.
%
```

```
% Note that this function calls the user-defined function
% Householder

if nargin < 2 || isempty(tol)
     tol = 1e-6;
end

if nargin < 3 || isempty(kmax)
     kmax = 50;
end

T = Householder(A);     % Call Householder
T(:,:,1) = T;

% QR factorization to reduce the off-diagonal entries of T
for m = 1:kmax,
    [Q, R] = qr(T(:,:,m));
    T(:,:,m+1)=R*Q;
    % Compare diagonals of two successive T matrices
    if norm(diag(T(:,:,m+1))-diag(T(:,:,m)))<tol,
       break;
    end
end

Tfinal = T(:,:,end);
T = T(:,:,1);
eigenvals = diag(Tfinal);
```

A Note on the Terminating Condition Used in "`HouseholderQR`"
The terminating condition employed here is based on the norm of the difference between two vectors whose components are the diagonal entries of two successive **T** matrices generated in the QR process. Although this works in most cases, there could be situations where the two aforementioned vectors have elements that are close to one another, hence meeting the tolerance condition, but the off-diagonal entries are not sufficiently small. One way to remedy this is to use a more firm terminating condition that inspects the ratio of the largest (magnitude) sub-diagonal entry to the smallest diagonal element. If this ratio is within the tolerance, the process is terminated. This will ensure that the sub-diagonal elements are considerably smaller than the diagonal ones. (See the Problem Set at the end of this chapter.)

Example 9.7: Householder's Method + QR Factorization

Consider Example 9.6 where a symmetric matrix was transformed into tridiagonal using Householder's method:

$$\mathbf{A} = \begin{bmatrix} 4 & 4 & 1 & 1 \\ 4 & 4 & 1 & 1 \\ 1 & 1 & 3 & 2 \\ 1 & 1 & 2 & 3 \end{bmatrix} \rightarrow \mathbf{T} = \begin{bmatrix} 4 & -4.2426 & 0 & 0 \\ -4.2426 & 5 & 1.4142 & 0 \\ 0 & 1.4142 & 4 & 0 \\ 0 & 0 & 0 & 1 \end{bmatrix}$$

Interestingly, $\mathbf{T}$ is in the block diagonal form (Chapter 1). Therefore, eigenvalues of $\mathbf{T}$ consist of the eigenvalues of the upper-left 3×3 block matrix, and a 1×1 block of 1. So, one eigenvalue ($\lambda_1 = 1$) is automatically decided. As a result, we now focus on the upper-left 3×3 block and find its eigenvalues. Note that this phenomenon does not generally occur and should be investigated on a case-by-case basis.

Therefore, we will proceed with the aforementioned 3×3 block matrix and repeatedly apply the QR factorization to this matrix. The process is initiated by setting

$$\mathbf{T}_0 = \begin{bmatrix} 4 & -4.2426 & 0 \\ -4.2426 & 5 & 1.4142 \\ 0 & 1.4142 & 4 \end{bmatrix}$$

The QR factorizations listed in Equation 9.19 are performed as follows. First, by Equation 9.23,

$$c_2 = 0.6860, \quad s_2 = -0.7276$$

so that

$$\mathbf{L}_2 = \begin{bmatrix} c_2 & s_2 & 0 \\ -s_2 & c_2 & 0 \\ 0 & 0 & 1 \end{bmatrix} \text{ and } \mathbf{L}_2\mathbf{T}_0 = \begin{bmatrix} 5.8310 & -6.5485 & -1.0290 \\ 0 & 0.3430 & 0.9701 \\ 0 & 1.4142 & 4 \end{bmatrix}$$

Next, the (3,2) entry of $\mathbf{L}_3\mathbf{L}_2\mathbf{T}_0$ is forced to be zero, and yields

$$-0.3430 s_3 + 1.4142 c_3 = 0 \Rightarrow \begin{matrix} c_3 = 0.2357 \\ s_3 = 0.9718 \end{matrix}$$

so that

$$\mathbf{L}_3 = \begin{bmatrix} 1 & 0 & 0 \\ 0 & c_3 & s_3 \\ 0 & -s_3 & c_3 \end{bmatrix} \text{ and}$$

$$\mathbf{R}_0 = \mathbf{L}_3\mathbf{L}_2\mathbf{T}_0 = \begin{bmatrix} 5.8310 & -6.5485 & -1.0290 \\ 0 & 1.4552 & 4.1160 \\ 0 & 0 & 0 \end{bmatrix}$$

Finally, letting $\mathbf{Q}_0 = \mathbf{L}_2^T\mathbf{L}_3^T$, we obtain

$$\mathbf{T}_1 = \mathbf{R}_0\mathbf{Q}_0 = \begin{bmatrix} 8.7647 & -1.0588 & 0 \\ -1.0588 & 4.2353 & 0 \\ 0 & 0 & 0 \end{bmatrix}$$

This tridiagonal matrix is also in a block diagonal form, including an upper-left 2×2 clock and a 1×1 block of 0. This implies one of its eigenvalues must be zero, hence $\lambda_2 = 0$. The remaining two eigenvalues of **A** are the eigenvalues of the upper-left 2×2 block matrix. We will proceed with the second iteration in Equation 9.19 using this 2×2 block matrix. So, we let

$$\mathbf{T}_1 = \begin{bmatrix} 8.7647 & -1.0588 \\ -1.0588 & 4.2353 \end{bmatrix}$$

Since the off-diagonal entries are still relatively large, matrix $\mathbf{T}_2$ must be formed, that is,

$$\mathbf{L}_2^{(1)} = \begin{bmatrix} c_2^{(1)} & s_2^{(1)} \\ -s_2^{(1)} & c_2^{(1)} \end{bmatrix} \quad \text{with} \quad \begin{aligned} c_2^{(1)} &= 0.9928 \\ s_2^{(1)} &= -0.1199 \end{aligned}$$

$$\text{so that} \quad \mathbf{R}_1 = \mathbf{L}_2^{(1)}\mathbf{T}_1 = \begin{bmatrix} 8.8284 & -1.5591 \\ 0 & 4.0777 \end{bmatrix}$$

Noting that $\mathbf{Q}_1 = \left[\mathbf{L}_2^{(1)}\right]^T$, we have

$$\mathbf{T}_2 = \mathbf{R}_1\mathbf{Q}_1 = \begin{bmatrix} 8.9517 & -0.4891 \\ -0.4891 & 4.0483 \end{bmatrix}$$

Finally, because the off-diagonal entries are considerably smaller in magnitude than the diagonal ones, the eigenvalues are approximately 9 and 4. Therefore, $\lambda(\mathbf{A}) = 0,1,4,9$. Executing the user-defined function `HouseholderQR` will confirm these results.

```
>> A=[4 4 1 1;4 4 1 1;1 1 3 2;1 1 2 3];
>> [T Tfinal eigenvals] =HouseholderQR(A)

T =    % Tridiagonal matrix generated by Householder
(see Example 9.6)

    4.0000    -4.2426     0.0000     0.0000
   -4.2426     5.0000     1.4142          0
   -0.0000     1.4142     4.0000     0.0000
    0.0000     0.0000     0.0000     1.0000
```

```
Tfinal =  % Tridiagonal matrix at the conclusion of QR process

    9.0000   -0.0008   -0.0000    0.0000
   -0.0008    4.0000   -0.0000   -0.0000
   -0.0000    0.0000    1.0000    0.0000
    0.0000    0.0000   -0.0000    0.0000

eigenvals =   % List of (estimated) eigenvalues of A

    9.0000
    4.0000
    1.0000
    0.0000
```

9.3.3 Transformation to Hessenberg Form (Nonsymmetric Matrices)

As mentioned at the outset of this section, Householder's method can also be applied to nonsymmetric matrices. Instead of a tridiagonal matrix, however, the outcome will be another special matrix known as the upper Hessenberg form

$$\mathbf{H} = \begin{bmatrix} h_{11} & h_{12} & h_{13} & \dots & h_{1n} \\ h_{21} & h_{22} & h_{23} & \dots & h_{2n} \\ & h_{32} & h_{33} & \dots & h_{3n} \\ & & \dots & \dots & \dots \\ & & & h_{n,n-1} & h_{nn} \end{bmatrix}$$

which is upper triangular plus the subdiagonal. In the second step of the process, repeated QR factorizations will be applied to **H** in order to obtain an upper triangular matrix. Since the eigenvalues of an upper triangular matrix are its diagonal entries, the eigenvalues of this final matrix, and hence of the original matrix, are along its diagonal.

The user-defined function `HouseholderQR`, which calls another user-defined function `Householder`, can be applied to any nonsymmetric matrix to accomplish this task.

Example 9.8: Nonsymmetric Matrix, Hessenberg Form

Consider the (nonsymmetric) matrix studied in Examples 9.3 and 9.4 of the last section:

$$\mathbf{A} = \begin{bmatrix} 4 & -3 & 3 & -9 \\ -3 & 6 & -3 & 11 \\ 0 & 8 & -5 & 8 \\ 3 & -3 & 3 & -8 \end{bmatrix}$$

```
>> A=[4 -3 3 -9;-3 6 -3 11;0 8 -5 8;3 -3 3 -8];
>> [H Hfinal eigenvals]=HouseholderQR(A)

H =          % Hessenberg form generated by Householder

    4.0000    -4.2426     3.8787    -8.1213
    4.2426    -5.0000     6.8995   -12.8995
    0.0000    -0.0000     4.6569    -9.6569
   -0.0000    -0.0000     1.6569    -6.6569

Hfinal =    % (Almost) upper triangular matrix at the
conclusion of QR process

   -5.0003     5.1960    13.4724   -16.2633
   -0.0002    -1.9997    -3.5347    -2.8588
    0.0000    -0.0000     3.0000    -1.1547
   -0.0000     0.0000     0.0000     1.0000

eigenvals =       % List of (estimated) eigenvalues of A

   -5.0003
   -1.9997
    3.0000
    1.0000
```

Note that `Hfinal` is not entirely upper triangular because we opted to use the default `tol=1e-6` in the user-defined function. A smaller value further eliminates the (2,1) entry in `Hfinal`.

PROBLEM SET

Power Method (Section 9.1)

In Problems 1 through 6, for each matrix $\mathbf{A}_{n\times n}$,

a. ✍ Starting with $\alpha_1 = 0$ and an $n \times 1$ initial vector $\mathbf{x}_0$ comprised of all ones, apply the power method to generate the scalars α_2 and α_3, and the normalized vectors $\mathbf{x}_2$ and $\mathbf{x}_3$.

b. Find the dominant eigenvalue and the corresponding eigenvector by executing the user-defined function `PowerMethod`.

1. $\mathbf{A} = \begin{bmatrix} 1 & 0 & 1 \\ 0 & 1 & 0 \\ 1 & 0 & 1 \end{bmatrix}$

2. $\mathbf{A} = \begin{bmatrix} 3 & -4 & -2 \\ -1 & 4 & 1 \\ 2 & -6 & -1 \end{bmatrix}$

3. $\mathbf{A} = \begin{bmatrix} 2 & -6 & -3 \\ 1 & 1 & -1 \\ -2 & -4 & 1 \end{bmatrix}$

4. $\mathbf{A} = \begin{bmatrix} 2 & -2 & 4 \\ -1 & 3 & 2 \\ -5 & 10 & -1 \end{bmatrix}$

5. $\mathbf{A} = \begin{bmatrix} 2 & 2 & -2 & 0 \\ -1 & -3 & 2 & -2 \\ 0 & -3 & 2 & -3 \\ 1 & 2 & -2 & 1 \end{bmatrix}$

6. $\mathbf{A} = \begin{bmatrix} 4 & 1 & 1 & 0 \\ 1 & 4 & 0 & 1 \\ 1 & 0 & 4 & 1 \\ 0 & 1 & 1 & 4 \end{bmatrix}$

In Problems 7 through 10 find all eigenvalues and eigenvectors of the matrix by applying the user-defined functions `PowerMethod` and/or `InvPowerMethod`. Use the initial vector $\mathbf{x}_0$ provided or the default, whichever works, in each application of these two functions.

7. $\mathbf{A} = \begin{bmatrix} 21 & -14 & 64 \\ -23 & 10 & -60 \\ -11 & 7 & -33 \end{bmatrix}, \quad \mathbf{x}_0 = \begin{Bmatrix} 1 \\ 0 \\ 1 \end{Bmatrix}$

8. $\mathbf{A} = \begin{bmatrix} 23 & -48 & 24 \\ 12 & -25 & 12 \\ 4 & -10 & 9 \end{bmatrix}, \quad \mathbf{x}_0 = \begin{Bmatrix} 1 \\ 0 \\ 1 \end{Bmatrix}$

9. $\mathbf{A} = \begin{bmatrix} -16 & -8 & 84 \\ -6 & -8 & 42 \\ -2 & -3 & 15 \end{bmatrix}, \quad \mathbf{x}_0 = \begin{Bmatrix} 0 \\ 1 \\ 1 \end{Bmatrix}$

10. $\mathbf{A} = \begin{bmatrix} -9 & -2 & 2 & 14 \\ 8 & 5 & -2 & -10 \\ 0 & -2 & 5 & -2 \\ -8 & -2 & 2 & 13 \end{bmatrix}, \quad \mathbf{x}_0 = \begin{Bmatrix} 1 \\ 0 \\ 0 \\ 0 \end{Bmatrix}$

Deflation Methods (Section 9.2)

In Problems 11 through 17, for each matrix **A**,

a. Find the dominant eigenvalue and corresponding eigenvector using `PowerMethod`. Use the default initial vector $\mathbf{x}_0$ unless otherwise specified.

b. Using Wielandt's deflation method, create a matrix $\mathbf{A}_1$, then deflate to a matrix $\mathbf{B}$, one size smaller than $\mathbf{A}$, whose eigenvalues are the remaining eigenvalues of $\mathbf{A}$.

11. $\mathbf{A} = \begin{bmatrix} 4 & -1 & 1 \\ -1 & 3 & -2 \\ 1 & -2 & 3 \end{bmatrix}$

12. $\mathbf{A} = \begin{bmatrix} 3 & -1 & 0 \\ -1 & 2 & -1 \\ 0 & -1 & 3 \end{bmatrix}$

13. $\mathbf{A} = \begin{bmatrix} 2 & -6 & -3 \\ 1 & 1 & -1 \\ -2 & -4 & 1 \end{bmatrix}$

14. $\mathbf{A} = \begin{bmatrix} 4 & -3 & 3 & -9 \\ -3 & 6 & -3 & 11 \\ 0 & 8 & -5 & 8 \\ 3 & -3 & 3 & -8 \end{bmatrix}$

15. $\mathbf{A} = \begin{bmatrix} 1 & 3 & -3 & 5 \\ 1 & 1 & 3 & -3 \\ 0 & -2 & 6 & -2 \\ -1 & 3 & -3 & 7 \end{bmatrix}$

16. $\mathbf{A} = \begin{bmatrix} -7 & 1 & -1 & 9 \\ 4 & 0 & 1 & -5 \\ 0 & -1 & 2 & -1 \\ -4 & 1 & -1 & 6 \end{bmatrix}, \quad \mathbf{x}_0 = \begin{Bmatrix} 1 \\ 0 \\ 1 \\ 0 \end{Bmatrix}$

17. $\mathbf{A} = \begin{bmatrix} -21 & -22 & 8 \\ 31 & 26 & -2 \\ 19 & 22 & -10 \end{bmatrix}$

18. Power method can handle repeated dominant eigenvalues, that is, two dominant eigenvalues with the same magnitude and same sign. The repeated nature of a dominant eigenvalue only becomes apparent after the successive applications of power and deflation methods. For the following matrix, calculate the dominant eigenvalue and its eigenvector by applying the

user-defined function PowerMethod, and then use this information to deflate to a smaller matrix, ultimately finding all three eigenvalues.

$$\mathbf{A} = \begin{bmatrix} 4 & 5 & 5 \\ 0 & 2 & -2 \\ 0 & -5 & -1 \end{bmatrix}$$

19. Repeat Problem 18 for the following matrix. Use the given initial vector when applying PowerMethod:

$$\mathbf{A} = \begin{bmatrix} -5 & -6 & -7 \\ 0 & -4 & 1 \\ 0 & 6 & 1 \end{bmatrix}, \quad \mathbf{x}_0 = \begin{Bmatrix} 1 \\ 0 \\ 0 \end{Bmatrix}$$

20. Write a user-defined function with function call

```
eigenvals=Deflation_Wielandt(A, x0, tol, kMax)
```

that uses Wielandt's deflation method to find all the eigenvalues of a matrix **A**. It must call the user-defined function PowerMethod to estimate the dominant eigenvalue and corresponding eigenvector of **A**, deflate to a smaller size matrix, and repeat the process until it reaches a 2×2 matrix. At that point, the MATLAB function eig must be used to find the eigenvalues of the 2×2 matrix. The function must return a list of all eigenvalues of matrix **A**. The input arguments have the same default values as in PowerMethod. Apply Deflation_Wielandt (with default inputs) to the matrix **A** in Example 9.4.

Note: A key part of the process is determining the first nonzero component of the eigenvector $\mathbf{v}_1$ in each step. But since PowerMethod generates this vector, it is only an approximation and as such, a component whose true value is zero will appear to have a very small value, but not exactly zero. Thus, your function must view any component with magnitude less than 10^{-4} as zero.

Householder Tridiagonalization and QR Factorization Methods (Section 9.3)

In Problems 21 through 25 find all eigenvalues of the matrix by executing the user-defined function HouseholderQR, and verify the results using MATLAB function eig.

21. $\mathbf{A} = \begin{bmatrix} 10 & 4 & 1 & 3 \\ 4 & 10 & 3 & 1 \\ 1 & 3 & 10 & 4 \\ 3 & 1 & 4 & 10 \end{bmatrix}$

22. $\mathbf{A} = \begin{bmatrix} 1 & -1 & 2 & 0 & 5 \\ -1 & 3 & 0 & 0 & -3 \\ 2 & 0 & 4 & -1 & 0 \\ 0 & 0 & -1 & 0 & 2 \\ 5 & -3 & 0 & 2 & 1 \end{bmatrix}$

23. $\mathbf{A} = \begin{bmatrix} -1 & 2 & 0 & 3 & 1 \\ 2 & 2 & -2 & 0 & 4 \\ 0 & -2 & 3 & -1 & 1 \\ 3 & 0 & -1 & 6 & -3 \\ 1 & 4 & 1 & -3 & 0 \end{bmatrix}$

24. $\mathbf{A} = \begin{bmatrix} 2 & -1 & 0 & 1 \\ -1 & 5 & 2 & -3 \\ 0 & 2 & -4 & 1 \\ 1 & -3 & 1 & 3 \end{bmatrix}$

25. $\mathbf{A} = \begin{bmatrix} -2 & 1 & 0 & -2 \\ 4 & 1 & 0 & 3 \\ -5 & -5 & -3 & -2 \\ 0 & 0 & 0 & -1 \end{bmatrix}$

26. Write a user-defined function with syntax

```
[T Tfinal eigenvals] = HouseholderQR_New(A, tol, kmax)
```

that uses Householder's method and the successive QR factorization process to find all eigenvalues of a matrix, as follows. Modify `HouseholderQR` by altering the terminating condition: the iterations must stop when the ratio of the largest (magnitude) subdiagonal entry to the smallest (magnitude) diagonal element is less than the prescribed tolerance. All input parameters have the same default as before. Apply this function to the matrix in Problem 25.

10

Numerical Solution of Partial Differential Equations

Partial differential equations (PDEs) play an important role in several areas of engineering, ranging from fluid mechanics, heat transfer, and applied mechanics to electromagnetic theory. Since it is generally much more difficult to find a closed-form solution for PDEs than it is for ordinary differential equations, they are usually solved numerically. In this chapter, we present numerical methods for the solution of PDEs, in particular, those that describe some fundamental problems in engineering applications, including the Laplace's equation, the heat equation, and the wave equation.

10.1 Introduction

A PDE is an equation involving a function (dependent variable) of at least two independent variables, and its partial derivatives. A PDE is of the order n if the highest derivative is of the order n. If a PDE is of the first degree in the dependent variable and its partial derivatives, it is called linear. Otherwise, it is called nonlinear. If a PDE is such that each of its terms involves either the dependent variable or its partial derivatives, it is called homogeneous. Otherwise, it is nonhomogeneous.

Notation: If $u = u(x,y)$, then we use the following brief notations for partial derivatives:

$$u_x = \frac{\partial u}{\partial x}, \quad u_{xx} = \frac{\partial^2 u}{\partial x^2}, \quad u_{xy} = \frac{\partial^2 u}{\partial y \partial x}$$

The number of spatial coordinates (not time t) determines the dimension of a PDE. For instance, a PDE with dependent variable $u = u(x,y,z)$ is three-dimensional. If $u = u(x,t)$ is the dependent variable, then the PDE is one-dimensional.

In particular, we consider a class of linear, second-order PDEs that appear in the form

$$au_{xx} + 2bu_{xy} + cu_{yy} = f(x,y,u,u_x,u_y) \tag{10.1}$$

A PDE in the form of Equation 10.1 is called elliptic if $b^2 - ac < 0$. Examples of elliptic PDEs include the two-dimensional Laplace's equation, $u_{xx} + u_{yy} = 0$, and the two-dimensional Poisson's equation, $u_{xx} + u_{yy} = f(x,y)$. A parabolic PDE satisfies $b^2 - ac = 0$. The one-dimensional (1D) heat equation, $u_t = \alpha^2 u_{xx}$ (α = constant > 0), is an example of a parabolic PDE. Finally, a PDE is called hyperbolic if $b^2 - ac > 0$. The 1D wave equation, $u_{tt} = \alpha^2 u_{xx}$ (α = constant > 0), is a well-known example of a hyperbolic PDE. Note that in our examples for parabolic and hyperbolic PDEs, the variable t has replaced y. In applications, when an elliptic PDE is involved, a boundary-value problem needs to be solved. Those involving parabolic and hyperbolic types require the solution of a boundary-initial-value problem. We first present the numerical solution of elliptic PDEs.

10.2 Elliptic PDEs

When an elliptic equation is to be solved in a specific region in the xy-plane, and the unknown function is prescribed along the boundary of the region, we have a Dirichlet problem. For instance, in the case of two-dimensional heat flow in steady state, temperature $u(x,y)$ is known along the boundary of the region. On the other hand, a Neumann problem refers to a boundary-value problem where the normal derivative of u, that is, $u_n = \partial u/\partial n$, is given on the boundary of the region. Note that along a vertical edge of a region, u_n is simply $u_x = \partial u/\partial x$, and along a horizontal edge, it is $u_y = \partial u/\partial y$. The mixed problem refers to the situation where u is specified on certain parts of the boundary, and u_n is specified on the others.

10.2.1 Dirichlet Problem

As a fundamental Dirichlet problem, we consider the solution of the two-dimensional Poisson's equation

$$u_{xx} + u_{yy} = f(x,y) \tag{10.2}$$

in the rectangular region shown in Figure 10.1, where $u(x,y)$ is assumed to be known along the boundary. The idea is to define a mesh size h and construct a grid by drawing equidistant vertical and horizontal lines of distance h. These lines are called grid lines, and the points at which they intersect are known as mesh points. Those mesh points that happen to be located on the boundary are called boundary points. Mesh points that lie inside the region are called interior mesh points. The goal is to approximate the solution u at the interior mesh points.

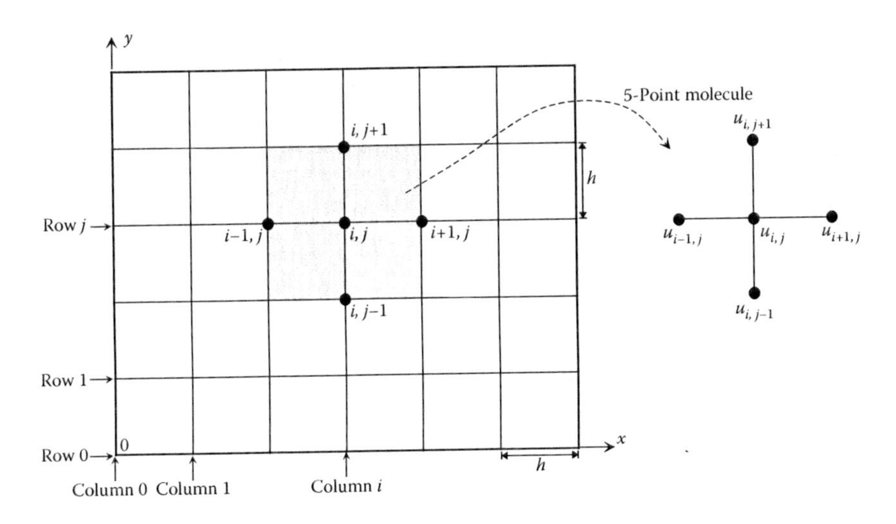

FIGURE 10.1
Grid for a rectangular region and a five-point molecule.

Let us denote the typical mesh point (x,y) by (ih,jh), simply labeled as (i,j) in Figure 10.1. The value of u at that point is then denoted by u_{ij}. Likewise, $f(x,y)$ is represented by f_{ij}. Approximating the second-order partial derivatives in Equation 10.1 with three-point central difference formulas (Section 6.2), we find

$$\frac{u_{i-1,j} - 2u_{ij} + u_{i+1,j}}{h^2} + \frac{u_{i,j-1} - 2u_{ij} + u_{i,j+1}}{h^2} = f_{ij}$$

that simplifies to

$$u_{i-1,j} + u_{i+1,j} + u_{i,j-1} + u_{i,j+1} - 4u_{ij} = h^2 f_{ij} \tag{10.3}$$

This is called the difference equation for Poisson's equation, which provides a relation between the solution u at (i,j) and four adjacent points. Similarly, for the Laplace's equation, $u_{xx} + u_{yy} = 0$, we find

$$u_{i-1,j} + u_{i+1,j} + u_{i,j-1} + u_{i,j+1} - 4u_{ij} = 0 \tag{10.4}$$

that is known as the difference equation for Laplace's equation.

In both cases, the application of the difference equation results in a linear system of algebraic equations, where the number of unknowns—and equations—is the number of interior mesh points generated by the grid. Assuming there are n interior mesh points in the region, this linear system is in the form $\mathbf{Au} = \mathbf{b}$, where $\mathbf{A}_{n\times n}$ is the coefficient matrix, $\mathbf{u}_{n\times 1}$ is the vector of the unknowns, and $\mathbf{b}_{n\times 1}$ is composed of known quantities. When at least one of the adjacent points in the molecule (Figure 10.1) is a boundary point, the value of u provided by the boundary condition is available and is ultimately moved to the right side of the equation, hence becoming part of vector **b**. In addition to the boundary points, in the case of Poisson's equation, the terms $h^2 f_{ij}$ will also be included in **b**. In practice, a large number of mesh points are needed for better accuracy, causing the coefficient matrix **A** to be large. This matrix is also sparse, with at most five nonzero entries in each of its rows. A linear system $\mathbf{Au} = \mathbf{b}$ with a large, sparse coefficient matrix is normally solved numerically via an indirect method such as the Gauss–Seidel iterative method (Chapter 4).

The user-defined function `DirichletPDE` uses the difference-equation approach outlined above to numerically solve the Poisson's equation—or Laplace's equation—in a rectangular region with the values of the unknown solution available on the boundary. The function returns the approximate solution at the interior mesh points, as well as the values at the boundary points in a pattern that resembles the gridded region. Additionally, it returns the three-dimensional (3D) plot of the results.

```
function U=DirichletPDE(x,y,f,uleft,uright,ubottom,utop)
%
%  DirichletPDE numerically solves an elliptic PDE with
%  Dirichlet boundary conditions over a rectangular
%  region.
%
%    U=DirichletPDE(x,y,f,uleft,uright,ubottom,utop) where
%
%      x is the 1-by-m vector of mesh points in the x
%      direction,
%      y is the n-by-1 vector of mesh points in the y
%      direction,
%      f is the inline function defining the forcing
%      function which is in terms of x and y, namely,
%      f(x,y),
%      ubottom(x),utop(x),uright(y),uleft(y) are the
%      functions defining the boundary conditions,
%
%      U is the solution at the interior mesh points.
m=size(x,2); n=size(y,1); N=(m-2)*(n-2);
A=diag(-4*ones(N,1));        % Create diagonal matrix
A=A+diag(diag(A,n-2)+1,n-2);    % Add n-2 diagonal
A=A+diag(diag(A,2-n)+1,2-n);    % Add 2-n diagonal
d1=ones(N-1,1);      % Create vector of ones
```

```
d1(n-2:n-2:end)=0;        % Insert zeros
A=A+diag(d1,1);           % Add upper diagonal
A=A+diag(d1,-1);          % Add lower diagonal
[X Y]=meshgrid(x(2:end-1),y(end-1:-1:2)); % Create mesh
h=x(2)-x(1);
%Define boundary conditions

for i=2:m-1,
    utopv(i-1)=utop(x(i));
    ubottomv(i-1)=ubottom(x(i));
end
for i=1:n,
    uleftv(i)=uleft(y(n+1-i));
    urightv(i)=uright(y(n+1-i));
end
% Build vector b

b=0;    % Initialize vector b

for i=1:N,
    b(i)=h^2*f(X(i),Y(i));
end

b(1:n-2:N)=b(1:n-2:N)-utopv;
b(n-2:n-2:N)=b(n-2:n-2:N)-ubottomv;
b(1:n-2)=b(1:n-2)-uleftv(2:n-1);
b(N-(n-3):N)=b(N-n+3:N)-urightv(2:n-1);

u=A\b';               % Solve the system
U=reshape(u,n-2,m-2);
U=[utopv;U;ubottomv];
U=[uleftv' U urightv'];
[X Y]=meshgrid(x,y(end:-1:1));
surf(X,Y,U);      % 3D plot of the numerical results
xlabel('x');ylabel('y');
```

Example 10.1: Dirichlet Problem

The Dirichlet problem in Figure 10.2 describes the steady-state temperature distribution inside a rectangular plate of length 1 and width 2. Three of the sides are kept at zero temperature, while the lower edge has a temperature profile of $\sin(\pi x/2)$. Using a mesh size of $h = 0.5$, construct a grid and find the approximate values of u at the interior mesh points, and calculate the relative errors associated with these approximate values. The exact solution is given by

$$u(x,y) = \frac{1}{\sinh(\pi/2)} \sin\frac{\pi x}{2} \sinh\frac{\pi(1-y)}{2}$$

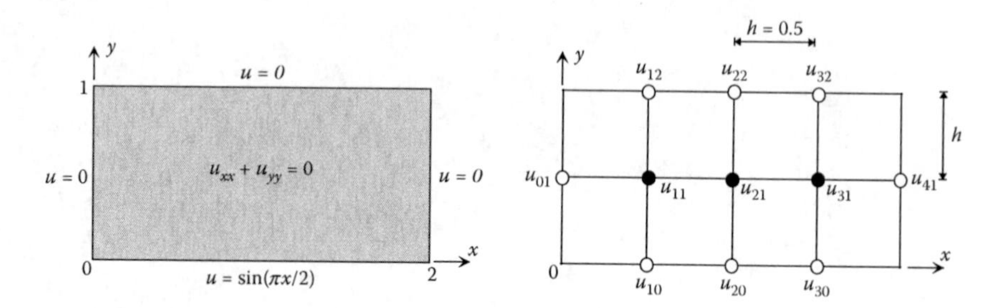

FIGURE 10.2
Dirichlet problem in Example 10.1.

SOLUTION

There are three interior mesh points and eight boundary points on the grid. Therefore, the difference equation, Equation 10.4, must be applied three times, once at each interior mesh point. As a result, we have

$$(i = 1, j = 1) \quad u_{01} + u_{10} + u_{21} + u_{12} - 4u_{11} = 0$$

$$(i = 2, j = 1) \quad u_{11} + u_{20} + u_{31} + u_{22} - 4u_{21} = 0$$

$$(i = 3, j = 1) \quad u_{21} + u_{30} + u_{41} + u_{32} - 4u_{31} = 0$$

Included in these equations are the values at the boundary points

$$u_{12} = u_{22} = u_{32} = u_{01} = u_{41} = 0, \quad u_{10} = 0.7071 = u_{30}, \quad u_{20} = 1$$

Inserting these into the system of equations, we find

$$\begin{aligned} -4u_{11} + u_{21} \quad &= -0.7071 \\ u_{11} - 4u_{21} + u_{31} &= -1 \\ u_{21} - 4u_{31} &= -0.7071 \end{aligned} \quad \overset{\text{In matrix form}}{\Rightarrow} \quad \begin{bmatrix} -4 & 1 & 0 \\ 1 & -4 & 1 \\ 0 & 1 & -4 \end{bmatrix} \begin{Bmatrix} u_{11} \\ u_{21} \\ u_{31} \end{Bmatrix} = \begin{Bmatrix} -0.7071 \\ -1 \\ -0.7071 \end{Bmatrix}$$

The solution of this system yields $u_{11} = 0.2735 = u_{31}$ and $u_{21} = 0.3867$. The exact values at these points are calculated as $u_{11} = 0.2669 = u_{31}$ and $u_{21} = 0.3775$, recording relative errors of 2.47% and 2.44%, respectively. The estimates turned out reasonably accurate considering the large mesh size that was used. Switching to a smaller mesh size of $h = 0.25$, for example, generates a grid that includes 21 interior mesh points and 20 boundary points. The ensuing linear system then comprises 21 equations and 21 unknowns, whose solutions are more accurate than those obtained here.

Executing the user-defined function `DirichletPDE` confirms the earlier numerical results. Note that the plot has been suppressed.

```
>> x=0:0.5:2; % x must be 1-by-m
>> y=0:0.5:1; y=y'; % y must be n-by-1
>> f=inline('0','x','y');
```

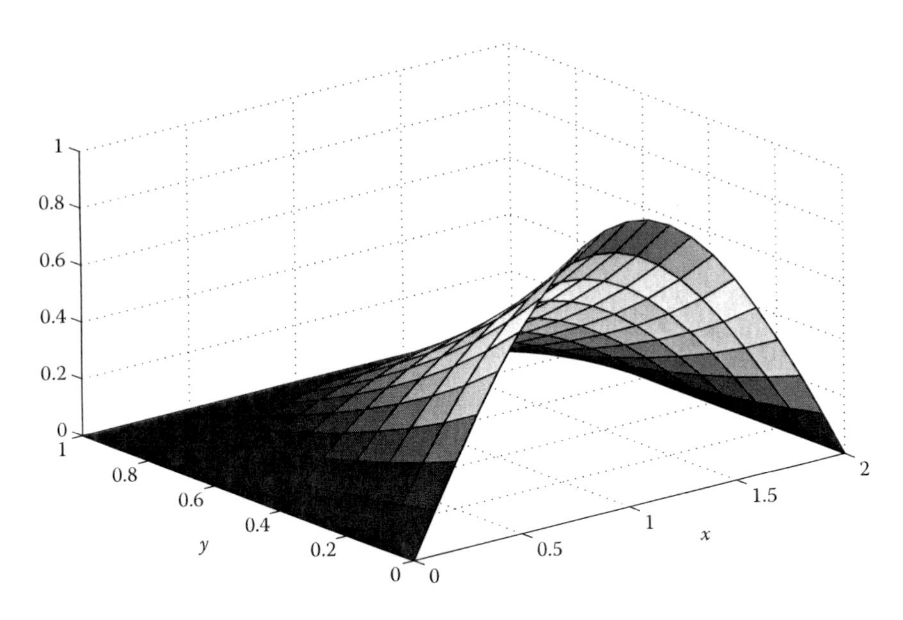

FIGURE 10.3
Steady-state temperature distribution in Example 10.1, using $h = 0.1$.

```
>> ubottom = inline('sin(pi*x/2)');
>> utop = inline('0','x'); uleft = inline('0','y');
uright = inline('0','y');
>> U = DirichletPDE(x,y,f,uleft,uright,ubottom,utop)

U =
         0         0         0         0         0
         0    0.2735    0.3867    0.2735         0
         0    0.7071    1.0000    0.7071         0
```

The three shaded values are the solution estimates at the interior mesh points, which agree with those obtained earlier. All other values correspond to the boundary points in the grid used. For plotting purposes, we reduce the mesh size to $h = 0.1$ and execute the function a second time, this time suppressing the numerical results with plot shown in Figure 10.3.

```
>> x = 0:0.1:2;
>> y = 0:0.1:1; y = y';
>> f = inline('0','x','y');
>> ubottom = inline('sin(pi*x/2)');
>> utop = inline('0','x'); uleft = inline('0','y');
uright = inline('0','y');
>> U = DirichletPDE(x,y,f,uleft,uright,ubottom,utop)
```

10.2.2 Alternating Direction-Implicit Methods

In the foregoing analysis, application of the difference equation for either Poisson's equation or Laplace's equation at the interior mesh points led to a linear system of equations whose coefficient matrix had at most five nonzero entries in each row. A small mesh size results in a large, sparse coefficient matrix. Although these types of matrices are desirable, the computations can be made even more efficient if they are tridiagonal (Section 4.3), namely, if they have at most three nonzero entries in each row.

The goal is thus to develop a scheme that leads to a system of equations with a tridiagonal coefficient matrix. We will present the idea while focusing on the Laplace's equation in a rectangular region. Suppose a mesh size h generates N internal mesh points per row and M internal mesh points per column. Equation 10.4 takes into account the five elements of the five-point molecule (Figure 10.1) all at once. That is, the elements $u_{i-1,j}$, u_{ij}, $u_{i+1,j}$ along the jth row and $u_{i,j-1}$, u_{ij}, $u_{i,j+1}$ along the ith column, with u_{ij} as the common element. Since we are aiming for a tridiagonal matrix, we write Equation 10.4 as

$$u_{i-1,j} - 4u_{ij} + u_{i+1,j} = -u_{i,j-1} - u_{i,j+1} \tag{10.5}$$

so that the members on the left side belong to the jth row and those on the right side belong to the ith column. Equation 10.4 may also be rewritten as

$$u_{i,j-1} - 4u_{ij} + u_{i,j+1} = -u_{i-1,j} - u_{i+1,j} \tag{10.6}$$

with the left-side terms belonging to the ith column and the right-side terms to the jth row. Alternating direction-implicit (ADI) methods use this basic idea to solve the Dirichlet problem iteratively. A complete iteration step consists of two halves. In the first half, Equation 10.5 is applied in every row in the grid. In the second half, Equation 10.6 is applied in every column of the grid. The most commonly used ADI method is the one proposed by Peaceman and Rachford, sometimes referred to as Peaceman–Rachford alternating direction-implicit method (PRADI).

10.2.2.1 Peaceman–Rachford Alternating Direction-Implicit Method

Choose the arbitrary starting value $u_{ij}^{(0)}$ at each interior mesh point (i,j). The first iteration has two halves. In the first half, update the values of u_{ij} row by row, in a manner suggested by Equation 10.5

jth row $(j = 1, 2, \ldots, M)$

$$u_{i-1,j}^{(1/2)} - 4u_{ij}^{(1/2)} + u_{i+1,j}^{(1/2)} = -u_{i,j-1}^{(0)} - u_{i,j+1}^{(0)}, \quad i = 1, 2, \ldots, N \tag{10.7}$$

Note that some of the u values are the known boundary values, which are not affected by the starting values and remain unchanged throughout the process. For each fixed j, one row, Equation 10.7 produces N equations. Since there are M rows, a total of MN equations will be generated. This system has a tridiagonal coefficient matrix, by design, which can then be solved efficiently using the Thomas method (Section 4.3). The solution at this stage represents the half-updated values with the superscript of (1/2).

In the second half, the values of $u_{ij}^{(1/2)}$ will be updated column by column, as suggested by Equation 10.6

$$i\text{th column } (i = 1, 2, \ldots, N)$$

$$u_{i,j-1}^{(1)} - 4u_{ij}^{(1)} + u_{i,j+1}^{(1)} = -u_{i-1,j}^{(1/2)} - u_{i+1,j}^{(1/2)}, \quad j = 1, 2, \ldots, M \tag{10.8}$$

For each fixed i, one column, Equation 10.8 produces M equations. Once again, the values at the boundary points are not affected and remain unchanged. Since there are N columns, a total of MN equations will be generated. This system also has a tridiagonal coefficient matrix, which can be solved efficiently using the Thomas method. The solution at this stage represents the updated values with the superscript of (1). This completes the first iteration. The second iteration has two halves. In the first half

$$j\text{th row } (j = 1, 2, \ldots, M)$$

$$u_{i-1,j}^{(3/2)} - 4u_{ij}^{(3/2)} + u_{i+1,j}^{(3/2)} = -u_{i,j-1}^{(1)} - u_{i,j+1}^{(1)}, \quad i = 1, 2, \ldots, N,$$

generates a system of MN equations with a tridiagonal coefficient matrix. In the second half

$$i\text{th column } (i = 1, 2, \ldots, N)$$

$$u_{i,j-1}^{(2)} - 4u_{ij}^{(2)} + u_{i,j+1}^{(2)} = -u_{i-1,j}^{(3/2)} - u_{i+1,j}^{(3/2)}, \quad j = 1, 2, \ldots, M$$

that also generates a system of MN equations with a tridiagonal coefficient matrix. The solution gives the updated values with the superscript of (2). The procedure is repeated until convergence is observed. This, of course, requires a terminating condition. One reasonable terminating condition is as follows. Assemble the values at the interior mesh points into a matrix, with the same configuration as the grid. If the norm of the difference between two successive matrices is less than a prescribed tolerance, terminate the iterations.

The user-defined function `PRADI` uses the Peaceman–Rachford ADI method to numerically solve the Poisson's equation—or Laplace's equation—in a rectangular region with the values of the unknown solution available on the boundary. The function returns the approximate solution at the interior mesh points, as well as the values at the boundary points in a pattern that resembles the gridded region. Additionally, it returns the 3D plot of the results.

```
function [U, k] = PRADI(x,y,f,uleft,uright,ubottom,utop,
tol,kmax)
%
% PRADI numerically solves an elliptic PDE with
% Dirichlet boundary conditions over a rectangular
% region using the Peaceman-Rachford alternating
% direction implicit method.
%
%  [U,k] = PRADI(x,y,f,uleft,uright,ubottom,utop,tol,
%  kmax) where
%
%     x is the 1-by-m vector of mesh points in the
%     x direction,
%     y is the n-by-1 vector of mesh points in the
%     y direction,
%     f is the inline function defining the forcing
%     function,
%     ubottom,uleft,utop,uright are the functions
%     defining the boundary conditions,
%     tol is the tolerance used for convergence
%     (default = 1e-4),
%     kmax is the maximum number of iterations
%     (default = 50),
%
%     U is the solution at the mesh points,
%     k is the number of full iterations needed to meet
%     the tolerance.
%
% Note: The default starting value at all mesh points
% is 0.5.
%
if nargin<9 || isempty(kmax), kmax = 50; end
if nargin<8 || isempty(tol), tol = 1e-4; end
[X Y] = meshgrid(x(2:end-1),y(2:end-1));
% Create messh grid
m = size(X,2); n = size(X,1); N = m*n;
u = 0.5*ones(n,m);           % Starting values
h = x(2)-x(1);      % Mesh size
% Define boundary conditions
for i = 2:m+1,
    utopv(i-1) = utop(x(i));
    ubottomv(i-1) = ubottom(x(i));
end
  for i = 1:n+2,
      uleftv(i) = uleft(y(i));
      urightv(i) = uright(y(i));
  end
```

```
U=[ubottomv;u;utopv]; U=[uleftv' U urightv'];
% Generate matrix A1 (first half) and A2 (second half).
A=diag(-4*ones(N,1));
d1=diag(A,1)+1; d1(m:m:N-1)=0;
d2=diag(A,-1)+1; d2(n:n:N-1)=0;
A2=diag(d2,1)+diag(d2,-1)+A;
A1=diag(d1,1)+diag(d1,-1)+A;
U1=U;
for i=1:N, % Initialize vector b
    b0(i)=h^2*f(X(i),Y(i));
end
b0=reshape(b0,n,m);
for k=1:kmax,
      % First half
  b=b0-U1(1:end-2,2:end-1)-U1(3:end,2:end-1);
  b(:,1)=b(:,1)-U(2:end-1,1);
  b(:,end)=b(:,end)-U(2:end-1,end);
  b=reshape(b',N,1);
  u=ThomasMethod(A1,b);
  % Tridiagonal system - Thomas method
  u=reshape(u,m,n);
  U1=[U(1,2:end-1);u';U(end,2:end-1)];
  U1=[U(:,1) U1 U(:,end)];
      % second half
  b=b0-U1(2:end-1,1:end-2)-U1(2:end-1,3:end);
  b(1,:)=b(1,:)-U(1,2:end-1);
  b(end,:)=b(end,:)-U(end,2:end-1);
  b=reshape(b,N,1);
  u=ThomasMethod(A2,b);
  % Tridiagonal system - Thomas method
  u=reshape(u,n,m);
  U2=[U(1,2:end-1);u;U(end,2:end-1)];
  U2=[U(:,1) U2 U(:,end)];
  if norm(U2-U1,inf)<=tol, break, end;
  U1=U2;
end
[X Y]=meshgrid(x,y);
U=U1;

for i=1:n+2,
    W(i,:)=U(n-i+3,:);
    YY(i)=Y(n-i+3);
end
    U=W; Y=YY;
    surf(X,Y,U);
xlabel('x');ylabel('y');
```

Example 10.2: PRADI Method

For the Dirichlet problem described in Figure 10.4

a. Perform one complete step of the PRADI method using $h = 1$ and the starting values of 0.5 for all interior mesh points.
b. Solve the Dirichlet problem using the user-defined function `PRADI` with default parameter values.

SOLUTION

a. *First half*: There are two rows and two columns in the grid, and a total of four interior mesh points. Equation 10.7 is first applied in the first row ($j = 1$). Since there are two mesh points in this row, Equation 10.7 is applied twice:

$$j = 1 \quad \begin{matrix} (i=1) & u_{01} - 4u_{11}^{(1/2)} + u_{21}^{(1/2)} = -u_{10} - u_{12}^{(0)} \\ (i=2) & u_{11}^{(1/2)} - 4u_{21}^{(1/2)} + u_{31} = -u_{20} - u_{22}^{(0)} \end{matrix} \tag{10.9}$$

Note that we have omitted superscripts for boundary values because they remain unchanged. We next apply Equation 10.7 at the two mesh points along the second row ($j = 2$):

$$j = 2 \quad \begin{matrix} (i=1) & u_{02} - 4u_{12}^{(1/2)} + u_{22}^{(1/2)} = -u_{11}^{(0)} - u_{13} \\ (i=2) & u_{12}^{(1/2)} - 4u_{22}^{(1/2)} + u_{32} = -u_{21}^{(0)} - u_{23} \end{matrix} \tag{10.10}$$

Combining Equations 10.9 and 10.10, and using the available boundary values as well as the starting values, we arrive at

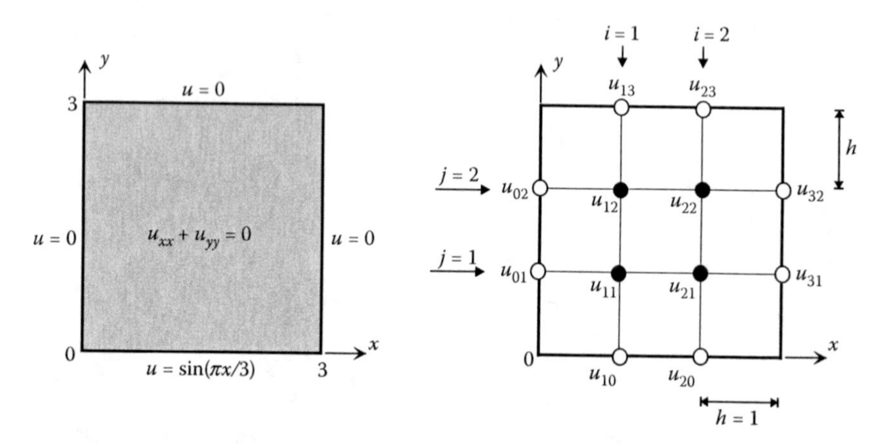

FIGURE 10.4
Dirichlet problem in Example 10.2.

$$\begin{bmatrix} -4 & 1 & 0 & 0 \\ 1 & -4 & 0 & 0 \\ 0 & 0 & -4 & 1 \\ 0 & 0 & 1 & -4 \end{bmatrix} \begin{Bmatrix} u_{11}^{(1/2)} \\ u_{21}^{(1/2)} \\ u_{12}^{(1/2)} \\ u_{22}^{(1/2)} \end{Bmatrix} = \begin{Bmatrix} -1.3660 \\ -1.3660 \\ -0.5 \\ -0.5 \end{Bmatrix} \overset{\text{Solve the tridiagonal system}}{\Rightarrow} \begin{array}{l} u_{11}^{(1/2)} = 0.4553 = u_{21}^{(1/2)} \\ u_{12}^{(1/2)} = 0.1667 = u_{22}^{(1/2)} \end{array}$$

Second half: Equation 10.8 is applied in the first column ($i = 1$). Since there are two mesh points in this column, it is applied twice:

$$i = 1 \quad \begin{array}{ll} (j = 1) & u_{10} - 4u_{11}^{(1)} + u_{12}^{(1)} = -u_{01} - u_{21}^{(1/2)} \\ (j = 2) & u_{11}^{(1)} - 4u_{12}^{(1)} + u_{13} = -u_{02} - u_{22}^{(1/2)} \end{array} \tag{10.11}$$

We next apply Equation 10.8 at the two mesh points along the second column ($i = 2$):

$$i = 2 \quad \begin{array}{ll} (j = 1) & u_{20} - 4u_{21}^{(1)} + u_{22}^{(1)} = -u_{11}^{(1/2)} - u_{31} \\ (j = 2) & u_{21}^{(1)} - 4u_{22}^{(1)} + u_{23} = -u_{12}^{(1/2)} - u_{32} \end{array} \tag{10.12}$$

Combining Equations 10.11 and 10.12, and using the available boundary values as well as the updated values from the previous half iteration, we arrive at

$$\begin{bmatrix} -4 & 1 & 0 & 0 \\ 1 & -4 & 0 & 0 \\ 0 & 0 & -4 & 1 \\ 0 & 0 & 1 & -4 \end{bmatrix} \begin{Bmatrix} u_{11}^{(1)} \\ u_{12}^{(1)} \\ u_{21}^{(1)} \\ u_{22}^{(1)} \end{Bmatrix} = \begin{Bmatrix} -1.3213 \\ -0.1667 \\ -1.3213 \\ -0.1667 \end{Bmatrix} \overset{\text{Solve the tridiagonal system}}{\Rightarrow} \begin{array}{l} u_{11}^{(1)} = 0.3635 = u_{21}^{(1)} \\ u_{12}^{(1)} = 0.1325 = u_{22}^{(1)} \end{array}$$

b. Executing the user-defined function `PRADI` (with default parameters) yields the solution estimates at the interior mesh points. Note that the plot has been suppressed.

```
>> x=0:1:3;
>> y=0:1:3; y=y';
>> f=inline('0','x','y');
>> ubottom=inline('sin(pi*x/3)');
>> utop=inline('0','x'); uleft=inline('0','y');
uright=inline('0','y');
>> [U, k]=PRADI(x,y,f,uleft,uright,ubottom,utop)

U =

        0          0          0     0
        0     0.1083     0.1083     0
        0     0.3248     0.3248     0
        0     0.8660     0.8660     0

k =
     5    % Five iterations required for convergence
```

Therefore, convergence occurs after five iterations. The numerical results obtained in (a) can be verified by letting the function PRADI perform only one iteration. Setting kmax = 1 and executing the function results in

```
>> [U, k] = PRADI(x,y,f,uleft,uright,ubottom,utop,1e-4,1)
U =
         0         0         0         0
         0    0.1325    0.1325         0
         0    0.3635    0.3635         0
         0    0.8660    0.8660         0

k =
     1
```

The numerical values for $u_{11}^{(1)}$, $u_{21}^{(1)}$, $u_{12}^{(1)}$, $u_{22}^{(1)}$ agree with those in (a).

10.2.3 Neumann Problem

In the formulation of the Neumann problem (Figure 10.5), the values of the normal derivatives of u are prescribed along the boundary. As before, we are focusing on the Laplace's equation in a rectangular region. In solving the Dirichlet problem, the objective was to find the solution estimates at all the interior mesh points by taking advantage of the known values of u along the boundary. In solving the Neumann problem, u is no longer available on the boundary; hence, the boundary points become part of the unknown vector.

Suppose the difference equation, Equation 10.4, is applied at the (3,3) mesh point:

$$\underbrace{u_{23} + u_{32} - 4u_{33}}_{\text{Interior mesh points}} + \underbrace{u_{43} + u_{34}}_{\text{Boundary points}} = 0 \tag{10.13}$$

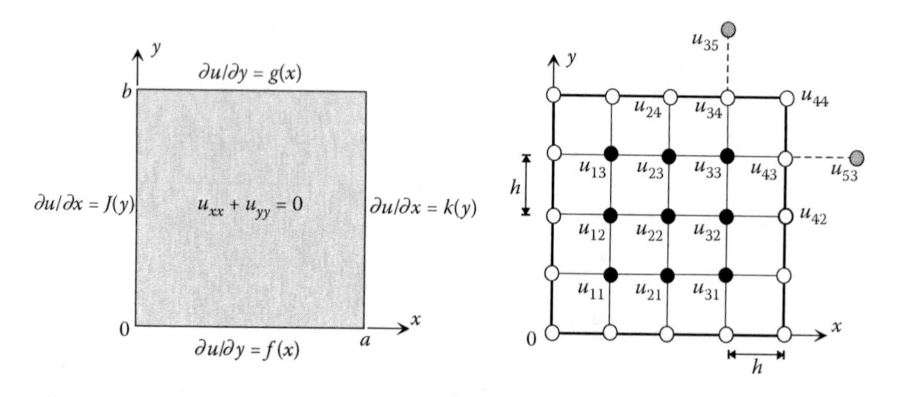

FIGURE 10.5
Neumann problem.

The three interior mesh points are part of the unknown vector. Since the boundary values u_{43} and u_{34} are not available, they must also be included in the unknown vector. Similar situations arise when we apply Equation 10.4 at near-boundary points. However, inclusion of all boundary points that lie on the grid in the unknown vector substantially increases the size of the unknown vector. For example, in Figure 10.5, there are 25 unknowns: 9 in the interior and 16 on the boundary. But since Equation 10.4 is applied at the interior mesh points, there are as many equations as there are interior mesh points. In Figure 10.5, for example, there will only be 9 equations, while there are 25 unknowns. Therefore, several more equations are needed to completely solve the ensuing system of equations.

To generate these additional (auxiliary) equations, we apply Equation 10.4 at each of the marked boundary points. Since we are currently concentrating on u_{43} and u_{34}, we apply Equation 10.4 at these two points:

$$\begin{aligned} u_{24} + u_{44} + u_{35} + u_{33} - 4u_{34} &= 0 \\ u_{42} + u_{44} + u_{53} + u_{33} - 4u_{43} &= 0 \end{aligned} \tag{10.14}$$

In Equation 10.14, there are two quantities that are not part of the grid and need to be eliminated: u_{53} and u_{35}. We will do this by extending the grid beyond the boundary of the region, and using the information on the vertical and horizontal boundary segments they are located on. At the (3,4) boundary point, we have access to $\partial u/\partial y = g(x)$. Let us call it g_{34}. Applying the two-point central difference formula (Section 6.2) at (3,4), we have

$$\left.\frac{\partial u}{\partial y}\right|_{(3,4)} = g_{34} = \frac{u_{35} - u_{33}}{2h} \overset{\text{Solve for } u_{35}}{\Longrightarrow} u_{35} = u_{33} + 2hg_{34}$$

Similarly, at (4,3)

$$\left.\frac{\partial u}{\partial x}\right|_{(4,3)} = k_{43} = \frac{u_{53} - u_{33}}{2h} \overset{\text{Solve for } u_{53}}{\Longrightarrow} u_{53} = u_{33} + 2hk_{43}$$

Substitution of these two relations into Equation 10.14 creates two new equations that involve only the interior mesh points and the boundary points. To proceed with this approach, we need to assume that the Laplace's equation is valid beyond the rectangular region, at least in the exterior area that contains the newly produced points such as u_{53} and u_{35}. If we continue with this strategy, we will end up with a system containing as many equations as the total number of interior mesh points and the boundary points. In Figure 10.5, for instance, the system will consist of 25 equations and 25 unknowns.

10.2.3.1 Existence of Solution for Neumann Problem

Existence of solution for Neumann problem entirely depends on the nature of the normal derivatives $u_n = \partial u/\partial n$ prescribed along different portions of the boundary. In fact, no solution is possible unless the line integral of the normal derivative taken over the boundary is zero:

$$\int_C \frac{\partial u}{\partial n}\, ds = 0 \tag{10.15}$$

Example 10.3: Neumann Problem

Consider the Neumann problem described in Figure 10.6 where the grid is constructed with $h = 1$. Using the approach outlined above, 12 unknowns are generated: 2 interior mesh points, 10 boundary points.

First, the condition for existence of solution, Equation 10.15, must be examined. Let the bottom edge of the region be denoted by C_1, and continuing counterclockwise, let the rest of the edges be denoted by C_2, C_3, and C_4. Then, it can be shown that (verify)

$$\int_C \frac{\partial u}{\partial n}\, ds = \int_{C_1} \frac{\partial u}{\partial n}\, ds + \int_{C_2} \frac{\partial u}{\partial n}\, ds + \int_{C_3} \frac{\partial u}{\partial n}\, ds + \int_{C_4} \frac{\partial u}{\partial n}\, ds$$

$$= \int_0^3 x\, dx + \int_0^2 y\, dy - \int_0^3 (x+1)\, dx - \int_0^2 (y-1)\, dy = -1 \neq 0$$

Therefore, the problem described in Figure 10.6 does not have a solution. In fact, if we had proceeded with the strategy introduced earlier, we would have obtained a 12×12 system of equations with a singular coefficient matrix.

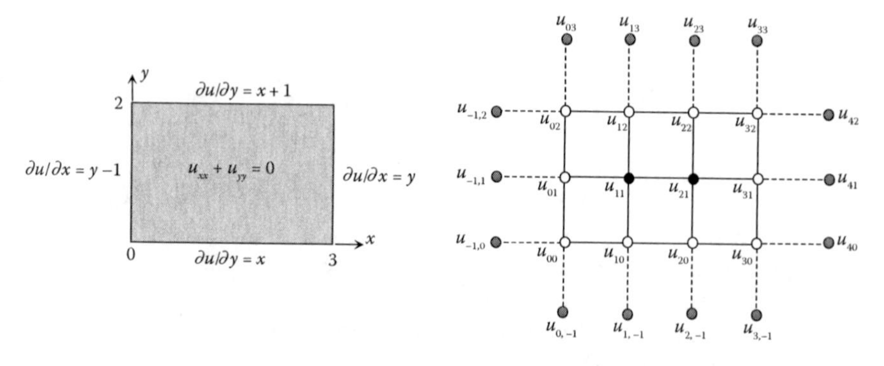

FIGURE 10.6
Neumann problem in Example 10.3.

10.2.4 Mixed Problem

The mixed problem refers to the case when u is prescribed along some portions of the boundary, while $u_n = \partial u/\partial n$ is known on the other portions. The numerical solution of these types of problems involves the same idea as in the case of the Neumann problem, with a lower degree of complexity. This is because u_n is dealt with only on certain segments of the boundary; hence, the region does not need to be entirely extended. Consequently, the linear system to be solved is not as large either.

Example 10.4: Mixed Problem

Solve the mixed problem described in Figure 10.7 using the grid with $h = 1$.

SOLUTION

Since the Dirichlet boundary conditions are prescribed along the left, lower, and upper edges, no extension of region is needed there. The only extension pertains to the (3,1) mesh point, where u itself is unknown, resulting in the creation of u_{41}. Application of Equation 10.4 at the two interior mesh points, as well as at (3,1), yields

$$\begin{array}{ll} (1,1) & u_{01} + u_{10} + u_{21} + u_{12} - 4u_{11} = 0 \\ (2,1) & u_{11} + u_{20} + u_{31} + u_{22} - 4u_{21} = 0 \\ (3,1) & u_{21} + u_{30} + u_{41} + u_{32} - 4u_{31} = 0 \end{array} \qquad (10.16)$$

To eliminate u_{41}, we use the two-point central difference formula at (3,1):

$$\left.\frac{\partial u}{\partial x}\right|_{(3,1)} = \left[2y\right]_{(3,1)} = 2 = \frac{u_{41} - u_{21}}{2} \overset{\text{Solve for } u_{41}}{\Longrightarrow} u_{41} = u_{21} + 4$$

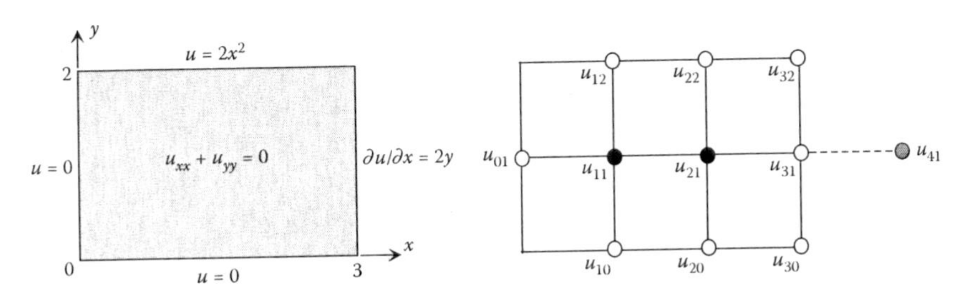

FIGURE 10.7
Mixed problem in Example 10.4.

Substitution of $u_{41} = u_{21} + 4$, as well as the boundary values provided by boundary conditions, into Equation 10.16 yields

$$\begin{aligned} &u_{21} + 2 - 4u_{11} = 0 \\ &u_{11} + u_{31} + 8 - 4u_{21} = 0 \\ &u_{21} + (u_{21} + 4) + 18 - 4u_{31} = 0 \end{aligned} \quad \overset{\text{Matrix form}}{\Rightarrow} \quad \begin{bmatrix} -4 & 1 & 0 \\ 1 & -4 & 1 \\ 0 & 2 & -4 \end{bmatrix} \begin{Bmatrix} u_{11} \\ u_{21} \\ u_{31} \end{Bmatrix} = \begin{Bmatrix} -2 \\ -8 \\ -22 \end{Bmatrix} \quad \overset{\text{Solve}}{\Rightarrow} \quad \begin{aligned} u_{11} &= 1.5769 \\ u_{21} &= 4.3077 \\ u_{31} &= 7.6538 \end{aligned}$$

10.2.5 More Complex Regions

Up to now, we have investigated boundary-value problems for elliptic PDEs in regions with relatively simple shapes, specifically, rectangular regions. As a result, it was possible to suitably adjust the grid so that some of the mesh points are located on the boundary of the region. But, in many applications, the geometry of the problem is not as simple as a rectangle and, as a result, the boundary of the region crosses the grid at points that are not mesh points.

Consider, for example, the problem of solving Laplace's equation, $u_{xx} + u_{yy} = 0$, in the region shown in Figure 10.8a. The region has an irregular boundary in that the curved portion intersects the grid at points A and B, neither of which is a mesh point. The points M and Q are treated as before because each has four adjacent mesh points that are located on the grid. But, a point such as P must be treated differently since two of its adjacent points, A and B, are not on the grid. Therefore, the objective is to derive expressions for $u_{xx}(P)$ and $u_{yy}(P)$, at mesh point P, to form a new difference equation for the Laplace's equation. Assume that A is located at a distance of αh to the right of P, and B is at a distance of βh above P. Write the Taylor's series expansion for

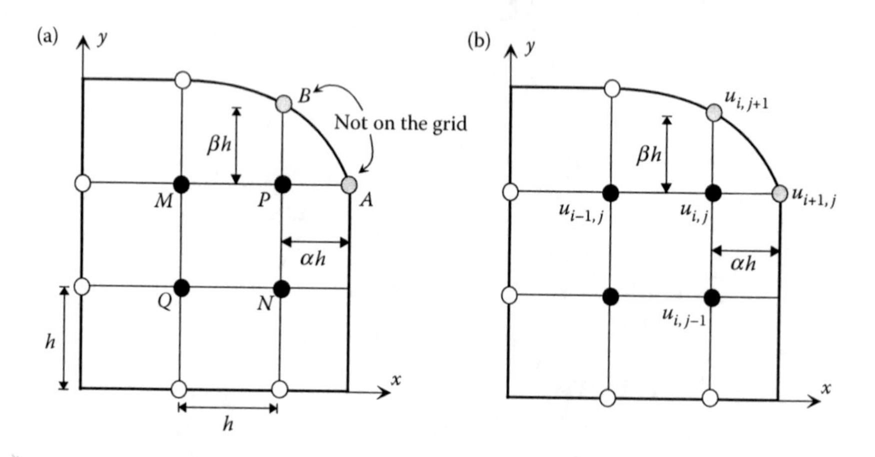

FIGURE 10.8
Region with an irregular boundary: (a) grid, (b) mesh points.

u at the four points A, B, M, and N about the point P. For example, $u(M)$ and $u(A)$ are expressed as

$$u(M) = u(P) - h\frac{\partial u(P)}{\partial x} + \frac{1}{2!}h^2\frac{\partial^2 u(P)}{\partial x^2} - \cdots \tag{10.17}$$

$$u(A) = u(P) + (\alpha h)\frac{\partial u(P)}{\partial x} + \frac{1}{2!}(\alpha h)^2\frac{\partial^2 u(P)}{\partial x^2} + \cdots \tag{10.18}$$

Multiply Equation 10.17 by α and add the result to Equation 10.18, while neglecting the terms involving h^2 and higher:

$$u(A) + \alpha u(M) \cong (\alpha + 1)u(P) + \frac{h^2}{2}\alpha(\alpha + 1)\frac{\partial^2 u(P)}{\partial x^2}$$

Solving for $\partial^2 u(P)/\partial x^2$, we find

$$\frac{\partial^2 u(P)}{\partial x^2} = \frac{2}{h^2}\left[\frac{1}{\alpha(\alpha+1)}u(A) + \frac{1}{\alpha+1}u(M) - \frac{1}{\alpha}u(P)\right] \tag{10.19}$$

Similarly, expanding $u(N)$ and $u(B)$ about P, and proceeding as above, yields

$$\frac{\partial^2 u(P)}{\partial y^2} = \frac{2}{h^2}\left[\frac{1}{\beta(\beta+1)}u(B) + \frac{1}{\beta+1}u(N) - \frac{1}{\beta}u(P)\right] \tag{10.20}$$

Addition of Equations 10.19 and 10.20 gives

$$u_{xx}(P) + u_{yy}(P) = \frac{2}{h^2}\left[\frac{1}{\alpha(\alpha+1)}u(A) + \frac{1}{\beta(\beta+1)}u(B) + \frac{1}{\alpha+1}u(M) + \frac{1}{\beta+1}u(N) - \left(\frac{1}{\alpha} + \frac{1}{\beta}\right)u(P)\right] \tag{10.21}$$

If the Laplace's equation is solved, then Equation 10.21 yields

$$\frac{2}{\alpha(\alpha+1)}u(A) + \frac{2}{\beta(\beta+1)}u(B) + \frac{2}{\alpha+1}u(M) + \frac{2}{\beta+1}u(N) - \frac{2(\alpha+\beta)}{\alpha\beta}u(P) = 0 \tag{10.22}$$

With the usual notations involved in Figure 10.8b, the difference equation is obtained by rewriting Equation 10.22 as

$$\frac{2}{\alpha(\alpha+1)}u_{i+1,j} + \frac{2}{\beta(\beta+1)}u_{i,j+1} + \frac{2}{\alpha+1}u_{i-1,j} + \frac{2}{\beta+1}u_{i,j-1} - \frac{2(\alpha+\beta)}{\alpha\beta}u_{i,j} = 0 \quad (10.23)$$

The case of Poisson's equation can be handled in a similar manner. Equation 10.23 is applied at any mesh point that has at least one adjacent point not located on the grid. In Figure 10.8a, for example, that would be points P and N. For the points M and Q, we simply apply Equation 10.4, as before, or we equivalently apply Equation 10.23 with $\alpha = 1 = \beta$.

Example 10.5: Irregular Boundary

Solve $u_{xx} + u_{yy} = 0$ in the region shown in Figure 10.9 subject to the given boundary conditions. The slanting segment of the boundary obeys $y = -\frac{2}{3}x + 2$.

SOLUTION

On the basis of the grid shown in Figure 10.9, Equation 10.4 can be applied at mesh points (1,1), (2,1), and (1,2) because all four neighboring points at those mesh points are on the grid. Using the boundary conditions provided, the resulting difference equations are

$$\begin{aligned} 1 + 4 + u_{12} + u_{21} - 4u_{11} &= 0 \\ u_{11} + 7 + 9.5 + u_{22} - 4u_{21} &= 0 \\ 1 + u_{11} + \tfrac{1}{3} + u_{22} - 4u_{12} &= 0 \end{aligned} \quad (10.24)$$

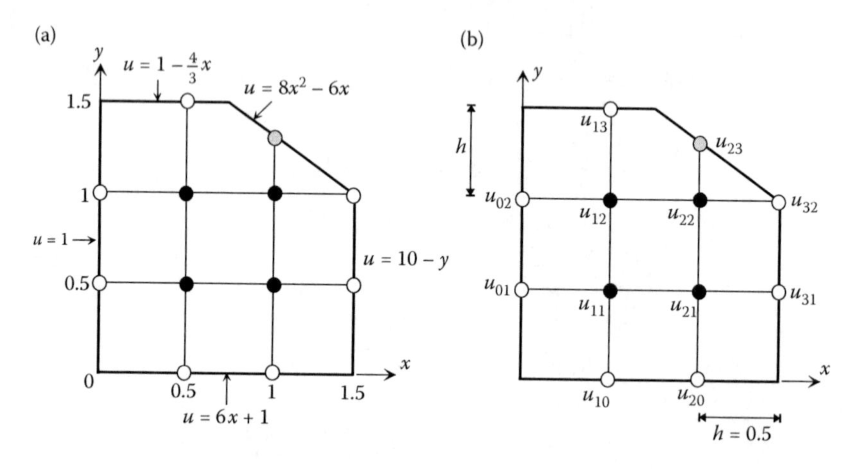

FIGURE 10.9
Example 10.5: (a) grid, (b) mesh points.

However, at (2,2), we need to use Equation 10.23. From the equation of the slanting segment, we find that point A is located at a vertical distance of $\frac{1}{3}$ from the (2,2) mesh point. Noting that $h = \frac{1}{2}$, we find $\beta = \frac{2}{3}$. On the other hand, $\alpha = 1$ since the neighboring point to the right of (2,2) is a mesh point. With these, Equation 10.23 yields

$$9 + \frac{2}{\frac{2}{3}\cdot\frac{5}{3}}(2) + u_{12} + \frac{2}{\frac{5}{3}}u_{21} - \frac{2(1+\frac{2}{3})}{\frac{2}{3}}u_{22} = 0$$

Combining this with Equation 10.24 and simplifying

$$\begin{bmatrix} -4 & 1 & 1 & 0 \\ 1 & -4 & 0 & 1 \\ 1 & 0 & -4 & 1 \\ 0 & 1.2 & 1 & -5 \end{bmatrix}\begin{Bmatrix} u_{11} \\ u_{21} \\ u_{12} \\ u_{22} \end{Bmatrix} = \begin{Bmatrix} -5 \\ -16.5 \\ -1.3333 \\ -12.6 \end{Bmatrix} \Rightarrow \begin{matrix} u_{11} = 3.3354 \\ u_{21} = 6.0666 \\ u_{12} = 2.2749 \\ u_{22} = 4.4310 \end{matrix}$$

10.3 Parabolic PDEs

In this section, we present two techniques for the numerical solution of parabolic PDEs: finite-difference (FD) method and Crank–Nicolson (CN) method. In both cases, the objective remains the same as before, namely, derive a suitable difference equation that can be used to generate solution estimates at mesh points. Contrary to the numerical solution of elliptic PDEs, there is no assurance that the difference equations for parabolic equations converge, regardless of the grid size. In these situations, it turns out that convergence can be guaranteed as long as some additional conditions are imposed.

10.3.1 Finite-Difference (FD) Method

The 1D heat equation, $u_t = \alpha^2 u_{xx}$ (α = constant > 0), is the simplest model of a physical system involving a parabolic PDE. Specifically, consider a laterally insulated wire of length L with its ends kept at zero temperature, and subjected to the initial temperature along the wire prescribed by $f(x)$. The boundary-initial-value problem at hand is

$$u_t = \alpha^2 u_{xx} \ (\alpha = \text{constant} > 0), \quad 0 \le x \le L, \ t \ge 0 \tag{10.25}$$

$$u(0,t) = 0 = u(L,t) \tag{10.26}$$

$$u(x,0) = f(x) \tag{10.27}$$

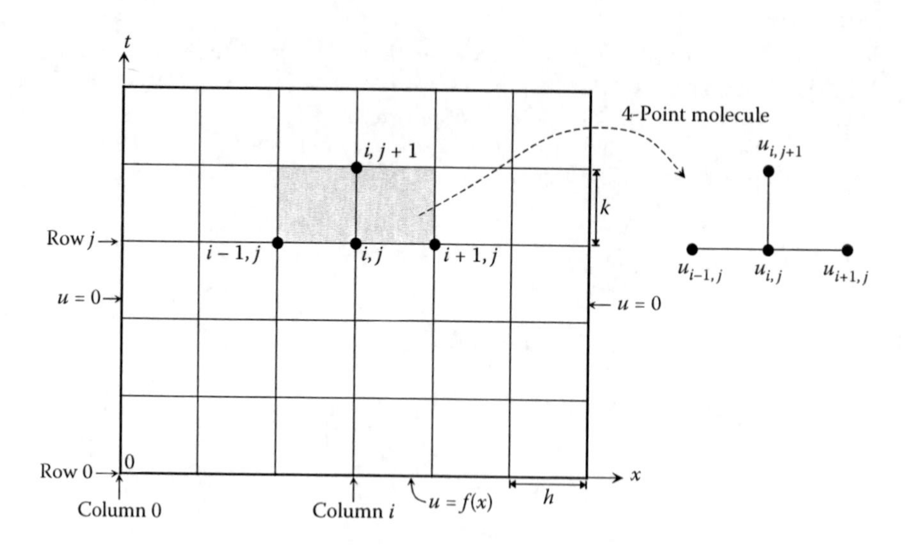

FIGURE 10.10
Region and grid used for solving the 1D heat equation.

Figure 10.10 shows that a grid is constructed using a mesh size of h in the x-direction and a mesh size of k in the t-direction. As usual, the partial derivatives in Equation 10.25 must be replaced by their FD approximations. The term u_{xx} is approximated by the three-point central difference formula (Section 6.2). For the term u_t, however, we must use the two-point forward-difference formula, as opposed to the more accurate central difference formula. This is because starting at $t = 0$ (the x-axis), we are only able to progress in the positive direction of the t-axis and we have no knowledge of $u(t)$ when $t < 0$. Consequently, Equation 10.25 yields

$$\frac{1}{k}(u_{i,j+1} - u_{ij}) = \frac{\alpha^2}{h^2}(u_{i-1,j} - 2u_{i,j} + u_{i+1,j}) \tag{10.28}$$

Referring to the four-point molecule in Figure 10.10, knowing $u_{i-1,j}$, $u_{i,j}$, and $u_{i+1,j}$, we can find $u_{i,j+1}$ at the higher level on the t-axis, via

$$u_{i,j+1} = \left[1 - \frac{2k\alpha^2}{h^2}\right]u_{i,j} + \frac{k\alpha^2}{h^2}(u_{i-1,j} + u_{i+1,j})$$

that can be simplified to

$$u_{i,j+1} = (1 - 2r)u_{ij} + r(u_{i-1,j} + u_{i+1,j}), \quad r = \frac{k\alpha^2}{h^2} \tag{10.29}$$

and is known as the difference equation for the 1D heat equation using the FD method.

10.3.1.1 Stability and Convergence of the FD Method

What is meant by convergence is that the approximate solution tends to the actual solution as the computational grid gets very fine, that is, as $h,k \to 0$. A numerical method is said to be stable if any small changes in the initial data result in small changes in the subsequent data, or errors such as round-off introduced at any time remain bounded throughout. It can then be shown that the FD method outlined in Equation 10.29 is both stable and convergent if*

$$r = \frac{k\alpha^2}{h^2} \le \frac{1}{2} \tag{10.30}$$

and is unstable and divergent when $r > \frac{1}{2}$.

The user-defined function Heat1DFD uses the FD approach to solve the 1D heat equation subject to zero-boundary conditions and a prescribed initial condition. The function returns a failure message if the ratio $r = k\alpha^2/h^2 > \frac{1}{2}$. The function returns the approximate solution at the interior mesh points, as well as the values at the boundary points in a pattern that resembles the gridded region rotated 90° clockwise.

```
function u = Heat1DFD(t,x,u,alpha)
%
%  Heat1DFD numerically solves the one-dimensional heat
%  equation, with zero boundary conditions, using the
%  finite-difference method.
%
%   u = Heat1DFD(t,x,u,alpha) where
%
%     t is the row vector of times to compute,
%     x is the column vector of x positions to compute,
%     u is the column vector of initial temperatures for
%     each value in x,
%     alpha is a given parameter of the PDE,
%
%     u is the solution at the mesh points.

u = u(:);       % u must be a column vector
k = t(2)-t(1);
h = x(2)-x(1);
r = (alpha/h)^2*k;
```

* For details, refer to R.L. Burden and J.D. Faires, *Numerical Analysis*. 3rd edition, Prindle, Weber, and Schmidt, 1985.

```
if r>0.5
  warning('Method is unstable and divergent. Results will
be inaccurate.')
end

i=2:length(x)-1;
for j=1:length(t)-1,
    u(i,j+1) = (1-2*r)*u(i,j) +r*(u(i-1,j)+u(i+1,j));
end
```

Example 10.6: Finite-Difference Method: 1D Heat Equation

Consider a laterally insulated wire of length $L = 1$ and $\alpha = 0.5$, whose ends are kept at zero temperature, subjected to the initial temperature $f(x) = 10\sin\pi x$. Compute the approximate values of temperature, $u(x,t)$, $0 \le x \le 1$, and $0 \le t \le 0.5$, at mesh points generated by $h = 0.25$ and $k = 0.1$. All parameter values are in consistent physical units. The exact solution is given as

$$u(x,t) = (10\sin \pi x)e^{-0.25\pi^2 t}$$

SOLUTION

We first calculate

$$r = \frac{k\alpha^2}{h^2} = \frac{(0.1)(0.5)^2}{(0.25)^2} = 0.4 < \frac{1}{2}$$

that signifies stability and convergence for the FD method described by Equation 10.29. With $r = 0.4$, Equation 10.29 reduces to

$$u_{i,j+1} = 0.2u_{ij} + 0.4(u_{i-1,j} + u_{i+1,j}) \tag{10.31}$$

and will help generate approximate solutions at the mesh points marked in Figure 10.11. The first application of Equation 10.31 is at the (1,0) position so that

$$u_{11} = 0.2u_{10} + 0.4(u_{00} + u_{20}) \underset{\substack{\text{boundary} \\ \text{values}}}{\overset{\text{using}}{=}} 0.2(0.7071) + 0.4(0 + 10) = 5.4142$$

It is next applied at (2,0) to find u_{21}, and at (3,0) to find u_{31}. This way, the values at the three mesh points along the time row $j = 1$ are all

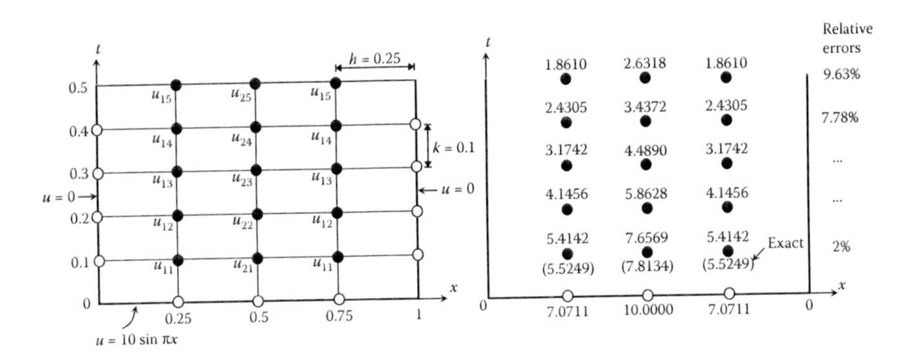

FIGURE 10.11
Grid and numerical results in Example 10.6.

determined. Subsequently, Equation 10.31 is applied at the three mesh points on the current row ($j = 1$) to find the values at the next time level, $j = 2$, and so on. Continuing with this strategy, we will generate the approximate values shown in Figure 10.11. It is interesting to note that at the first time row, we are experiencing relative errors of around 2%, but this grows to about 9.63% by the time we arrive at the fifth time level.

The numerical results obtained in this manner can be readily verified by executing the user-defined function `Heat1DFD`:

```
>> t=0:0.1:0.5;
>> x=0:0.25:1; x=x';
>> u=10.*sin(pi*x);
>> u=Heat1DFD(t,x,u,0.5)
u =
         0         0         0         0         0         0
    7.0711    5.4142    4.1456    3.1742    2.4304    1.8610
   10.0000    7.6569    5.8627    4.4890    3.4372    2.6318
    7.0711    5.4142    4.1456    3.1742    2.4304    1.8610
    0.0000         0         0         0         0         0
```

10.3.2 Crank–Nicolson (CN) Method

The condition $r = k\alpha^2/h^2 \leq \frac{1}{2}$, required by the FD method for stability and convergence, could lead to serious computational problems. For example, suppose $\alpha = 1$ and $h = 0.2$. Then, $r \leq \frac{1}{2}$ imposes $k \leq 0.02$, requiring too many time steps. Moreover, reducing the mesh size h by half to $h = 0.1$ increases the number of time steps by a factor of 4. Generally, to decrease r, we must either decrease k or increase h. Decreasing k forces additional time levels to be generated that increases the amount of computations. Increasing h causes a reduction of accuracy.

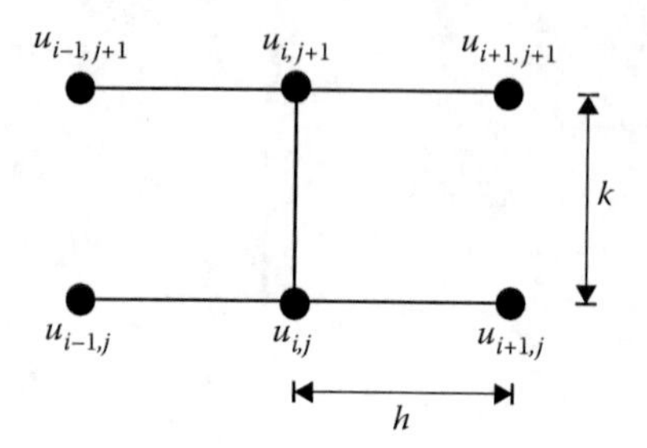

FIGURE 10.12
Six-point molecule used in CN method.

CN method offers a technique for solving the 1D heat equation with no restriction on the ratio $r = k\alpha^2/h^2$. The idea behind the CN method is to employ a six-point molecule (Figure 10.12) as opposed to the four-point molecule (Figure 10.10) used with the FD method.

To derive the difference equation associated with this method, we consider Equation 10.28. On the right-hand side, we write two expressions similar to that inside parentheses: one for the jth time row, one for the $(j+1)$st time row. Multiply each by $\alpha^2/2h^2$, and add them to obtain

$$\frac{1}{k}(u_{i,j+1} - u_{ij}) = \frac{\alpha^2}{2h^2}(u_{i-1,j} - 2u_{i,j} + u_{i+1,j}) + \frac{\alpha^2}{2h^2}(u_{i-1,j+1} - 2u_{i,j+1} + u_{i+1,j+1}) \tag{10.32}$$

Multiply Equation 10.32 by $2k$ and let $r = k\alpha^2/h^2$ as before, and rearrange the outcome so that the three terms associated with the higher time row [$(j+1)$st row] appear on the left side of the equation. The result is

$$2(1+r)u_{i,j+1} - r(u_{i-1,j+1} + u_{i+1,j+1}) = 2(1-r)u_{ij} + r(u_{i-1,j} + u_{i+1,j}), \quad r = \frac{k\alpha^2}{h^2} \tag{10.33}$$

This is called the difference equation for the 1D heat equation using the CN method. Starting with the 0th time level ($j = 0$), each time Equation 10.33 is applied, the three values on the right side, $u_{i-1,j}$, u_{ij}, $u_{i+1,j}$, are available from the initial temperature $f(x)$, while the values at the higher time level ($j = 1$) are unknown. If there are n nonboundary mesh points in each row, the ensuing system of equations to be solved is $n \times n$ with a tridiagonal coefficient matrix. Solving the system yields the values of u at the mesh points along the $j = 1$ row. Repeating the procedure leads to the approximate values of u at all desired mesh points.

The user-defined function `Heat1DCN` uses the CN method to solve the 1D heat equation subject to zero-boundary conditions and a prescribed initial condition. The function returns the approximate solution at the interior mesh points, as well as the values at the boundary points in a pattern that resembles the gridded region rotated 90° clockwise.

```
function u=Heat1DCN(t,x,u,alpha)
%
%  Heat1DCN numerically solves the one-dimensional heat
%  equation, with zero boundary conditions, using the
%  Crank-Nicolson method.
%
%   u=Heat1DCN(t,x,u,alpha) where
%
%     t is the row vector of times to compute,
%     x is the column vector of x positions to compute,
%     u is the column vector of initial temperatures for
%     each value in x,
%     alpha is a given parameter of the PDE,
%
%     u is the solution at the mesh points.
u=u(:);    % u must be a column vector
k=t(2)-t(1); h=x(2)-x(1); r=(alpha/h)^2*k;

% Compute A
n=length(x);
A=diag(2*(1+r)*ones(n-2,1));
A=A+diag(diag(A,-1)-r,-1);
A=A+diag(diag(A,1)-r, 1);

% Compute B
B=diag(2*(1-r)*ones(n-2,1));
B=B+diag(diag(B,-1)+r,-1);
B=B+diag(diag(B,1)+r,1);

C=A\B;

i=2:length(x)-1;
for j=1:length(t)-1,
    u(i,j+1)=C*u(i,j);
end
```

Example 10.7: CN Method: 1D Heat Equation

Consider the temperature distribution problem outlined in Example 10.6. Find the approximate values of $u(x,t)$ at the mesh points, and compare with the actual values as well as those obtained using the FD method.

SOLUTION

With $r = 0.4$, Equation 10.33 is written as

$$2.8u_{i,j+1} - 0.4(u_{i-1,j+1} + u_{i+1,j+1}) = 1.2u_{ij} + 0.4(u_{i-1,j} + u_{i+1,j}) \quad (10.34)$$

Applying Equation 10.34 at the $j = 0$ level, we find

$$\begin{aligned} 2.8u_{11} - 0.4(u_{01} + u_{21}) &= 1.2u_{10} + 0.4(u_{00} + u_{20}) \\ 2.8u_{21} - 0.4(u_{11} + u_{31}) &= 1.2u_{20} + 0.4(u_{10} + u_{30}) \\ 2.8u_{31} - 0.4(u_{21} + u_{41}) &= 1.2u_{30} + 0.4(u_{20} + u_{40}) \end{aligned}$$

Substituting the values from the boundary and initial conditions yields

$$\begin{bmatrix} 2.8 & -0.4 & 0 \\ -0.4 & 2.8 & -0.4 \\ 0 & -0.4 & 2.8 \end{bmatrix} \begin{Bmatrix} u_{11} \\ u_{21} \\ u_{31} \end{Bmatrix} = \begin{Bmatrix} 12.4853 \\ 17.6569 \\ 12.4853 \end{Bmatrix} \underset{\text{tridiagonal system}}{\overset{\text{Solve}}{\Longrightarrow}} \begin{matrix} u_{11} = 5.5880 = u_{31} \\ u_{21} = 7.9026 \end{matrix}$$

Next, Equation 10.34 is applied at the $j = 1$ level:

$$\begin{bmatrix} 2.8 & -0.4 & 0 \\ -0.4 & 2.8 & -0.4 \\ 0 & -0.4 & 2.8 \end{bmatrix} \begin{Bmatrix} u_{12} \\ u_{22} \\ u_{32} \end{Bmatrix} = \begin{Bmatrix} 9.8666 \\ 13.9535 \\ 9.8666 \end{Bmatrix} \underset{\text{tridiagonal system}}{\overset{\text{Solve}}{\Longrightarrow}} \begin{matrix} u_{12} = 4.4159 = u_{32} \\ u_{22} = 6.2451 \end{matrix}$$

This procedure is repeated until the approximate values of u at the mesh points along the $j = 5$ row are calculated. Execution of the user-defined function `Heat1DCD` yields

```
>> t=0:0.1:0.5;
>> x=0:0.25:1; x=x';
>> u=10.*sin(pi*x);
>> u=Heat1DCN(t,x,u,0.5)

u =

         0         0         0         0         0         0
    7.0711    5.5880    4.4159    3.4897    2.7578    2.1794
   10.0000    7.9026    6.2451    4.9352    3.9001    3.0821
    7.0711    5.5880    4.4159    3.4897    2.7578    2.1794
    0.0000         0         0         0         0         0
```

These numerical results, together with the associated relative errors, are shown in Figure 10.13. Comparing the relative errors with those in Example 10.6, it is evident that the CN method produces more accurate estimates. Note that the values returned by the CN method overshoot the actual values, while those generated by the FD method (Figure 10.11) undershoot the exact values.

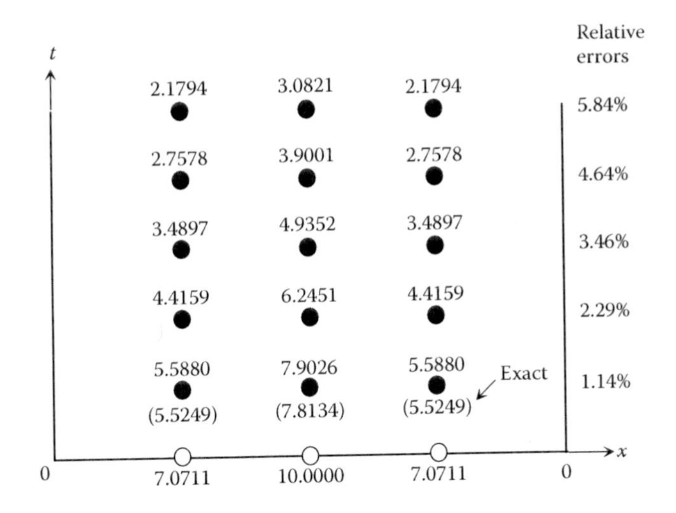

FIGURE 10.13
Grid and numerical results in Example 10.7.

10.3.2.1 CN Method versus FD Method

In Examples 10.6 and 10.7, the value of r satisfied the condition $r \le \frac{1}{2}$ so that both FD and CN were successfully applied, with CN yielding more accurate estimates. There are other situations where r does not satisfy $r \le \frac{1}{2}$; hence, the FD method cannot be implemented. To implement FD, we must reduce the value of r, which is possible by either increasing h or decreasing k, which causes reduction in accuracy or increase in computations. The following example demonstrates that even with a substantial increase in the number of time steps—to assure stability and convergence of FD—the values provided by FD are not any more accurate than those provided by CN.

Example 10.8: CN versus FD

Consider the temperature distribution problem studied in Examples 10.6 and 10.7. Assume $h = 0.25$.

a. Use the CN method with $r = 1$ to find the temperature at the mesh points u_{11}, u_{21}, and u_{31} in the first time row.
b. Since $r = 1$ does not satisfy the condition of $r \le \frac{1}{2}$, pick $r = 0.25$, for example, and $h = 0.25$ as before, and apply the FD method to find the values at the points u_{11}, u_{21}, and u_{31} in (a). Considering that the number of time steps has quadrupled, decide whether FD generates more accurate results than CN.

SOLUTION

a. We first note that

$$r = \frac{k\alpha^2}{h^2} = 1 \underset{h=0.25}{\overset{\alpha=0.5}{\Rightarrow}} k = r\left(\frac{h}{\alpha}\right)^2 = 0.25$$

With $r = 1$, Equation 10.33 becomes

$$4u_{i,j+1} - u_{i-1,j+1} - u_{i+1,j+1} = u_{i-1,j} + u_{i+1,j}$$

Applying this equation with $j = 0$, we find

$$\begin{aligned} 4u_{11} - 0 - u_{21} &= 0 + 10 \\ 4u_{21} - u_{11} - u_{31} &= 2(7.0711) \\ 4u_{31} - u_{21} - 0 &= 10 + 0 \end{aligned} \Rightarrow \begin{bmatrix} 4 & -1 & 0 \\ -1 & 4 & -1 \\ 0 & -1 & 4 \end{bmatrix} \begin{Bmatrix} u_{11} \\ u_{21} \\ u_{31} \end{Bmatrix} = \begin{Bmatrix} 10 \\ 14.1422 \\ 10 \end{Bmatrix} \Rightarrow \begin{matrix} u_{11} = u_{31} = 3.8673 \\ u_{21} = 5.4692 \end{matrix}$$

b. We note that

$$r = \frac{k\alpha^2}{h^2} = 0.25 \underset{h=0.25}{\overset{\alpha=0.5}{\Rightarrow}} k = r\left(\frac{h}{\alpha}\right)^2 = 0.0625$$

Therefore, the step size along the t-axis has been reduced from $k = 0.25$ to $k = 0.0625$, implying that four time-step calculations are required to find the values of u_{11}, u_{21}, and u_{31} in (a). With $r = 0.25$, Equation 10.29 reduces to

$$u_{i,j+1} = 0.5u_{ij} + 0.25(u_{i-1,j} + u_{i+1,j})$$

Proceeding as always, the solution estimates at the desired mesh points will be calculated. It should be mentioned that with the new, smaller step size, $k = 0.0625$, the old u_{11}, u_{21}, and u_{31} in (a) are now labeled u_{14}, u_{24}, and u_{34}. The computed values are

$$\begin{aligned} u_{14} &= u_{34} = 3.7533 \\ u_{24} &= 5.3079 \end{aligned}$$

The numerical results obtained in (a) and (b) are summarized in Figure 10.14 where it is readily observed that although FD used four times as many time levels as CN, the accuracy of the results by CN is still superior.

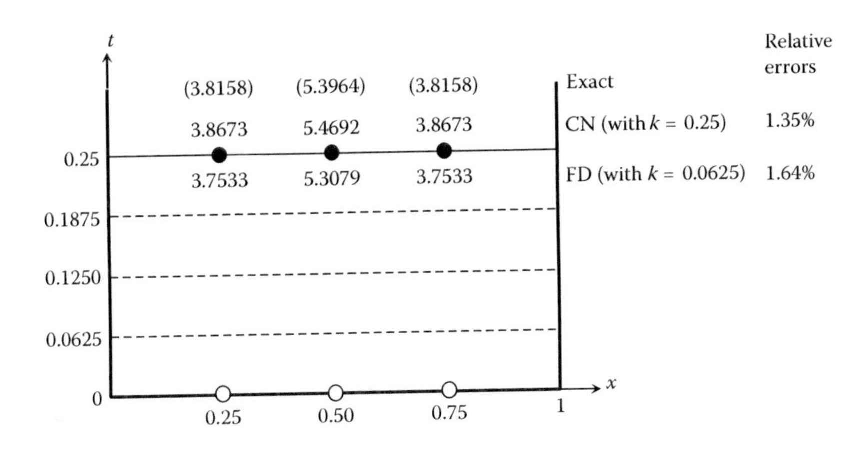

FIGURE 10.14
Accuracy and efficiency of CN versus FD.

10.4 Hyperbolic PDEs

The 1D wave equation, $u_{tt} = \alpha^2 u_{xx}$ (α = constant > 0), is the simplest model of a physical system involving a hyperbolic PDE. Specifically, consider an elastic string of length L fixed at both ends. Assuming the string is driven by only initial displacement $f(x)$ and initial velocity $g(x)$, the free vibration of the string is governed by the boundary-initial-value problem

$$u_{tt} = \alpha^2 u_{xx} \ (\alpha = \text{constant} > 0), \quad 0 \le x \le L, \quad t \ge 0 \tag{10.35}$$

$$u(0,t) = 0 = u(L,t) \tag{10.36}$$

$$u(x,0) = f(x), \quad u_t(x,0) = g(x) \tag{10.37}$$

Figure 10.15 shows that a grid is constructed using a mesh size of h in the x-direction and a mesh size of k in the t-direction. The terms u_{xx} and u_{tt} in Equation 10.35 will be replaced by three-point central difference approximations (Section 6.2). Consequently, Equation 10.35 yields

$$\frac{1}{k^2}(u_{i,j-1} - 2u_{ij} + u_{i,j+1}) = \frac{\alpha^2}{h^2}(u_{i-1,j} - 2u_{i,j} + u_{i+1,j}) \tag{10.38}$$

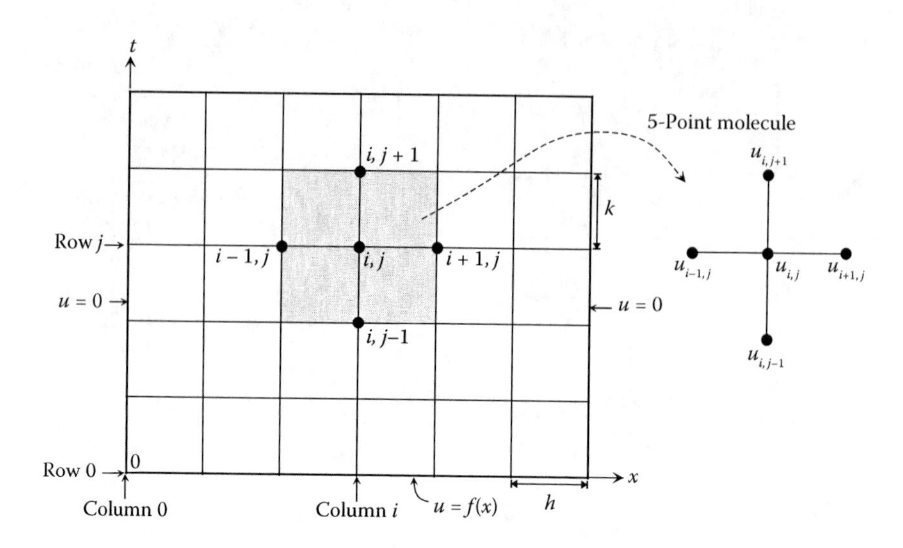

FIGURE 10.15
Region and grid used for solving the 1D wave equation.

Multiplying by k^2, letting $\tilde{r} = (k\alpha/h)^2$, and solving for $u_{i,j+1}$, we find

$$u_{i,j+1} = -u_{i,j-1} + 2(1-\tilde{r})u_{ij} + \tilde{r}(u_{i-1,j} + u_{i+1,j}), \quad \tilde{r} = \left(\frac{k\alpha}{h}\right)^2 \tag{10.39}$$

that is known as the difference equation for the 1D wave equation using the FD method. It can be shown that the numerical method described by Equation 10.39 is stable and convergent if $\tilde{r} \le 1$.

10.4.1 Starting the Procedure

Applying Equation 10.39 along the $j = 0$ level, we have

$$u_{i,1} = -u_{i,-1} + 2(1-\tilde{r})u_{i,0} + \tilde{r}(u_{i-1,0} + u_{i+1,0}) \tag{10.40}$$

The quantities $u_{i,0}$, $u_{i-1,0}$, and $u_{i+1,0}$ are available from the initial displacement, but $u_{i,-1}$ is not yet known. To find $u_{i,-1}$, we use the information on the initial velocity $u_t(x,0) = g(x)$. Let $x_i = ih$ and $g_i = g(x_i)$. Using the central difference formula for $u_t(x_i,0)$

$$\frac{u_{i,1} - u_{i,-1}}{2k} = g_i \quad \overset{\text{Solve for } u_{i,-1}}{\Rightarrow} \quad u_{i,-1} = u_{i,1} - 2kg_i$$

Inserting this into Equation 10.40

$$u_{i,1} = (1-\tilde{r})u_{i,0} + \frac{1}{2}\tilde{r}(u_{i-1,0} + u_{i+1,0}) + kg_i \tag{10.41}$$

In summary, the FD approximation for the 1D wave equation is implemented as follows. First, apply Equation 10.41 using the knowledge of initial displacement and velocity. This gives the values of u along the first time step, $j = 1$. From this point onward, apply Equation 10.39 to find u at the mesh points on the higher time levels.

The user-defined function `Wave1DFD` uses the FD approach to solve the 1D wave equation subject to zero-boundary conditions and prescribed initial displacement and velocity. The function returns a failure message if $r = (k\alpha/h)^2 > 1$. The function returns the approximate solution at the interior mesh points, as well as the values at the boundary points in a pattern that resembles the gridded region rotated 90° clockwise.

```
function u=Wave1DFD(t,x,u,ut,alpha)
%
% Wave1DFD numerically solves the one-dimensional wave
% equation, with zero boundary conditions, using the
% finite-difference method.
%
%   u=Wave1DFD(t,x,u,ut,alpha) where
%
%      t is the row vector of times to compute,
%      x is the column vector of x positions to compute,
%      u is the column vector of initial displacements
%      for each value in x,
%      ut is the column vector of initial velocities for
%      each value in x,
%      alpha is a given parameter of the PDE,
%
%      u is the solution at the mesh points.

u=u(:);        % u must be a column vector
ut=ut(:);      % ut must be a column vector

k=t(2)-t(1);
h=x(2)-x(1);
r=(k*alpha/h)^2;

if r>1
    warning('Method is unstable and divergent. Results
will be inaccurate.')
end
```

```
i=2:length(x)-1;
u(i,2) = (1-r)*u(i,1) +r/2*(u(i-1,1) +u(i+1,1)) +k*ut(i);
for j=2:length(t)-1,
    u(i,j+1) =-u(i,j-1) +2*(1-r)*u(i,j) +r*(u(i-1,j) +
u(i+1,j));
end
```

Example 10.9: Free Vibration of an Elastic String

Consider an elastic string of length $L=2$ with $\alpha=1$, fixed at both ends. Suppose the string is subjected to an initial displacement $f(x)=5\sin(\pi x/2)$ and zero initial velocity, $g(x)=0$. Using $h=0.4=k$, find the displacement $u(x,t)$ of the string for $0 \le x \le L$ and $0 \le t \le 2$. All parameters are in consistent physical units. The exact solution is given by

$$u(x,t) = 5\sin\frac{\pi x}{2}\cos\frac{\pi t}{2}$$

SOLUTION

We first calculate

$$\tilde{r} = \left(\frac{k\alpha}{h}\right)^2 = 1$$

To find the values of u at the $j=1$ level ($t=0.4$), we apply Equation 10.41. But since $g_i=0$ and $\tilde{r}=1$, Equation 10.41 simplifies to

$$u_{i,1} = \frac{1}{2}(u_{i-1,0} + u_{i+1,0}), \quad i = 1,2,3,4$$

so that

$$u_{11} = \frac{1}{2}(u_{00} + u_{20}),\ u_{21} = \frac{1}{2}(u_{10} + u_{30}),\ u_{31} = \frac{1}{2}(u_{20} + u_{40}),\ u_{41} = \frac{1}{2}(u_{30} + u_{50})$$

This way, the estimates at all four interior mesh points along $j=1$ are determined. For higher time level, we use Equation 10.39 that simplifies to

$$u_{i,j+1} = -u_{i,j-1} + u_{i-1,j} + u_{i+1,j}, \quad i,j = 1,2,3,4 \tag{10.42}$$

To find the four estimates on the $j = 2$ level, for example, we fix $j = 1$ in Equation 10.42, vary $i = 1,2,3,4$, and use the boundary values, as well as those obtained previously on the $j = 1$ level. As a result, we find u_{12}, u_{22}, u_{32}, and u_{42}. Continuing in this manner, estimates at all desired mesh points will be calculated; see Figure 10.16. These results can be readily verified by executing the user-defined function `Wave1DFD`. We can also use this function to plot the variations of u versus x for fixed values of time. That is, plot the values obtained in each time level in Figure 10.16 versus x. For plotting purposes, we will use a smaller increment of 0.1 for both t and x. Note that the increments h and k are constrained by the condition $(k\alpha/h)^2 \leq 1$.

```
>> t=0:0.1:2;
>> x=0:0.1:2; x=x';
>> u=5.*sin(pi*x/2);
>> ut=zeros(length(x),1);
>> u=Wave1DFD(t,x,u,ut,1);
>> plot(x,u)                    % Figure 10.17
```

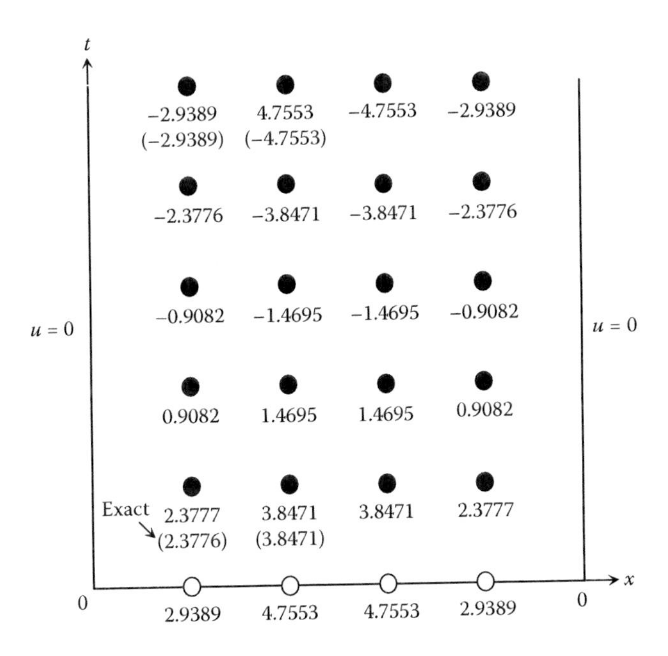

FIGURE 10.16
Grid and numerical results in Example 10.9.

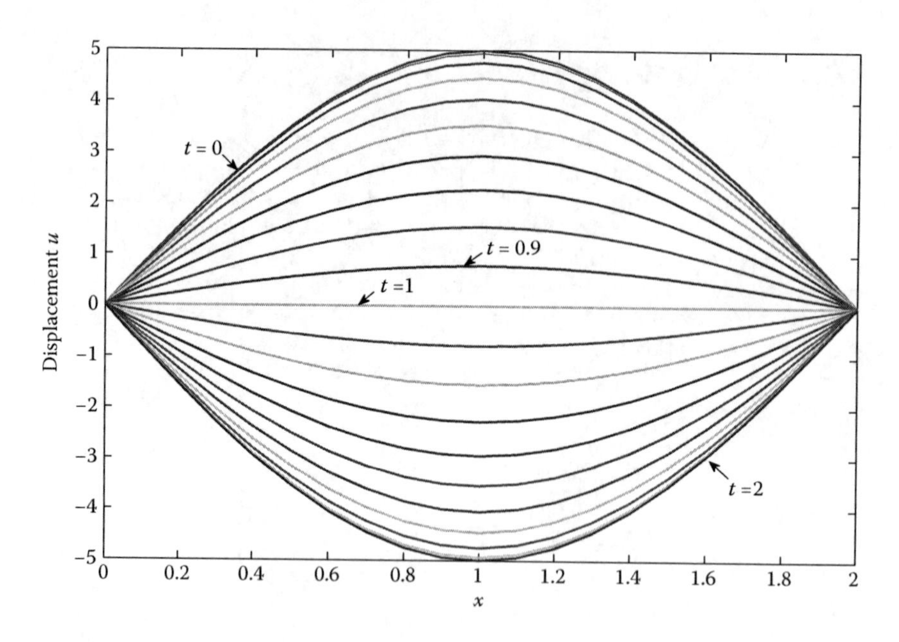

FIGURE 10.17
Displacement of the elastic string for various fixed values of time.

PROBLEM SET

Elliptic Partial Differential Equations (Section 10.2)

Dirichlet Problem (Laplace's Equation)

In Problems 1 through 4

a. Solve the Dirichlet problem described in Figure 10.18 with the specified boundary conditions.

b. Confirm the results by executing the user-defined function `DirichletPDE`. Suppress the plot.

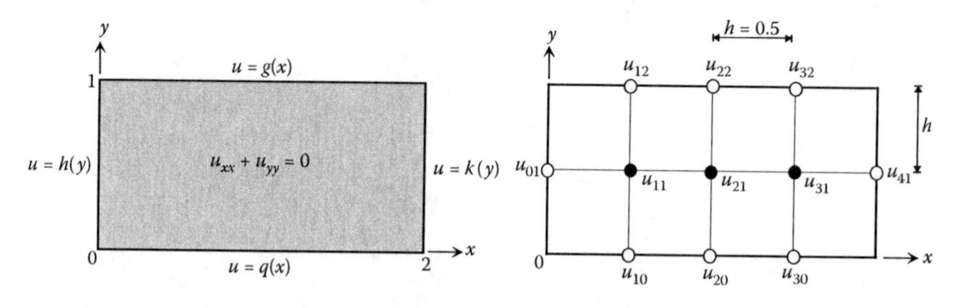

FIGURE 10.18
Dirichlet problem in Problems 1 through 4.

c. Reduce the mesh size by half and execute `DirichletPDE` to solve the problem. Compare the numerical results at the original three interior mesh points with those obtained in (a) and (b).

1. $q(x) = 85x(2 - x)$, $g(x) = 0$, $h(y) = 0$, $k(y) = 0$
2. $q(x) = 50\sin(\pi x/2)$, $g(x) = 100x(2 - x)$, $h(y) = 0$, $k(y) = 0$
3. $q(x) = 10x(2 + x)$, $g(x) = 0$, $h(y) = 0$, $k(y) = 80(1 - y^2)$
4. $q(x) = 9$, $g(x) = 12$, $h(y) = 5y^2 - 2y + 9$, $k(y) = 2y^2 + y + 9$

In Problems 5 through 8

a. Solve the Dirichlet problem described in Figure 10.19 with the specified boundary conditions.

b. Confirm the results by executing the user-defined function `DirichletPDE`. Suppress the plot.

c. Reduce the mesh size by half and execute `DirichletPDE` to solve the problem. Compare the numerical results at the original four interior mesh points with those obtained in (a) and (b).

5. $a = 3$, $h = 1$, $q(x) = x$, $g(x) = 3$, $h(y) = y$, $k(y) = 3$
6. $a = 3$, $h = 1$, $q(x) = 100$, $g(x) = 178 - 17x - x^3$,
   $h(y) = 3y^3 - y + 100$, $k(y) = 100$
7. $a = 9$, $h = 3$, $q(x) = 50\sin^2(\pi x/9)$, $g(x) = x^2 + x$,
   $h(y) = 0$, $k(y) = y(y + 1)$

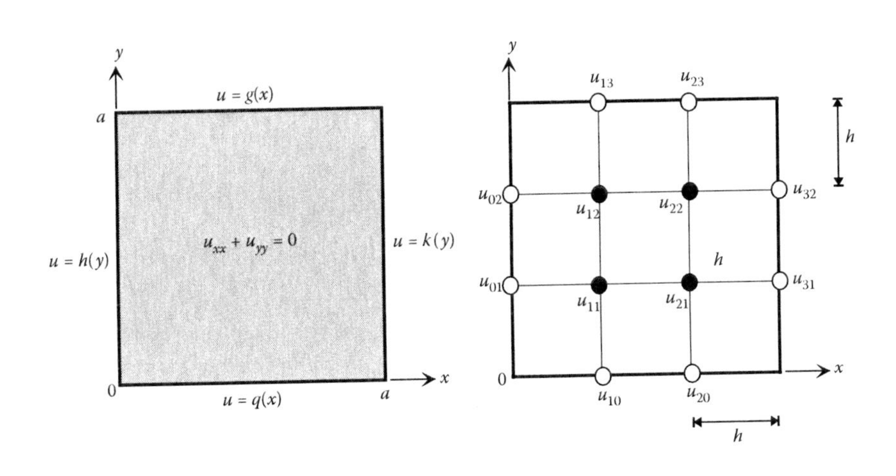

FIGURE 10.19
Region and grid used in Problems 5 through 8.

8. $a = 9$, $h = 3$, $q(x) = 0$, $g(x) = 18 - 2x$, $h(y) = 18\sin(\pi y/18)$, $k(y) = 20\left[\sin(\pi y/9) + 3\sin(\pi y/3)\right]$

Dirichlet Problem (Poisson's Equation)

In Problems 9 through 14

a. ✍ Solve the Poisson's equation in the specified region subject to the given boundary conditions.

b. Confirm the results by executing the user-defined function `DirichletPDE`. Suppress the plot.

c. Reduce the mesh size by half and execute `DirichletPDE` to solve the problem. Compare the numerical results at the original four interior mesh points with those obtained in (a) and (b).

9. $u_{xx} + u_{yy} = (x-3)(2y-1)^2$, same region, grid, and boundary conditions as in Problem 5.

10. $u_{xx} + u_{yy} = 10x^2 - 25y^3$, same region, grid, and boundary conditions as in Problem 6.

11. $u_{xx} + u_{yy} = 0.1y\sin(\pi x/9)$, same region, grid, and boundary conditions as in Problem 7.

12. $u_{xx} + u_{yy} = x + y$, same region, grid, and boundary conditions as in Problem 8.

13. $u_{xx} + u_{yy} = y - x + 1$, same region and grid as in Figure 10.19 with $a = 1.2$ and boundary conditions described by

$$q(x) = x^2,\ g(x) = 1.2 + 1.4x,\ h(y) = y,\ k(y) = 1.44 + y^2$$

14. $u_{xx} + u_{yy} = \sin \pi x$, same region and grid as in Figure 10.19 with $a = 1.8$ and boundary conditions described by

$$q(x) = x,\ g(x) = 1.8,\ h(y) = y,\ k(y) = 1.8$$

PRADI Method

15. Referring to Example 10.2

a. ✍ Perform the second iteration to calculate the estimates $u_{11}^{(2)}, u_{21}^{(2)}, u_{12}^{(2)}, u_{22}^{(2)}$. Compare the corresponding relative errors with those at the completion of one iteration step.

b. Verify the results of (a) using the user-defined function `PRADI`.

c. Determine the number of iterations required to meet the (default) tolerance.

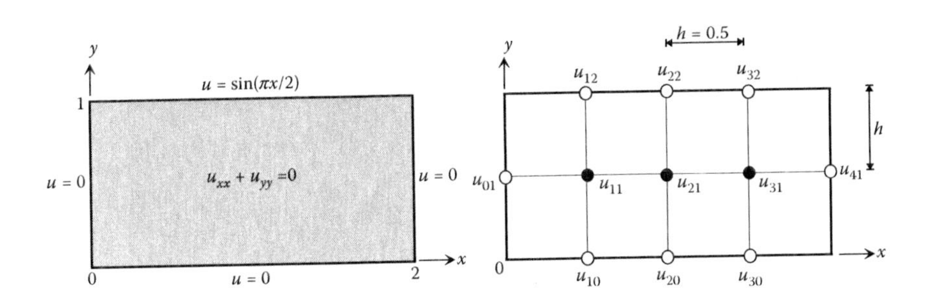

FIGURE 10.20
Region and grid in Problem 16.

16. Consider the Dirichlet problem described in Figure 10.20.

    a. ✍ Perform one iteration step of the PRADI method using starting values of 0.4 for the interior mesh points.

    b. Verify the numerical results of (a) using the user-defined function `PRADI`.

✍ In Problems 17 through 22, perform one complete iteration step of the PRADI method using the given starting values for the interior mesh points in the specified region and subject to the given boundary conditions.

17. $u_{xx} + u_{yy} = 0$ in the region, and with the boundary conditions, described in Problem 5; starting values of 2.
18. $u_{xx} + u_{yy} = 0$ in the region, and with the boundary conditions, described in Problem 6; starting values of 110.
19. $u_{xx} + u_{yy} = 0$ in the region, and with the boundary conditions, described in Problem 7; starting values of 22.
20. $u_{xx} + u_{yy} = x + y$ in the region, and with the boundary conditions, described in Problem 8; starting values of –10.
21. Problem 13; starting values of 0.6.
22. Problem 14; starting values of 0.8.

Mixed Problem

✍ In Problems 23 through 30, solve the mixed problem in the given region subject to the boundary conditions provided. In all cases, use a mesh size of $h = 1$.

23. For this problem, use the following figure.

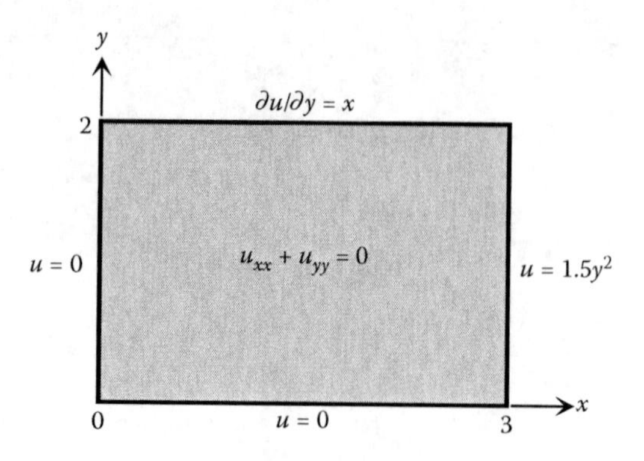

24. For this problem, use the following figure.

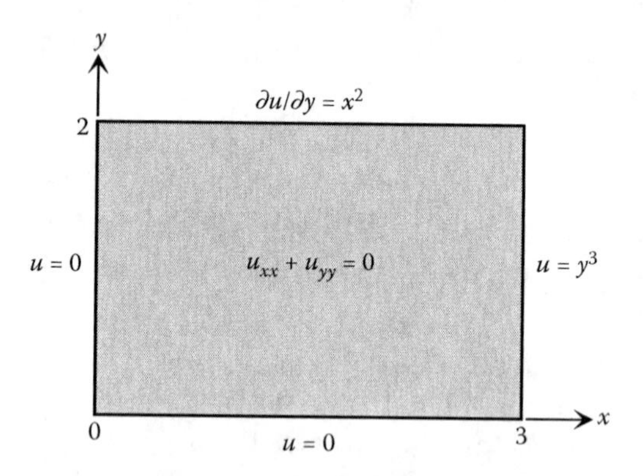

25. For this problem, use the following figure.

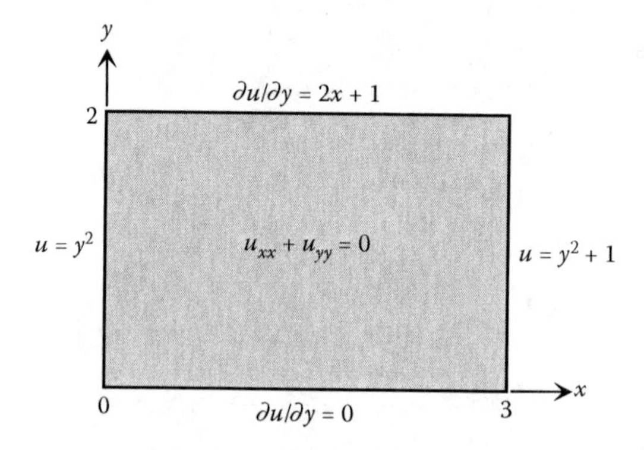

26. For this problem, use the following figure.

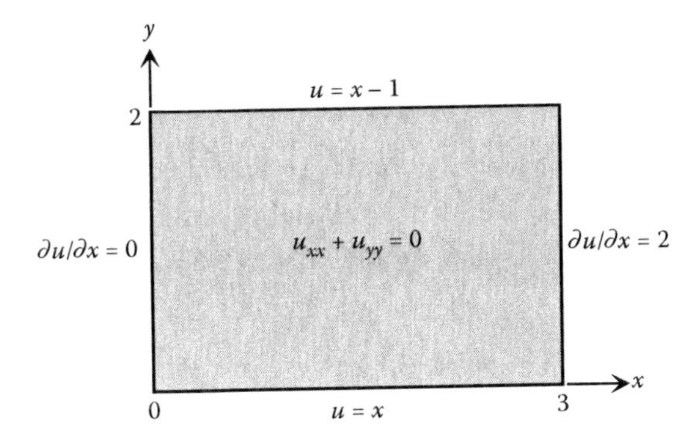

27. For this problem, use the following figure.

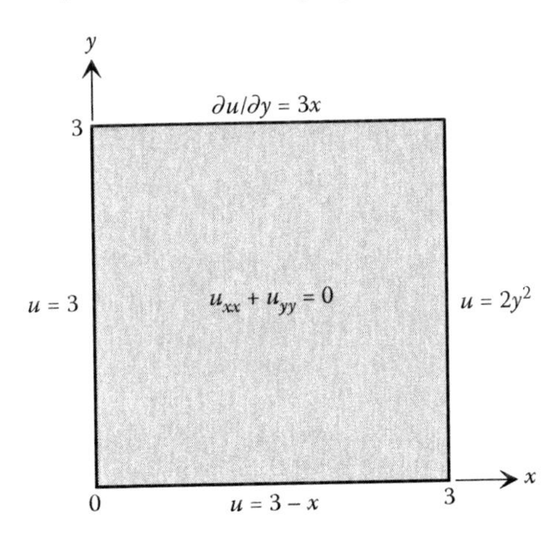

28. For this problem, use the following figure.

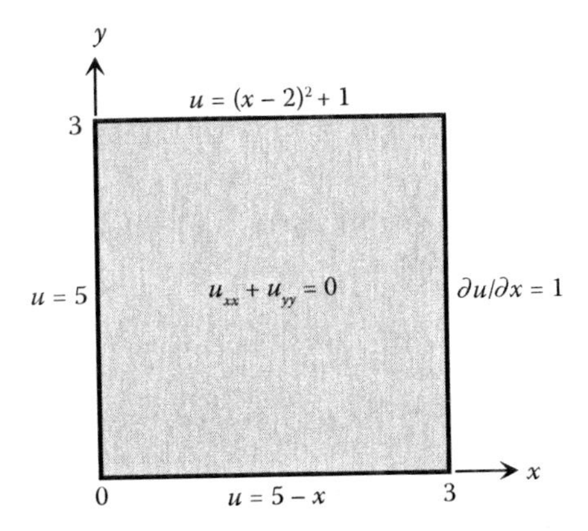

29. For this problem, use the following figure ($f(x,y) = \frac{1}{3}x^2 + \frac{1}{2}y^2$).

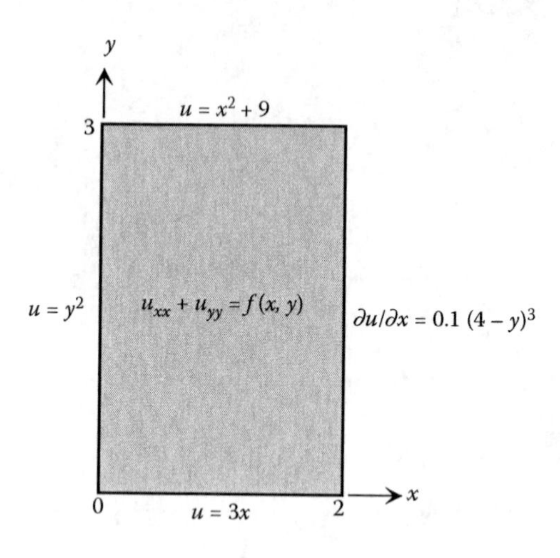

30. For this problem, use the following figure ($f(x,y) = 0.2xy$).

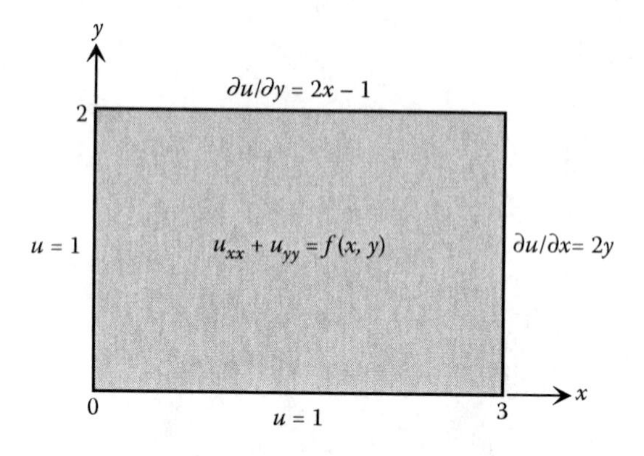

More Complex Regions

31. ✍ Solve $u_{xx} + u_{yy} = 0$ in the region shown in the following figure subject to the indicated boundary conditions. The curved portion of the boundary obeys $x^2 + y^2 = 41$.

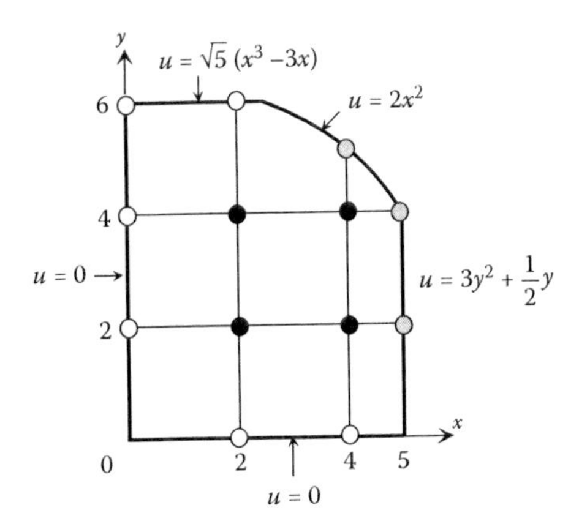

32. ✍ Solve $u_{xx} + u_{yy} = 0$ in the region shown in the following figure subject to the indicated boundary conditions. The slanted portion of the boundary obeys $y + x = 8$.

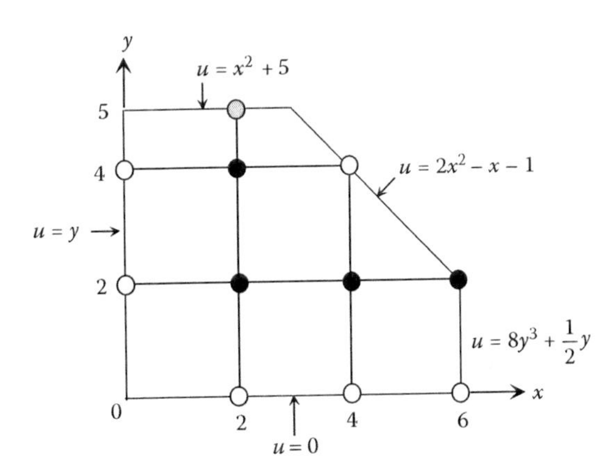

33. ✍ Solve $u_{xx} + u_{yy} = 0.5x^2y$ in the region shown in the figure in Problem 32 above subject to the indicated boundary conditions. The slanted portion of the boundary obeys $y + x = 8$.
34. ✍ Solve $u_{xx} + u_{yy} = 0$ in the region shown in the following figure subject to the indicated boundary conditions. The curved portion of the boundary obeys $(x-7)^2 + (y-9)^2 = 16$.

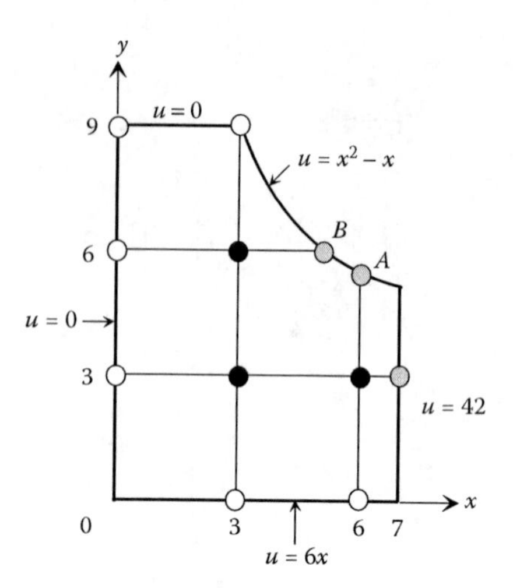

Parabolic Partial Differential Equations (Section 10.3)

FD Method

35. Consider a laterally insulated wire of length $L = 2$ and $\alpha = 1$, whose ends are kept at zero temperature, subjected to the initial temperature

$$f(x) = \begin{cases} x & \text{if } 0 \le x \le 1 \\ 2 - x & \text{if } 1 \le x \le 2 \end{cases}$$

   a. ✍ Using the FD method, compute the estimated values of temperature, $u(x,t)$, $0 \le x \le 2$, $0 \le t \le 0.5$, at the mesh points generated by $h = 0.50$ and $k = 0.125$. Also calculate the relative error at each time level. The exact solution is given by $u_{\text{Exact}}(x,t) \cong (8/\pi^2)\sin(\pi x/2)\, e^{-\pi^2 t/4}$.

   b. Confirm the numerical results of (a) using `Heat1DFD`.

36. Consider a laterally insulated wire of length $L = 2$ and $\alpha = 1$, whose ends are kept at zero temperature, subjected to the initial temperature $f(x) = 10\sin(\pi x/2)$.

   a. ✍ Using the FD method, find the estimated values of temperature, $u(x,t)$, $0 \le x \le 2$, $0 \le t \le 0.16$, at the mesh points generated by $h = 0.4$ and $k = 0.04$. Also calculate the relative error at each time level. The exact solution is given by $u_{\text{Exact}}(x,t) = 10\sin(\pi x/2)\, e^{-\pi^2 t/4}$.

   b. Confirm the numerical results of (a) using `Heat1DFD`.

37. ✍ Reconsider the wire in Problem 35. Assuming that all information is unchanged except for $k = 0.0625$, calculate the temperature estimate $u(0.5,0.125)$, and compare the corresponding relative error with that obtained in Problem 35.

38. ✍ Reconsider the wire in Problem 36. Assuming that all information is unchanged except for $k = 0.02$, calculate the temperature estimate $u(0.4,0.04)$, and compare the corresponding relative error with that obtained in Problem 36.

39. Consider a laterally insulated wire of length $L = 1$ and $\alpha = 1$, whose ends are kept at zero temperature, subjected to the initial temperature $f(x) = x(1-x)$.

    a. ✍ Using the FD method, find the estimated values of temperature, $u(x,t)$, $0 \le x \le 1$, $0 \le t \le 0.04$, at the mesh points generated by $h = 0.2$ and $k = 0.01$.

    b. Confirm the numerical results of (a) using `Heat1DFD`.

40. Write a user-defined function with function call `u = Heat1DFD_gen(t,x,u,alpha,q,g)` that uses the FD method to solve the 1D heat equation subjected to general boundary conditions. All arguments are as in `Heat1DFD`, while the two new parameters `q` and `g` are inline functions, respectively, representing $u(0,t)$ and $u(L,t)$. Using `Heat1DFD_gen` find the temperature estimates for a laterally insulated wire of length $L = 1$ and $\alpha = 1$, subjected to the initial temperature $f(x) = x(1-x)$ and boundary conditions $u(0,t) = 0$, $u(1,t) = t$, $0 < t < 0.05$. Construct a grid using $h = 0.2$, $k = 0.01$.

41. Using the user-defined function `Heat1DFD_gen` (see Problem 40), find the temperature estimates for a laterally insulated wire of length $L = 3$ and $\alpha = 0.8$, subjected to the initial temperature $f(x) = \sin \pi x$ and boundary conditions $u(0,t) = 1$, $u(1,t) = 1$, $0 < t < 0.3$. Construct a grid using $h = 0.3$, $k = 0.05$.

42. Write a user-defined function with function call `u = Heat1DFD_insul_ends(t,x,u,alpha)` that uses the FD method to solve the 1D heat equation subjected to insulated ends, that is, $u_x(0,t) = 0$ and $u_x(L,t) = 0$. Note that the difference equation must be applied at the boundary mesh points as well since the estimates at these points are now part of the unknown vector. Use the central difference formulas to approximate the first partial derivative at these points. All arguments are as in `Heat1DFD`. Using `Heat1DFD_insul_ends` find the temperature estimates for a laterally insulated wire of length $L = 1$ and $\alpha = 1$, subjected to the initial temperature $f(x) = x(1-x)$ and insulated ends. Construct a grid using $h = 0.2$, $k = 0.01$, and assume $0 < t < 0.05$.

CN Method

43. Consider a laterally insulated wire of length $L = 2$ and $\alpha = 1$, whose ends are kept at zero temperature, subjected to the initial temperature

$$f(x) = \begin{cases} x & \text{if } 0 \le x \le 1 \\ 2 - x & \text{if } 1 \le x \le 2 \end{cases}$$

   a. ✍ Using the CN method, compute the estimated values of temperature, $u(x,t)$, $0 \le x \le 2$, $0 \le t \le 0.25$, at the mesh points generated by $h = 0.50$ and $k = 0.125$. Calculate the relative error at each time level, and compare with those corresponding to the FD method (see Problem 35). The exact solution is given in closed form as $u_{\text{Exact}}(x,t) \cong (8/\pi^2)\sin(\pi x/2)\,e^{-\pi^2 t/4}$.

   b. Confirm the numerical results of (a) using `Heat1DCN`.

44. Consider a laterally insulated wire of length $L = 1$ and $\alpha = 0.5$, whose ends are kept at zero temperature, subjected to the initial temperature $f(x) = \sin \pi x + \sin 2\pi x$.

   a. ✍ Using the CN method, compute the estimated values of temperature, $u(x,t)$, $0 \le x \le 1$, $0 \le t \le 0.25$, at the mesh points generated by $h = 0.25$ and $k = 0.125$.

   b. Confirm the numerical results of (a) using `Heat1DCN`.

45. Consider a laterally insulated wire of length $L = 3$ and $\alpha = 1.5$, whose ends are kept at zero temperature, subjected to the initial temperature $f(x) = 0.5x(3 - x)$.

   a. ✍ Using the CN method, compute the estimated values of temperature, $u(x,t)$, $0 \le x \le 3$, $0 \le t \le 0.25$, at the mesh points generated by $h = 0.75$ and $k = 0.125$.

   b. Confirm the numerical results of (a) using `Heat1DCN`.

46. Repeat Problem 45 for an initial temperature of $f(x) = \sin(\pi x/3) + \sin 2\pi x$.

47. Consider a laterally insulated wire of length $L = 0.5$ and $\alpha = 1$, with ends kept at zero temperature, subjected to the initial temperature $f(x) = x(1 - 2x)$. Let $h = 0.1$.

   a. Find the temperature estimates, $u(x,t)$, $0 \le x \le 0.5$, $0 \le t \le 0.04$, using the `Heat1DCN` with $k = 0.01$.

   b. Reduce the time step size by half and apply `Heat1DFD` to find the estimates at the mesh points. Compare the numerical results by the two methods at the time level $t = 0.01$.

48. Consider a laterally insulated wire of length $L = 1$ and $\alpha = 0.7071$, with ends kept at zero temperature, subjected to the initial temperature $f(x) = 1 - \cos 2\pi x$. Let $h = 0.25$.

   a. Find the temperature estimates, $u(x,t)$, $0 \le x \le 1$, $0 \le t \le 0.5$, using the `Heat1DCN` with $r = 1$.

   b. Adjust the time step size so that $r = 0.5$, and apply `Heat1DFD` to find the estimates at the mesh points. Compare the numerical results by the two methods at the time level $t = 0.125$.

49. Consider a laterally insulated wire of length $L = 2$ and $\alpha = 1$, subjected to the initial temperature $f(x) = 3x(2 - x)$ and boundary conditions $u(0,t) = 1$, $u(2,t) = t$, $0 < t < 0.25$. Construct a grid using $h = 0.5$, $k = 0.125$. Find the temperature estimates at the mesh points using the CN method.

50. Write a user-defined function with function call `u = Heat1DCN_gen(t,x,u,alpha,q,g)` that uses the CN method to solve the 1D heat equation subjected to general boundary conditions. All arguments are as in `Heat1DCN`, while the two new parameters `q` and `g` are inline functions, respectively, representing $u(0,t)$ and $u(L,t)$. Using `Heat1DCN_gen` find the temperature estimates at the mesh points for the problem formulated in Problem 49.

Hyperbolic Partial Differential Equations (Section 10.4)

In Problems 51 through 56, an elastic string of length L with constant α, fixed at both ends, is considered. The string is subjected to initial displacement $f(x)$ and initial velocity $g(x)$.

a. Using the FD method with the indicated mesh sizes h and k, find the estimates for displacement $u(x,t)$ for $0 \le x \le L$ and the given time interval.

b. Confirm the numerical results of (a) using `Wave1DFD`.

51. $L = 1$, $\alpha = 1$, $f(x) = 6 \sin 2\pi x$, $g(x) = 0$, $h = 0.2 = k$, $0 \le t \le 0.4$
52. $L = 0.5$, $\alpha = 1$, $f(x) = 10x(1 - 2x)$, $g(x) = 0$, $h = 0.1 = k$, $0 \le t \le 0.3$
53. $L = 0.5$, $\alpha = 1$, $f(x) = 0$, $g(x) = -\sin(2\pi x)$, $h = 0.1 = k$, $0 \le t \le 0.2$
54. $L = 1$, $\alpha = 4$, $f(x) = 0$, $g(x) = 90x^2$, $h = 0.2$, $k = 0.05$, $0 \le t \le 0.15$
55. $L = 1$, $\alpha = 2$, $f(x) = x(1 - x)$, $g(x) = \sin \pi x$, $h = 0.2$, $k = 0.1$, $0 \le t \le 0.3$
56. $L = 1$, $\alpha = 2$, $f(x) = x(1 - x)$, $g(x) = 10x$, $h = 0.2$, $k = 0.1$, $0 \le t \le 0.3$

Bibliography

1. Burden, R.L. and Faires, J.D., *Numerical Analysis*. 9th edition, Brooks Cole, Independence, KY, 2010.
2. Craig, J.J., *Introduction to Robotics: Mechanics and Control*. 3rd edition, Prentice Hall, Englewood Cliffs, NJ, 2004.
3. Esfandiari, R.S., *Applied Mathematics for Engineers*. 4th edition, Atlantis, Los Angeles, CA, 2007.
4. Esfandiari, R.S., *MATLAB Manual for Advanced Engineering Mathematics*. Atlantis, Los Angeles, CA, 2007.
5. Esfandiari, R.S. and Lu, B., *Modeling and Analysis of Dynamic Systems*. CRC Press, Boca Raton, FL, 2010.
6. Golub, G.H. and Van Loan, C.F., *Matrix Computations*. 3rd edition, The Johns Hopkins University Press, Baltimore, MD, 1996.
7. Henrici, P., *Elements of Numerical Analysis*. John Wiley & Sons, New York, NY, 1966.
8. Johnson, L.W. and Riess, R.D., *Numerical Analysis*. Addison-Wesley, Reading, MA, 1977.
9. Kelly, S.G., *Fundamentals of Mechanical Vibrations*. 2nd edition, McGraw-Hill, New York, NY, 2000.
10. Li, J. and Chen, Y.-T., *Computational Partial Differential Equations Using MATLAB*. CRC Press, Boca Raton, FL, 2008.
11. Mathews, J.H. and Fink, K.K., *Numerical Methods Using MATLAB*. 4th edition, Pearson, Upper Saddle River, NJ, 2004.
12. Meirovitch, L., *Fundamentals of Vibrations*. Waveland Press, Long Grove, IL, 2010.
13. Noble, B. and Daniel, J.W., *Applied Linear Algebra*. 3rd edition, Pearson, Upper Saddle River, NJ, 1987.
14. Rao, S.S., *Mechanical Vibrations*. 5th edition, Prentice Hall, Englewood Cliffs, NJ , 2010.
15. Smith, W.A., *Elementary Numerical Analysis*. Brady, Upper Saddle River, NJ, 1986.
16. Spong, M.W. and Vidyasagar, M., *Robot Dynamics and Control*. Wiley, New York, NY, 1989.
17. Strang, G., *Linear Algebra and Its Applications*. 4th edition, Brooks Cole, Independence, KY, 2005.
18. Strikwerda, J.C., *Finite Difference Schemes and Partial Differential Equations*. 2nd edition, SIAM, Philadelphia, PA, 2007.
19. Vidyasagar, M., *Nonlinear Systems Analysis*. 2nd edition, SIAM, Philadelphia, PA, 2002.
20. Weaver, W., Timoshenko, S.P., and Young, D.H., *Vibration Problems in Engineering*. 5th edition, Wiley-Interscience, New York, NY, 1990.

Index

D

E

F

G

H

I

J

K

L

T

U

V

W